Probability Theory and Stochastic Modelling

Volume 71

The **Stochastic Modelling and Probability Theory** series is a merger and continuation of Springer's two well established series Stochastic Modelling and Applied Probability and Probability and Its Applications series. It publishes research monographs that make a significant contribution to probability theory or an applications domain in which advanced probability methods are fundamental. Books in this series are expected to follow rigorous mathematical standards, while also displaying the expository quality necessary to make them useful and accessible to advanced students as well as researchers. The series covers all aspects of modern probability theory including

- Gaussian processes
- Markov processes
- Random Fields, point processes and random sets
- Random matrices
- Statistical mechanics and random media
- Stochastic analysis

as well as applications that include (but are not restricted to):

- Branching processes and other models of population growth
- Communications and processing networks
- Computational methods in probability and stochastic processes, including-simulation
- Genetics and other stochastic models in biology and the life sciences
- Information theory, signal processing, and image synthesis
- Mathematical economics and finance
- Statistical methods (e.g. empirical processes, MCMC)
- Statistics for stochastic processes
- Stochastic control
- Stochastic models in operations research and stochastic optimization
- Stochastic models in the physical sciences

More information about this series at http://www.springer.com/series/13205

Oleg Klesov

Limit Theorems for Multi-Indexed Sums of Random Variables

Springer

Oleg Klesov
Department of Mathematical Analysis
and Probability Theory
National Technical University of Ukraine
"Kiev Polytechnic Institute"
Kyiv
Ukraine

ISSN 2199-3130 ISSN 2199-3149 (electronic)
ISBN 978-3-662-44387-3 ISBN 978-3-662-44388-0 (eBook)
DOI 10.1007/978-3-662-44388-0

Library of Congress Control Number: 2014945956

Mathematics Subject Classification (2010): 60F15, 60F05, 60F10, 60E15, 60E07, 60E10, 60F20

Springer Heidelberg New York Dordrecht London

Printed on acid-free paper

Springer is part of Springer Science+Business Media (www.springer.com)

Dedicated to my parents, Lidiya and Ivan

Preface

It would not be an exaggeration to say that the history of probability theory is the story of limit theorems, notably the law of large numbers, the law of small numbers and the central limit theorem. The names of Jakob Bernoulli, Abraham de Moivre, Pierre-Simon Laplace and Siméon Denis Poisson are forever linked to their pioneering limit theorems for the Bernoulli scheme and its simplest generalizations.

After these classical works, probability theory remained largely the theory of limit theorems. Invaluable contributions were made to this theory in the late nineteenth and early twentieth centuries by the outstanding Russian mathematicians P.L. Chebyshev, A.A. Markov and A.M. Lyapunov.

Limit theorems continued to be the most popular topic of probability theory in the first half of the twentieth century, the foundations of which were established by E. Borel, A.N. Kolmogorov, A.Ya. Khintchine, P. Lévy and W. Feller. It became clear in the twentieth century that the problems concerning Bernoulli trials have natural analogues which constitute a more general theory of sums of independent random variables. The monograph by B.V. Gnedenko and A.N. Kolmogorov, published in 1949, is the "final chord" in the area of research concerning the weak convergence of distribution functions related to sums of independent random variables.

The second half of the twentieth century was marked by outstanding achievements within the framework of limit problems for sums of independent random variables: many old problems were solved and new directions appeared. In particular, necessary and sufficient conditions for the law of the iterated logarithm and the strong law of large numbers with an arbitrary normalization were found, and functional limit theorems (often called invariance principles) were discovered.

Despite the undoubted achievements of the theory of limit theorems, its main problem for almost sure convergence remains unresolved, namely to find necessary and sufficient conditions for

$$\mathrm{P}(S_n > b_n \text{ i. o.}) = 0 \quad \text{and} \quad \mathrm{P}(S_n > b_n \text{ i.o.}) = 1$$

as well as for

$$P(S_n < b_n \text{ i.o.}) = 0 \text{ and } P(S_n < b_n \text{ i.o.}) = 1,$$

where $\{S_n\}$ is a sequence of cumulative sums of independent identically distributed random variables, $\{b_n\}$ is a sequence of real numbers, and "i. o." abbreviates "infinitely often". Note that the strong law of large numbers and the law of the iterated logarithm mentioned above can obviously be embedded into this setting.

The natural logic of development of limit theorems in probability theory has led to many generalizations of classical results. The first of these generalizations deals with a continuous index t instead of a discrete index n. The wealth of problems and results in this direction are not discussed here because it is virtually impossible to do so within the framework of a single monograph.

At the same time there appeared other generalizations. One of them, closely related to the theory of stochastic processes, concerns limit theorems for sums of random elements taking values in abstract linear spaces. This line of research also cannot be described in a few pages; even a simple summary of various problems and related results requires a significant amount of space.

Another generalization arose by replacing the rule of "accumulation" of random variables. The classical theory uses summation $\sum$ for this, while multiplication $\prod$ or maximum max can also be used. Each of these "substitutions" led to an independent theory with many publications. Of special interest is the case of the operation $\prod$ for random matrices instead of random variables.

Yet another generalization occurs if we drop one of the key assumptions of the classical theory, namely the condition that the terms are infinitesimal negligible. Many dozens of publications related to this topic have appeared in the past few decades.

A special place is occupied by a generalization where independent random variables are replaced by those with different schemes of dependence. This generalization has produced an almost endless stream of results and publications; hundreds of pages in the literature are devoted to various problems for dependent random variables. We will not discuss limit theorems for dependent random variables here, since even a brief listing of results would take too much space.

Instead, we will focus on a different area of limit theorems, which arises if the indices of the random variables belong to a more general space than $\mathbf{N}$ or $\mathbf{R}$, namely we study limit theorems for random variables that depend on several discrete arguments. The set of indices we are interested in is denoted by $\mathbf{N}^d$ where d is the dimension of the indices. In this case, we say that random variables depend on a *multiple index* or *multi-index*. The members of the space are denoted by $\mathbf{k}, \mathbf{m}, \mathbf{n}, \dots$.

Thus we consider *multi-indexed* independent random variables and study limit theorems for them. The main focus is on almost sure convergence. Any family of random variables that depend on the indices belonging to the space $\mathbf{N}^d$ is called a *random field*.

The convergence of random fields can be defined in various ways depending on what is meant by "$\mathbf{n} \to \infty$" in the case of $d > 1$. We are mostly interested in the following two modes of convergence:

$$(n_1, \ldots, n_d) \to \infty \Longleftrightarrow \min(n_1, \ldots, n_d) \to \infty$$

or

$$(n_1, \ldots, n_d) \to \infty \Longleftrightarrow \max(n_1, \ldots, n_d) \to \infty$$

The first is called min-convergence, while the second is called max-convergence. In a sense, min-convergence and max-convergence are respectively the weakest and strongest modes of convergence.

A brief survey of the development of the theory of limit theorems for multi-indexed sums of random variables is given in Chap. 1.

Contents of the Book

Following a "telegraphic style", we briefly discuss, chapter by chapter, the contents of the book.

Chapter 1 is an informal introduction to the theory of limit theorems for multi-indexed sums of independent random variables. We describe what this theory is and what the main differences are between the results for $d = 1$ and $d > 1$. The chapter also contains a short history of the development of the subject over the last 50 years.

Chapter 2 contains a number of inequalities for distributions and moments of multi-indexed sums of independent random variables. There is a fundamental difference between the cases $d = 1$ and $d > 1$, which is caused by the absence of a complete ordering of the space $\mathbf{N}^d$ if $d > 1$. We describe several ways to overcome this obstacle in the proof of maximal inequalities for multi-indexed sums. The results of Chap. 2 are used throughout the book.

A classical problem concerning the set of limit distributions in the case $d > 1$ is solved in Chap. 3. We find necessary and sufficient conditions for weak convergence for both modes of convergence, min and max. Although these conditions do not differ in form, there is actually a big difference between them. We exhibit this difference in Chap. 4 when discussing the law of large numbers for both modes of convergence.

Chapter 5 studies almost sure convergence of multi-indexed series of independent random variables. An unexpected aspect of the case $d > 1$ is that the general term of a convergent series does not necessarily approach 0. In the case $d > 1$, in contrast to $d = 1$, the criterion of almost sure convergence of a series of independent terms is expressed in terms of convergence of four numerical series. In some situations, however, this result can be reduced to the classical form by using only three series.

Chapter 6 is closely related to the preceding chapter and discusses the almost sure boundedness of multi-indexed series. We note a phenomenon that occurs if $d > 1$: a convergent series is not necessarily bounded.

Chapter 7 is also related to Chap. 5 and studies the rate of convergence of multi-indexed series. Here we have a new phenomenon for $d > 1$: if a series converges, then its tails are not necessarily convergent. The converse is also true, namely the tails of a divergent series may converge.

The strong law of large numbers for multi-indexed sums of independent random variables is studied in Chap. 8. One of the unexpected obstacles is the lack of an analogue of Kronecker's lemma in the case $d > 1$, which requires new approaches and methods.

Chapter 9 continues the investigation of the strong law of large numbers under an additional assumption that the random variables are identically distributed. The general results of Chap. 8 are used. The peculiarity of the case $d > 1$ is that the strong law of large numbers may hold for some "exotic" normalizations $b(\mathbf{n})$ that have no analogues for $d = 1$. If, for example, $d = 2$, then one of those normalizations is $b(m, n) = m\sqrt{n}$.

Chapter 10 is devoted to yet another classical result, the law of the iterated logarithm. Necessary and sufficient conditions for the law of the iterated logarithm are found in the case of identically distributed random variables. We face a challenging problem when studying the law of the iterated logarithm for non-identically distributed random variables, namely the normalization is not universal for non-identically distributed random variables even if the second moments are finite; it strongly depends on the structure of the $\mathtt{var}\,[S(\mathbf{n})]$.

We study the asymptotic behaviour of renewal functions and processes constructed from random walks with multi-dimensional time in Chap. 11. A new approach to the definition of renewal processes is presented. New methods for developing the results are also introduced.

The existence of moments of supremums of weighted multi-indexed sums of independent identically distributed random variables is established in Chap. 12. Some applications of these results are exhibited: we deal with new forms of the strong law of large numbers and the law of the iterated logarithm that have no analogues in the case $d = 1$.

We study so-called complete convergence in Chap. 13. The results on complete convergence are traditionally used to describe the rate of convergence in the law of large numbers and to express necessary and sufficient conditions for the strong law of large numbers and the law of the iterated logarithm. Several other applications are also known.

Some related references are given at the end of each chapter.

Appendix A discusses min-, max- and other modes of convergence in the space $\mathbf{N}^d$. In particular, all necessary notations can be found in Appendix. Various other results, definitions and notation for the space $\mathbf{N}^d$ are also given. Specific features of the case $d > 1$ and the difference between this case and $d = 1$ are discussed. In general, Appendix A is a very dense introduction to elementary analysis in the space of multi-indices $\mathbf{N}^d$.

I would greatly appreciate feedback from the readers, who are invited to contact me via e-mail address klesov@matan.kpi.ua.

Kyiv, May 2014 Oleg Klesov

Acknowledgments

I was introduced to the topic of limit theorems for multi-indexed sums in 1975 by M.I. Yadrenko when I was a student in the Mechanics and Mathematics Faculty at Kyiv Taras Shevchenko University. One of the results of my student diploma was an analogue of the Kolmogorov three series theorem for almost sure convergence of multi-indexed series of independent random variables.

At that time, the graduates of the Department of Probability Theory were usually invited to deliver a talk at A.V. Skorokhod's seminar to explain their results before the official defence procedure took place. I also had a chance to deliver a talk at one of the Skorokhod seminars. During the talk it appeared to me that A.V. Skorokhod was listening with interest to what I was saying, but when I had finished, his verdict was unexpected: "... *you are using the wrong definition* ..." he said. I must confess that I did not quite understand his comment, since all the definitions used in my talk were taken from "classical" textbooks and seemed absolutely correct. I was too shy to ask Anatoly Vladimirovich what he had meant by "wrong definition". Perhaps he soon forgot the remark; but in contrast, for many years I pondered over it, trying to get closer to his opinion. In fact, several chapters of this book contain different responses to Skorokhod's remark (or rather, responses to my modern interpretation of his remark).

A little later M.I. Yadrenko suggested that I continue to study this topic and later prepare a candidate dissertation (Ph.D. thesis). Before presenting the thesis to the Academic Council of the Mechanics and Mathematics Faculty, Michael Iosifovich sent me to Leningrad to deliver a talk at V.V. Petrov's well-known seminar. It was there that I felt for the first time that my topic was of interest to people other than myself and my thesis supervisor. At the seminar, I met Alexander Martikainen and Thomas Mikosh who became (and still are, I believe) my best friends and colleagues. After the talk, Valentin Vladimirovich Petrov gave me a lot of time to discuss existing results and further generalizations. On behalf of the so-called *leading organization* he sent a positive report on my thesis to the Academic Council.

At that time, the rule was that three reports were necessary for the defence. One should come from the *leading organization* (Leningrad University in my case) and the other two should be given by the so-called "official opponents" chosen by the Academic Council. A.Ya. Dorogovtsev and V.V. Buldygin had been chosen to serve as official opponents for the defence of my thesis in January 1981. Anatoly Yakovlevich in his speech noted, among other things, that the results of the thesis ought to constitute a larger volume and that the author should publish a book. I did not even try to implement his advice, because my guess was that there were too many obstacles to publish a book in the Soviet Union. Only now am I ready to come back to his idea, but the current version of the book has almost nothing common with what I would have published at that time.

Valeriy Vladimirovich was much more emotional in his speech at the defence procedure and also advised me to continue scientific investigations on this topic. Since then, I have repeatedly discussed various aspects of limit theorems with him. Much later, it was he who insisted that I should also defend my doctoral dissertation (habilitation). My opinion at that time was that the habilitation was not necessary for me, but now I am grateful to him for his persistence in implementing the idea.

There were three "official opponents" for a defence of any doctoral dissertation, A.V. Ivanov, Yu.V. Kozachenko and A.I. Martikainen in my case. From their reports, I realized that the text of my dissertation was not only far from perfect, but also that some of the results deserve attention. Since then, I have revised the text (and, I hope, improved it) and now convey it to the readers.

I also wish to acknowledge other colleagues with whom I have discussed some of the ideas and results of this book, namely Paul Doukhan, Istvan Fazekas, Makhamat Gafurov, Allan Gut, Karl-Heinz Indlekofer, Imre Kátai, Deli Li, Andrew Rosalsky, Zdzislaw Rychlik, Josef Steinebach and Ulrich Stadtmüller. My scientific discussions with them had a great (permanent and stimulating) effect on my work in this monograph.

I sincerely appreciate the creative communications and useful scientific discussions with all of my colleagues mentioned above. I take this opportunity to express my sincere gratitude to all of them.

Contents

Chapter 1
Some Remarks on the Theory of Limit Theorems for Multi-Indexed Sums

There exists an (unjustified) point of view that all limit theorems for multi-indexed sums are straightforward consequences of the corresponding classical results for cumulative sums and thus the case of multi-indices is non-independent. We shortly discuss this controversial claim. First we consider in more detail one of the results for multi-indexed sums that indeed follows easily from its classical counterpart (this is an example that supports the above mentioned point of view). Then we explain why this is not a universal phenomenon.

1.1 Is It True That All Results for Multi-Indexed Sums Follow From Their Classical Counterparts?

The relation

$$\frac{\zeta_n}{n} \xrightarrow{\mathsf{P}} \mu, \qquad n \to \infty \tag{1.1}$$

is called the *law of large numbers* for independent identically distributed random variables $\xi_k, k \geq 1$, where $\zeta_n = \xi_1 + \cdots + \xi_n$ and μ is a non-random constant. The Khintchine criterion of the law of large numbers is well known; for the case of $\mu = 0$, it can be written as

$$\lim_{m\to\infty} m[F(-m) + 1 - F(m)] = 0, \qquad \lim_{m\to\infty} \int_{-m}^{m} x\, dF(x) = 0,$$

where F is the distribution function of the random variable ξ_1 (see Theorem 4.2).

Now consider a family of independent identically distributed random variables $\{X(k,l), k \geq 1, l \geq 1\}$ that depend on two discrete parameters k and l. The double sums $S(m,n)$ are defined by

O. Klesov, *Limit Theorems for Multi-Indexed Sums of Random Variables*,
Probability Theory and Stochastic Modelling 71, DOI 10.1007/978-3-662-44388-0_1

$$S(m,n) = \sum_{k=1}^{m}\sum_{l=1}^{n} X(k,l), \qquad m \ge 1, \quad n \ge 1. \tag{1.2}$$

By analogy with the classical case, we say that the *law of large numbers* holds for double sums $\{S(m,n)\}$ if

$$\frac{S(m,n)}{mn} \xrightarrow{\mathsf{P}} \mu$$

as m and n increase to infinity. In other words, the law of large numbers for double sums is a statement that the probabilities

$$\mathsf{P}(|S(m,n) - \mu mn| \ge \varepsilon mn)$$

tend to 0 for all $\varepsilon > 0$ as both indices m and n increase to infinity.

Along with $\{X(k,l)\}$ consider an auxiliary sequence of independent random variables $\{\xi_i\}$ whose distribution function is the same as that of $X(1,1)$; as above, we denote by ζ_n the partial sums of random variables $\{\xi_n\}$. For $k = mn$, the distribution function of the random variable $S(m,n)/mn$ is the same as that of ζ_k/k and thus the law of large numbers for $\{S(m,n)\}$ holds if and only if it holds for $\{\zeta_n\}$. Therefore the law of large numbers for double sums follows explicitly from that for usual cumulative sums.

It may appear that such a reasoning can be applied to *any* limit theorem for double sums but actually this is not the case at all. We try to explain why this is not a universal phenomenon. The reasoning above is based on two properties, namely:

(i) the distribution function of the random variable $S(m,n)$ is determined by the product mn;
(ii) the centering and normalizing sequences, μmn and mn, are also determined by the product mn.

If at least one of these properties fails, then the method described above cannot be used. For example, property (i) fails if random variables $X(k,l)$ are not identically distributed. Property (ii) fails if the normalizing sequence is $m^a n^b, a \ne b$, say, instead of mn.

Moreover, we will see immediately that the above reasoning fails for almost sure convergence even for the normalizing constants mn.

1.2 Is It True That All Classical Methods Apply in the Proofs for Multi-Indexed Sums?

A less sceptical point of view concerning the limit theorems for multi-indexed sums agrees that the results for multi-indexed sums do not always follow from their classical counterparts but the classical methods can easily be adapted to the case of multi-indices and thus deriving results for multi-indexed sums is routine.

The working example which "supports" this point of view concerns the *strong law of large numbers*. Recall that the relation

$$\frac{\zeta_n}{n} \xrightarrow{\text{a.s.}} \mu, \qquad n \to \infty,$$

is called the strong law of large numbers for the sequence of independent identically distributed random variables $\{\xi_k, k \ge 1\}$ (here "a.s." abbreviates the expression "almost surely"). We retain the notation for cumulative sums ζ_n, expectation μ, and distribution function F introduced above. Kolmogorov's criterion for the strong law of large numbers is the existence of the first moment

$$\int_{-\infty}^{\infty} x\,dF(x) = \mu \tag{1.3}$$

(see Theorem 9.2).

It is natural to treat the relation

$$\frac{S(m,n)}{mn} \xrightarrow{\text{a.s.}} \mu \tag{1.4}$$

as m and n increase to infinity as an analog of Kolmogorov's strong law of large numbers for double sums.

According to the reasoning in Sect. 1.1, it is also reasonable to conjecture that condition (1.3) is sufficient for both strong laws of large numbers (1.1) and (1.4). To "support" this conjecture assume that condition (1.3) does indeed hold. The convergence (1.4) in the case of independent identically distributed random variables $\{X(k,l), k \ge 1, l \ge 1\}$ can be "proved" by using the Cauchy convergence criterion for fields depending on two indices (see Sect. A.3)

Consider two arbitrary increasing sequences of positive integers $\{m_k, k \ge 1\}$ and $\{n_k, k \ge 1\}$. Then

$$\frac{S(m_k, n_k)}{m_k n_k}$$

is an usual subsequence of random variables. Putting $i_k = m_k n_k$ and $\zeta_{i_k} = S(m_k, n_k)$ we see that $(\zeta_{i_k}, k \ge 1) \subseteq (\zeta_n, n \ge 1)$. Since

$$\frac{\zeta_{i_k}}{i_k} = \frac{S(m_k, n_k)}{m_k n_k}, \tag{1.5}$$

the classical Kolmogorov strong law of large numbers implies that the limit of the left-hand side of (1.5) exists almost surely, since the first moment is finite. This also means that the limit of the right-hand side of (1.5) exists almost surely. Now the sequences $\{m_k\}$ and $\{n_k\}$ are arbitrary and thus the Cauchy convergence criterion for sequences depending on two indices "implies" that the limit of the right-hand side of (1.4) exists almost surely.

Note, however, that the limit of sequence (1.5) exists on a random event Ω_1. The probability of this event does indeed equal one, but Ω_1 depends on the sequences $\{m_k\}$ and $\{n_k\}$. Thus one cannot expect that the limit for (1.4) exists outside the random event $\cap\Omega_1$, where the intersection is considered with respect to all increasing sequences $\{m_k\}$ and $\{n_k\}$.

The above "proof" is destroyed by the following observation. The set of increasing sequences is uncountable and thus the random event $\cap\Omega_1$ may have probability zero and thus the implication (1.3)$\Rightarrow$(1.4) cannot be accepted. Moreover, we will see later that, in contrast to the classical case, the existence of the first moment does not always imply that the probability of the event $\cap\Omega_1$ equals one (see Theorem 9.4).

The example just discussed shows that the results for multi-indexed sums do not always follow from their classical counterparts.

1.3 Is It True That Any Classical Method Can Be Used in the Case of Multi-Indexed Sums?

One of the "correct" proofs of the strong law of large numbers for multi-indexed sums mentioned above is a return to Kolmogorov's original proof together with a result concerning asymptotic behavior in the Dirichlet divisor problem. This is a reason to take an even less sceptical point of view of the limit theorems for multi-indexed sums. This point of view is that all of the results for multi-indexed sums can be obtained by using the classical schemes of proof with (possibly) some necessary technical changes to take into account the features (if any) of the multi-index case.

This point of view is also wrong in many settings. We show this for another strong law of large numbers for non-identically distributed random variables due to Kolmogorov (see Theorem 8.1). According to this result, if

$$\sum_{n=1}^{\infty} \frac{\mathsf{var}\,[\xi_n]}{b_n^2} < \infty$$

for independent random variables $\{\xi_n, n \geq 1\}$ and an increasing sequence of non-random numbers $\{b_n, n \geq 1\}$, then

$$\frac{\zeta_n - \mathsf{E}\,[\zeta_n]}{b_n} \xrightarrow{\text{a.s.}} 0.$$

The classical Kolmogorov proof that can be found in every advanced textbook on probability theory consists of two steps and is based on his criterion for the almost sure convergence of series of independent terms in combination with the Kronecker lemma.

This approach does not work in the case of multi-indexed sums. The simplest way to see this is to look at Example A.2 showing that the Kronecker lemma fails for multi-indexed sums.

Even more essential is that the classical Kolmogorov result on the almost sure convergence of series of independent random variables also fails for multi-indexed sums. Example 5.1 shows that there are independent random variables $\{X(k, l)\}$ such that the double series $\sum X(i, j)$ converges almost surely but the series $\sum \mathsf{P}(|X(i, j)| \geq c)$ diverges *for all* $c > 0$ in contrast to the classical case where the counterpart of the above series *converges* for all $c > 0$. Therefore both steps in the classical proof fail in the multi-index case and the Kolmogorov method cannot be used to prove the strong law of large numbers for multi-indexed sums at all.

1.4 Is It True That the Classical Ideas Are Suitable for Multi-Indexed Sums?

Given that the classical methods do not always work in the case of multi-indexed sums, can we nevertheless expect that the classical tools can be configured in such a way that they would allow one to prove the results for the multi-index case, too? This is hopeless.

Actually, not only do the *methods* of the classical theory fail in the case of multi-indexed sums, even the main *ideas* are not always applicable to the multi-index case. For example, one of the fundamental ideas of the classical theory is to use the "first exit time". For instance, the proof of the Lévy inequality for the maximum of cumulative sums of independent symmetric terms relies on the random variable $\tau = \tau(\omega)$, the *first* exit time of a random walk from the interval $(-\infty, x]$:

$$\tau = \min\{n : S_n > x\}.$$

Since the elements of the space $\mathbf{N}^d$ are not linearly ordered, an analog of the random variable τ written as

$$\tau = \min\{(m, n) : S(m, n) > x\}$$

does not exist for $d \geq 2$ (e.g., what is the first exit time from the interval $(-\infty, x]$ if $S(1, 1) \leq x$ but $S(1, 2) > x$ and $S(2, 1) > x$?). This, unfortunately, means that all the nice proofs for $d = 1$ that use τ are useless in the case $d > 1$.

1.5 Is It True That the Classical Conditions Are Suitable for Multi-Indexed Sums?

We have already observed that the existence of the first moment,

$$\int_{-\infty}^{\infty} |x| \, dF(x) < \infty$$

is not a sufficient condition for the strong law of large numbers in the case of multi-indexed sums in contrast to the case of cumulative sums. We shall prove that a necessary and sufficient condition for the strong law of large numbers is

$$\int_{-\infty}^{\infty} |x|(\log^+ |x|)^{d-1}\, dF(x) < \infty$$

in the case of d-tuple sums (see Theorem 9.4). Substituting $d = 1$ here one obtains the preceding (classical!) condition for the strong law of large numbers. This is not surprising and moreover this kind of "continuity" of a condition with respect to the dimension of multi-indices might be expected for all other results.

As for the other "obvious" conclusions about the case $d > 1$, the latter statement concerning "continuity" is also wrong: several cases show the difference in conditions between the cases $d = 1$ and $d > 1$. For example, the criterion for the law of the iterated logarithm for cumulative sums of independent identically distributed random variables found by Hartman and Wintner [156] and proved by Strassen [385] is the existence of the second moment,

$$\mathsf{E}\left[X^2\right] < \infty. \tag{1.6}$$

We show in Theorem 10.9 that the criterion for the same result in the case of multi-indexed sums is Wichura's condition

$$\mathsf{E}\left[X^2 \frac{(\log^+ |X|)^{d-1}}{\log^+ \log^+ |X|}\right] < \infty,$$

where d is the dimension of the multi-indices. Substituting $d = 1$ here we do not get the preceding (classical!) condition (1.6). In other words, Wichura's criterion for the law of the iterated logarithm for $d > 1$ does not turn into the Hartman–Wintner–Strassen condition if $d = 1$.

1.6 Is It True That the Classical Constants Do not Change in the Case of Multi-Indexed Sums?

One can continue to extend the list of the differences between the cases $d = 1$ and $d > 1$. But, what has been already said, is, perhaps, sufficient to agree that the limit theorems for multi-indexed sums are worthy of attention and that they need other methods as compared to the case of cumulative sums. Nevertheless we point out yet another feature of multi-indexed sums.

The role of various inequalities in probability theory is well known. Inequalities often include some constants; finding optimal values of those constants is a challenging question. The optimal value of some constants are still not known; for example, the Berry–Esseen inequality includes a constant whose calculated values have decreased markedly over the years, from the original value of 7.59 in 1942 to 0.4748 in 2012.

There are also other inequalities where the constant are optimal from the beginning. For instance, the Lévy inequality says that

$$\mathsf{P}\left(\max_{k\le n} S_k \ge x\right) \le 2\mathsf{P}(S_n \ge x),$$

where $x > 0$ and $\{S_k\}$ are cumulative sums of independent symmetric random variables. The constant 2 on the right-hand side is optimal, since

$$\mathsf{P}\left(\max_{0\le k\le n} S_k \ge x\right) = 2\mathsf{P}(S_n > x) + \mathsf{P}(S_n = x) \qquad (1.7)$$

if the terms $\{X_k\}$ have the symmetric Bernoulli distribution:

$$\mathsf{P}(X_k = -1) = \mathsf{P}(X_k = 1) = \frac{1}{2}.$$

Equality (1.7) for non-integer x proves the optimality of the constant 2 in the Lévy inequality.

We will see below that the constants in the most important results from the classical theory of limit theorems change in their counterparts for multi-indexed sums. For example, the constant 1 in the law of the iterated logarithm becomes $\sqrt{d}$, the constant 2 in the Lévy inequality becomes 2^d and so on. The precise values of the constants for the latter two examples are particularly important. For the sake of clarity, we briefly discuss the constant for the Lévy inequality. If $d = 2$, then a counterpart of the Lévy inequality for double sums reads as follows

$$\mathsf{P}\left(\max_{\substack{k\le m\\ l\le n}} S(k,l) \ge x\right) \le 4\mathsf{P}(S(m,n) \ge x). \qquad (1.8)$$

Is the constant 4 on the right-hand side optimal? To answer this question, we turn to the so-called *invariance principle*.

If the second moment exists, then the invariance principle implies that

$$\mathsf{P}\left(\max_{k\le n} |S_k| \ge x\sqrt{n}\right) \to \mathsf{P}\left(\sup_{0\le t\le 1} |w(t)| \ge x\right), \qquad n \to \infty,$$

where w is a standard Wiener process. The probability on the right-hand side is evaluated in a closed form with the help of the *reflection principle*,

$$\mathsf{P}\left(\sup_{0\le t\le 1}|w(t)|\ge x\right)=2\mathsf{P}(|w(1)|\ge x).$$

In particular,

$$\lim_{x\to\infty}\frac{1}{2\Psi(x)}\mathsf{P}\left(\sup_{0\le t\le 1}|w(t)|\ge x\right)=1,$$

where Ψ is the tail of the distribution function of the absolute value of a standard Gaussian random variable. This consideration allows us to conclude that, for large n and x,

$$\mathsf{P}\left(\max_{k\le n}|S_k|\ge x\sqrt{n}\right)\approx\mathsf{P}\left(\sup_{0\le t\le 1}|w(t)|\ge x\right)\approx 2\Psi(x).$$

The invariance principle can also be proved for $d=2$,

$$\mathsf{P}\left(\max_{\substack{k\le m\\ l\le n}}|S(k,l)|\ge x\sqrt{mn}\right)\to\mathsf{P}\left(\sup_{\substack{0\le s\le 1\\ 0\le t\le 1}}|w_2(s,t)|\ge x\right),\qquad m,n\to\infty,$$

where w_2 is an analog of the Wiener process for the two dimensional time called the *Chentsov–Yeh field*. The precise distribution of $\max|w_2|$, not known yet, is an important characteristic in many statistical procedures. Luckily one can find its asymptotic behavior, namely

$$\lim_{x\to\infty}\frac{1}{4\Psi(x)}\mathsf{P}\left(\sup_{\substack{0\le s\le 1\\ 0\le t\le 1}}|w_2(s,t)|\ge x\right)=1.$$

The optimality (in the sense explained above) of the constant 4 in inequality (1.8) is derived in the same way as above.

1.7 Some Classical Results Do not Have Counterparts for Multi-Indexed Sums

One cannot expect to construct a theory for $d>1$ which is fully compatible with the classical theory. Some results for $d=1$ cannot be extended to the case $d>1$. An example of this kind is related to the Petrov classes of functions (see [366], Chapter IX).

Assume that $\{X_n\}$ are independent random variables with certain finite moments (for definiteness, let us assume that $\mathsf{E}[X_k]=0$, $k\ge 1$, and that the second moments exist). Then there exists a class of functions Ψ_c such that

$$\lim_{n\to\infty} \frac{S_n}{\psi(B_n)} = 0 \qquad \text{a. s.}$$

for every $\psi \in \Psi_c$, where $\{S_n\}$ are cumulative sums of random variables $\{X_k\}$ and $B_n = \mathsf{E}\left[S_n^2\right]$. It is important for this result that the class of functions Ψ_c is universal, that is, it is independent of the univariate distributions of the sequence $\{X_k\}$.

The situation changes drastically in the case of multi-indexed sums. Example A.5 shows for double sums that *for any given* function ψ there exists a field of independent random variables $\{X(k,l);\, k \geq 1, l \geq 1\}$ with zero expectations and finite second moments such that

$$\lim_{\substack{m\to\infty\\ n\to\infty}} \frac{S(m,n)}{\psi(B(m,n))} \neq 0 \qquad \text{a. s.,}$$

where $\{S(m,n)\}$ are double sums of the independent random variables $\{X(k,l)\}$ and $B(m,n) = \mathsf{E}\left[S^2(m,n)\right]$.

Of course, Example A.5 does not mean that one cannot find a function ψ for a given field $\{X(k,l)\}$ such that

$$\lim_{\substack{m\to\infty\\ n\to\infty}} \frac{S(m,n)}{\psi(B(m,n))} = 0 \qquad \text{a. s.}$$

Such a function does indeed exist! On the other hand, Example A.5 means that, even for this exceptional function, there exists a further field $\{X'(k,l)\}$ for which the latter property does not hold.

Therefore there is no universal Petrov class of functions ψ for $d > 1$.

1.8 Some Results for Multi-Indexed Sums Do not Have Counterparts in the Classical Case

The classical theory of limit theorems for $d = 1$ can be described by attaching a name to each of its result. Then the counterparts for $d > 1$ get the same names with an indication that they belong to the theory for $d > 1$ ("the law of large numbers for $d > 1$", say). It may appear from what was said in the preceding section that the set of names for $d > 1$ is a subset of that for $d = 1$. This again is a wrong impression. There are examples of the opposite, where a result for multi-indexed sums does not have an analog in the classical theory.

Consider, for example, the field

$$\frac{S(m,n)}{m\sqrt{n}}, \qquad m \geq 1, \quad n \geq 1,$$

where the double sums $S(m,n)$ are defined by (1.2) and $\{X(k,l)\}$ are independent identically distributed random variables with zero mean and finite second moment. Is it possible that

$$\frac{S(m,n)}{m\sqrt{n}} \xrightarrow{\mathsf{P}} 0? \tag{1.9}$$

If the limit is considered along the first coordinate, then this is obvious (see the law of large numbers (1.1)). But if the limit is considered with respect to the second coordinate, then the latter result would contradict the central limit theorem. We will, nevertheless, see in Chap. 4 that relation (1.9) holds under some assumptions as both $m \to \infty$ and $n \to \infty$. Nothing of this kind can be expected for $d = 1$.

Other examples of the same kind are the following two limit properties

$$\lim_{m\to\infty} \frac{1}{m} \sup_{n\geq 1} \left| \frac{S(m,n)}{n} \right| = 0 \qquad \text{a. s.}$$

and

$$\limsup_{m\to\infty} \frac{1}{\sqrt{m \log\log m}} \sup_{n\geq 1} \left| \frac{S(m,n)}{\sqrt{n \log\log n}} \right| = \sqrt{2} \qquad \text{a. s.,}$$

to be considered in Chap. 12.

Obviously, the classical theory does not have any results of this kind, since they both essentially use two parameters.

Therefore the classical theory suggests the direction of study for multi-indexed sums (for example, the law of large numbers or the law of the iterated logarithm etc.), but in most settings its ideas, methods, results, conditions, and universal constants can be used only after appropriate changes.

1.9 Some History of the Theory of Limit Theorems for Multi-Indexed Sums

The history of limit theorems for multi-indexed sums of random variables can be divided into two conventional stages.

1.9.1 First Stage

The pioneering studies that can be attributed to the first stage of probabilistic limit theorems for multi-indexed sums began in ergodic theory. Wiener [418] considered transformations T preserving a measure μ on some measurable space (Ω, μ). He studied d-tuple square sums, $d \geq 1$,

$$S_m[f] = \sum_{n_1=1}^{m} \cdots \sum_{n_d=1}^{m} f\left(T^{n_1+\cdots+n_d}\omega\right)$$

for functions $f \in \mathbf{L}_1(\Omega, \mu)$ and proved that the limit

$$\lim_{m\to\infty} \frac{S_m[f]}{m^d}$$

exists for μ-almost all $\omega \in \Omega$. Note that the normalizing sequence depends on the dimension d but the class of functions for which the above limit exists is the same as in the case of the usual cumulative sums studied in classical ergodic theory.

Dunford [79] and Zygmund [440] considered a similar problem but for multiple rectangular sums with different "right upper corner" coordinates, that is

$$S_{m_1,\dots,m_d}[f] = \sum_{n_1=1}^{m_1} \cdots \sum_{n_d=1}^{m_d} f\left(T^{n_1+\dots+n_d}\omega\right).$$

The class of functions f for which the limit

$$\lim_{\substack{m_1\to\infty\\ \dots \\ m_d\to\infty}} \frac{S_{m_1,\dots,m_d}[f]}{m_1\dots m_d}$$

exists for μ-almost all $\omega \in \Omega$, as proved by Dunford [79] and Zygmund [440], is such that

$$\int_{\Omega} |f(\omega)| \left(\log^+ |f(\omega)|\right)^{d-1} d\mu < \infty. \tag{1.10}$$

Note that the class of functions possessing property (1.10) is narrower than the class $\mathbf{L}_1(\Omega, \mu)$ of Wiener's theorem. It became apparent later that condition (1.10) cannot be improved in the general case.

The studies in ergodic theory done by Wiener, Dunford, and Zygmund, as seen from the probabilistic point of view, deal with weakly stationary random fields

$$X(n_1, \dots, n_d) = T^{n_1+\dots+n_d} X_0,$$

where T is a shift operator, and their sums

$$S(m_1, \dots, m_d) = \sum_{n_1=1}^{m_1} \cdots \sum_{n_d=1}^{m_d} X(n_1, \dots, n_d).$$

In the language of probability theory, the Wiener, Dunford, and Zygmund results mentioned above are the strong laws of large numbers for weakly stationary fields.

During the 1940s and 1950s only a few papers were devoted to limit theorems for random fields. They nevertheless contributed to an accumulation of new settings and problems that later became more popular. For example, Bochner [27] and Krickeberg [262] considered martingales depending on a parameter belonging to a partially ordered set ($\mathbf{N}^d$ is clearly an example of such a set and sums of independent random variables are examples of martingales).

An interest in purely probabilistic problems for random variables with multi-dimensional indices began to grow, starting in the 1960s. For example, Chernoff and Teicher [52] studied the limit distributions of

$$W_n = \min_{1 \le i \le n} \max_{1 \le j \le n} X(i, j), \qquad Z_n = \max_{1 \le i \le n} \min_{1 \le j \le n} X(i, j),$$

where $X(i, j)$ are independent and identically distributed random variables. This direction was further developed by experts in game theory and in applied science; several applications are still developing (various maximin schemes, asymptotic reliability of parallel-sequential systems etc.).

1.9.2 Second Stage

Wichura [419] was, perhaps, the first to prove results in the classical theory of limit theorems for multi-indexed sums of independent random variables by using purely probabilistic tools. His main aim was to prove the invariance principle for random variables indexed with a multi-index. His approach was to reduce the problem for multi-indexed sums to a corresponding problem for cumulative sums but for random vectors of the corresponding dimension.

Wichura's [419] paper and, particularly, the survey paper by Pyke [345] had a great impact on the investigation of limit theorems for random variables with multi-indices. Soon after Pyke's paper, Smythe [373] found the first probabilistic proof of the strong law of large numbers for multi-indexed sums of independent identically distributed random variables. Turning back to the language of ergodic theory used by Wiener, Dunford and Zygmund, one can say that Smythe was able to contribute the following result to the ergodic theorem for multi-parameter groups of transformations preserving a measure: condition (1.10) is necessary and sufficient for such an ergodic theorem in the class of transformations related to independent identically distributed random variables. Smythe's proof is based on the machinery of reversed martingales. It was observed later that such a technique is not applicable to other normalizing sequences.

The papers by Wichura, Pyke and Smythe manifested the second stage of the theory of limit theorems for multi-indexed sums of random variables. Various counterparts of many classical limit theorems for $d > 1$ have been proved since then.

The further history of each topic considered in the monograph is discussed in more detail in the corresponding chapter. Here we use a plain "telegraphic style" just to mention the main authors, and their papers, that influenced the development of the theory.

Wichura [420] studied the Strassen form of the law of the iterated logarithm for d-tuple sums of independent identically distributed random variables. He found a necessary and sufficient condition for the law of the iterated logarithm. As mentioned above, his condition is remarkable because if we substitute $d = 1$ into Wichura's condition, it does not transform into the classical Hartman–Wintner–Strassen condition.

Smythe [374] turned to general limit theorems for sums of independent random variables indexed with elements of a partially ordered set that is more general than $\mathbf{N}^d$. In addition to almost sure convergence, he also studied complete convergence.

Shorack and Smythe [366] obtained the Hájek–Rényi type inequality for moments of the maximum of multi-indexed sums of random variables. This result allows one to extend the strong law of large numbers to a larger class of normalizations. An even wider class is obtained with the help of a corresponding inequality in [211].

In his first paper devoted to limit theorems for random variables with multi-indices, Gut [128] studied the convergence of reversed martingales. He was aware that this approach cannot be used to prove exact results for independent identically distributed random variables and so in [129] he started using another, let us say "classical", method that allowed him to prove the strong law of large numbers for Marcinkiewicz–Zygmund normalizations. Necessary and sufficient conditions for the strong law of large numbers for even more general normalizations have been found in [221]. In the paper [130], A. Gut studied conditions for the existence of certain moments of the suprema of multi-indexed sums of independent identically distributed random variables. In [132] he suggested using the Hoffman-Jørgensen inequality for complete convergence. This essentially shortened the proofs and made them transparent.

In his dissertation [114], Gabriel made an observation that if the indices of multi-indexed sums are considered in a subset of $\mathbf{N}^d$ rather than in the whole space $\mathbf{N}^d$, then conditions for limit theorems become closer to the classical ones. Gut [133] proved this general rule in the case of the strong law of large numbers if the set where the indices vary coincides with a sector $\theta n_i \le n_j \le \theta^{-1} n_i$, $i \ne j$, for some $0 < \theta < 1$. It turned out that, like the classical case, the existence of the first moment is equivalent to the strong law of large numbers if the indices belong to that restricted domain.

Gabriel [115] considered the almost sure convergence of multi-indexed series and convergence of d-parametric martingales; he generalized the Kolmogorov three series theorem to the case of multi-indexed sums. It has later become clear that the three numerical series involved in the Kolmogorov criterion for $d = 1$ should be supplemented by a fourth series [210], [232].

Bounds for the distributions and moments of the maximum of cumulative sums is a powerful tool in the study of almost sure convergence if $d = 1$. The corresponding bounds are even more important for the theory of multi-indexed sums, since other tools may not work in this case.

The Lévy inequality for the maximum of multi-indexed sums of independent symmetric random variables is a rich source for several generalizations used in many limit theorems and has been considered by many authors. Zimmerman [433] seems to be the first who proved the Lévy inequality for double sums. The general case of $d \ge 1$ is considered by Paranjape and Park [335] (no proof is given by them). A more general inequality generalizing Petrov's inequality [337] for quantiles is obtained in [215].

The Prokhorov–Loève form of the strong law of large numbers for multi-indexed sums is studied in [218]. Mikosch [315] and Mikosch and Norvaisa [317] considered a similar problem (also for the Banach-valued random variables). Note that the search for conditions for the Prokhorov–Loève strong law of large numbers for multi-indexed sums is not yet complete, since not all normalizations fit the conditions in the papers mentioned above. The reason for this, explained by Martikainen [308], is that a certain structural property is not universal for normalizations if $d > 1$, while any increasing normalization possesses this property if $d = 1$. Martikainen [308] was able to overcome some difficulties in this problem and approached the solution in the general case of $d > 1$.

The Kolmogorov form of the law of the iterated logarithm for multi-indexed sums of independent bounded terms is proved in [215] (again, the Martikainen characteristic plays a crucial role here). The method of proof in [215] is classical with the necessary changes related to the unexpected obstacle that a "geometrically growing" subsequence is a rather complicated object in the case $d > 1$. This problem is overcome in [215] by assuming the field of variances of multi-indexed sums has a special structure. Martikainen [307] showed that if this assumption is dropped, then one has to change the normalization in the law of the iterated logarithm: his characteristic of the normalizing field substitutes the log log function (see Sect. A.8). The law of the iterated logarithm for stable terms is considered in [212].

The Erdös–Rényi law of large numbers for multi-indexed sums is obtained by Steinebach in [381].

The necessity of studying the convergence of series of large, moderate, and small deviations such as

$$\sum_{\mathbf{n}\in\mathbf{N^d}} w(\mathbf{n})\mathsf{P}(|\mathbf{S}(\mathbf{n})| \geq \varepsilon\mathbf{b}(\mathbf{n}))$$

for various classes of fields $\{w(\mathbf{n}), \mathbf{n} \in \mathbf{N^d}\}$ and $\{b(\mathbf{n}), \mathbf{n} \in \mathbf{N^d}\}$ is often explained by the needs of the so-called complete convergence. The first to study such series for $d > 1$ were Smythe [374] and Gut [129].

Renewal theory is closely connected to asymptotic properties of sums of random variables. The renewal function constructed from a random walk with multidimensional time, that is the series

$$\sum_{\mathbf{n}\in\mathbf{N^d}} \mathsf{P}(S(\mathbf{n}) < \mathbf{x}),$$

is studied by Ney and Wainger [330], Maejima and Mori [300], Galambos and Kátai [119, 120], Galambos et al. [118]. The renewal process in the case $d > 1$ is studied in [245].

We also mention some results by Pyke et al. [10, 20, 21] concerning the strong law of large numbers, the law of the iterated logarithm, and the functional central limit theorem for partial sums constructed from random variables indexed by subsets of $\mathbf{R}^d$.

1.10 Comments

1.10.1 Books Devoted to Limit Theorems

The literature devoted to limit theorems in probability theory is immense. Every advanced textbook contains several such theorems, with certain details. Among the books completely devoted to limit theorems we would like to mention the monographs by Gnedenko and Kolmogorov [126], Petrov [336, 338], Stout [383], Chow and Teicher [57], Hall and Heyde [151], Révész [352], Csörgő and Révész [65], and Gut [138].

1.10.2 Monographs Devoted to Random Fields

At the present time, there are no monographs devoted to limit theorems for multi-indexed sums. On the other hand, there are several books where various particular cases of these theorems are discussed in the context of the theory of random functions. To mention a few, these are Adler and Taylor [8], Azaïs and Wshebor [13], Ivanov and Leonenko [185], Khoshnevisan [206], Yadrenko [422].

Chapter 2
Maximal Inequalities for Multi-Indexed Sums of Independent Random Variables

Bounds for distributions and moments of sums of independent random variables play a crucial role in the proofs of limit theorems (especially for almost sure convergence). One of the first results of this kind is *Chebyshev's inequality*.

Theorem 2.1 (P. L. Chebyshev) *Let* $\{X_k\}$ *be independent random variables,* $n \geq 1$, $S_n = X_1 + \cdots + X_n$. *Assume that, for all* $k \leq n$,

$$\mathsf{E}[X_k] = 0, \quad \mathsf{E}\left[X_k^2\right] = \sigma_k^2 < \infty.$$

Then for all $x > 0$

$$\mathsf{P}(|S_n| \geq x) \leq \frac{\sigma_1^2 + \cdots + \sigma_n^2}{x^2}.$$

This result holds even if the maximum of cumulative sums substitutes S_n. The corresponding result is called *Kolmogorov's inequality*.

Theorem 2.2 (A. N. Kolmogorov) *Let* $\{X_k\}$ *be independent random variables,* $n \geq 1$, *and* $S_n = X_1 + \cdots + X_n$. *Assume that, for all* $k \leq n$,

$$\mathsf{E}[X]_k = 0, \quad \mathsf{E}[X]_k^2 = \sigma_k^2 < \infty.$$

Then for all $x > 0$

$$\mathsf{P}\left(\max_{k \leq n} |S_k| \geq x\right) \leq \frac{\sigma_1^2 + \cdots + \sigma_n^2}{x^2}. \tag{2.1}$$

Kolmogorov's inequality can also be proved for weighted sums S_k/b_k (in this form, it is called *Hájek–Rényi's inequality*).

Theorem 2.3 (J. Hájek, A. Rényi) *Let* $\{X_k\}$ *be independent random variables,* $n \geq 1$, *and* $S_n = X_1 + \cdots + X_n$. *Assume that, for all* $k \leq n$,

O. Klesov, *Limit Theorems for Multi-Indexed Sums of Random Variables*,
Probability Theory and Stochastic Modelling 71, DOI 10.1007/978-3-662-44388-0_2

$$\mathsf{E}[X_k] = 0, \quad \mathsf{E}\left[X_k^2\right] = \sigma_k^2 < \infty.$$

Also let $\{b_k\}$ *be an increasing sequence of positive numbers. Then for all* $x > 0$

$$\mathsf{P}\left(\max_{k \le n} \frac{|S_k|}{b_k} \ge x\right) \le \frac{1}{x^2} \sum_{k=1}^{n} \frac{\sigma_k^2}{b_k^2}. \tag{2.2}$$

Another generalization of Theorem 2.2, called *Doob's inequality*, provides a bound for the second moment of the maximum of cumulative sums.

Theorem 2.4 (J. Doob) *Let* $\{X_k\}$ *be independent random variables,* $n \ge 1$, *and* $S_n = X_1 + \cdots + X_n$. *Assume that, for all* $k \le n$,

$$\mathsf{E}[X_k] = 0, \quad \mathsf{E}\left[X_k^2\right] = \sigma_k^2 < \infty.$$

Then

$$\mathsf{E}\left[\max_{k \le n} S_k^2\right] \le 4(\sigma_1^2 + \cdots + \sigma_n^2). \tag{2.3}$$

Yet another classical result is *Lévy's inequality* for symmetric random variables.

Theorem 2.5 (P. Lévy) *Let* $\{X_k\}$ *be independent symmetric random variables. Then*

$$\mathsf{P}\left(\max_{k \le n} S_k \ge x\right) \le 2\mathsf{P}(S_n \ge x) \tag{2.4}$$

for all $x > 0$.

The methods used to prove Theorems 2.2–2.5 for $d = 1$ are similar and based on the *first time* when the sequence $S_1, \ldots, S_n$ crosses a level x from below (in other words, enters an interval $[x, \infty])$. Since the "first time" when a field $\{S(\mathbf{k}), \mathbf{k} \le \mathbf{n}\}$ crosses a level x is not well defined for $d > 1$, the methods of proof of the above inequalities also change for $d > 1$.

The analogs of the above inequalities for multi-indexed sums are collected and proved in this chapter. These results will be used in the later chapters of the book.

2.1 Estimates for Moments of Multi-Indexed Sums

In this section, we consider a sequence $\{X_k, 1 \le k \le n\}$ of independent random variables and a field $\{X(\mathbf{n}), \mathbf{n} \in \mathbf{N}^d\}$ of independent random variables. Some of the results presented below also hold for certain classes of dependent random variables.

We will indicate all such cases where the random variables $\{X_k\}$ or $\{X(\mathbf{n})\}$ are not necessarily independent but still admit the corresponding inequality.

Let $n > 1$ and put $S_1 = X_1$, $S_n = X_1 + \cdots + X_n$. Assume that $r > 0$ and

$$\mathsf{E}\left[|X_k|^r\right] < \infty, \quad 1 \le k \le n.$$

Similar assumptions are made concerning the random fields $\{X(\mathbf{n}), \mathbf{n} \in \mathbf{N}^d\}$, that is we assume that

$$\mathsf{E}\left[|X(\mathbf{k})|^r\right] < \infty, \quad \mathbf{k} \le \mathbf{n}.$$

2.1.1 The Bahr–Esseen and Dharmadhikari–Jogdeo Inequalities

The triangle inequality implies that

$$|S_n|^r \le |X_1|^r + \cdots + |X_n|^r$$

for $0 < r \le 1$. Passing to the expectations on both sides of the latter inequality we get

$$\mathsf{E}\left[|S_n|^r\right] \le \sum_{k=1}^{n} \mathsf{E}\left[|X_k|^r\right]. \tag{2.5}$$

Inequality (2.5) holds for $r = 2$, as well, if $\mathsf{E}[X_k] = 0$, $1 \le k \le n$. Moreover, it becomes an equality in this case. The case $1 < r < 2$ is more delicate.

Theorem 2.6 (B. Bahr, C.-G. Esseen) *Let X_k, $k \le n$, be independent random variables with $\mathsf{E}[X_k] = 0$, $k \le n$. If $1 \le r \le 2$, then*

$$\mathsf{E}\left[|S_n|^r\right] \le \left(2 - \frac{1}{n}\right) \sum_{k=1}^{n} \mathsf{E}\left[|X_k|^r\right]. \tag{2.6}$$

For convenience, we refer to the constant on the right-hand side of inequalities (2.5) and (2.6) as the *Bahr–Esseen constant*. Thus the Bahr–Esseen constant equals 1 if $0 < r \le 1$ or $r = 2$ and $2 - \frac{1}{n}$ if $1 < r < 2$.

The case $r > 2$ is the most complicated one.

Theorem 2.7 (S. W. Dharmadhikari, K. Jogdeo) *Let X_k, $k \le n$, be independent random variables with $\mathsf{E}[X_k] = 0$, $k \le n$. If $r \ge 2$, then*

$$\mathsf{E}\left[|S_n|^r\right] \le J_r n^{\frac{r}{2}-1} \sum_{k=1}^{n} \mathsf{E}\left[|X_k|^r\right], \tag{2.7}$$

where

$$J_r = \frac{r(r-1)}{2} \max\left(1, 2^{r-3}\right)\left(1 + \frac{2}{r} K_{2m}^{(r-2)/2m}\right), \quad K_{2m} = \sum_{i=1}^{m} \frac{i^{2m-1}}{(i-1)!},$$

and where the number m is defined by the condition $2m \le r < 2m+2$.

For convenience, we refer to J_r as the *Dharmadhikari–Jogdeo constant.*

Remark 2.1 Any attempt to prove inequality (2.7) by "elementary" reasoning (with the help of Hölder's inequality, for example) is doomed to failure. Indeed, let $r > 2$ and p be such that $\frac{1}{p} + \frac{1}{r} = 1$. Then Hölder's inequality implies that

$$\left|\sum_{k=1}^{n} X_k\right| \le n^{1/p}\left(\sum_{k=1}^{n} |X_k|^r\right)^{1/r},$$

whence

$$\mathsf{E}\left[|S_n|^r\right] \le n^{r/p} \sum_{k=1}^{n} \mathsf{E}\left[|X_k|^r\right].$$

This bound is worse than (2.7) as far as the order of growth of n is concerned, since

$$\frac{r}{p} = r - 1 > \frac{r}{2} - 1.$$

The preceding inequalities can be rewritten for multi-indexed sums as follows.

Theorem 2.8 *Let* $\{X(\mathbf{n})\}$ *be a field of independent random variables. Let* $\mathfrak{L} \subset \mathbf{N}^d$ *be an arbitrary finite subset. Let either* $0 < r < 1$ *or* $r \ge 1$ *and* $\mathsf{E}[X_k] = 0$ *for all* $\mathbf{k} \in \mathfrak{L}$. *Then*

$$\mathsf{E}\left[\left|\sum_{\mathbf{k}\in\mathfrak{L}} X(\mathbf{k})\right|^r\right] \le C_r \sum_{\mathbf{k}\in\mathfrak{L}} \mathsf{E}\left[|X(\mathbf{k})|^r\right], \quad 0 < r \le 2, \tag{2.8}$$

$$\mathsf{E}\left[\left|\sum_{\mathbf{k}\in\mathfrak{L}} X(\mathbf{k})\right|^r\right] \le C_r \left(\operatorname{card}(\mathfrak{L})\right)^{\frac{r}{2}-1} \sum_{\mathbf{k}\in\mathfrak{L}} \mathsf{E}\left[|X(\mathbf{k})|^r\right], \quad r > 2, \tag{2.9}$$

where C_r *equals the Bahr–Esseen constant for* $0 < r \le 2$ *and the Dharmadhikari–Jogdeo constant for* $r > 2$, *that is*

$$C_r = \begin{cases} 1, & 0 < r \le 1 \text{ or } r = 2, \\ 2, & 1 < r < 2, \\ J_r, & r > 2. \end{cases} \tag{2.10}$$

2.1.2 Rosenthal's Inequality

Let $r \geq 2$. Another bound for the expectation $\mathsf{E}\left[|S_n|^r\right]$ is written in terms of the moments of the random variables $\{X_k\}$ only and without any multiplicative terms such as $n^{\frac{r}{2}-1}$.

Theorem 2.9 (H. Rosenthal) *Let X_k, $k \leq n$, be independent random variables. Assume that $r \geq 2$ and*

$$\mathsf{E}\left[X_k\right] = 0, \quad 1 \leq k \leq n.$$

Then

$$\mathsf{E}\left[|S_n|^r\right] \leq R_r \left[\sum_{k=1}^{n} \mathsf{E}\left[|X_k|^r\right] + B_n^{r/2}\right], \tag{2.11}$$

where $B_n = \sum_{k=1}^{n} \mathsf{E}\left[X_k^2\right]$, $R_r = 2^{r^2}$.

For convenience, we refer to R_r as the *Rosenthal constant*.

Rosenthal's inequality for multi-indexed sums is written as follows.

Theorem 2.10 *Let $\{X(\mathbf{n})\}$ be a field of independent random variables. Assume that $r \geq 2$. Let $\mathfrak{L} \subset \mathbf{N}^d$ be an arbitrary finite set and let*

$$\mathsf{E}\left[X(\mathbf{k})\right] = 0, \quad \mathbf{k} \in \mathfrak{L}.$$

Then

$$\mathsf{E}\left[\left|\sum_{\mathbf{k}\in\mathfrak{L}} X(\mathbf{k})\right|^r\right] \leq R_r \left[\sum_{\mathbf{k}\in\mathfrak{L}} \mathsf{E}\left[|X(\mathbf{k})|^r\right] + B^{r/2}(\mathfrak{L})\right], \tag{2.12}$$

where

$$B(\mathfrak{L}) = \sum_{\mathbf{k}\in\mathfrak{L}} \mathsf{E}\left[X^2(\mathbf{k})\right].$$

2.1.3 The Marcinkiewicz–Zygmund Inequality

Yet another bound for moments of sums of independent random variables is due to Marcinkiewicz and Zygmund.

Theorem 2.11 (J. Marcinkiewicz, A. Zygmund) *Let X_i, $1 \leq i \leq n$, be independent random variables. Let $r \geq 2$. Assume that $\mathsf{E}\left[X_i\right] = 0$ and $\mathsf{E}\left[|X_i|^r\right] < \infty$, $1 \leq i \leq n$. Then*

$$Z_r' \mathsf{E}\left[\left(\sum_{i=1}^{n} X_i^2\right)^{r/2}\right] \leq \mathsf{E}\left[|S_n|^r\right] \leq Z_r'' \mathsf{E}\left[\left(\sum_{i=1}^{n} X_i^2\right)^{r/2}\right], \tag{2.13}$$

where Z_r' and Z_r'' are universal constants that depend only on r.

We refer to the constants Z_r' and Z_r'' as the *upper and lower Marcinkiewicz–Zygmund constants.*

An analog of inequality (2.13) for multi-indexed sums is given below.

Theorem 2.12 *Let $\{X(\mathbf{n})\}$ be a field of independent random variables. Let $\mathfrak{L} \subset \mathbf{N}^d$ be a finite set and $r \geq 2$. Assume that $\mathsf{E}\left[|X(\mathbf{k})|^r\right] < \infty$ and $\mathsf{E}[X(\mathbf{k})] = 0$ for all $\mathbf{k} \in \mathfrak{L}$. Then*

$$Z_r'\mathsf{E}\left[\left(\sum_{\mathbf{k}\in\mathfrak{L}} X^2(\mathbf{k})\right)^{r/2}\right] \leq \mathsf{E}\left[\left|\sum_{\mathbf{k}\in\mathfrak{L}} X(\mathbf{k})\right|^r\right] \leq Z_r''\mathsf{E}\left[\left(\sum_{\mathbf{k}\in\mathfrak{L}} X^2(\mathbf{k})\right)^{r/2}\right], \tag{2.14}$$

where Z_r' and Z_r'' are the upper and lower Marcinkiewicz–Zygmund constants.

2.1.4 Skorokhod's Inequality

An interesting generalization of inequality (2.11) can be obtained from *Skorokhod's inequality.*

Theorem 2.13 (A. V. Skorokhod) *Let $\{X_i\}$ be independent bounded random variables, $|X_i| \leq 1$. Assume that $\mathsf{P}(|S_n| \geq a) \leq 1/8e$ for some a. Then*

$$\mathsf{E}\left[|S_n|^r\right] \leq \mathscr{S}_r(a+1)^r \tag{2.15}$$

for any positive integer r, where $\mathscr{S}_r$ is a universal constant that depends only on r.

We refer to $\mathscr{S}_r$ as the *Skorokhod constant.*

Remark 2.2 One can estimate the number a in Theorem 2.13 with the help of the Markov–Chebyshev inequality. Indeed, if $t \geq 1$, then $a = \left(8e\mathsf{E}\left[|S_n|^t\right]\right)^{1/t}$ fits the restriction in Theorem 2.13. For this a, we get from inequality (2.15)

$$\mathsf{E}\left[|S_n|^r\right] \leq \mathscr{S}_r\left(\left(8e\mathsf{E}\left[|S_n|^t\right]\right)^{1/t} + 1\right)^r.$$

Thus if n is such that $\mathsf{E}\left[|S_n|^t\right] \geq 1/8e$, then

$$\mathsf{E}\left[|S_n|^r\right] \leq 2^r\mathscr{S}_r(8e)^{r/t}\left(\mathsf{E}\left[|S_n|^t\right]\right)^{r/t}. \tag{2.16}$$

The restriction $|X_i| \leq 1, i \leq n$, in Theorem 2.13 can easily be replaced by a more general one $|X_i| \leq b$, $i \leq n$, for some constant $b > 0$. This is proved by passing to the random variables $X_k' \stackrel{\text{def}}{=} X_k/b$ and applying inequality (2.16) for X_k'. It is worth mentioning that inequality (2.16) does not change at all in this case.

If $r \le t$, then inequality (2.16) is too rough, since

$$\left(\mathsf{E}\left[|S_n|^r\right]\right)^{1/r} \le \left(\mathsf{E}\left[|S_n|^t\right]\right)^{1/t}$$

by *Lyapunov's inequality*. Otherwise, namely if $r > t$, then (2.16) is a new result.

Put, for example, $t = 2$ in inequality (2.16). Then, for $r > 2$,

$$\mathsf{E}\left[|S_n|^r\right] \le \mathscr{S}_r' \left(\sum_{i=1}^n \mathsf{E}\left[X_i^2\right]\right)^{r/2}$$

if $\mathsf{E}\left[S_n^2\right] \ge 1/8e$, where $\mathscr{S}_r' = 2^r \mathscr{S}_r (8e)^{r/2}$. The latter inequality, in contrast to Rosenthal's inequality (2.11), does not contain higher moments of the random variables $\{X_i\}$ on the right-hand side.

Inequality (2.16) can be used for values other than $t = 2$. For example, let $1 \le t \le 2$. Assuming that $\mathsf{E}[X_i] = 0, i \le n$, we derive from the Bahr–Esseen inequality (2.6) that

$$\mathsf{E}\left[|S_n|^r\right] \le \mathscr{S}_{r,t} \left(\sum_{i=1}^n \mathsf{E}\left[|X_i|^t\right]\right)^{r/t},$$

where $\mathscr{S}_{r,t} = 2^r (16e)^{r/t} \mathscr{S}_r$.

Skorokhod's inequality for multi-indexed sums reads as follows.

Theorem 2.14 *Let $\{X(\mathbf{n})\}$ be a field of independent random variables. Let $\mathfrak{L} \subset \mathbf{N}^d$ be an arbitrary finite set and $r \ge 2$ be a positive integer, $1 \le t \le 2$. Also let $\mathsf{E}[X(\mathbf{k})] = 0$ and $|X(\mathbf{k})| \le b$ almost surely for some non-random constant $b > 0$ and for all $\mathbf{k} \in \mathfrak{L}$. If $\mathsf{E}\left[\left|\sum_{\mathbf{k}\in\mathfrak{L}} X(\mathbf{k})\right|^t\right] \ge 1/8e$, then*

$$\mathsf{E}\left[\left|\sum_{\mathbf{k}\in\mathfrak{L}} X(\mathbf{k})\right|^r\right] \le \mathscr{S}_{r,t} \left(\sum_{\mathbf{k}\in\mathfrak{L}} \mathsf{E}\left[|X(\mathbf{k})|^t\right]\right)^{r/t}, \tag{2.17}$$

where $\mathscr{S}_{r,t}$ is the universal constant defined above.

Remark 2.3 It may appear that the assumption that the terms in the sum are bounded restricts the possible applications of Skorokhod's inequality. This is not the case for problems related to the almost sure asymptotic behavior of normalized sums of independent random variables, since a common trick in the proofs of such results is to truncate the random variables so that after a point we are dealing with bounded terms and Skorokhod's inequality applies.

Finally we note that the restriction $\mathsf{E}\left[|S_n|^r\right] \ge 1/8e$ is not critical at all. Instead one can assume that

$$\theta \stackrel{\text{def}}{=} \inf_n \mathsf{E}\left[|S_n|^r\right] > 0.$$

In this case, the constant $\mathscr{S}_{r,t}$ in (2.17) changes and depends on θ.

2.2 Maximal Inequalities for Distributions of Multi-Indexed Sums

2.2.1 A Generalization of Petrov's Inequality

Let $0 < q < 1$. Recall that a number $\varkappa_q = \varkappa_q(X)$ is called a *q-quantile* of a random variable X if

$$\mathsf{P}(X \le \varkappa_q) \ge q, \quad \mathsf{P}(X \ge \varkappa_q) \ge 1 - q.$$

A q-quantile is called a *median* if $q = \frac{1}{2}$.

Lemma 2.1 describes some key properties of quantiles.

Lemma 2.1 *Every random variable either possesses a unique q-quantile or the set of its q-quantiles coincides with a closed interval in the real line.*

If $\varkappa_q$ is a q-quantile of a random variable X, then $-\varkappa_q$ is a $(1-q)$-quantile of the random variable $-X$.

The following is a known inequality for $d = 1$ due to V. V. Petrov.

Theorem 2.15 (V. V. Petrov) *Let $0 < q < 1$. If $X_1, \dots, X_n$ are independent random variables, then*

$$\mathsf{P}\left(\max_{k\le n}[S_k - \varkappa_q(S_k - S_n)] \ge x\right) \le q^{-1}\mathsf{P}(S_n \ge x) \tag{2.18}$$

for all $x \in \mathbf{R}$.

We refer to (2.18) as *Petrov's inequality*.

Our aim in this section is to generalize inequality (2.18) to the case $d > 1$. Let $\mathbf{n} \in \mathbf{N}^d$ be a fixed multi-index. Given $\mathbf{k} \le \mathbf{n}$, introduce $d+1$ multi-indices $\mathbf{k}_0, \dots, \mathbf{k}_d$: the "end" members of this sequence are defined by

$$\mathbf{k}_0 = \mathbf{k}, \quad \mathbf{k}_d = \mathbf{n}.$$

The construction is complete if $d = 1$. Let $d \ge 2$. If $0 < j < d$ is given, then the first j coordinates of the multi-index $\mathbf{k}_j$ coincide with the corresponding coordinates of the element $\mathbf{n}$, while the last $d - j$ coordinates of $\mathbf{k}_j$ coincide with those of the multi-index $\mathbf{k}$:

$$\mathbf{k}_j = (n_1, \dots, n_j, k_{j+1}, \dots, k_d), \quad 1 \le j \le d-1. \tag{2.19}$$

The definition of the multi-indices $\mathbf{k}_1$ and $\mathbf{k}_{d-1}$ include the expressions "$n_1, \dots, n_1$" and "$k_d, \dots, k_d$", which are to be understood as just "n_1" and "k_d", respectively.

The procedure of creating a multi-index $\mathbf{k}_j$ from multi-indices $\mathbf{k} \le \mathbf{n}$ can be repeated several times. For example, $\left(\mathbf{k}_0\right)_1 = (k_1, k_2, \dots, k_d)_1 = (n_1, k_2, \dots, k_d)$, $\left(\mathbf{k}_1\right)_1 = (n_1, k_2, \dots, k_d)_1 = (n_1, k_2, \dots, k_d) = \mathbf{k}_1$,

$$(\mathbf{k}_1)_2 = (n_1, k_2, \ldots, k_d)_2 = (n_1, n_2, k_3, \ldots, k_d)$$

etc.

Let $\{X(\mathbf{k}), \mathbf{k} \le \mathbf{n}\}$ be independent random variables. As usual, their sums are denoted by $\{S(\mathbf{k}), \mathbf{k} \le \mathbf{n}\}$. Put

$$\Gamma(\mathbf{k}) = \sum_{j=1}^{d} \varkappa_q(S(\mathbf{k}_{j-1}) - S(\mathbf{k}_j)). \tag{2.20}$$

Note that the number $\Gamma(\mathbf{k})$ depends not only on $\mathbf{k}$, but also on the multi-index $\mathbf{n}$. In fact, $\Gamma(\mathbf{k}) = \Gamma(\mathbf{k}, \mathbf{n})$.

Theorem 2.16 *Let* $0 < q < 1$. *If* $\{X(\mathbf{k}), \mathbf{k} \le \mathbf{n}\}$ *are independent random variables, then*

$$\mathsf{P}\left(\max_{\mathbf{k} \le \mathbf{n}} [S(\mathbf{k}) - \Gamma(\mathbf{k})] \ge x \right) \le q^{-d} \mathsf{P}(S(\mathbf{n}) \ge x) \tag{2.21}$$

for all $x \in \mathbf{R}$, *where one can choose arbitrary quantiles* $\varkappa_q(S(\mathbf{k}_{j-1}) - S(\mathbf{k}_j))$ *to form the number* $\Gamma(\mathbf{k})$.

Proof Inequality (2.21) coincides with Petrov's inequality (2.18) if $d = 1$. We prove (2.21) for $d \ge 2$ by induction on the dimension of the space $\mathbf{N}^d$. Assume that Theorem 2.16 holds for all spaces $\mathbf{N}^b$ with $b < d$ and let us prove it for $b = d$.

Let $\mathbf{n} = (n_1, \ldots, n_d)$. Given $1 \le j \le d$, we recursively define the sequence of random variables $\{\mathscr{M}_j(\ell), \ell \le n_j\}$ and the random number I_j, namely if the random variables $\{\mathscr{M}_j(\ell), \ell \le n_j\}$ are already defined for a given $1 \le j \le d$, then the random number I_j is chosen such that

$$I_j = \begin{cases} \min\{\ell : \mathscr{M}_j(\ell) \ge x\}, & \text{if } \max_{\ell \le n_j} \mathscr{M}_j(\ell) \ge x, \\ 0, & \text{if } \max_{\ell \le n_j} \mathscr{M}_j(\ell) < x. \end{cases} \tag{2.22}$$

The sequences $\{\mathscr{M}_j(\ell), \ell \le n_j\}$ are defined as follows: for $j = 1$,

$$\mathscr{M}_1(\ell) = \max_{\substack{k_2 \le n_2 \\ \cdots \\ k_d \le n_d}} [S(\ell, k_2, \ldots, k_d) - \Gamma(\ell, k_2, \ldots, k_d)];$$

for $2 \le j < d$ (clearly, this case happens only if $d \ge 3$),

$$\begin{aligned} \mathscr{M}_j(\ell) = \max_{\substack{k_{j+1} \le n_{j+1} \\ \cdots \\ k_d \le n_d}} & [S(I_1, \ldots, I_{j-1}, \ell, k_{j+1}, \ldots, k_d) \\ & - \Gamma(I_1, \ldots, I_{j-1}, \ell, k_{j+1}, \ldots, k_d)]; \end{aligned}$$

and finally for $j = d$,

$$\mathscr{M}_d(\ell) = S(I_1, \dots, I_{d-1}, \ell) - \Gamma(I_1, \dots, I_{d-1}, \ell).$$

In the definition of $\mathscr{M}_2(\ell)$ and $\mathscr{M}_{d-1}(\ell)$, the expressions "$I_1, \dots, I_1$" and "$k_d, \dots, k_d$" are to be understood as "I_1" and "k_d", respectively. The sequences $\{\mathscr{M}_j(\ell), \ell \le n_j\}$ are well defined if we agree that $\Gamma(\mathbf{k}) = 0$ and $S(\mathbf{k}) = 0$ in the case where at least one of the coordinates of the element $\mathbf{k}$ equals 0.

Then we introduce the random events $\{\mathscr{I}(\mathbf{k}), \mathbf{k} \le \mathbf{n}\}$ and $\{E(\mathbf{k}), \mathbf{k} \le \mathbf{n}\}$ as follows:

$$\begin{aligned}\mathscr{I}(\mathbf{k}) &= \{\omega : I_1 = k_1, \dots, I_d = k_d\},\\ E(\mathbf{k}) &= \{\omega : S(\mathbf{k}_1) - S(\mathbf{k}_0) - \varkappa_{1-q}(S(\mathbf{k}_1) - S(\mathbf{k}_0)) \ge 0\}.\end{aligned}$$

The events $\mathscr{I}(\mathbf{k}), \mathbf{k} \le \mathbf{n}$, are disjoint, that is $\mathscr{I}(\mathbf{k}) \cap \mathscr{I}(\mathbf{m}) = \varnothing$ if $\mathbf{k} \ne \mathbf{m}$. Moreover

$$\bigcup_{\mathbf{k} \le \mathbf{n}} \mathscr{I}(\mathbf{k}) = \left\{ \max_{\mathbf{k} \le \mathbf{n}} [S(\mathbf{k}) - \Gamma(\mathbf{k})] \ge x \right\}.$$

Note that the random event $\mathscr{I}(\mathbf{k})$ for every $\mathbf{k} \le \mathbf{n}$ is defined in terms of random variables $X(\mathbf{m})$ whose indices $\mathbf{m} = (m_1, \dots, m_d)$ are such that $m_1 \le k_1$, while the random event $E(\mathbf{k})$ is expressed in terms of the random variables $X(\mathbf{m})$ whose indices are such that $m_1 > k_1$. Thus $\mathscr{I}(\mathbf{k})$ and $E(\mathbf{k})$ are pairwise independent for all $\mathbf{k} \le \mathbf{n}$. Moreover, the definition of a $(1-q)$-quantile implies that

$$\mathsf{P}(E(\mathbf{k})) \ge q \quad \text{for all} \quad \mathbf{k} \le \mathbf{n}.$$

Thus

$$q\mathsf{P}\left(\max_{\mathbf{k} \le \mathbf{n}} [S(\mathbf{k}) - \Gamma(\mathbf{k})] \ge x \right) = q \sum_{\mathbf{k} \le \mathbf{n}} \mathsf{P}(\mathscr{I}(\mathbf{k})) \le \sum_{\mathbf{k} \le \mathbf{n}} \mathsf{P}(\mathscr{I}(\mathbf{k}) \cap E(\mathbf{k})).$$

For $\omega \in \mathscr{I}(\mathbf{k})$, we have $S(\mathbf{k}) \ge \Gamma(\mathbf{k}) + x$. Recall $\mathbf{k}_0 = \mathbf{k}$ and thus

$$S(\mathbf{k}_1) - \varkappa_{1-q}(S(\mathbf{k}_1) - S(\mathbf{k}_0)) \ge \Gamma(\mathbf{k}) + x$$

for $\omega \in \mathscr{I}(\mathbf{k}) \cap E(\mathbf{k})$. Since

$$\varkappa_{1-q}(S(\mathbf{k}_1) - S(\mathbf{k}_0)) + \Gamma(\mathbf{k}) = \Gamma(\mathbf{k}_1),$$

we get $S(\mathbf{k}_1) - \Gamma(\mathbf{k}_1) \ge x$ for $\omega \in \mathscr{I}(\mathbf{k}) \cap E(\mathbf{k})$. Therefore

$$
\begin{aligned}
q\mathsf{P}\left(\max_{\mathbf{k}\le\mathbf{n}}[S(\mathbf{k})-\Gamma(\mathbf{k})]\ge x\right)
&\le \sum_{\mathbf{k}\le\mathbf{n}} \mathsf{P}(\mathscr{I}(\mathbf{k}), S(\mathbf{k}_1)-\Gamma(\mathbf{k}_1)\ge x)\\
&\le \sum_{\mathbf{k}\le\mathbf{n}} \mathsf{P}\Bigg(\mathscr{I}(\mathbf{k}), \max_{\substack{m_2\le n_2\\ \dots\\ m_d\le n_d}}[S(\mathbf{m}_1)-\Gamma(\mathbf{m}_1)]\ge x\Bigg)\\
&\le \mathsf{P}\Bigg(\max_{\substack{k_2\le n_2\\ \dots\\ k_d\le n_d}}[S(\mathbf{k}_1)-\Gamma(\mathbf{k}_1)]\ge x\Bigg).
\end{aligned}
\tag{2.23}
$$

Now we use the inductive assumption. Put

$$
X'(\widetilde{\mathbf{k}})=\sum_{\ell=1}^{n_d} X(\ell,\widetilde{\mathbf{k}}),\quad S'(\widetilde{\mathbf{k}})=\sum_{\widetilde{\mathbf{m}}\le\widetilde{\mathbf{k}}} X'(\widetilde{\mathbf{m}}),
$$

where $\widetilde{\mathbf{k}}=(k_2,\dots,k_d)$ and $\widetilde{\mathbf{m}}=(m_2,\dots,m_d)$. The random variables $\{X'(\widetilde{\mathbf{k}}), \widetilde{\mathbf{k}}\le\widetilde{\mathbf{n}}\}$ are independent. Using the sums $S'(\widetilde{\mathbf{k}})$, we define the numbers $\Gamma'(\widetilde{\mathbf{k}})$ according to (2.20) but for elements $\widetilde{\mathbf{k}}$ of the space $\mathbf{N}^{d-1}$ rather than for elements of the space $\mathbf{N}^d$. By the inductive assumption,

$$
\mathsf{P}\left(\max_{\widetilde{\mathbf{k}}\le\widetilde{\mathbf{n}}}[S'(\widetilde{\mathbf{k}})-\Gamma'(\widetilde{\mathbf{k}})]\ge x\right)\le q^{-(d-1)}\mathsf{P}(S'(\widetilde{\mathbf{n}})\ge x).
$$

It is clear that $S'(\widetilde{\mathbf{k}})=S(\mathbf{k}_1)$, $\Gamma'(\widetilde{\mathbf{k}})=\Gamma(\mathbf{k}_1)$, and $S'(\widetilde{\mathbf{n}})=S(\mathbf{n})$. Thus (2.23) implies

$$
q\mathsf{P}\left(\max_{\mathbf{k}\le\mathbf{n}}[S(\mathbf{k})-\Gamma(\mathbf{k})]\ge x\right)\le q^{-(d-1)}\mathsf{P}(S(\mathbf{n})\ge x)
$$

and this completes the proof. □

Corollary 2.1 *Assume that there are a constant $c\ge 0$ and q-quantiles of the random variables $S(\mathbf{k}_{j-1})-S(\mathbf{k}_j)$ such that*

$$
\varkappa_q(S(\mathbf{k}_{j-1})-S(\mathbf{k}_j))\le c \quad \textit{for all} \quad \mathbf{k}\le\mathbf{n} \;\; \textit{and} \;\; 1\le j\le d.
$$

Then

$$
\mathsf{P}\left(\max_{\mathbf{k}\le\mathbf{n}} S(\mathbf{k})\ge x\right)\le q^{-d}\mathsf{P}(S(\mathbf{n})\ge x-dc) \tag{2.24}
$$

for all $x\in\mathbf{R}$.

Proof Since $\Gamma(\mathbf{k}) \le dc$, we get

$$\left\{\omega\colon \max_{\mathbf{k}\le\mathbf{n}} S(\mathbf{k}) \ge x\right\} \subseteq \left\{\omega\colon \max_{\mathbf{k}\le\mathbf{n}}[S(\mathbf{k}) - \Gamma(\mathbf{k})] \ge x - dc\right\}$$

for all $x \in \mathbf{R}$. Thus inequality (2.24) follows from (2.21). □

Corollary 2.2 *If for some constants $c \ge 0$ and $q > 0$*

$$\mathsf{P}(S(\mathbf{k}_{j-1}) - S(\mathbf{k}_j) \le c) \ge q \quad \textit{for all } j = 1, \dots, d \ \textit{ and } \mathbf{k} \le \mathbf{n},$$

then inequality (2.24) *holds for all $x \in \mathbf{R}$.*

To prove Corollary 2.2 we need the following auxiliary results.

Lemma 2.2 *Let $0 < q < 1$. If $\mathsf{P}(X \le a) \ge q$ for some $a \in \mathbf{R}$, then there exists a quantile $\varkappa_q(X)$ such that $\varkappa_q(X) \le a$. If $\mathsf{P}(X \ge b) \ge q$, then there exists a quantile $\varkappa_{1-q}(X)$ such that $\varkappa_{1-q}(X) \ge b$.*

Proof of Lemma 2.2 We prove only the first assertion of the lemma (the second one is proved similarly). Let $\mathsf{P}(X \le a) \ge q$. Then $\mathsf{P}(X > a) \le 1 - q$. Let $\varkappa_q^*(X)$ be the minimal q-quantile of the random variable X. We have

$$\mathsf{P}(X \ge \varkappa_q^*(X)) \ge 1 - q \ge \mathsf{P}(X > a).$$

This implies that either $\varkappa_q^*(X) \le a$ or $\mathsf{P}(a < X < \varkappa_q^*(X)) = 0$. In the latter case every number from the interval $(a, \varkappa_q^*(X))$ is a q-quantile of the random variable X, that is $\varkappa_q^*(X)$ is not the minimal q-quantile. Thus $\varkappa_q^*(X) \le a$. □

Proof of Corollary 2.2 Lemma 2.2 implies that there exist q-quantiles such that $\varkappa_q(S(\mathbf{k}_{j-1}) - S(\mathbf{k}_j)) \le c$ for all $\mathbf{k} \le \mathbf{n}$ and $1 \le j \le d$. Thus Corollary 2.2 follows from Corollary 2.1 with the same quantiles. □

2.2.2 Lévy's Inequality

Lévy's inequality is a particular case of inequality (2.18) corresponding to $q = \frac{1}{2}$, that is to medians. Put

$$M(\mathbf{k}) = \sum_{j=1}^{d} \operatorname{med}(S(\mathbf{k}_{j-1}) - S(\mathbf{k}_j)),$$

where $\operatorname{med}(\xi)$ is a median of a random variable ξ.

Corollary 2.3 *Let $\{X(\mathbf{k}), \mathbf{k} \le \mathbf{n}\}$ be independent random variables. Then*

$$\mathsf{P}\left(\max_{\mathbf{k}\le\mathbf{n}} [S(\mathbf{k}) - M(\mathbf{k})] \ge x \right) \le 2^d \mathsf{P}(S(\mathbf{n}) \ge x) \tag{2.25}$$

for all $x \in \mathbf{R}$.

Proof Inequality (2.25) follows from inequality (2.21) with $q = \frac{1}{2}$. □

Corollary 2.4 *Let $\{X(\mathbf{k}), \mathbf{k} \le \mathbf{n}\}$ be independent symmetric random variables. Then*

$$\mathsf{P}\left(\max_{\mathbf{k}\le\mathbf{n}} S(\mathbf{k}) \ge x \right) \le 2^d \mathsf{P}(S(\mathbf{n}) \ge x), \tag{2.26}$$

$$\mathsf{P}\left(\max_{\mathbf{k}\le\mathbf{n}} |S(\mathbf{n})| \ge x \right) \le 2^d \mathsf{P}(|S(\mathbf{n})| \ge x) \tag{2.27}$$

for all $x \in \mathbf{R}$.

Proof One of the medians $\mu(S(\mathbf{k}_{j-1}) - S(\mathbf{k}_j))$ equals zero in the case of symmetric random variables $\{X(\mathbf{k})\}$ for all $j = 1, \dots, d$ and $\mathbf{k} \le \mathbf{n}$. Thus inequality (2.26) follows directly from (2.25) where one takes the zero medians. Applying (2.26) to the random variables $\{-X(\mathbf{k}), \mathbf{k} \le \mathbf{n}\}$ we obtain (2.27). □

Remark 2.4 Corollary 2.4 generalizes Lévy's inequality (2.4) to the case $d > 1$. The constant 2^d on the right-hand side is optimal for $d = 1$. This follows by considering the case of Bernoulli random variables $X_1, \dots, X_n$ and non-integer x, since both sides of inequality (2.4) coincide in this case. The question of whether or not the constant 2^d is optimal for $d > 1$ is not as simple, however the answer is positive in this case, too.

According to Wichura's invariance principle [419]

$$\frac{1}{n^{d/2}} S([nt_1], \dots, [nt_d]) \quad \Longrightarrow \quad W(t_1, \dots, t_d),$$

in the space $D[0, 1]^d$, where W is the so-called Chentsov–Yeh random field. In particular,

$$\lim_{n\to\infty} \mathsf{P}\left(\frac{1}{n^{d/2}} \max_{\mathbf{k}\le(n,\dots,n)} S(\mathbf{k}) \ge x \right) = \mathsf{P}\left(\sup_{t_1,\dots,t_d\in[0,1]^d} W(t_1, \dots, t_d) \ge x \right).$$

It is shown in Chap. 2 of [341] that

$$\mathsf{P}\left(\sup_{t_1,\dots,t_d\in[0,1]^d} W(t_1, \dots, t_d) \ge x \right) \sim 2^d \mathsf{P}(W(1, \dots, 1) \ge x) \qquad \text{as } x \to \infty.$$

This means that the inequality

$$\mathsf{P}\left(\max_{\mathbf{k}\le\mathbf{n}} S(\mathbf{n}) \ge x\right) \le c\mathsf{P}(S(\mathbf{n}) \ge x)$$

does not hold for any $c < 2^d$ if x and $n_1, \dots, n_d$ are sufficiently large.

On the other hand it is still unknown whether or not the latter inequality with $c < 2^d$ may hold for small x, or small $n_1, \dots, n_d$, or in the case where the second moments do not exist.

2.2.3 *Ottaviani's Inequality*

One of the achievements of probability theory in the first half of the 20th century is Lévy's theorem on the equivalence of almost sure convergence and convergence in probability of a series of independent random variables (Theorem 5.2 is an analog of Lévy's theorem for multi-indexed sums). One of the proofs of Levy's theorem is based on the following result, known as *Ottaviani's inequality*.

Theorem 2.17 (G. Ottaviani) *If*

$$\mathsf{P}(|S_k - S_n| \ge \varepsilon) \le \alpha, \qquad \textit{for all} \quad k = 0, \dots, n-1,$$

for some $\varepsilon > 0$ *and* $0 < \alpha < 1$, *then*

$$\mathsf{P}\left(\max_{k\le n} |S_k| \ge 2\varepsilon\right) \le \frac{\alpha}{1-\alpha}. \tag{2.28}$$

One can prove a generalization of Ottaviani's inequality for $d > 1$ with the help of inequality (2.21).

Corollary 2.5 *Assume that*

$$\mathsf{P}(S(\mathbf{k}_{j-1}) - S(\mathbf{k}_j) > \varepsilon) \le \alpha \qquad \textit{for all} \quad j = 1, \dots, d \;\; \textit{and} \quad \mathbf{k} \le \mathbf{n}, \tag{2.29}$$

for some $\varepsilon > 0$ *and* $0 < \alpha < 1$, *where the multi-indices* $\mathbf{k}_j$ *are defined by equality* (2.19). *Then*

$$\mathsf{P}\left(\max_{\mathbf{k}\le\mathbf{n}} S(\mathbf{k}) \ge (d+1)\varepsilon\right) \le (1-\alpha)^{-d}\mathsf{P}(S(\mathbf{n}) \ge \varepsilon). \tag{2.30}$$

Moreover, if

$$\mathsf{P}(|S(\mathbf{k}_{j-1}) - S(\mathbf{k}_j)| > \varepsilon) \le \alpha \quad \textit{for all } j = 1, \dots, d \;\; \textit{and} \quad \mathbf{k} \le \mathbf{n},$$

then

$$\mathsf{P}\left(\max_{\mathbf{k}\le\mathbf{n}} |S(\mathbf{k})| \ge (d+1)\varepsilon\right) \le (1-\alpha)^{-d}\mathsf{P}(|S(\mathbf{n})| \ge \varepsilon). \tag{2.31}$$

Proof The assumption of the corollary implies that

$$\mathsf{P}(S(\mathbf{k}_{j-1}) - S(\mathbf{k}_j) \le \varepsilon) \ge 1-\alpha \qquad \text{for all} \quad j=1,\dots,d \ \text{ and } \quad \mathbf{k}\le\mathbf{n}.$$

Thus Corollary 2.2 with $c=\varepsilon$, $x=\varepsilon$, and $q=1-\alpha$ implies inequality (2.30).

The same reasoning for random variables $-X(\mathbf{k})$ yields

$$\mathsf{P}\left(\max_{\mathbf{k}\le\mathbf{n}}[-S(\mathbf{k})] \ge (d+1)\varepsilon\right) \le (1-\alpha)^{-d}\mathsf{P}(-S(\mathbf{n}) \ge \varepsilon).$$

Since

$$\left\{\omega\colon \max_{\mathbf{k}\in A} S(\mathbf{k}) \ge z\right\} \cup \left\{\omega\colon \max_{\mathbf{k}\in A}[-S(\mathbf{k})] \ge z\right\} = \left\{\omega\colon \max_{\mathbf{k}\in A} |S(\mathbf{k})| \ge z\right\}$$

for all $A \subseteq \mathbf{N}^d$ and $z>0$, inequality (2.31) is also proved. □

The most useful variant of Corollary 2.5 is given below.

Corollary 2.6 *Assume that condition* (2.29) *holds for some* $\varepsilon>0$ *and* $0<\alpha<\frac{1}{d}$. *Then*

$$\mathsf{P}\left(\max_{\mathbf{k}\le\mathbf{n}} |S(\mathbf{k})| \ge (d+1)\varepsilon\right) \le \frac{\alpha}{1-d\alpha}. \tag{2.32}$$

Proof Corollary 2.6 follows from Corollary 2.5 in view of the Bernoulli inequality: $1-dt \le (1-t)^d$, $0\le t\le 1$. □

2.2.4 Kolmogorov's Inequality for Probabilities

The proof of the law of the iterated logarithm for $d=1$ is based on the so-called *Kolmogorov inequality*.

Theorem 2.18 (A. N. Kolmogorov) *If* $\mathsf{E}[X_i]=0$ *and* $\mathsf{E}\left[X_i^2\right]<\infty$ *for all* $i=1,\dots,n$, *then*

$$\mathsf{P}\left(\max_{k\le n} S_n \ge x\right) \le 2\mathsf{P}\left(S_n \ge x-\sqrt{2B_n}\right) \tag{2.33}$$

for all $x\in\mathbf{R}$, *where* $B_n=\mathsf{E}\left[S_n^2\right]$.

An analogous inequality for the case $d>1$ can be derived from Corollary 2.2.

Corollary 2.7 *Let* $\mathsf{E}[X(\mathbf{k})] = 0$ *and* $\mathsf{E}\left[X^2(\mathbf{k})\right] = 0$ *for all* $\mathbf{k} \le \mathbf{n}$. *Put* $B(\mathbf{n}) = \mathsf{E}\left[S^2(\mathbf{n})\right]$. *Then*

$$\mathsf{P}\left(\max_{\mathbf{k}\le\mathbf{n}} S(\mathbf{k}) \ge x\right) \le 2^d \mathsf{P}\left(S(\mathbf{n}) \ge x - d\sqrt{2B(\mathbf{n})}\right), \tag{2.34}$$

$$\mathsf{P}\left(\max_{\mathbf{k}\le\mathbf{n}} |S(\mathbf{k})| \ge x\right) \le 2^d \mathsf{P}\left(|S(\mathbf{n})| \ge x - d\sqrt{2B(\mathbf{n})}\right) \tag{2.35}$$

for all $x \in \mathbf{R}$.

An even more general result can be proved for which Corollary 2.7 is the particular case corresponding to $q = \frac{1}{2}$ and $r = 2$.

Corollary 2.8 *Let* $\mathbf{n} \in \mathbf{N}^d$, $0 < q < 1$, *and* $r > 0$. *Assume that*

$$\mathsf{E}\left[|X(\mathbf{k})|^r\right] < \infty \quad \textit{for all } \mathbf{k} \le \mathbf{n}.$$

Put

$$M_r(\mathbf{n}) = \sum_{\mathbf{k}\le\mathbf{n}} \mathsf{E}\left[|X(\mathbf{k})|^r\right].$$

If $0 < r \le 1$, *then*

$$\mathsf{P}\left(\max_{\mathbf{k}\le\mathbf{n}} S(\mathbf{k}) \ge x\right) \le q^{-d} \mathsf{P}\left(S(\mathbf{n}) \ge x - d(1-q)^{-1/r} M_r^{1/r}(\mathbf{n})\right) \tag{2.36}$$

for all $x \in \mathbf{R}$.
If $1 \le r \le 2$ *and* $\mathsf{E}[X(\mathbf{k})] = 0$ *for all* $\mathbf{k} \le \mathbf{n}$, *then*

$$\mathsf{P}\left(\max_{\mathbf{k}\le\mathbf{n}} S(\mathbf{k}) \ge x\right) \le q^{-d} \mathsf{P}\left(S(\mathbf{n}) \ge x - d(1-q)^{-1/r} \left(c_r M_r(\mathbf{n})\right)^{1/r}\right) \tag{2.37}$$

for all $x \in \mathbf{R}$, *where*

$$c_r = \begin{cases} 1, & \text{for } r = 1 \text{ or } r = 2, \\ 2, & \text{for } 1 < r < 2. \end{cases}$$

Finally, if $r \ge 1$, *then*

$$\mathsf{P}\left(\max_{\mathbf{k}\le\mathbf{n}} S(\mathbf{k}) \ge x\right) \le q^{-d} \mathsf{P}\left(S(\mathbf{n}) \ge x - d(1-q)^{-1/r} \left(\mathsf{E}\left[|S(\mathbf{n})|^r\right]\right)^{1/r}\right) \tag{2.38}$$

for all $x \in \mathbf{R}$.

Proof For any $t > 0$,

$$\mathsf{P}\left(|S(\mathbf{k}_{j-1}) - S(\mathbf{k}_j)| \geq t(1-q)^{-1/r}\right) \leq (1-q)\frac{\mathsf{E}\left[|S(\mathbf{k}_{j-1}) - S(\mathbf{k}_j)|^r\right]}{t^r} \tag{2.39}$$

by the Chebyshev–Markov inequality for all $j = 1, \dots, d$ and $\mathbf{k} \leq \mathbf{n}$. To prove (2.36) we choose $t = (M_r(\mathbf{n}))^{1/r}$ and use the inequality

$$\mathsf{E}\left[|S(\mathbf{k}_{j-1}) - S(\mathbf{k}_j)|^r\right] \leq M_r(\mathbf{n})$$

that results from $|x+y|^r \leq |x|^r + |y|^r$ for $0 < r \leq 1$.

To prove (2.37) we choose $t = (c_r M_r(\mathbf{n}))^{1/r}$ and use the inequality

$$\mathsf{E}\left[|S(\mathbf{k}_{j-1}) - S(\mathbf{k}_j)|^r\right] \leq c_r M_r(\mathbf{n})$$

that follows from the Bahr–Esseen inequality (2.8).

Finally, to prove (2.38) we choose $t = \left(\mathsf{E}\left[|S(\mathbf{n})|^r\right]\right)^{1/r}$ and use the inequality

$$\mathsf{E}\left[|S(\mathbf{k}_j) - S(\mathbf{k}_{j-1})|^r\right] \leq \mathsf{E}\left[|S(\mathbf{n})|\right]^r,$$

which is a consequence of the martingale property of the sequence $\zeta_j = S(k_j)$.

In either case,

$$\mathsf{P}\left(S(\mathbf{k}_j) - S(\mathbf{k}_{j-1}) \geq -t(1-q)^{-1/r}\right) \geq q$$

and thus Corollary 2.8 follows from Corollary 2.2. □

2.3 Maximal Inequalities

2.3.1 A Generalization of Kolmogorov's Inequality for Moments

An analog of inequality (2.1) for $d > 1$ can be proved by using the results of the preceding section. Moreover, this method allows us to obtain upper bounds not only for probabilities but also for moments as in Theorem 2.4.

Corollary 2.9 *Let* $\mathbf{n} \in \mathbf{N}^d$ *and* $r > 0$. *If* $\{X(\mathbf{k}), \mathbf{k} \leq \mathbf{n}\}$ *are independent symmetric random variables such that* $\mathsf{E}\left[|X(\mathbf{k})|^r\right] < \infty$ *for all* $\mathbf{k} \leq \mathbf{n}$, *then*

$$\mathsf{E}\left[\max_{\mathbf{k}\leq\mathbf{n}} |S(\mathbf{k})|^r\right] \leq 2^d \mathsf{E}\left[|S(\mathbf{n})|^r\right]. \tag{2.40}$$

Proof First we multiply inequality (2.27) by rx^{r-1} and integrate the result in the interval $[0, \infty)$. Using the representation for the moment of the random variable ξ

$$\mathsf{E}\left[|\xi|^r\right] = r\int_0^\infty x^{r-1}\mathsf{P}(|\xi| \ge x)\,dx$$

(see Sect. A.19, we prove inequality (2.40). □

Corollary 2.10 *Let* $\mathbf{n} \in \mathbf{N}^d$ *and* $r > 0$. *Let* $\{X(\mathbf{k}), \mathbf{k} \le \mathbf{n}\}$ *be independent random variables such that* $\mathsf{E}\left[|X(\mathbf{k})|^r\right] < \infty$ *for all* $\mathbf{k} \le \mathbf{n}$. *Then*

$$\mathsf{E}\left[\max_{\mathbf{k}\le\mathbf{n}} |S(\mathbf{k})|^r\right] \le \sum_{\mathbf{k}\le\mathbf{n}} \mathsf{E}\left[|X(\mathbf{k})|^r\right] \tag{2.41}$$

in the case when $0 < r \le 1$.

If $r \ge 1$ *and* $\mathsf{E}[X(\mathbf{k})] = 0$ *for all* $\mathbf{k} \le \mathbf{n}$, *then*

$$\mathsf{E}\left[\max_{\mathbf{k}\le\mathbf{n}} |S(\mathbf{k})|^r\right] \le 2^{d+r}\mathsf{E}\left[|S(\mathbf{n})|^r\right]. \tag{2.42}$$

Moreover, under the same assumptions

$$\mathsf{E}\left[\max_{\mathbf{k}\le\mathbf{n}} |S(\mathbf{k})|^r\right] \le 2^{d+r+1}\sum_{\mathbf{k}\le\mathbf{n}} \mathsf{E}\left[|X(\mathbf{k})|^r\right] \tag{2.43}$$

for $1 \le r \le 2$ *and*

$$\mathsf{E}\left[\max_{\mathbf{k}\le\mathbf{n}} |S(\mathbf{k})|^r\right] \le 2^{d+r} J_r |\mathbf{n}|^{-1+r/2}\sum_{\mathbf{k}\le\mathbf{n}} \mathsf{E}\left[|X(\mathbf{k})|^r\right] \tag{2.44}$$

for $r \ge 2$, *where* J_r *is the Dharmadhikari–Jogdeo constant* (2.9).

Remark 2.5 As seen from the proof below, the constant on the right-hand side of (2.43) equals 2^{d+2} if $r = 2$. In fact, this constant can be decreased up to 2^{d+1} in this case.

Proof The inequality (2.41) for $0 < r \le 1$ follows from the bound

$$|S(\mathbf{k})|^r \le \sum_{\mathbf{m}\le\mathbf{n}} \mathsf{E}\left[|S(\mathbf{m})|^r\right] \qquad \text{for all } \mathbf{k} \le \mathbf{n}.$$

To prove inequality (2.42) for $r \ge 1$, we first apply inequality (2.40) for symmetrized $\{S^{(s)}(\mathbf{k}), \mathbf{k} \le \mathbf{n}\}$:

$$\mathsf{E}\left[\max_{\mathbf{k}\le\mathbf{n}} |S^{(s)}(\mathbf{k})|^r\right] \le 2^d\mathsf{E}\left[|S^{(s)}(\mathbf{n})|^r\right].$$

Then we estimate the right-hand side by the Hölder inequality

$$\mathsf{E}\left[|S^{(s)}(\mathbf{n})|^r\right] \le 2^r \mathsf{E}\left[|S(\mathbf{n})|^r\right].$$

Note that if $r = 2$, then, in fact, the constant on the right-hand side equals 2.

To complete the proof of inequality (2.42), we denote by $\mathfrak{F}$ the σ-algebra generated by the random variables $\{X(\mathbf{k}), \mathbf{k} \le \mathbf{n}\}$. Then we deduce from Jensen's inequality for conditional expectations that

$$\begin{aligned}\mathsf{E}\left[\max_{\mathbf{k}\le\mathbf{n}} |S^{(s)}(\mathbf{k})|^r\right] &= \mathsf{E}\left[\mathsf{E}\left[\max_{\mathbf{k}\le\mathbf{n}} |S^{(s)}(\mathbf{k})|^r/\mathfrak{F}\right]\right] \ge \mathsf{E}\left[\max_{\mathbf{k}\le\mathbf{n}} \mathsf{E}\left[|S^{(s)}(\mathbf{k})|^r/\mathfrak{F}\right]\right]\\ &\ge \mathsf{E}\left[\max_{\mathbf{k}\le\mathbf{n}} \left|\mathsf{E}\left[S^{(s)}(\mathbf{k})/\mathfrak{F}\right]\right|^r\right] = \mathsf{E}\left[\max_{\mathbf{k}\le\mathbf{n}} |S(\mathbf{k})|^r\right],\end{aligned}$$

since $\mathsf{E}[S(\mathbf{k})] = 0$.

Finally, to prove inequalities (2.43) and (2.44), we use (2.42) and apply the Bahr–Esseen inequality (2.8) if $1 \le r \le 2$ or the Dharmadhikari–Jogdeo inequality (2.9) if $r \ge 2$. □

Kolmogorov's inequality (2.1) follows from Corollary 2.10 with $r = 2$ and Chebyshev's inequality.

Corollary 2.11 *Let $\mathbf{n} \in \mathbf{N}^d$ and let $\{X(\mathbf{k}), \mathbf{k} \le \mathbf{n}\}$ be independent random variables such that $\mathsf{E}[X(\mathbf{k})] = 0$ and $\mathsf{E}\left[X^2(\mathbf{k})\right] < \infty$ for all $\mathbf{k} \le \mathbf{n}$. Then*

$$\mathsf{P}\left(\max_{\mathbf{k}\le\mathbf{n}} |S(\mathbf{k})| \ge x\right) \le \frac{2^{d+1}}{x^2} \sum_{\mathbf{k}\le\mathbf{n}} \mathsf{E}\left[X^2(\mathbf{k})\right] \tag{2.45}$$

for all $x > 0$.

Further bounds for the expectation of the maximum of sums of independent random variables can be derived from inequality (2.42). For example, the following result follows from Rosenthal's inequality (2.12).

Corollary 2.12 *Let $\mathbf{n} \in \mathbf{N}^d$, $r \ge 2$, and let $\{X(\mathbf{k}), \mathbf{k} \le \mathbf{n}\}$ be independent random variables such that $\mathsf{E}[X(\mathbf{k})] = 0$ and $\mathsf{E}\left[|X(\mathbf{k})|^r\right] < \infty$ for all $\mathbf{k} \le \mathbf{n}$. Then*

$$\mathsf{E}\left[\max_{\mathbf{k}\le\mathbf{n}} |S(\mathbf{k})|^r\right] \le 2^{d+r} R_r \left[\sum_{\mathbf{k}\le\mathbf{n}} \mathsf{E}\left[|X(\mathbf{k})|^r\right] + (\mathsf{var}\,[S(\mathbf{n})])^{r/2}\right], \tag{2.46}$$

where R_r is the Rosenthal constant.

2.4 A Generalization of the Hájek–Rényi Inequality

Two methods can be used to establish a generalization of the Hájek–Rényi inequality (2.2) for $d > 1$. The first allows us to obtain results for normalizing fields with non-negative increments (see Definition A.11). The second method works for a wider class of monotone normalizing fields and has the advantage that it leads to better constants in the bounds. We show how the first method works in the proof of the Hájek–Rényi version of Kolmogorov's inequality.

2.4.1 First Method

Let $\mathbf{n} \in \mathbf{N}^d$ and let a field $\{b(\mathbf{k}), \mathbf{k} \le \mathbf{n}\}$ have non-negative increments. Consider a field $\{X(\mathbf{k}), \mathbf{k} \le \mathbf{n}\}$ of independent random variables such that $\mathsf{E}[X(\mathbf{k})] = 0$ and $\mathsf{E}\left[X^2(\mathbf{k})\right] < \infty$ for all $\mathbf{k} \le \mathbf{n}$. According to Proposition A.8,

$$\max_{\mathbf{k}\le\mathbf{n}} \left|\frac{S(\mathbf{k})}{b(\mathbf{k})}\right| \le 2^d \max_{\mathbf{k}\le\mathbf{n}} \left|\sum_{\mathbf{m}\le\mathbf{k}} \frac{X(\mathbf{m})}{b(\mathbf{m})}\right|. \tag{2.47}$$

Now we square both sides of the latter inequality and pass to the mathematical expectation. Then we apply inequality (2.43) with $r = 2$ to the right-hand side:

$$\mathsf{E}\left[\max_{\mathbf{k}\le\mathbf{n}} \left|\sum_{\mathbf{m}\le\mathbf{k}} \frac{X(\mathbf{m})}{b(\mathbf{m})}\right|^2\right] \le 2^{d+1} \sum_{\mathbf{k}\le\mathbf{n}} \frac{\mathsf{E}\left[X^2(\mathbf{k})\right]}{b^2(\mathbf{k})}. \tag{2.48}$$

Combining all the above results with the Chebyshev inequality we get, for $x > 0$,

$$\mathsf{P}\left(\max_{\mathbf{k}\le\mathbf{n}} \left|\frac{S(\mathbf{k})}{b(\mathbf{k})}\right| \ge x\right) \le \frac{2^{2d+1}}{x^2} \sum_{\mathbf{k}\le\mathbf{n}} \frac{\mathsf{E}\left[X^2(\mathbf{k})\right]}{b^2(\mathbf{k})}.$$

This is an analog of inequality(2.2) for $d > 1$.

Inequality (2.47) holds in the general case, that is not only for independent random variables. On the other hand, the second step of the proof of inequality (2.48) requires the property of independence.

2.4.2 Second Method

The second method allows us to study monotone normalizing fields $\{b(\mathbf{k}), \mathbf{k} \le \mathbf{n}\}$ (see Definition A.11). Recall that the property of monotonicity of a field is less restrictive than the property of non-negativity of its increments (see Proposition A.7). Moreover, the second method also works for dependent random variables.

Let $\mathbf{n} \in \mathbf{N}^d$, $\{X(\mathbf{k}), \mathbf{k} \le \mathbf{n}\}$ be random variables, and let $\{b(\mathbf{k}), \mathbf{k} \le \mathbf{n}\}$ be an increasing field of positive real numbers such that $b(\mathbf{1}) > 0$. Let $c > 1$ and let the sets A_t, $t \ge 0$, be defined as follows:

$$A_t = \{\mathbf{k} \le \mathbf{n} : b(\mathbf{k}) \le c^t b(\mathbf{1})\}. \tag{2.49}$$

It is clear that $A_0 = \{\mathbf{k} : b(\mathbf{k}) = b(\mathbf{1})\}$.

Let $r > 0$ and let $\{\lambda(\mathbf{k}), \mathbf{k} \le \mathbf{n}\}$ be a field of non-negative numbers for which

$$\mathsf{E}\left[\max_{\mathbf{k}\in A_t} |S(\mathbf{k})|\right]^r \le \sum_{\mathbf{k}\in A_t} \lambda(\mathbf{k}) \qquad \text{for all } t \ge 0. \tag{2.50}$$

A field $\{\lambda(\mathbf{k}), \mathbf{k} \le \mathbf{n}\}$ satisfying condition (2.50) exists for every field of random variables $\{X(\mathbf{k}), \mathbf{k} \le \mathbf{n}\}$. Indeed, put

$$g(t) = \mathsf{E}\left[\max_{\mathbf{k}\in A_t} |S(\mathbf{k})|^r\right]$$

and, for every $t \ge 0$, choose an arbitrary point $\mathbf{k}_t \in A_t \setminus A_{t-1}$ if the set $A_t \setminus A_{t-1}$ is non-empty. It is obvious that the field

$$\lambda(\mathbf{k}) = \begin{cases} g(t), & \text{if } \mathbf{k} = \mathbf{k}_t, \\ 0, & \text{if } \mathbf{k} \ne \mathbf{k}_t \end{cases}$$

satisfies condition (2.50).

Note that the choice of $\{\lambda(\mathbf{k})\}$ described above is not unique. Moreover, this choice is not optimal in many cases. Below we will consider other fields $\{\lambda(\mathbf{k})\}$ as well.

Theorem 2.19 *Let $\mathbf{n} \in \mathbf{N}^d$, $r > 0$, and let $\{b(\mathbf{k}), \mathbf{k} \le \mathbf{n}\}$ be an increasing field of positive real numbers. If condition* (2.50) *holds for some $c > 1$ and some non-negative numbers $\{\lambda(\mathbf{k}), \mathbf{k} \le \mathbf{n}\}$, then*

$$\mathsf{E}\left[\max_{\mathbf{k}\le\mathbf{n}} \left|\frac{S(\mathbf{k})}{b(\mathbf{k})}\right|^r\right] \le \frac{c^{2r}}{c^r - 1} \sum_{\mathbf{k}\le\mathbf{n}} \frac{\lambda(\mathbf{k})}{b^r(\mathbf{k})}. \tag{2.51}$$

Proof Put $B_0 = A_0$ and $B_t = A_t \setminus A_{t-1}$ for $t \ge 1$, where the sets A_t are defined in (2.49). Note that the sets B_t are empty for sufficiently large t. Moreover,

$$\bigcup_{i=0}^{t} B_i = A_t, \qquad \bigcup_{t=0}^{\infty} B_t = \{\mathbf{k} : \mathbf{k} \le \mathbf{n}\}.$$

It is clear that

$$\mathsf{E}\left[\max_{\mathbf{k}\le\mathbf{n}}\left|\frac{S(\mathbf{k})}{b(\mathbf{k})}\right|^r\right] \le \sum_{t=0}^{\infty}\mathsf{E}\left[\max_{\mathbf{k}\in B_t}\left|\frac{S(\mathbf{k})}{b(\mathbf{k})}\right|^r\right] \le \left(\frac{c}{b_1}\right)^r\sum_{t=0}^{\infty}c^{-tr}\mathsf{E}\left[\max_{\mathbf{k}\in A_t}|S(\mathbf{k})|^r\right],$$

where $b_1 = b(\mathbf{1})$. The assumptions of the theorem imply that

$$\sum_{t=0}^{\infty}c^{-tr}\mathsf{E}\left[\max_{\mathbf{k}\in A_t}|S(\mathbf{k})|^r\right] \le \sum_{t=0}^{\infty}c^{-tr}\sum_{i=0}^{t}\sum_{\mathbf{k}\in B_i}\lambda(\mathbf{k}) \le \sum_{i=0}^{\infty}\sum_{\mathbf{k}\in B_i}\lambda(\mathbf{k})\sum_{t=i}^{\infty}c^{-tr}$$
$$= \frac{c^r}{c^r-1}\sum_{i=0}^{\infty}c^{-ir}\sum_{\mathbf{k}\in B_i}\lambda(\mathbf{k}) \le b_1^r\frac{c^r}{c^r-1}\sum_{\mathbf{k}\le\mathbf{n}}\frac{\lambda(\mathbf{k})}{b^r(\mathbf{k})}.$$

The latter two bounds complete the proof of the theorem. □

Corollary 2.13 *Let* $\mathbf{n} \in \mathbf{N}^d$, $r > 0$, *and let* $\{b(\mathbf{k}), \mathbf{k} \le \mathbf{n}\}$ *be an increasing field of positive numbers. If condition* (2.50) *holds for some non-negative numbers* $\lambda(\mathbf{k})$, $\mathbf{k} \le \mathbf{n}$, *and all* $c > 1$, *then*

$$\mathsf{E}\left[\max_{\mathbf{k}\le\mathbf{n}}\left|\frac{S(\mathbf{k})}{b(\mathbf{k})}\right|^r\right] \le 4\sum_{\mathbf{k}\le\mathbf{n}}\frac{\lambda(\mathbf{k})}{b^r(\mathbf{k})}. \tag{2.52}$$

Proof Corollary 2.13 follows from Theorem 2.19, since

$$\min_{c>1}\frac{c^{2r}}{c^r-1} = 4.$$

□

The assumptions of Corollary 2.13 can easily be checked for independent random variables.

Corollary 2.14 *Let* $\mathbf{n} \in \mathbf{N}^d$ *and* $r > 0$. *Let* $\{X(\mathbf{k}), \mathbf{k} \le \mathbf{n}\}$ *be a field of independent random variables such that* $\mathsf{E}\left[|X(\mathbf{k})|^r\right] < \infty$ *for all* $\mathbf{k} \le \mathbf{n}$. *Let* $\{b(\mathbf{k}), \mathbf{k} \le \mathbf{n}\}$ *be an increasing field of positive real numbers. Then*

$$\mathsf{E}\left[\max_{\mathbf{k}\le\mathbf{n}}\left|\frac{S(\mathbf{k})}{b(\mathbf{k})}\right|^r\right] \le 4\sum_{\mathbf{k}\le\mathbf{n}}\frac{\mathsf{E}\left[|X(\mathbf{k})|^r\right]}{b^r(\mathbf{k})} \tag{2.53}$$

if $0 < r \le 1$.

If $1 \le r \le 2$ *and* $\mathsf{E}\left[X(\mathbf{k})\right] = 0$ *for all* $\mathbf{k} \le \mathbf{n}$, *then*

$$\mathsf{E}\left[\max_{\mathbf{k}\le\mathbf{n}}\left|\frac{S(\mathbf{k})}{b(\mathbf{k})}\right|^r\right] \le 2^{d+r+3}\sum_{\mathbf{k}\le\mathbf{n}}\frac{\mathsf{E}\left[|X(\mathbf{k})|^r\right]}{b^r(\mathbf{k})} \tag{2.54}$$

(*the constant on the right-hand side can be decreased up to* 2^{d+3} *if* $r = 2$).

Finally, if $r \geq 2$ *and* $\mathsf{E}[X(\mathbf{k})] = 0$ *for all* $\mathbf{k} \leq \mathbf{n}$, *then*

$$\mathsf{E}\left[\max_{\mathbf{k}\leq\mathbf{n}}\left|\frac{S(\mathbf{k})}{b(\mathbf{k})}\right|^r\right] \leq 2^{d+r+2} J_r |\mathbf{n}|^{-1+r/2} \sum_{\mathbf{k}\leq\mathbf{n}} \frac{\mathsf{E}\left[|X(\mathbf{k})|^r\right]}{b^r(\mathbf{k})}, \qquad (2.55)$$

where J_r *is the Dharmadhikari–Jogdeo constant* (2.9).

Proof We check the assumptions of Corollary 2.13. For a fixed $c > 1$, define the sets A_t according to (2.49). Now fix $t \geq 0$ and define the random variables

$$X'(\mathbf{k}) = \begin{cases} X(\mathbf{k}), & \mathbf{k} \in A_t, \\ 0, & \mathbf{k} \notin A_t, \end{cases}$$

their sums $S'(\mathbf{k}) = \sum_{\mathbf{m}\leq\mathbf{k}} X'(\mathbf{m})$, $\mathbf{m} \leq \mathbf{n}$, and the field

$$\lambda(\mathbf{k}) = \begin{cases} \mathsf{E}\left[|X(\mathbf{k})|^r\right], & 0 < r \leq 1, \\ 2^{d+1+r}\mathsf{E}\left[|X(\mathbf{k})|^r\right], & 1 < r < 2, \\ 2^{d+1}\mathsf{E}\left[|X(\mathbf{k})|^2\right], & r = 2, \\ 2^{d+r} J_r |\mathbf{n}|^{-1+r/2}\mathsf{E}\left[|X(\mathbf{k})|^r\right], & r > 2. \end{cases}$$

Note that $S(\mathbf{k}) = S'(\mathbf{k})$ for $\mathbf{k} \in A_t$ and thus

$$\max_{\mathbf{k}\in A_t}\left|\frac{S(\mathbf{k})}{b(\mathbf{k})}\right| \leq \max_{\mathbf{k}\leq\mathbf{n}}\left|\frac{S'(\mathbf{k})}{b(\mathbf{k})}\right|.$$

Applying Corollary 2.10 to the sum $S'(\mathbf{k})$ we prove condition (2.50). Now Corollary 2.14 follows from Corollary 2.13. □

Remark 2.6 Even in the case when $d = 1$, the method described above allows us to obtain new results. One example of this kind is based on the monotonicity of the sequence $\mathsf{E}\left[|S_n|^r\right]$ if the random variables $\{X_k\}$ are independent. Let $r > 1$ and put $S_0 = 0$ and

$$\lambda_k = \left(\frac{r}{r-1}\right)^r \left(\mathsf{E}\left[|S_k|^r\right] - \mathsf{E}\left[|S_{k-1}|^r\right]\right), \qquad k \geq 1.$$

Then Doob's inequality for martingales implies that

$$\mathsf{E}\left[\max_{k\leq n} |S_k|^r\right] \leq \left(\frac{r}{r-1}\right)^r \mathsf{E}\left[|S_n|^r\right] = \sum_{k\leq n} \lambda_k.$$

Thus Corollary 2.13 yields

$$\mathsf{E}\left[\max_{k\le n}|S_k|^r\right]\le 4\left(\frac{r}{r-1}\right)^r\sum_{1\le k\le n}\frac{\mathsf{E}\left[|S_k|^r\right]-\mathsf{E}\left[|S_{k-1}|^r\right]}{b_k^r}\tag{2.56}$$

for all monotone sequences $\{b_k, k\le n\}$. The right-hand side of inequality (2.56) for $r=2$ coincides, up to a multiplicative factor, with that of the Hájek–Rényi inequality. Note that (2.56) is more general, since it holds for moments, while the original Hájek–Rényi inequality is valid only for probabilities. For other r, inequality (2.56) may even be a better estimate of the probabilities involved in the Hájek–Rényi inequality.

Example 2.1 Let, for example, $r=4$, $b_k=k$, and let $X_1,\dots,X_n$ be Gaussian random variables such that

$$\mathsf{E}[X_k]=0,\qquad \mathsf{E}\left[S_k^2\right]=\frac{k^2}{\ln(3k)}.$$

The Hájek–Rényi inequality provides the bound

$$\mathsf{P}\left(\max_{k\le n}\left|\frac{S_k}{k}\right|\ge x\right)\le\frac{1}{x^2}\sum_{k=1}^n\frac{\mathsf{E}\left[X_k^2\right]}{k^2}\asymp\ln\ln n.\tag{2.57}$$

On the other hand,

$$\mathsf{E}\left[S_k^4\right]-\mathsf{E}\left[S_{k-1}^4\right]\asymp\frac{k^3}{\ln^2 k}$$

and inequality (2.56) for $r=4$ means that

$$\mathsf{P}\left(\max_{k\le n}\left|\frac{S_k}{k}\right|\ge x\right)\le\frac{16}{x^2}\sum_{k=1}^n\frac{\mathsf{E}\left[S_k^4\right]-\mathsf{E}\left[S_{k-1}^4\right]}{k^4}\asymp 1.$$

Clearly the latter bound is asymptotically better than the bound in (2.57).

2.4.3 The Hájek–Rényi Inequality for some Classes of Dependent Random Variables

Just to demonstrate how general the second method is, we provide below some maximal inequalities for the three particular cases of orthogonal, martingale, and homogeneous random variables.

2.4.3.1 The Hájek–Rényi Inequality for Orthogonal Random Variables

Throughout this section, log stands for $\log_2$.

A field $\{X(\mathbf{k}), \mathbf{k} \le \mathbf{n}\}$ of random variables is called *orthogonal* if $\mathsf{E}\left[X^2(\mathbf{k})\right] < \infty$ for all $\mathbf{k} \le \mathbf{n}$ and

$$\mathsf{E}\left[X(\mathbf{k})X(\mathbf{l})\right] = 0 \qquad \text{for all } \mathbf{k} \ne \mathbf{l}.$$

Members of an orthogonal field are called *orthogonal random variables.*

Corollary 2.15 *Let* $\mathbf{n} \in \mathbf{N}^d$ *and let* $\{X(\mathbf{k}), \mathbf{k} \le \mathbf{n}\}$ *be orthogonal random variables. Assume that* $\{b(\mathbf{k}), \mathbf{k} \le \mathbf{n}\}$ *is an increasing field of positive real numbers. Then*

$$\mathsf{E}\left[\max_{\mathbf{k}\le\mathbf{n}}\left|\frac{S(\mathbf{k})}{b(\mathbf{k})}\right|^2\right] \le 4\left(\log(2n_1)\right)^2 \ldots \left(\log(2n_d)\right)^2 \sum_{\mathbf{k}\le\mathbf{n}} \frac{\mathsf{E}\left[X^2(\mathbf{k})\right]}{b^2(\mathbf{k})}. \tag{2.58}$$

Proof Moricz [318] proved the following analog of the Menschoff–Rademacher maximal inequality for orthogonal random variables in the case of $d > 1$:

$$\mathsf{E}\left[\max_{\mathbf{k}\le\mathbf{n}} S^2(\mathbf{k})\right] \le \left(\log(2n_1)\right)^2 \ldots \left(\log(2n_d)\right)^2 \sum_{\mathbf{k}\le\mathbf{n}} \mathsf{E}\left[X^2(\mathbf{k})\right].$$

The rest of the proof is the same as that of Corollary 2.14. □

2.4.3.2 The Hájek–Rényi Inequality for Martingales

Cairoli [43] established several examples showing that some classical inequalities for maximums of submartingales with discrete time are not valid in the case of *submartingales with multi-dimensional time*. Below we show that, despite his examples, some other inequalities can be generalized to the case of $d > 1$, too.

Definition 2.1 Let $\{\mathfrak{F}(\mathbf{k}), \mathbf{k} \le \mathbf{n}\}$ be a family of σ-algebras such that

$$\mathfrak{F}(\mathbf{k}) \subseteq \mathfrak{F}(\mathbf{l}) \qquad \text{for all } \mathbf{k} \le \mathbf{n}.$$

A field of random variables $\{X(\mathbf{k}), \mathbf{k} \le \mathbf{n}\}$ is called a *martingale difference* with respect to $\{\mathfrak{F}(\mathbf{k}), \mathbf{k} \le \mathbf{n}\}$ if $X(\mathbf{k})$ is a $\mathfrak{F}(\mathbf{k})$-measurable random variable for all $\mathbf{k} \le \mathbf{n}$ and

$$\mathsf{E}\left[S(\mathbf{k})/\mathfrak{F}(\mathbf{l})\right] = S(\min\{k_1, l_1\}, \ldots, \min\{k_d, l_d\}) \qquad \text{for all } \mathbf{k}, \mathbf{l} \le \mathbf{n}.$$

The field $\{S(\mathbf{k}), \mathbf{k} \le \mathbf{n}\}$ is called a *martingale* with respect to $\{\mathfrak{F}(\mathbf{k}), \mathbf{k} \le \mathbf{n}\}$.

Clearly, when $d = 1$, Definition 2.1 defines a usual martingale sequence with respect to an increasing sequence of σ-algebras.

Let $r > 1$. Cairoli [43] proved the following analog of Doob's inequality for $d > 1$:

$$\mathsf{E}\left[\max_{\mathbf{k}\le\mathbf{n}} |S(\mathbf{k})|^r\right] \le \left(\frac{r}{r-1}\right)^{dr} \mathsf{E}\left[|S(\mathbf{n})|^r\right]. \tag{2.59}$$

The right-hand side of this inequality is not always a field with non-negative increments for $d > 1$.

Indeed, this is the case if, for example, $r = 2$. But, in general, this inequality cannot be used to establish a Hájek–Rényi type inequality for martingales if $d > 1$. On the other hand, $\mathsf{E}\left[|S(\mathbf{n})|^r\right]$ can be estimated with the help of the Bahr–Esseen or Dharmadhikari–Jogdeo inequality and the result can already be used to derive a Hájek–Rényi inequality.

Since

$$\zeta_i = \sum_{k_1=1}^{n_1} \cdots \sum_{k_{d-1}=1}^{n_{d-1}} \sum_{k_d=1}^{i} X(\mathbf{k})$$

is a usual martingale with respect to the σ-algebras $\mathfrak{F}(n_1, \dots, n_{d-1}, i)$, $i \le n_d$, we use induction with respect to the dimension of the spaces $\mathbf{N}^d$ and prove that the Bahr–Esseen inequality [16] for martingales with one-dimensional indices implies that

$$\mathsf{E}\left[|S(\mathbf{n})|^r\right] \le 2^d \sum_{\mathbf{k}\le\mathbf{n}} \mathsf{E}\left[|X(\mathbf{k})|^r\right] \tag{2.60}$$

if $1 \le r \le 2$, while the Dharmadhikari–Jogdeo inequality [73] for usual martingales implies that

$$\mathsf{E}\left[|S(\mathbf{n})|^r\right] \le J_r^d |\mathbf{n}|^{-1+r/2} \sum_{\mathbf{k}\le\mathbf{n}} \mathsf{E}\left[|X(\mathbf{k})|^r\right] \tag{2.61}$$

if $r > 2$.

Corollary 2.16 *Let $\mathbf{n} \in \mathbf{N}^d$, $r > 1$, and let $\{X(\mathbf{k}), \mathbf{k} \le \mathbf{n}\}$ be a martingale difference with respect to a family of increasing σ-algebras. Let $\{b(\mathbf{k}), \mathbf{k} \le \mathbf{n}\}$ be an increasing field of positive numbers. Then*

$$\mathsf{E}\left[\max_{\mathbf{k}\le\mathbf{n}} \left|\frac{S(\mathbf{k})}{b(\mathbf{k})}\right|^r\right] \le 4 \left(\frac{r}{r-1}\right)^{dr} \sum_{\mathbf{k}\le\mathbf{n}} \frac{\mathsf{E}\left[|X(\mathbf{k})|^r\right]}{b^r(\mathbf{k})} \tag{2.62}$$

if $0 < r \le 1$;

$$\mathsf{E}\left[\max_{\mathbf{k}\le\mathbf{n}} \left|\frac{S(\mathbf{k})}{b(\mathbf{k})}\right|^r\right] \le 2^{d+2} \left(\frac{r}{r-1}\right)^{dr} \sum_{\mathbf{k}\le\mathbf{n}} \frac{\mathsf{E}\left[|X(\mathbf{k})|^r\right]}{b^r(\mathbf{k})} \tag{2.63}$$

if $1 < r \le 2$; *and*

$$\mathsf{E}\left[\max_{\mathbf{k}\le\mathbf{n}}\left|\frac{S(\mathbf{k})}{b(\mathbf{k})}\right|^r\right] \le 4J_r^d\left(\frac{r}{r-1}\right)^{dr}|\mathbf{n}|^{-1+r/2}\sum_{\mathbf{k}\le\mathbf{n}}\frac{\mathsf{E}\left[|X(\mathbf{k})|^r\right]}{b^r(\mathbf{k})} \tag{2.64}$$

if $r > 2$, *where* J_r *is the Dharmadhikari–Jogdeo constant* (2.9).

Proof Let $c > 1$. The proof of Corollary 2.16 for $0 < r \le 1$ coincides with that of Corollary 2.14.

Now let $r > 1$. Define the sets A_t, $t \ge 0$, according to the rule (2.49). Then we apply Lemma 2.4 given below for all $\mathbf{k}_0 \notin A_t$ and prove that the field

$$X'(\mathbf{k}) = \begin{cases} X(\mathbf{k}), & \mathbf{k} \in A_t, \\ 0, & \mathbf{k} \notin A_t, \end{cases}$$

is a martingale difference with respect to the σ-algebras $\{\mathfrak{F}(\mathbf{k}), \mathbf{k} \le \mathbf{n}\}$. Now we use inequalities (2.60) and (2.61) for random variables $\{X'(\mathbf{k})\}$. As a result, we prove that condition (2.50) holds for $\{X(\mathbf{k})\}$ with the corresponding numbers $\{\lambda(\mathbf{k})\}$. Inequality (2.63) follows from (2.59), Corollary 2.13, and the Bahr–Esseen inequality (2.60), while (2.64) follows from (2.59), Corollary 2.13, and the Dharmadhikari–Jogdeo inequality (2.61). □

Lemma 2.3 *Let* $\{X(\mathbf{k}), \mathbf{k} \le \mathbf{n}\}$ *be a martingale difference with respect to a family of increasing* σ*-algebras* $\{\mathfrak{F}(\mathbf{k}), \mathbf{k} \le \mathbf{n}\}$. *Then, for all* $\mathbf{k}, \mathbf{l} \le \mathbf{n}$,

$$\mathsf{E}\left[X(\mathbf{k})/\mathfrak{F}(\mathbf{l})\right] = \begin{cases} X(\mathbf{k}), & \mathbf{k} \le \mathbf{l}, \\ 0, & \mathbf{k} \not\le \mathbf{l}. \end{cases} \tag{2.65}$$

Proof Let $\mathbf{k} = (k_1, \dots, k_d)$, $\mathbf{l} = (l_1, \dots, l_d)$. As usual, we also let $S(\mathbf{l}) = \sum_{\mathbf{k}\le\mathbf{l}} X(\mathbf{k})$. Then we use the representation

$$X(\mathbf{k}) = \sum_{\boldsymbol{\varepsilon}\in\mathscr{E}_d}(-1)^{\varepsilon_1+\dots+\varepsilon_d}S(\mathbf{k}-\boldsymbol{\varepsilon})$$

(see Proposition A.5), where $\boldsymbol{\varepsilon}$ is a vector $(\varepsilon_1, \dots, \varepsilon_d)$ and each of its coordinates ε_i is either equal to 0 or 1. We deduce from this representation that

$$\mathsf{E}\left[X(\mathbf{k})/\mathfrak{F}(\mathbf{l})\right] = \sum_{\boldsymbol{\varepsilon}\in\mathscr{E}_d}(-1)^{\varepsilon_1+\dots+\varepsilon_d}\mathsf{E}\left[S(\mathbf{k}-\boldsymbol{\varepsilon})/\mathfrak{F}(\mathbf{l})\right].$$

If $\mathbf{k} \le \mathbf{l}$, then $\mathsf{E}\left[S(\mathbf{k}-\boldsymbol{\varepsilon})/\mathfrak{F}(\mathbf{l})\right] = S(\mathbf{k}-\boldsymbol{\varepsilon})$ for all $\boldsymbol{\varepsilon} \in \mathscr{E}_d$ and thus (2.65) holds.

Otherwise, that is if $\mathbf{k} \not\le \mathbf{l}$, then there is a non-empty set $J \subseteq \{1, \dots, d\}$ such that $k_j > l_j$ for $j \in J$ and $k_j \le l_j$ for $j \notin J$. Note that if $J = \{1, \dots, d\}$, then there is no $j \in J$ for which $k_j \le l_j$ and thus the latter condition does not hold at all.

First we consider the case of $J = \{1, \ldots, d\}$, that is, the case of $k_1 > l_1, \ldots, k_d > l_d$. Then $\mathbf{l} \le \mathbf{k} - \boldsymbol{\varepsilon}$, whence $\mathsf{E}\left[S(\mathbf{k} - \boldsymbol{\varepsilon}) / \mathfrak{F}(\mathbf{l}) \right] = S(\mathbf{l})$ by Definition 2.1. Hence

$$\mathsf{E}\left[X(\mathbf{k})/\mathfrak{F}(\mathbf{l})\right] = S(\mathbf{l}) \sum_{\boldsymbol{\varepsilon} \in \mathscr{E}_d} (-1)^{\varepsilon_1 + \cdots + \varepsilon_d} = 0.$$

Now we consider the case of a set $J = \{1, \ldots, c\}$, $1 \le c < d$, that is, the case where $k_1 > l_1, \ldots, k_c > l_c$ and $k_{c+1} \le l_{c+1}, \ldots, k_d \le l_d$. Then

$$\mathsf{E}\left[S(\mathbf{k} - \boldsymbol{\varepsilon}) / \mathfrak{F}(\mathbf{l}) \right] = S(l_1, \ldots, l_c, k_{c+1} - \varepsilon_{c+1}, \ldots, k_d - \varepsilon_d),$$

whence

$$\begin{aligned} \mathsf{E}\left[X(\mathbf{k})/\mathfrak{F}(\mathbf{l})\right] = {} & \sum_{\mathscr{E}_{1,\ldots,c}} (-1)^{\varepsilon_1 + \cdots + \varepsilon_c} \\ & \sum_{\mathscr{E}_{c+1,\ldots,d}} (-1)^{\varepsilon_{c+1} + \cdots + \varepsilon_d} S(l_1, \ldots, l_c, k_{c+1} - \varepsilon_{c+1}, \ldots, k_d - \varepsilon_d) = 0, \end{aligned}$$

where the sums are formed with respect to the sets $\mathscr{E}_{1,\ldots,c}$ and $\mathscr{E}_{c+1,\ldots,d}$ consisting of those elements $(\varepsilon_1, \ldots, \varepsilon_c)$ and $(\varepsilon_{c+1}, \ldots, \varepsilon_d)$, respectively, whose coordinates equal either 0 or 1.

The case of a general set J is considered similarly to the case of $J = \{1, \ldots, c\}$ in view of symmetry. The lemma is completely proved. □

Lemma 2.4 *Let $\{X(\mathbf{k}), \mathbf{k} \le \mathbf{n}\}$ be a martingale difference with respect to a family of increasing σ-algebras $\{\mathfrak{F}(\mathbf{k}), \mathbf{k} \le \mathbf{n}\}$ and let $\mathbf{k}_0 \le \mathbf{n}$ be fixed. Put $S(\mathbf{k}) = \sum_{\mathbf{l} \le \mathbf{k}} X(\mathbf{l})$ and*

$$X'(\mathbf{k}) = \begin{cases} X(\mathbf{k}), & \mathbf{k} \ne \mathbf{k}_0, \\ 0, & \mathbf{k} = \mathbf{k}_0, \end{cases} \qquad S'(\mathbf{m}) = \sum_{\mathbf{k} \le \mathbf{m}} X'(\mathbf{k}), \quad \mathbf{m} \le \mathbf{n}.$$

Then $\{S'(\mathbf{k}), \mathbf{k} \le \mathbf{n}\}$ is a martingale with respect to the σ-algebras $\{\mathfrak{F}(\mathbf{k}), \mathbf{k} \le \mathbf{n}\}$.

Proof It is clear that

$$S'(\mathbf{k}) = \begin{cases} S(\mathbf{k}), & \mathbf{k}_0 \not\le \mathbf{k}, \\ S(\mathbf{k}) - X(\mathbf{k}_0), & \mathbf{k}_0 \le \mathbf{k}. \end{cases}$$

If $\mathbf{k}_0 \not\le \min\{\mathbf{k}, \mathbf{l}\}$ and $\mathbf{k}_0 \not\le \mathbf{k}$, then

$$\mathsf{E}\left[S'(\mathbf{k})/\mathfrak{F}(\mathbf{l})\right] = \mathsf{E}\left[S(\mathbf{k})/\mathfrak{F}(\mathbf{l})\right] = S(\min\{\mathbf{k}, \mathbf{l}\}) = S'(\min\{\mathbf{k}, \mathbf{l}\}).$$

On the other hand, if $\mathbf{k}_0 \le \min\{\mathbf{k}, \mathbf{l}\}$, then $\mathbf{k}_0 \le \mathbf{k}$ and

$$\mathsf{E}\left[S'(\mathbf{k})/\mathfrak{F}(\mathbf{l})\right] = \mathsf{E}\left[S(\mathbf{k}) - X(\mathbf{k}_0)/\mathfrak{F}(\mathbf{l})\right] = S(\min\{\mathbf{k}, \mathbf{l}\}) - X(\mathbf{k}_0) = S'(\min\{\mathbf{k}, \mathbf{l}\})$$

by Lemma 2.3. Finally, if $\mathbf{k}_0 \not\leq \min\{\mathbf{k}, \mathbf{l}\}$ and $\mathbf{k}_0 \leq \mathbf{k}$, then $\mathbf{k}_0 \not\leq \mathbf{l}$ and thus

$$\mathsf{E}\left[S'(\mathbf{k})/\mathfrak{F}(\mathbf{l})\right] = \mathsf{E}\left[S(\mathbf{k}) - X(\mathbf{k}_0)/\mathfrak{F}(\mathbf{l})\right] = S(\min\{\mathbf{k}, \mathbf{l}\}) = S'(\min\{\mathbf{k}, \mathbf{l}\})$$

again by Lemma 2.3. The proof is complete. □

2.4.3.3 The Hájek–Rényi Inequality for Quadratic Forms

A random quadratic form is a particular case of a martingale for $d = 2$.

Assume that $\{\xi_i, i \leq n\}$ are independent random variables such that $\mathsf{E}[\xi_i] = 0$ and $\mathsf{E}\left[\xi_i^2\right] < \infty$ for all $i \leq n$. Let $\{a_{ij}, i, j \leq n\}$ be some collection of real numbers. Put

$$X(i, j) = a_{ij}\left(\xi_i\xi_j - \mathsf{E}\left[\xi_i\xi_j\right]\right), \qquad i, j \leq n,$$
$$S(m_1, m_2) = \sum_{i_1=1}^{m_1}\sum_{i_2=1}^{m_2} X(i_1, i_2).$$

Then $\{X(i, j); (i, j) \leq (n, n)\}$ is a martingale-difference with respect to the family of σ-algebras

$$\mathfrak{F}(k, l) = \sigma\{X(i, j); (i, j) \leq (k, l)\} = \sigma(\xi_i, i \leq \max\{k, l\}).$$

Further, let $\{b(i, j); (i, j) \leq (n, n)\}$ be an increasing field of positive real numbers. Put $\sigma_i^2 = \mathsf{E}\left[\xi_i^2\right]$. Corollary 2.16 for $r = 2$ implies that

$$\begin{aligned}
&\mathsf{E}\left[\max_{(k,l)\leq(n_1,n_2)}\left|\frac{1}{b(k, l)}\sum_{i=1}^{k}\sum_{j=1}^{l} a_{ij}\left(\xi_i\xi_j - \mathsf{E}\left[\xi_i\xi_j\right]\right)\right|^2\right] \\
&\quad \leq 2^{2d+3}\mathsf{E}\left[S^2(n_1, n_2)\right] \leq 2^{2d+3}\left[\sum_{i=1}^{\min\{n_1,n_2\}} \frac{a_{ii}^2}{b^2(i, i)}\mathsf{E}\left[\left(\xi_i^2 - \sigma_i^2\right)^2\right]\right. \\
&\qquad \left. + \sum_{i=1}^{n_1}\sum_{j=1}^{n_2} \frac{a_{ij}^2}{b^2(i, j)}\sigma_i^2\sigma_j^2\right]
\end{aligned}$$

for all $n_1, n_2 \leq n$.

2.4.3.4 The Hájek–Rényi Inequality for Homogeneous Random Fields

A random field $\{X(\mathbf{k}), \mathbf{k} \leq \mathbf{n}\}$ is called *homogeneous* if there exists a function R depending on d arguments such that

$$\mathsf{E}\left[X(\mathbf{k})\right] = \text{const}, \qquad \mathsf{E}\left[X(\mathbf{k})X(\mathbf{l})\right] = R(\mathbf{k}-\mathbf{l}) \qquad \text{for all} \quad \mathbf{k}, \mathbf{l} \le \mathbf{n}.$$

In what follows we assume that $\mathsf{E}\left[X(\mathbf{k})\right] = 0$ for $\mathbf{k} \le \mathbf{n}$. It is easy to show that

$$\mathsf{E}\left[S^2(\mathbf{n})\right] \le |\mathbf{n}| \sum_{\mathbf{0} \le \mathbf{k} \le \mathbf{n}-\mathbf{1}} |R(\mathbf{k})|.$$

Since the right-hand side of the latter inequality is of a super-additive structure, Moricz's inequality [318] implies that

$$\mathsf{E}\left[\max_{\mathbf{k} \le \mathbf{n}} |S(\mathbf{k})|^2\right] \le (\log(2n_1))^2 \dots (\log(2n_d))^2\, |\mathbf{n}| \sum_{\mathbf{0} \le \mathbf{k} \le \mathbf{n}-\mathbf{1}} |R(\mathbf{k})|.$$

This bound together with Corollary 2.13 yields an analog of the Hájek–Rényi inequality for homogeneous fields.

Corollary 2.17 *Let* $\mathbf{n} \in \mathbf{N}^d$ *and let* $\{X(\mathbf{k}), \mathbf{k} \le \mathbf{n}\}$ *be a homogeneous field. Further let* $\{b(\mathbf{k}), \mathbf{k} \le \mathbf{n}\}$ *be an increasing field of positive real numbers. Then*

$$\mathsf{E}\left[\max_{\mathbf{k} \le \mathbf{n}} \left|\frac{S(\mathbf{k})}{b(\mathbf{k})}\right|^2\right] \le 4\,(\log(2n_1))^2 \dots (\log(2n_d))^2\, |\mathbf{n}| \sum_{\mathbf{0} \le \mathbf{k} \le \mathbf{n}-\mathbf{1}} \frac{|R(\mathbf{k})|}{b^2(\mathbf{k})}. \tag{2.66}$$

2.5 Comments

Theorem 2.1 was stated (without proof) by Bienayme in 1853; Chebyshev independently proved it in 1867. Theorem 2.2 was proved by Kolmogorov [254]; its generalization, Theorem 2.3, is mentioned by Hájek and Rényi [150]. Theorem 2.4 is a corollary of Doob's inequality [77] for martingales. Theorem 2.5 is obtained by Lévy [283] in his studies of the convergence of sums of independent random variables to the stable laws.

Section 2.1 The Bahr–Esseen inequality (2.6) is proved in [16]; the Dharmadhikari–Jogdeo inequality (2.7) in [73]; Rosenthal's inequality (2.11) in [360]; and the Marcinkiewicz–Zygmund inequality (2.11) in [301]. Theorem 2.15 is due to Skorokhod [371].

Section 2.2 Theorem 2.15 is due to Petrov [337]. Lemma 2.2 can be found, for example, in Petrov's monograph [338]. Theorem 2.16 is proved by Petrov [337] in the case $d = 1$ (see inequality (2.18)). Corollary 2.4 for $d = 1$ is proved by Lévy [283], see [215] for the case $d > 1$. Inequality (2.26) is mentioned by Zimmerman [433] in the case $d = 2$; the same inequality for $d > 1$ is given by Paranjape and Park [335] without proof.

Theorem 2.17 is proved by Ottaviani [333]; its analog for $d > 1$ (Corollary 2.6) is obtained in [210] using another method. A similar result is given by N. Etemadi in [89] for $d = 1$ and in [92] for $d > 1$.

The idea to use Corollary 2.2 to prove inequality (2.33) is proposed by Petrov [337] for $d = 1$. Corollary 2.8 is proved by Petrov [337] for the case of $d = 1$ and $r = 2$ (the general case is obtained in [215]). All other results of this section for $d > 1$ are proved in [215].

Section 2.3 Inequality (2.43) for $r = 2$ is due to Wichura [419] with the constant 4^d instead of 2^{d+2} on the right-hand side. All other results for $d > 1$ are obtained in [211].

Christofides [59] and Christofides and Serfling [62] describe an application of inequality (2.44) for generalized U-statistics. Inequality (2.45) is used by Shcherbakova [365] to study the almost sure limit properties of increments of random fields.

Section 2.4 The first method of proof for the Hájek–Rényi type inequalities for $d > 1$ is proposed by Shorack and Smythe [366]; the second method is presented in [211]. Etemadi [92] obtains an inequality of type (2.54) for fields with independent increments.

The relationship between inequalities like (2.2) and the strong law of large numbers is clear. It is shown in [98] and [234] that the strong law of large numbers is always a consequence of such an inequality (Khoshnevisan [205] discusses a similar idea for multi-indexed sums of independent identically distributed random variables). This approach was later used by many authors for various types of dependence: Wang et al. [411] study the so-called d-demimartingales; Wang and Jin [414] reduce the complete convergence of multi-indexed sums to the corresponding inequalities for weighted sums; Hung and Tien [173] show some applications of the Hájek–Rényi type inequalities to Banach-valued martingale fields.

The literature on Hájek–Rényi type inequalities for various dependence schemes is rather extensive. Below is a short list of relevant schemes: pairwise independent random variables (Etemadi [91]), pairwise NQ dependent random variables (Yang et al. [424]), linearly dependent NQD random variables (Hu et al. [170]), AANA sequences (Wang et al. [407]), sequences with φ-mixing (Wang et al. [409]) or with ψ-mixing (Yang et al. [410]), mixingales (Wang et al. [413]), demimartingales (Wang and Hu [406]).

Several generalizations of martingales are known in the case of sequences depending on multi-indices (Cairoli and Walsh [46], Wong and Zakai [421], Zakai [430]). Definition 2.1 was introduced by Shorack and Smythe [366] (they also prove inequality (2.59)). Below is a list of papers where Kolmogorov and Hájek–Rényi inequalities have been discovered for multi-parameter martingales: Li [285], Fazekas [95], Sung [390], He [157], and Christofides and Serfling [60, 61].

Inequalities for maximums of multi-indexed sums of random variables with different types of dependence have been proved by Kryzhanovskaya [264], Borodikhin [29], Lagodowski [271], Kurbanmuradov and Sabelfeld [269], Shashkin [364], and Bakhtin and Bulinskii [17].

Chapter 3
Weak Convergence of Multi-Indexed Sums of Independent Random Variables

Denote by $\mathbb{C}(F)$ the set of points of continuity of a function F. A sequence of distribution functions $\{F_n\}$ (as well as the sequence of corresponding random variables $\{\zeta_n\}$) is said to *weakly converge* to a function F if

$$\lim_{n\to\infty} F_n(x) = F(x) \quad \text{for all } x \in \mathbb{C}(F).$$

We denote the weak convergence of a sequence of distribution functions $\{F_n\}$ to F by $F_n \xrightarrow{\mathscr{D}} F$ or $F_n(x) \xrightarrow{\mathscr{D}} F(x)$.

The main classical results of probability theory can easily be stated in terms of the weak convergence of some distribution functions. For example, the law of large numbers, central limit theorem, and Poisson theorem (law of rare events) are particular examples of the weak convergence of distribution functions.

The theory of the weak convergence of distribution functions of sums $\zeta_n = \sum_{k\le k_n} \xi_{nk}$ of infinitesimal negligible independent random variables $\{\xi_{nk}\}$ can be viewed as a complete theory for the classical case of $d = 1$ (see [126]). Recall that random variables $\{\xi_{nk}, k \le k_n\},\ n \ge 1$ are called *infinitesimal negligible* if

$$\lim_{n\to\infty} \max_{k\le k_n} \mathsf{P}(|\xi_{nk}| \ge \varepsilon) = 0 \quad \text{for all} \quad \varepsilon > 0. \tag{3.1}$$

There exists an extension of the classical weak convergence theory where the assumption that the random variables are infinitesimal negligible is not imposed (see [439]). Below we restrict our attention to the classical case of infinitesimal negligible random variables only.

The general theory of the weak convergence of multi-indexed sums of independent random variables is similar to its counterpart in the case where $d = 1$ and, in a certain sense, follows from the theory of cumulative sums. Nevertheless, a number of particular cases for $d > 1$ are rather unexpected and this also excuses the development of the theory for $d > 1$. Some such cases are considered in Chap. 4.

O. Klesov, *Limit Theorems for Multi-Indexed Sums of Random Variables*, Probability Theory and Stochastic Modelling 71, DOI 10.1007/978-3-662-44388-0_3

3.1 Main Definitions for Multi-Indexed Sums

The definition of a usual triangular array $\{\xi_{nk}, k \le k_n\}$, $n \ge 1$, for $d = 1$ uses two parameters. Similarly, the definition of a scheme of series for fields also requires two vector parameters whose dimensions, in the general case, are different.

Definition 3.1 A family of random variables $\{\xi_{nk}, k \le k_n\}$, $n \ge 1$ is called a *sequence of series* if the random variables $\{\xi_{nk}, k \le k_n\}$ are independent for all $n \in \mathbf{N}$.

Definition 3.2 Let d and r be two positive integers. Also let $\mathbf{k}(\mathbf{n}) \in \mathbf{N}^r$ for every $\mathbf{n} \in \mathbf{N}^d$. A family of random variables

$$\{X(\mathbf{n}, \mathbf{k}), \mathbf{k} \le \mathbf{k}(\mathbf{n})\}, \quad \mathbf{n} \in \mathbf{N}^d,$$

is called a *(d, r)-field of series* (equivalently, a *(d, r)-series*) if the random variables $\{X(\mathbf{n}, \mathbf{k}); \mathbf{k} \le \mathbf{k}(\mathbf{n})\}$ are independent for every fixed $\mathbf{n} \in \mathbf{N}^d$.

Remark 3.1 Any $(1, 1)$-series is a classical sequence of series of independent random variables in the sense of Definition 3.1. For other values of d and r, a field of series is a more complicated object that, however, can be reduced to the case of $r = 1$ when studying weak convergence. Indeed, for all (d, r)-fields of series, one can construct a $(d, 1)$-series such that the sums in both series are identical. To construct such a $(d, 1)$-series, one can place the random variables in each (d, r)-series one by one according to any linear order. Therefore one can restrict the consideration to the case of $(d, 1)$-series if only the distributions of sums $S(\mathbf{n})$ matter (see (3.2)).

Consider the sums of random variables in each series:

$$S(\mathbf{n}) = \sum_{\mathbf{k} \le \mathbf{k}(\mathbf{n})} X(\mathbf{n}, \mathbf{k}), \qquad \mathbf{n} \in \mathbf{N}^d. \tag{3.2}$$

In this chapter we study the following classical questions concerning the multi-indexed sums $S(\mathbf{n})$:

(a) what functions are weak limits of the distribution functions of sums $S(\mathbf{n})$?
(b) if a certain function F is a weak limit of distribution functions of sums $S(\mathbf{n})$, then what are the conditions to be imposed on a (d, r)-field of series in order that their sums $S(\mathbf{n})$ weakly converge to F?

As in the classical case, problem (a) has an obvious solution if no extra restriction is imposed, namely every distribution function is a weak limit of distribution functions of sums $S(\mathbf{n})$. By analogy with the case $d = 1$ we introduce a restriction that the "role" of every term in the sum $S(\mathbf{n})$ is asymptotically negligible (cf. condition (3.1)).

Definition 3.3 The relation

$$\lim(\max) \max_{\mathbf{k}\le\mathbf{k}(\mathbf{n})} \mathsf{P}(|X(\mathbf{n},\mathbf{k})| \ge \varepsilon) = 0 \quad \text{for all} \quad \varepsilon > 0 \tag{3.3}$$

is called the condition of *infinitesimal negligibility for the* $\lim(\max)$*-convergence* of distribution functions of a (d, r)-series.

Definition 3.4 The relation

$$\lim(\min) \max_{\mathbf{k}\le\mathbf{k}(\mathbf{n})} \mathsf{P}(|X(\mathbf{n},\mathbf{k})| \ge \varepsilon) = 0 \quad \text{for all} \quad \varepsilon > 0 \tag{3.4}$$

is called the condition of the *infinitesimal negligibility for the* $\lim(\min)$*-convergence* of distribution functions of a (d, r)-series.

3.2 Limit Laws in the Scheme of Series

We recall some classical definitions and results for the case $d = 1$. The distribution function of a random variable ξ is sometimes denoted by Law (ξ). The set of points of continuity of a distribution function F is denoted by $\mathbb{C}(F)$.

Definition 3.5 A field of distribution functions $\{F(\mathbf{n};\cdot)\}$ is called *weakly convergent in the sense of* $\lim(\max)$*-convergence* to the function $F(\cdot)$ if

$$\lim(\max) F(\mathbf{n}; x) = F(x)$$

at each point $x \in \mathbb{C}(F)$. In the case of the $\lim(\max)$-convergence of distribution functions $F(\mathbf{n};\cdot)$ to the limit $F(\cdot)$ we write $F(\mathbf{n};\cdot) \xrightarrow[\max]{\mathscr{D}} F(\cdot)$.

It is sometimes more convenient to express the weak convergence in terms of the random variables rather than in terms of their distribution functions. If a field of distribution functions $\{F(\mathbf{n};\cdot)\}$ of random variables $\xi(\mathbf{n})$, $\mathbf{n} \in \mathbf{N}^d$, weakly converges in the sense of $\lim(\max)$-convergence to the distribution function $F(\cdot)$ of a random variable ξ, then we also write

$$\xi(\mathbf{n}) \xrightarrow[\max]{\mathscr{D}} \xi, \quad \text{or} \quad \xi(\mathbf{n}) \xrightarrow[\max]{\mathscr{D}} F, \quad \text{or} \quad F(\mathbf{n};\cdot) \xrightarrow[\max]{\mathscr{D}} \xi.$$

Definition 3.6 A field of distribution functions $\{F(\mathbf{n};\cdot)\}$ is said to *weakly converge in the sense of* $\lim(\min)$*-convergence* to a function $F(\cdot)$ if

$$\lim(\min) F(\mathbf{n}; x) = F(x)$$

at each point of continuity $x \in \mathbb{C}(F)$. In the case of the lim(min)-convergence of distribution functions $F(\mathbf{n};\cdot)$ to a limit $F(\cdot)$, we write $F(\mathbf{n};\cdot) \xrightarrow[\min]{\mathscr{D}} F(\cdot)$.

If a field of distribution functions $\{F(\mathbf{n};\cdot)\}$ of random variables $\xi(\mathbf{n})$ weakly converges in the sense of lim(min)-convergence to the distribution function $F(\cdot)$ of a random variable ξ, then we also write

$$\xi(\mathbf{n}) \xrightarrow[\min]{\mathscr{D}} \xi, \quad \text{or} \quad \xi(\mathbf{n}) \xrightarrow[\min]{\mathscr{D}} F, \quad \text{or} \quad F(\mathbf{n};\cdot) \xrightarrow[\min]{\mathscr{D}} \xi.$$

3.2.1 The Case d = 1

The set of all characteristic functions is denoted by $\mathfrak{H}$. It is well known that if the infinitesimal negligibility condition (3.1) holds, then the solution of problem (a) above is the set of infinitely divisible distributions $\mathfrak{D}$ in the case of $d = 1$ (see [126] Chap. 4, or [336], Chap. 4). Recall that a distribution function F is called *infinitely divisible* if, for all $n \geq 1$, its characteristic function f admits the representation

$$f(t) = \left[f_n(t)\right]^n, \qquad t \in \mathbf{R}, \tag{3.5}$$

where f_n is a certain characteristic function. We also say that a characteristic function f is infinitely divisible if its distribution function is infinitely divisible.

For the sake of completeness, we recall a theorem of Khintchine on the class of limit distributions for the weak convergence of distribution functions of sums of independent random variables.

Theorem 3.1 (A.Ya. Khintchine (1937)) *Consider the sums*

$$\zeta_n = \sum_{k=1}^{k_n} \xi_{nk}$$

constructed from a $(1, 1)$*-sequence of series* $\{\xi_{nk}, k \leq k_n\}$, $n \geq 1$, *that satisfies the infinitesimal negligibility condition* (3.1) *and let* $k_n \to \infty$ *as* $n \to \infty$. *If the weak limit of distribution functions of sums* ζ_n *exists, then it is infinitely divisible. Moreover, each infinitely divisible distribution function is the weak limit of distribution functions of sums of a certain sequence of series.*

3.2.2 The Case of (d, r)-Series

Denote the set of weak lim(max)-limits (or lim(min)-limits) of distribution functions of sums $S(\mathbf{n})$ of random variables satisfying condition (3.3) (or condition (3.4)) by $\mathfrak{D}_d^{\max}$ (by $\mathfrak{D}_d^{\min}$, respectively).

Theorem 3.2 *Let* $\{X(\mathbf{n}, \mathbf{k}), \mathbf{k} \le \mathbf{k}(\mathbf{n})\}$, $\mathbf{n} \in \mathbf{N}^d$, *be a* (d, r)*-field of series satisfying condition* (3.3). *Assume that* $\lim(\max)|\mathbf{k}(\mathbf{n})| = \infty$.[1] *If the distributions of sums* $S(\mathbf{n})$ *have a* $\lim(\max)$*-limit, then the limit is infinitely divisible.*

Moreover, for an arbitrary infinite divisible distribution function F, *there exists a* (d, r)*-field of series satisfying condition* (3.3) *for which the distributions of sums* $S(\mathbf{n})$ *weakly* $\lim(\max)$*-converge to* F. *In other words,*

$$\mathfrak{D}_d^{\max} = \mathfrak{D}.$$

Theorem 3.3 *Let* $\{X(\mathbf{n}, \mathbf{k}), \mathbf{k} \le \mathbf{k}(\mathbf{n})\}$, $\mathbf{n} \in \mathbf{N}^d$, *be a* (d, r)*-field of series satisfying condition* (3.4). *Assume that* $\lim(\min)|\mathbf{k}(\mathbf{n})| = \infty$. *If the distributions of sums* $S(\mathbf{n})$ *have a* $\lim(\min)$*-limit, then the limit is infinitely divisible.*

Moreover, for an arbitrary infinite divisible distribution function F, *there exists a* (d, r)*-field of series satisfying condition* (3.4) *for which the distributions of sums* $S(\mathbf{n})$ *weakly* $\lim(\min)$*-converge to* F. *In other words,*

$$\mathfrak{D}_d^{\min} = \mathfrak{D}.$$

Proof of Theorem 3.2 Without loss of generality, we may restrict the consideration to the case $r = 1$ (see Remark 3.1). The statement of the theorem is well known for $d = 1$ (see Theorem 3.1), thus we turn to the case $d > 1$. Assume that a distribution function F is a weak limit of distribution functions of sums (3.2) where the dimension of indices $\mathbf{k}$ equals 1. Then, by the definition of the $\lim(\max)$ convergence, the sequence of distribution functions corresponding to random variables $\{\zeta_m, m \ge 1\}$ also weakly converges to F, where

$$\zeta_m = S(m, \underbrace{1, \dots, 1}_{(d-1)}) = \sum\nolimits_{k=1}^{k_m} \xi_{mk}, \qquad \xi_{mk} \stackrel{\text{def}}{=} X(m, \underbrace{1, \dots, 1}_{(d-1)}; k),$$
$$k_m = |\mathbf{k}(m, \underbrace{1, \dots, 1}_{(d-1)})|.$$

It is clear that $\{\xi_{mk}; k \le k_m; m \ge 1\}$ is a $(1, 1)$-sequence of series. Moreover, it follows from (3.3) that this series satisfies the classical infinitesimal negligibility condition (3.1), that is ζ_m are cumulative sums of a $(1, 1)$-sequence of series of random variables satisfying the infinitesimal negligibility condition. Moreover, the distribution functions of cumulative sums ζ_m weakly converge to the limit F. Thus Theorem 3.1 implies that the function F is infinitely divisible, that is $\mathfrak{D}_d^{\max} \subseteq \mathfrak{D}$.

Now we prove the converse inclusion $\mathfrak{D} \subseteq \mathfrak{D}_d^{\max}$. Let F be an infinitely divisible distribution function. We construct a $(d, 1)$-field of series for which condition (3.3) holds and the distribution functions of sums $S(\mathbf{n})$ weakly converge to F.

According to Theorem 3.1, there exists a $(1, 1)$-sequence of series of infinitesimal negligible random variables $\{\xi_{mk}; k \le k_m\}$, $m \ge 1$, for which the distribution

[1] Recall that $|\mathbf{m}| = m_1 \dots m_d$.

functions of sums ζ_m weakly converge to F. Put $\Xi_m = \{\xi_{mk}; k \le k_m\}$, $m \ge 1$, that is Ξ_m is the mth series of independent random variables. Then we define the $(d, 1)$-sequence of series $\{\Xi(\mathbf{n}), \mathbf{n} \in \mathbf{N}^d\}$ by putting $\Xi(\mathbf{n}) = \Xi_{\max(n_1,\dots,n_d)}$. Note that "many" series of the field $\{\Xi(\mathbf{n}), \mathbf{n} \in \mathbf{N}^d\}$ are identical but the random variables in each series $\Xi(\mathbf{n})$ are independent. It is clear that the field of series $\{\Xi(\mathbf{n}), \mathbf{n} \in \mathbf{N}^d\}$ satisfies the infinitesimal negligibility condition (3.3).

Moreover, the distribution function $F(\mathbf{n}; \cdot)$ of the sum $S(\mathbf{n})$ of the random variables in the series $\Xi(\mathbf{n})$ coincides with the distribution function F_m of the sum ζ_m in the series Ξ_m for $m = \max(n_1, \dots, n_d)$. Thus the weak lim(max)-limit of the field $\{F(\mathbf{n}; \cdot)\}$ coincides with the weak limit of F_m. This means that $\mathfrak{D} \subseteq \mathfrak{D}_d^{\max}$. Theorem 3.2 is proved. □

Proof of Theorem 3.3 Since lim(min)-convergence follows from lim(max)-convergence, we have $\mathfrak{D}_d^{\max} \subseteq \mathfrak{D}_d^{\min}$ and thus Theorem 3.2 implies that $\mathfrak{D} \subseteq \mathfrak{D}_d^{\min}$.

To prove the converse inclusion $\mathfrak{D}_d^{\min} \subseteq \mathfrak{D}$ assume that $F \in \mathfrak{D}_d^{\min}$. This means that there exists some (d, r)-field of series $\{\Xi(\mathbf{n}), \mathbf{n} \in \mathbf{N}^d\}$ of random variables satisfying the infinitesimal negligibility condition (3.4) and that the distribution functions of sums $S(\mathbf{n})$ weakly lim(min)-converge to the distribution function F. For every $n \ge 1$, consider the series

$$\Xi\big(\underbrace{n, \dots, n}_{d}\big) = \big\{X\big(\underbrace{n, \dots, n}_{d}; \mathbf{k}\big), \mathbf{k} \le \mathbf{k}\big(\underbrace{n, \dots, n}_{d}\big)\big\}.$$

We place the random variables of this series one by one according to a linear order and denote the resulting random variables by ξ_{nk}, $k \le k_n$, where

$$k_n \stackrel{\text{def}}{=} \big|\mathbf{k}\big(\underbrace{n, \dots, n}_{d}\big)\big|.$$

It is clear that the sequence of series $\{\xi_{nk}, k \le k_n\}$, $n \ge 1$, satisfies the infinitesimal negligibility condition (3.1) and that the distribution functions of sums $\zeta_n \stackrel{\text{def}}{=} \sum_{k=1}^{k_n} \xi_{nk}$ weakly converge to F. This means that $\mathfrak{D}_d^{\min} \subseteq \mathfrak{D}$. Theorem 3.3 is proved. □

3.2.3 Centered Sums

As for the case $d = 1$, the limit theorems can be studied for centered sums

$$\sum_{\mathbf{k} \le \mathbf{k}(\mathbf{n})} X(\mathbf{n}, \mathbf{k}) - c(\mathbf{n}), \tag{3.6}$$

too, where $\{c(\mathbf{n})\}$ is a field of real numbers.

Theorem 3.4 *The set of distribution functions that appear as the limits of centered sums* (3.6), *where the random variables* $\{X(\mathbf{n},\mathbf{k}),\mathbf{k}\le\mathbf{k}(\mathbf{n})\}$, $\mathbf{n}\in\mathbf{N}^d$, *satisfy the infinitesimal negligibility condition* (3.3) *in the case of* lim(max)*-convergence or condition* (3.4) *in the case of* lim(min)*-convergence, coincides with the set of infinitely divisible distribution functions* $\mathfrak{D}$.

Remark 3.2 A more general condition than (3.3) or (3.4), called the *condition of infinitesimal stability*, is also studied in the theory of weak convergence of sums of independent random variables, namely

$$\max_{\mathbf{k}\le\mathbf{k}(\mathbf{n})}\mathsf{P}(|X(\mathbf{n},\mathbf{k})-l(\mathbf{n},\mathbf{k})|\ge\varepsilon)\to 0\quad\text{for all}\quad\varepsilon>0. \tag{3.7}$$

where $\{l(\mathbf{n},\mathbf{k}),\mathbf{k}\le\mathbf{k}(\mathbf{n})\}$, $\mathbf{n}\in\mathbf{N}^d$, is a certain family of real numbers. Condition (3.7) is considered for both lim(max)-convergence and lim(min)-convergence. As in the case of the infinitesimal condition, the set of weak limits of distribution functions of sums $S(\mathbf{n})$ coincides with the set of infinitely divisible distribution functions provided that the random variables $\{X(\mathbf{n},\mathbf{k})\}$ are infinitesimal stable (in other words, if they satisfy condition (3.7)).

3.3 Conditions for Weak Convergence to a Given Infinitely Divisible Law

Simple additional reasoning allows us to obtain necessary and sufficient conditions for the convergence of distribution functions of sums $S(\mathbf{n})$ to a given infinitely divisible distribution function. Recall that a characteristic function f of an arbitrary infinitely divisible distribution function admits the so-called *Lévy–Khintchine representation*:

$$f(t)=\exp\left\{i\gamma t+\int_{-\infty}^{\infty}\left(e^{itx}-1-\frac{itx}{1+x^2}\right)\frac{1+x^2}{x^2}\,dG(x)\right\}, \tag{3.8}$$

where γ is a real constant and $G(x)$ is a non-decreasing bounded function.[2] Note that G is called the *spectral function* corresponding to the characteristic function f.

Recall that the notation $G(\mathbf{n})\rightrightarrows G$ for lim(max)-convergence means that the field of functions $\{G(\mathbf{n};\cdot)\}$ weakly lim(max)-converges to G and, moreover, $\lim(\max)G(\mathbf{n};-\infty)=G(-\infty)$ and $\lim(\max)G(\mathbf{n};\infty)=G(\infty)$ (a similar notation is introduced for lim(min)-convergence). Here and in what follows we put $H(\pm\infty)=\lim_{x\to\pm\infty}H(x)$ for an arbitrary function H if the limits exist (these limits obviously exist in the case of distribution functions).

[2] The expression in the integral equals $-t^2/2$ if $x=0$.

Theorem 3.5 *Let $F(x)$ be an infinitely divisible distribution function and let $f(t)$ be its characteristic function admitting the Lévy–Khintchine representation* (3.8). *Let*

$$\{X(\mathbf{n}, \mathbf{k}), \mathbf{k} \le \mathbf{k}(\mathbf{n})\}, \ \mathbf{n} \in \mathbf{N}^d, \tag{3.9}$$

be a (d, r)-field of series of independent random variables satisfying the infinitesimal negligibility condition (3.3). *Put*

$$F(\mathbf{n}, \mathbf{k}; x) = \mathsf{P}(X(\mathbf{n}, \mathbf{k}) < x). \tag{3.10}$$

In order that the distribution functions of the sums

$$\sum_{\mathbf{k} \le \mathbf{k}(\mathbf{n})} X(\mathbf{n}, \mathbf{k}) \tag{3.11}$$

weakly lim(max)*-converge to $F(\cdot)$ it is necessary and sufficient that the following conditions hold for the* lim(max)*-convergence*:

$$G(\mathbf{n}; \cdot) \rightrightarrows G(\cdot), \qquad \gamma(\mathbf{n}) \to \gamma, \tag{3.12}$$

where

$$G(\mathbf{n}; x) = \sum_{\mathbf{k} \le \mathbf{k}(\mathbf{n})} \int_{-\infty}^{x} \frac{y^2}{1+y^2} \, dF^*(\mathbf{n}, \mathbf{k}; x), \tag{3.13}$$

$$\gamma(\mathbf{n}) = \sum_{\mathbf{k} \le \mathbf{k}(\mathbf{n})} \left\{ a(\mathbf{n}, \mathbf{k}) + \int_{-\infty}^{\infty} \frac{x}{1+x^2} \, dF^*(\mathbf{n}, \mathbf{k}; x) \right\}, \tag{3.14}$$

$$a(\mathbf{n}, \mathbf{k}) = \int_{|x|<\tau} x \, dF(\mathbf{n}, \mathbf{k}; x), \tag{3.15}$$

$$F^*(\mathbf{n}, \mathbf{k}; x) = F(\mathbf{n}, \mathbf{k}; x + a(\mathbf{n}, \mathbf{k})). \tag{3.16}$$

Here τ is an arbitrary real number.

Theorem 3.6 *Let $F(x)$ be an infinitely divisible distribution function and let $f(t)$ be its characteristic function admitting the Lévy–Khintchine representation* (3.8), *where γ is a real number and $G(x)$ is a non-decreasing bounded function.*[3]

Let a (d, r)-field of series (3.9) *satisfy the infinitesimal negligibility condition* (3.4). *In order that the distribution functions of sums* (3.11) *weakly* lim(min)*-converge to $F(x)$ it is necessary and sufficient that conditions* (3.12) *hold for the* lim(min)*-convergence, where $F(\mathbf{n}, \mathbf{k}; x)$ are distribution functions* (3.10) *and*

[3] The expression in the integral (3.8) equals $-t^2/2$ if $x = 0$.

$G(\mathbf{n}; x)$, $\gamma(\mathbf{n})$, $a(\mathbf{n}, \mathbf{k})$, and $F^*(\mathbf{n}, \mathbf{k}; x)$ are defined by equalities (3.13)–(3.16), *respectively, and where τ is an arbitrary positive number.*

Remark 3.3 Theorems 3.5 and 3.6 coincide in the case of $d = 1$ (the proof of both results can be found in [336]). The proofs of these results for $d > 1$ are the same for the lim(max)-convergence and lim(min)-convergence. Thus we consider both proofs simultaneously.

Proof of Theorem 3.5 and 3.6 Let conditions (3.12) hold. Choose a sequence of monotone multi-indices $\{\mathbf{n}_i, i \ge 1\}$ such that $\max(n_{1i}, \dots, n_{di}) \to \infty$ as $i \to \infty$ in the case of Theorem 3.5 or $\min(n_{1i}, \dots, n_{di}) \to \infty$ as $i \to \infty$ in the case of Theorem 3.6. According to the Cauchy convergence criterion (see Proposition A.3), relations (3.12) imply that

$$G(\mathbf{n}_i) \rightrightarrows G, \qquad \gamma(\mathbf{n}_i) \to \gamma \tag{3.17}$$

as $i \to \infty$. Denote the (d, r)-field of series $\{X(\mathbf{n}, \mathbf{k}); \mathbf{k} \le \mathbf{k}(\mathbf{n})\}$ by $\Xi(\mathbf{n})$. Then the sequence $\Xi_i \stackrel{\text{def}}{=} \Xi(\mathbf{n}_i)$, $i \ge 1$, forms a $(1, r)$-field of series. In every series Ξ_i, we place the random variables one by one according to a linear order. This, in particular, means that Ξ_i, $i \ge 1$, is a $(1, 1)$-sequence of series. The sequence Ξ_i, $i \ge 1$, just defined possesses the following properties:

(a) each series is constituted of independent random variables;
(b) the random variables in the series Ξ_i satisfy the classical infinitesimal negligibility condition;
(c) the sums of random variables in each series Ξ_i equal the corresponding sums in the series $\Xi(\mathbf{n}_i)$;
(d) the functions $G(\mathbf{n}_i)$ and numbers $\gamma(\mathbf{n}_i)$ equal the functions G_i and numbers γ_i corresponding to the series Ξ_i.

Conditions (3.17) imply that

$$G_i \rightrightarrows G, \qquad \gamma_i \to \gamma. \tag{3.18}$$

According to the case $d = 1$ (see Theorem 3, Chap. 4 in [336]) this means that the distribution functions in the series Ξ_i weakly converge to F as $i \to \infty$. Since the sequence $\{\mathbf{n}_i, i \ge 1\}$ is arbitrary, the Cauchy convergence criterion (see Corollary A.1 in the case of Theorem 3.5 or Corollary A.2 in the case of Theorem 3.6), implies that the distribution functions of sums (3.11) weakly converge to F.

Now we prove the converse statement, namely we show that (3.12) follows from the weak convergence. Let the distribution functions of sums (3.11) weakly converge to an infinitely divisible distribution function F. Choose a sequence of multi-indices $\{\mathbf{n}_i, i \ge 1\}$ such that $\max(n_{1i}, \dots, n_{di}) \to \infty$ as $i \to \infty$ in the case of Theorem 3.5 or $\min(n_{1i}, \dots, n_{di}) \to \infty$ as $i \to \infty$ in the case of Theorem 3.6. According to the Cauchy convergence criterion (Corollary A.1 or Corollary A.2) the distribution functions of sums $\sum_{\mathbf{k} \le \mathbf{k}(\mathbf{n}_i)} X(\mathbf{n}_i, \mathbf{k})$ weakly converge to F. As before, we denote the (d, r)-series $\{X(\mathbf{n}, \mathbf{k}); \mathbf{k} \le \mathbf{k}(\mathbf{n})\}$ by $\Xi(\mathbf{n})$. Then $\{\Xi_i \stackrel{\text{def}}{=} \Xi(\mathbf{n}_i), i \ge 1\}$ forms a classical $(1, 1)$-sequence of series and the random variables constituting these series are infinitesimally negligible. The distribution functions F_i of sums in the series Ξ_i coincide with those of sums in the series $\Xi(\mathbf{n}_i)$ and thus F_i weakly converge. According to the case $d = 1$ (see Theorem 3, Chap. 4 in [336]) this means that condition (3.18) holds, where the functions G_i and numbers γ_i are defined with respect to the sums of random variables in the series Ξ_i. It is clear that $G_i = G(\mathbf{n}_i)$, $\gamma_i = \gamma(\mathbf{n}_i)$ and thus condition (3.17) holds. Since the sequence of multi-indices $\{\mathbf{n}_i, i \ge 1\}$ is arbitrary, the Cauchy convergence criterion (see Corollary A.1 or Corollary A.2) yields condition (3.12). Theorems 3.5 and 3.6 are proved. □

3.4 Weak Convergence of Cumulative Sums

In this section, we consider a particular case of a scheme of series. Having in mind the case $d = 1$, we use the term "cumulative" sums even when $d > 1$. First we recall the results in the classical case.

3.4.1 The Case d = 1

The following definition is attributed to P. Lévy.

Definition 3.7 We say that a distribution function F belongs to the class $\mathfrak{L}$ if there exists a sequence of independent random variables $\{X_n\}$ and two sequences of real numbers $\{b_n\}$ and $\{c_n\}$ such that $b_n > 0$, $n \ge 1$,

$$\lim_{n\to\infty} \max_{k\le n} \mathsf{P}(|X_k| \ge \varepsilon b_n) = 0 \qquad \text{for all} \quad \varepsilon > 0, \tag{3.19}$$

and

$$\operatorname{Law}\left(\frac{S_n}{b_n} - c_n\right) \xrightarrow{\mathscr{D}} F. \tag{3.20}$$

Condition (3.19) means that the random variables in the series

$$\left\{\frac{X_k}{b_n},\ k \le n\right\}, \qquad n \ge 1,$$

are infinitesimal, while condition (3.20) says that the distribution functions of normalized and centered cumulative sums $S_n = X_1 + \cdots + X_n$ weakly converge to the function F.

It is straightforward from (3.19)–(3.20) that if the distribution function F is not degenerate (that is, F does not have a jump of unit height), then

$$\lim_{n\to\infty} b_n = \infty, \qquad \lim_{n\to\infty} \frac{b_{n+1}}{b_n} = 1. \tag{3.21}$$

A characterization of characteristic functions of probability laws belonging to the class $\mathfrak{L}$ can be derived from condition (3.21) and the convergence (3.20).

Theorem 3.7 (Characterization in terms of characteristic functions) *A distribution function $F(x)$ with the characteristic function $f(t)$ belongs to the class $\mathfrak{L}$ if and only if, for any $0 < \alpha < 1$, there exists a characteristic function $f_\alpha(t)$ such that*

$$f(t) = f(\alpha t) f_\alpha(t).$$

Moreover, f_α is an infinitely divisible characteristic function.

Since condition (3.21) follows from (3.19) and (3.20), one can define the class $\mathfrak{L}$ in an equivalent way. The definition below is attributed to W. Feller.

Definition 3.8 We say that a distribution function F belongs to the class $\mathfrak{L}$ if there exist a sequence of independent random variables $\{X_n\}$ and two sequences of real numbers $\{b_n\}$, $b_n > 0$, $n \geq 1$, and $\{c_n\}$ satisfying condition (3.21) and such that the convergence (3.20) holds.

The proofs of the classical results for the class $\mathfrak{L}$ are exhibited in the literature in the following order:

1. first, we derive properties (3.21) from Definition 3.7;
2. then we obtain a characterization of the class $\mathfrak{L}$ (see Theorem 3.7) by using properties (3.21) and the convergence (3.20);
3. the remaining results concerning weak convergence are proved with the use of Theorem 3.7.

Theorem 3.7 provides yet another characterization of the class $\mathfrak{L}$ which is equivalent to the preceding two characterizations.

Definition 3.9 We say that a distribution function F belongs to the class $\mathfrak{L}$ if there exists a sequence of independent random variables $\{X_n\}$ such that

$$\operatorname{Law}\left(\frac{S_n}{n}\right) \xrightarrow{\mathscr{D}} F, \tag{3.22}$$

where, as usual, $S_n = X_1 + \cdots + X_n$.

To prove the equivalence of Definitions 3.8 and 3.9 it is sufficient to show that, for any distribution function F, condition (3.22) follows from (3.21) and (3.20). Denote by f the characteristic function of F. Theorem 3.7 means that, for any $k \geq 1$,

$$f_k(t) \stackrel{\text{def}}{=} \frac{f(kt)}{f((k-1)t)}$$

is a characteristic function of a random variable X_k. Thus f is the characteristic function of the random variable S_n/n and, in addition, condition (3.22) holds.

There are several other characterizations of the distribution functions belonging to the class $\mathfrak{L}$. Below we give two such characterizations. The first of them is given in terms of the spectral function G in the Lévy–Khintchine representation (3.8); while the second one is based on yet another property of the corresponding characteristic functions.

Theorem 3.8 (Characterization in terms of the spectral function) *A distribution function $F(x)$ with the characteristic function $f(t)$ belongs to the class $\mathfrak{L}$ if and only if, in the Lévy–Khintchine representation* (3.8),

(i) *G is a non-decreasing and bounded function;*
(ii) *$G(-\infty)=0$;*
(iii) *at each point $x\neq 0$, the left derivative $G'_{left}(x)$ and the right derivative $G'_{right}(x)$ both exist;*
(iv) *the functions*

$$\frac{1+x^2}{x^2}G'_{left}(x) \quad \textit{and} \quad \frac{1+x^2}{x^2}G'_{right}(x)$$

are non-increasing in the intervals $(-\infty,0)$ and $(0,\infty)$.

Theorem 3.9 (Kubik characterization) *Let K_1 be the class of functions*

$$G(x)=\begin{cases}0, & x<A,\\ a\ln\frac{1+A^2}{1+x^2}, & A\le x\le 0,\\ a\ln(1+A^2), & x>0,\end{cases}$$

that depend on two parameters $a\ge 0$ and $A\le 0$. Also let K_2 and K_3 be the classes of functions

$$G(x)=\begin{cases}0, & x\le 0,\\ b\ln(1+x^2), & 0<x\le B,\\ b\ln(1+B^2), & x>B;\end{cases} \qquad G(x)=\begin{cases}0, & x\le 0,\\ c, & x>0,\end{cases}$$

respectively. Note that the functions of the class K_2 depend on two parameters $b\ge 0$ and $B>0$, while those of K_3 depend on a unique parameter $c\ge 0$. Further, denote by $\mathcal{G}$ the class of distribution functions whose characteristic functions admit the Lévy–Khintchine (3.8) *representation with $G\in K_1\cup K_2\cup K_3$.*

Then the class $\mathfrak{L}$ consists of convolutions of a finite number of distribution functions of $\mathcal{G}$ and of the weak limits of these convolutions.

3.4.2 Urbanik Classes

A natural generalization of the property described in Theorem 3.7 is due to Urbanik and leads to some further important classes of characteristic functions.

The Urbanik classes are introduced recursively by starting with the family of all characteristic functions denoted by $\mathfrak{H}$. Let $\mathfrak{U}_0$ be the class of all infinitely divisible characteristic functions, that is $\mathfrak{U}_0 = \mathfrak{D}$. Put

$$\mathfrak{U}_1 = \left\{ f \in \mathfrak{H} : \frac{f(t)}{f(\alpha t)} \in \mathfrak{U}_0 \text{ for all } \quad \alpha \in (0, 1) \right\}.$$

According to Theorem 3.7, $\mathfrak{U}_1 = \mathfrak{L}$. All other Urbanik classes are introduced as follows:

$$\mathfrak{U}_k = \left\{ f \in \mathfrak{H} : \frac{f(t)}{f(\alpha t)} \in \mathfrak{U}_{k-1} \text{ for all } \quad \alpha \in (0, 1) \right\}.$$

It is known that all classes $\mathfrak{U}_k$ are different and moreover $\bigcap_{k \ge 1} \mathfrak{U}_k$ is a non-empty class.

Consider, for example, the case $k = 2$ in more detail. The characteristic property of the class $\mathfrak{U}_2$ can be written as follows: for any number $\alpha \in (0, 1)$, the function $g(t) \stackrel{\text{def}}{=} f(t)/f(\alpha t)$ belongs to the class $\mathfrak{L}$. Theorem 3.7 implies that

$$\frac{g(t)}{g(\beta t)} \quad \text{is a characteristic function for all} \quad \beta \in (0, 1)$$

whatever the number $\alpha \in (0, 1)$ defining g. In other words, using the definition of the function g,

$$\frac{f(t) f(\alpha\beta t)}{f(\alpha t) f(\beta t)} \in \mathfrak{H} \quad \text{for all} \quad \alpha, \beta \in (0, 1). \tag{3.23}$$

The characterization property of the class $\mathfrak{U}_k$ can be written similarly for every $k \ge 1$. This property is easy to write with the help of two sets A and B defined for an arbitrary collection of numbers $\alpha_1, \dots, \alpha_k \in (0, 1)$. Each of the sets A and B consists of 2^{k-1} numbers; each of the numbers constituting A and B equals one of the products $\alpha_{i_1} \dots \alpha_{i_j}$, where $i_1, \dots, i_j$ are different integers from $\{1, \dots, k\}$ (there is a "product" corresponding to the empty collection of α's; we agree that the expression $\alpha_{i_1} \dots \alpha_{i_j}$ equals 1 if the collection $\{i_1, \dots, i_j\}$ is empty).

The difference between A and B is that the numbers constituting A are products of an even number of terms, while the numbers constituting B are products of an odd number of terms. Note that the term with the empty collection of α's belongs to A.

Then the characterization of the class $\mathfrak{U}_k$ can be written as follows: $f \in \mathfrak{U}_k$ if and only if

$$\prod_{a\in A} f(at) \Big/ \prod_{b\in B} f(bt) \in \mathfrak{H} \quad \text{for all} \quad \alpha_1, \ldots, \alpha_k \in (0,1). \tag{3.24}$$

It is clear that condition (3.24) reduces to the condition of Theorem 3.7 if $k = 1$. It is also clear that (3.24) coincides with (3.23) if $k = 2$.

3.4.3 The Case $d > 1$

An analogous question about the limit distributions of "cumulative" sums can be asked for $d > 1$, too, for both lim(max)- and lim(min)-convergence. Each of the Definitions 3.7–3.9 has its analog in the case $d > 1$ which results in six classes of limit distributions and not all of them coincide when $d > 1$. To highlight a relationship between the class $\mathfrak{L}$ related to cumulative sums for $d = 1$ and the general problem for $d > 1$ we write $\mathfrak{L}_1$ instead of $\mathfrak{L}$.

Let $\{X(\mathbf{n}), \mathbf{n} \in \mathbf{N}^d\}$ be a field of independent random variables and let $\{b(\mathbf{n}), \mathbf{n} \in \mathbf{N}^d\}$ and $\{c(\mathbf{n}), \mathbf{n} \in \mathbf{N}^d\}$ be two fields of real numbers such that $b(\mathbf{n}) > 0$. We study the weak convergence of distribution functions of centered and normalized sums

$$\frac{1}{b(\mathbf{n})} \sum_{\mathbf{k}\le\mathbf{n}} X(\mathbf{k}) - c(\mathbf{n}). \tag{3.25}$$

The problem described above for sums (3.25) is a particular case for the scheme of (d, d)-series with $\mathbf{k}(\mathbf{n}) = \mathbf{n}$ for all $\mathbf{n} \in \mathbf{N}^d$ and

$$\left\{ X(\mathbf{n}, \mathbf{k}) \stackrel{\text{def}}{=} \frac{X(\mathbf{k})}{b(\mathbf{n})} - \frac{c(\mathbf{n})}{|\mathbf{n}|},\ \mathbf{k} \le \mathbf{n} \right\}, \quad \mathbf{n} \in \mathbf{N}^d.$$

3.4.4 Lévy–Khintchine Classes for $d > 1$

First we consider analogs of Definition 3.7 for lim(max)- and lim(min)-convergence. We use the superscript L to denote the corresponding classes resulting from the Lévy Definition 3.7 for $d = 1$.

Definition 3.10 We say that a distribution function F belongs to the class $\mathfrak{L}^L_{d;\max}$ if there exist a field of independent random variables $\{X(\mathbf{n})\}$ and two fields of non-random real numbers $\{b(\mathbf{n})\}$ and $\{c(\mathbf{n})\}$ such that

$$\operatorname{Law}\left(\frac{1}{b(\mathbf{n})} \sum_{\mathbf{k}\le\mathbf{n}} X(\mathbf{k}) - c(\mathbf{n}) \right) \xrightarrow[\max]{\mathscr{D}} F, \tag{3.26}$$

$b(\mathbf{n}) > 0$, $\mathbf{n} \in \mathbf{N}^d$, and

$$\lim(\max) \max_{\mathbf{k} \le \mathbf{n}} \mathsf{P}(|X(\mathbf{k})| \ge \varepsilon b(\mathbf{n})) = 0 \qquad \text{for all} \quad \varepsilon > 0. \tag{3.27}$$

Definition 3.11 We say that a distribution function F belongs to the class $\mathfrak{L}^L_{d;\min}$ if there exist a field of independent random variables $\{X(\mathbf{n})\}$ and two fields of non-random real numbers $\{b(\mathbf{n})\}$ and $\{c(\mathbf{n})\}$ such that

$$\operatorname{Law}\left(\frac{1}{b(\mathbf{n})} \sum_{\mathbf{k} \le \mathbf{n}} X(\mathbf{k}) - c(\mathbf{n})\right) \xrightarrow[\min]{\mathscr{D}} F, \tag{3.28}$$

$b(\mathbf{n}) > 0$, $\mathbf{n} \in \mathbf{N}^d$, and

$$\lim(\min) \max_{\mathbf{k} \le \mathbf{n}} \mathsf{P}(|X(\mathbf{k})| \ge \varepsilon b(\mathbf{n})) = 0. \tag{3.29}$$

Theorem 3.10 $\mathfrak{U}_d \subseteq \mathfrak{L}^L_{d;\max} \subseteq \mathfrak{L}_1 \subseteq \mathfrak{L}^L_{d;\min}$.

Proof First we prove the inclusion $\mathfrak{U}_d \subseteq \mathfrak{L}^L_{d;\max}$. Let $F \in \mathfrak{U}_d$. We denote the characteristic function of F by f. Recall that the set $\mathscr{E}_d$ consists of the vectors $\boldsymbol{\varepsilon} = (\varepsilon_1, \dots, \varepsilon_d)$ each coordinate of which is equal to either 0 or 1 (see Proposition A.5). For every $\mathbf{k} \in \mathbf{N}^d$, we define the sets

$$\begin{aligned} A &= \{\mathbf{a} : \mathbf{a} = \mathbf{k} - \boldsymbol{\varepsilon},\ \boldsymbol{\varepsilon} \in \mathscr{E}_d,\ \varepsilon_1 + \dots + \varepsilon_d \text{ even}\}, \\ B &= \{\mathbf{b} : \mathbf{b} = \mathbf{k} - \boldsymbol{\varepsilon},\ \boldsymbol{\varepsilon} \in \mathscr{E}_d,\ \varepsilon_1 + \dots + \varepsilon_d \text{ odd}\} \end{aligned} \tag{3.30}$$

and two functions

$$g(\mathbf{k}; t) = \prod_{\mathbf{a} \in A} f(|\mathbf{a}| t), \quad h(\mathbf{k}; t) = \prod_{\mathbf{b} \in B} f(|\mathbf{b}| t).$$

According to condition (3.24), $g(\mathbf{k}; t)/h(\mathbf{k}; t)$ is a characteristic function for all $\mathbf{k} \in \mathbf{N}^d$. Moreover,

$$\prod_{\mathbf{k} \le \mathbf{n}} \frac{g(\mathbf{k}; t)}{h(\mathbf{k}; t)} = f(|\mathbf{n}| t), \quad \mathbf{n} \in \mathbf{N}^d.$$

This means that if a random variable $X(\mathbf{k})$ has the characteristic function $g(\mathbf{k}; t)/h(\mathbf{k}; t)$ and if the random variables $\{X(\mathbf{n})\}$ are jointly independent, then the characteristic function of $S(\mathbf{n})/|\mathbf{n}|$ is $f(t)$.

To complete the proof of the inclusion $F \in \mathfrak{L}^L_{d;\max}$, it remains to show that condition (3.27) holds. Let $q : A \to B$ be a bijection such that the elements $\mathbf{a}$ and $q(\mathbf{a})$ differ only by one coordinate for all $\mathbf{a} \in A$. Then, for all $\mathbf{k} \le \mathbf{n}$,

$$\frac{g(\mathbf{k}; t/|\mathbf{n}|)}{h(\mathbf{k}; t/|\mathbf{n}|)} = \prod_{\mathbf{a}\in A} \frac{f\left(t|\mathbf{a}|/|\mathbf{n}|\right)}{f\left(t|\mathbf{b}|/|\mathbf{n}|\right)}, \quad \mathbf{b} = q(\mathbf{a}). \tag{3.31}$$

Since every characteristic function is uniformly continuous, we deduce that

$$\max_{\substack{\mathbf{a}\le\mathbf{n}\\ \mathbf{a}\in A}} \left| f\left(t|\mathbf{a}|/|\mathbf{n}|\right) - f\left(t|\mathbf{b}|/|\mathbf{n}|\right)\right| \underset{\max}{\longrightarrow} 0$$

for all finite intervals I uniformly with respect to $t \in I$, where $\mathbf{b} = q(\mathbf{a})$. Since f is an infinitely divisible characteristic function, it does not vanish in the real line. Thus

$$\max_{\substack{\mathbf{a}\le\mathbf{n}\\ \mathbf{a}\in A}} \left| \frac{f\left(t|\mathbf{a}|/|\mathbf{n}|\right)}{f\left(t|\mathbf{b}|/|\mathbf{n}|\right)} - 1\right| \underset{\max}{\longrightarrow} 0$$

for all finite intervals I uniformly with respect to $t \in I$ (here $\mathbf{b} = q(\mathbf{a})$). By (3.31), we see that

$$\max_{\mathbf{k}\le\mathbf{n}} \left| \frac{g(\mathbf{k}; t/|\mathbf{n}|)}{h(\mathbf{k}; t/|\mathbf{n}|)} - 1\right| \underset{\max}{\longrightarrow} 0$$

for all finite intervals I uniformly with respect to $t \in I$. In particular, this means that if $X(\mathbf{n}; \mathbf{k})$ is a random variable with the characteristic function $f(\mathbf{n}; \mathbf{k}; t) = g(\mathbf{k}; t/|\mathbf{n}|)/h(\mathbf{k}; t/|\mathbf{n}|)$, then

$$\max_{\mathbf{k}\le\mathbf{n}} |f(\mathbf{k}; \mathbf{n}; t) - 1| \underset{\max}{\longrightarrow} 0 \tag{3.32}$$

uniformly with respect to $t \in I$, where I is an arbitrary finite interval.

Lemma 3.1 *Let* $\{X(\mathbf{n}, \mathbf{k}), \mathbf{k} \le \mathbf{k}(\mathbf{n})\}$, $\mathbf{n} \in \mathbf{N}^d$, *be a* (d, d)*-series. Then condition* (3.32) *implies the infinitesimal negligibility condition* (3.27).

The proof of Lemma 3.1 does not differ from that for the classical case $d = 1$, thus we omit it here.

The reasoning given above together with Lemma 3.1 implies condition (3.27), that is $\mathfrak{U}_d \in \mathfrak{L}^L_{d;\max}$.

Now we prove the inclusion $\mathfrak{L}^L_{d;\max} \subseteq \mathfrak{L}_1$. If $F \in \mathfrak{L}^L_{d;\max}$, then there exist a field of independent random variables $\{X(\mathbf{n})\}$ and two fields of real numbers $\{b(\mathbf{n})\}$ and $\{c(\mathbf{n})\}$, $b(\mathbf{n}) > 0$, for which conditions (3.27)–(3.26) hold. Therefore, condition (3.20) holds for all random variables

$$S_m \stackrel{\text{def}}{=} \sum_{k=1}^{m} X(k, 1, \dots, 1)$$

and sequences $b_m \stackrel{\text{def}}{=} b(m, 1, \dots, 1)$ and $c_m \stackrel{\text{def}}{=} c(m, 1, \dots, 1)$. Since the random variables $X_k \stackrel{\text{def}}{=} X(k, 1, \dots, 1)$ are independent and satisfy the infinitesimal negligibility condition (3.19) and weak convergence (3.20), we obtain $F \in \mathfrak{L}_1$, that is $\mathfrak{L}_{d;\max}^L \subseteq \mathfrak{L}_1$.

Finally, we prove the inclusion $\mathfrak{L}_1 \subseteq \mathfrak{L}_{d;\min}^L$. Let $F \in \mathfrak{L}_1$. According to Definition 3.9, there exists a sequence of independent random variables $\{X_n\}$ for which condition (3.22) holds. Now we introduce the field of independent random variables $\{X(\mathbf{n})\}$ by putting

$$X(\mathbf{n}) = \begin{cases} X_{n_1}, & \text{if } n_1 = n_2 = \cdots = n_d, \\ 0, & \text{otherwise.} \end{cases}$$

Put $m \stackrel{\text{def}}{=} \min\{n_1, \ldots, n_d\}$. Then, for the field $b(\mathbf{n}) \stackrel{\text{def}}{=} m$, we have $S(\mathbf{n})/b(\mathbf{n}) = S_m/m$, whence

$$\frac{S(\mathbf{n})}{b(\mathbf{n})} \xrightarrow[\min]{\mathscr{D}} F.$$

To show that the infinitesimal negligibility condition (3.29) holds, note that (3.22) implies that

$$\max_{k \le n} \mathsf{P}(|X_k| \ge \varepsilon n) \to 0, \quad n \to \infty,$$

for all $\varepsilon > 0$. The latter result is proved above for the general case $d \ge 1$. In summary, this means $\mathfrak{L}_1 \subseteq \mathfrak{L}_{d;\min}^L$. □

3.4.5 Feller Classes for d > 1

Now we consider some analogues of Definition 3.8. We shall use the superscript F to indicate that a given class is defined using Feller's definition.

Definition 3.12 We say that a distribution function F belongs to the class $\mathfrak{L}_{d;\max}^F$ if there exist a field of independent random variables $\{X(\mathbf{n})\}$ and two fields of non-negative real numbers $\{b(\mathbf{n})\}$ and $\{c(\mathbf{n})\}$ such that $b(\mathbf{n}) > 0$, $\mathbf{n} \in \mathbf{N}^d$,

$$\lim(\max) b(\mathbf{n}) = \infty, \quad \lim(\max) \frac{b(\mathbf{n}+\mathbf{1})}{b(\mathbf{n})} = 1 \tag{3.33}$$

and if conditions (3.26) and (3.27) hold.

When $d = 1$, Definition 3.12 is a combination of Definitions 3.7 and 3.8, since it includes both the infinitesimal negligibility condition (3.27) and condition (3.33) describing a "slow enough" evolution of the normalizing field. As mentioned above, such a "mixture" of properties does not reduce the class of possible limit distributions if $d = 1$.

Definition 3.13 We say that a distribution function F belongs to the class $\mathfrak{L}^F_{d;\min}$ if there exist a field of independent random variables $\{X(\mathbf{n})\}$ and two fields of non-negative real numbers $\{b(\mathbf{n})\}$ and $\{c(\mathbf{n})\}$ such that $b(\mathbf{n}) > 0$, $\mathbf{n} \in \mathbf{N}^d$,

$$\lim(\min) b(\mathbf{n}) = \infty, \quad \lim(\min) \frac{b(\mathbf{n}+\mathbf{1})}{b(\mathbf{n})} = 1 \tag{3.34}$$

and if conditions (3.28) and (3.29) hold.

Theorem 3.11 $\mathfrak{U}_d \subseteq \mathfrak{L}^F_{d;\max} \subseteq \mathfrak{L}^F_{d;\min} = \mathfrak{L}_1$.

Proof The inclusion $\mathfrak{U}_d \subseteq \mathfrak{L}^F_{d;\max} \subseteq \mathfrak{L}_1 \subseteq \mathfrak{L}^F_{d;\min}$ is proved by the same reasoning as that in the proof of Theorem 3.10. It remains to show that $\mathfrak{L}^F_{d;\min} \subseteq \mathfrak{L}_1$.

Let $F \in \mathfrak{L}^F_{d;\min}$. Introduce the field of independent random variables $\{X(\mathbf{n})\}$ and two fields of real numbers $\{b(\mathbf{n})\}$ and $\{c(\mathbf{n})\}$ for which conditions (3.34) and (3.28)–(3.29) hold. Then

$$\operatorname{Law}\left(\frac{S(n,\dots,n)}{b(n,\dots,n)} - c(n,\dots,n)\right) \xrightarrow{\mathcal{D}} F.$$

Put

$$S_n = S(n,\dots,n), \quad b_n = b(n,\dots,n), \quad c_n = c(n,\dots,n).$$

Then

$$\operatorname{Law}\left(\frac{S_n}{b_n} - {}_n\right) \xrightarrow{\mathcal{D}} F, \quad b_n \to \infty, \quad \frac{b_{n+1}}{b_n} \to 1.$$

According to Definition 3.8, $F \in \mathfrak{L}_1$, that is $\mathfrak{L}^F_{d;\min} \subseteq \mathfrak{L}_1$. □

3.5 Weak Convergence of Cumulative Sums of Identically Distributed Random Variables

Let the random variables $\{X(\mathbf{n}), \mathbf{n} \in \mathbf{N}^d\}$ be independent and identically distributed. One of the basic results of the classical theory says that the set of limit distribution functions of normalized and centered sums of independent identically distributed random variables coincides with the set of stable distributions (see, for example, §33 in [126]). The case of $d > 1$ is treated below.

Definition 3.14 A distribution function is called *stable* if it is a weak limit of the distribution functions of sums

$$\frac{1}{b_n}\sum_{k=1}^{n} X_k - c_n, \tag{3.35}$$

where $\{X_k, k \geq 1\}$ are independent identically distributed random variables and where $b_n > 0$ and c_n are real numbers. The set of stable laws is denoted by $\mathfrak{S}$.

Definition 3.15 Denote by $\mathfrak{S}_d^{\max}$ the set of distribution functions (laws) that are weak limits for the distribution functions of sums

$$\frac{1}{b(\mathbf{n})} \sum_{\mathbf{k} \leq \mathbf{n}} X(\mathbf{k}) - c(\mathbf{n}) \tag{3.36}$$

in the sense of lim(max)-convergence, where $\{X(\mathbf{n}), \mathbf{n} \in \mathbf{N}^d\}$ are independent identically distributed random variables and where $b(\mathbf{n}) > 0$ and $c(\mathbf{n})$ are real numbers.

Definition 3.16 Denote by $\mathfrak{S}_d^{\min}$ the set of distribution functions (laws) that are weak limits for the distribution functions of sums (3.36) in the sense of lim(min)-convergence, where $\{X(\mathbf{n}), \mathbf{n} \in \mathbf{N}^d\}$ are independent identically distributed random variables and where $b(\mathbf{n}) > 0$ and $c(\mathbf{n})$ are real numbers.

It turns out that the classes $\mathfrak{S}_d^{\max}$ and $\mathfrak{S}_d^{\min}$ coincide with $\mathfrak{S}$ for all $d \geq 1$. For $d = 1$, this is a classical result.

Theorem 3.12 *For all* $d \geq 1$,

$$\mathfrak{S}_d^{\max} = \mathfrak{S}.$$

Theorem 3.13 *For all* $d \geq 1$,

$$\mathfrak{S}_d^{\min} = \mathfrak{S}.$$

We have used in the proofs of Theorems 3.12 and 3.13 the property that the normalizing field $\{b(\mathbf{n}), \mathbf{n} \in \mathbf{N}^d\}$ is unbounded, namely if the distribution functions of random variables (3.36) weakly converge to a limit distribution and if the common distribution of terms $X(\mathbf{k})$ is nondegenerate, then $\lim(\max) b(\mathbf{n}) = \infty$ in the case of lim(max)-convergence and $\lim(\min) b(\mathbf{n}) = \infty$ in the case of lim(min)-convergence (see Lemmas 3.3 and 3.4). This property of normalizing fields is a simple consequence of the following result.

Lemma 3.2 *Let* g *and* f *be characteristic functions such that* $|g(t)| \not\equiv 1$. *Let* $\{b_n, n \geq 1\}$ *be a sequence of positive numbers and* $\{l_n, n \geq 1\}$ *be a sequence of positive integers such that* $l_n \to \infty$. *If*

$$\left| g\left(\frac{t}{b_n}\right) \right|^{l_n} \to |f(t)|, \quad n \to \infty,$$

for all real numbers t, *then* $\lim_{n\to\infty} b_n = \infty$.

Proof of Lemma 3.2 Assume that $b_n \not\to \infty$. Then there exist a real number b_0, $|b_0| < \infty$, and a subsequence $\{n_k, k \geq 1\}$ for which $b_{n_k} \to b_0$ as $k \to \infty$. Choose a number $\delta > 0$ for which the function f does not have roots in the set $|t| \le \delta$ (such a number exists, since $f(0) = 1$ and f is a continuous function).

Recall that $|g(t)| \not\equiv 1$, thus one of the following two possibilities may occur:

(a) $|g(t)| < 1$ for all $t \ne 0$,
(b) $|g(t)| = 1$ only for a countable set of isolated points t

(see, for example, [57], Corollary 2, p 275).

Choose a point t_0 such that $|t_0| \le \delta$ and $|g(t_0/b_0)| < 1$. Then, on one hand,

$$\left| g\left(\frac{t_0}{b_{n_k}}\right)\right|^{l_{n_k}} \to 0, \qquad k \to \infty,$$

since g is continuous. On the other hand, $|g(t_0/b_{n_k})|^{l_{n_k}} \to |f(t_0)|$ by the assumption of the lemma, that is $f(t_0) = 0$. This contradiction proves that $b_n \to \infty$. The lemma is proved. □

Lemma 3.3 *Let g be a common characteristic function of each of the independent identically distributed random variables $X(\mathbf{n})$, $\mathbf{n} \in \mathbf{N}^d$, such that $g(t) \not\equiv 1$. If the distribution functions of the sums*

$$\frac{1}{b(\mathbf{n})} \sum_{\mathbf{k} \le \mathbf{n}} X(\mathbf{k}) - c(\mathbf{n}) \tag{3.37}$$

weakly lim(min)*-converge to a limit, then*

$$\lim(\min) b(\mathbf{n}) = \infty. \tag{3.38}$$

Lemma 3.4 *Let g be a common characteristic function of each of the independent identically distributed random variables $X(\mathbf{n})$, $\mathbf{n} \in \mathbf{N}^d$, such that $g(t) \not\equiv 1$. If the distribution functions of the sums* (3.37) *weakly* lim(max)*-converge to a limit, then*

$$\lim(\max) b(\mathbf{n}) = \infty. \tag{3.39}$$

Proof of Lemma 3.3 Consider an arbitrary sequence of multi-indices $\mathbf{n}_j \in \mathbf{N}^d$, $j \ge 1$, such that $\mathbf{n}_j = (n_{1j}, \dots, n_{dj})$ and $\min\{n_{1j}, \dots, n_{dj}\} \to \infty$ as $j \to \infty$. Then the distribution functions of the random variables

$$\zeta_j = \frac{1}{b(\mathbf{n}_j)} \sum_{\mathbf{k} \le \mathbf{n}_j} X(\mathbf{k}) - c(\mathbf{n}_j)$$

weakly converge to a limit as $j \to \infty$. By Helly's theorem (see, for example, Theorem 1, §13 in [126]), there exists a pointwise limit as $j \to \infty$ of the sequence of characteristic functions $h_{\zeta_j}(t) \stackrel{\text{def}}{=} (g(t/b(\mathbf{n}_j)))^{l_j} e^{-itc(\mathbf{n}_j)}$, where $l_j = |\mathbf{n}_j|$. This limit is also a characteristic function, and we denote it by f. Hence $|g(t/b(\mathbf{n}_j))|^{l_j} \to$

$|f(t)|$ as $j \to \infty$. Applying Lemma 3.2, we prove that $b(\mathbf{n}_j) \to \infty$ as $j \to \infty$. Since the sequence $\{\mathbf{n}_j\}$ is arbitrary, relation (3.38) follows from the Cauchy convergence criterion for lim(min)-convergence (see Corollary A.2). □

The proof of Lemma 3.4 is similar to that of Lemma 3.3. The only difference concerns the sequence of multi-indices $\mathbf{n}_j$, $j \geq 1$: it should be chosen such that $\max\{n_{1j}, \dots, n_{dj}\} \to \infty$ as $j \to \infty$ rather than $\min\{n_{1j}, \dots, n_{dj}\} \to \infty$ as $j \to \infty$ and thus one needs to apply the Cauchy convergence criterion for lim(max)-convergence (see Corollary A.1) instead of the Cauchy convergence criterion for lim(min)-convergence as in Lemma 3.3.

Proof of Theorem 3.12 First we prove that $\mathfrak{S} \subseteq \mathfrak{S}_d^{\max}$. Let a distribution function F belong to $\mathfrak{S}$. Then there are two sequences of real numbers $\{b_n, n \geq 1\}$, $b_n > 0$, and $\{c_n, n \geq 1\}$ and a sequence of independent identically distributed random variables $\{X_n, n \geq 1\}$ for which the distribution functions of sums (3.35) weakly converge to F.

Next we introduce the following two numerical fields $\{b(\mathbf{n})\}$ and $\{c(\mathbf{n})\}$ and a field of random variables $\{X(\mathbf{n}), \mathbf{n} \in \mathbf{N}^d\}$:

$$b(\mathbf{n}) = b_{|\mathbf{n}|}, \qquad c(\mathbf{n}) = c_{|\mathbf{n}|}, \qquad X(\mathbf{n}) \stackrel{d}{=} X_1.$$

An important additional requirement is that the random variables $\{X(\mathbf{n})\}$ are jointly independent. The symbol $\xi \stackrel{d}{=} \eta$ means that the random variables ξ and η are identically distributed. It is clear that $b(\mathbf{n}) > 0$ and

$$\operatorname{Law}\left(\frac{1}{b(\mathbf{n})}\sum_{\mathbf{k}\leq\mathbf{n}} X(\mathbf{k}) - c(\mathbf{n})\right) = \operatorname{Law}\left(\frac{1}{b_{|\mathbf{n}|}}\sum_{k=1}^{|\mathbf{n}|} X_k - c_{|\mathbf{n}|}\right)$$

for all $\mathbf{n} \in \mathbf{N}^d$. Thus the distribution functions of sums (3.36) weakly converge to F in the sense of lim(max)-convergence, that is $\mathfrak{S} \subseteq \mathfrak{S}_d^{\max}$.

Next we prove the inverse inclusion $\mathfrak{S}_d^{\max} \subseteq \mathfrak{S}$. Let $F \in \mathfrak{S}_d^{\max}$. Choose the three fields $\{b(\mathbf{n}), \mathbf{n} \in \mathbf{N}^d\}$, $\{c(\mathbf{n}), \mathbf{n} \in \mathbf{N}^d\}$ and $\{X(\mathbf{n}), \mathbf{n} \in \mathbf{N}^d\}$ such that $b(\mathbf{n}) > 0$, $c(\mathbf{n})$ are real numbers, and $X(\mathbf{n})$ are independent identically distributed random variables such that the distribution of sums (3.36) weakly converge to F in the sense of lim(max)-convergence. Then the distribution functions of the sums

$$\frac{1}{b(n, 1, \dots, 1)}\sum_{k=1}^{n} X(k, 1, \dots, 1) - c(n, 1, \dots, 1)$$

converge to F, as well. Put

$$b_n = b(n, 1, \dots, 1), \qquad c_n = c(n, 1, \dots, 1), \qquad X_n = X(n, 1, \dots, 1).$$

It is clear that the distribution functions of sums (3.35) for the fields just defined weakly converge to F, that is $\mathfrak{S}_d^{\max} \subseteq \mathfrak{S}$. The proof of Theorem 3.12 is complete. □

Proof of Theorem 3.13 The inclusion $\mathfrak{S} \subseteq \mathfrak{S}_d^{\min}$ is proved in the same way as the inclusion $\mathfrak{S} \subseteq \mathfrak{S}_d^{\max}$ in Theorem 3.12.

It remains to prove the inverse inclusion $\mathfrak{S}_d^{\min} \subseteq \mathfrak{S}$. Let $F \in \mathfrak{S}_d^{\min}$. Since degenerate distributions are stable, we assume in what follows that

$$F(x) \text{ is a nondegenerate distribution function,}$$

that is, there is no point $x_0 \in \mathbf{R}^1$ such that $F(x_0) = 0$ and $F(x_0 + 0) = 1$. Choose the fields $\{b(\mathbf{n}), \mathbf{n} \in \mathbf{N}^d\}$, $\{c(\mathbf{n}), \mathbf{n} \in \mathbf{N}^d\}$ and $\{X(\mathbf{n}), \mathbf{n} \in \mathbf{N}^d\}$ such that $b(\mathbf{n}) > 0$, $X(\mathbf{n})$ are independent identically distributed random variables, and the distribution functions of sums (3.36) weakly converge to F in the sense of lim(min)-convergence. We denote the common distribution function of random variables $X(\mathbf{n})$ by H. It is clear that the distribution functions of normalized and centered "square" sums

$$\frac{1}{b(n, n, \dots, n)} \sum_{k_1=1}^{n} \sum_{k_2=1}^{n} \cdots \sum_{k_d=1}^{n} X(k_1, k_2, \dots, k_d) - c(n, n, \dots, n) \tag{3.40}$$

weakly converge to F.

To prove the inclusion $F \in \mathfrak{S}$, we construct sequences of

(a) non-random numbers $\{b_n, n \geq 1\}$, $b_n > 0$, and $\{c_n, n \geq 1\}$;
(b) independent identically distributed random variables $\{X_n, n \geq 1\}$,

for which the distribution functions of sums (3.35) weakly converge to F. The construction of such sequences runs as follows. Considering the convergence of the distribution functions of random variables (3.40) we choose the sequences $\{b_{n^d}\}$ and $\{c_{n^d}\}$ by putting

$$b_{n^d} = b(n, n, \dots, n), \qquad c_{n^d} = c(n, n, \dots, n).$$

Now to define random variables $\{X_n, n \geq 1\}$ we first define the sets $\Gamma_1 = \Pi_1 = \{\mathbf{1}\}$ and

$$\Pi_n = \{\mathbf{k} \colon \mathbf{k} \leq (n, \dots, n)\} \quad \text{and} \quad \Gamma_n = \Pi_n \setminus \Pi_{n-1}, \qquad n > 1.$$

Now we write the random variables $\{X(\mathbf{n}), \mathbf{n} \in \mathbf{N}^d\}$ as follows: the unique random variable of the set Γ_1 is written first; then we write (in an arbitrary order) the random variables of the set Γ_2, then those of the set Γ_3 and so forth. It is clear that:

(a) $b_{n^d} > 0, n \geq 1$;
(b) the random variables $\{X_n\}$ are jointly independent and identically distributed;

(c) the distribution functions of normalized and centered sums

$$\frac{1}{b_{n^d}} \sum_{k=1}^{n^d} X_k - c_{n^d}$$

weakly converge to F.

The latter property is a consequence of the weak convergence of random variables (3.40), since

$$\sum_{k_1=1}^{n} \sum_{k_2=1}^{n} \cdots \sum_{k_d=1}^{n} X(k_1, k_2, \ldots, k_d) = \sum_{k=1}^{n^d} X_k.$$

Note, however, that the properties indicated above do not, in general, mean that the whole sequence of distribution functions of sums (3.35) weakly converges to F for the random variables just defined. Properties (a)–(c) only mean that the distribution functions corresponding to the subsequence $\{1, 2^d, 3^d, \ldots\}$ weakly converge. To prove the existence of sequences $\{b_n\}$ and $\{c_n\}$ for which the whole sequence converges we will make use of the following result.

Theorem 3.14 (W. Döblin and B. V. Gnedenko) *For every distribution function H, only one of the following three possibilities may occur:*

(i) *H does not belong to the domain of partial attraction of any probability law;*
(ii) *H belongs to the domain of partial attraction of an uncountable set of types;*
(iii) *H belongs to the domain of partial attraction of exactly one type.*

Recall that a distribution function H is said to belong to the domain of partial attraction of a law G if there are two sequences of real numbers $\{\beta_n, n \geq 1\}$, $\beta_n > 0$, and $\{\delta_n, n \geq 1\}$, an increasing subsequence of positive integers $\{k_n, n \geq 1\}$, and a sequence of independent identically distributed random variables $\{X_n, n \geq 1\}$ with distribution function H such that

$$\frac{1}{\beta_{k_n}} \sum_{k=1}^{k_n} X_k - \delta_{k_n} \xrightarrow{\mathscr{D}} G. \tag{3.41}$$

The collection of distribution functions $\{G(ax + b), a, b \in \mathbf{R}\}$ is called the *type of the law G*.

As has been already proved, the distribution function H of the random variables $X(\mathbf{n})$ belongs to the domain of partial attraction of the law F (see property (c) above). Thus case (i) in Theorem 3.14 does not happen. We now prove that case (ii) is also impossible.

Assume that, for some sequences $\{\beta_n\}$, $\{\delta_n\}$ and $\{k_n\}$, relation (3.41) holds. Our current goal is to prove that $G(x) = F(ax + b)$ for some $a, b \in \mathbf{R}$.

Since $\beta_{k_n} \to \infty$ by Lemma 3.2, the distribution functions of the sums

$$\frac{1}{\beta_{k_n}} \sum_{k=k_n+1}^{k_n+q} X_k$$

weakly converge to a degenerate distribution function whatever the number $q \ge 1$ is. This means that, for $q \ge 1$ and $\varepsilon > 0$,

$$f_n(\varepsilon, q) \stackrel{\text{def}}{=} \mathsf{P}\left(\left|\frac{1}{\beta_{k_n}} \sum_{k=k_n+1}^{k_n+q} X_k\right| \ge \varepsilon\right) \to 0, \quad n \to \infty. \tag{3.42}$$

Lemma 3.5 *Assume that positive functions $\{f_n(\varepsilon, q)\}$ are decreasing with respect to ε and that*

$$\lim_{n\to\infty} f_n(\varepsilon, q) = 0$$

for all $q \ge 1$ and $\varepsilon > 0$. Then there exists a sequence $\{q_n, n \ge 1\}$ such that $\lim_{n\to\infty} q_n = \infty$ and

$$\lim_{n\to\infty} f_n(\varepsilon, q_n) = 0$$

for all $\varepsilon > 0$.

Proof of Lemma 3.5 By the assumptions of the lemma, given $k \ge 1$, there exists a positive integer n_k such that

$$f_n\left(\frac{1}{2^k}, k\right) < \frac{1}{2^k} \qquad \text{for all} \quad n \ge n_k.$$

Now we put $q_n = k$ for all $n_k \le n < n_{k+1}$. We prove that $\{q_n\}$ is the sequence we are searching for. The property $\lim q_n = \infty$ is obvious. Next, given $\varepsilon > 0$, we find k_0 such that $2^{-k_0} \le \varepsilon < 2^{-k_0-1}$. If $n_k \le n < n_{k+1}$ and $k \ge k_0$, then

$$f_n(\varepsilon, q_n) = f_n(\varepsilon, k) \le f_n\left(\frac{1}{2^{k_0}}, k\right) \le f_n\left(\frac{1}{2^k}, k\right) \le \frac{1}{2^k}.$$

This proves the lemma. □

Using notation (3.42) and Lemma 3.5, we choose a sequence of positive integers $\{q_n, n \ge 1\}$ such that $q_n \to \infty$ as $n \to \infty$ and $f_n(\varepsilon, q_n) \to 0$ as $n \to \infty$ for all $\varepsilon > 0$. This means that

$$\frac{1}{\beta_{k_n}} \sum_{k=k_n+1}^{k_n+q_n} X_k \xrightarrow{\mathsf{P}} 0, \qquad n \to \infty. \tag{3.43}$$

Remark 3.4 Relation (3.43) is a particular case of the *weak law of large numbers in the scheme of series* for the random variables $\{X_{nk}, k_n + 1 \le k \le k_n + q_n\}$, $n \ge 1$, where $X_{nk} = X_k/\beta_{k_n}$. The infinitesimal negligibility condition (3.1) holds in this case, since the distribution functions of the random variables X_k are identical and $\beta_{k_n} \to \infty$. A criterion for the weak law of large numbers (3.43) in the scheme of series is written as a combination of the following three conditions:

$$\begin{aligned} &q_n \mathsf{P}(|X| \ge \varepsilon\beta_{k_n}) \to 0, \qquad n \to \infty, \\ &\frac{q_n}{\beta_{k_n}} \int\limits_{|x|<\tau\beta_{k_n}} x\,dH(x) \to 0, \qquad n \to \infty, \\ &\frac{q_n}{\beta_{k_n}^2} \int\limits_{|x|<\tau\beta_{k_n}} x^2\,dH(x) \to 0, \qquad n \to \infty, \end{aligned} \tag{3.44}$$

for all $\varepsilon > 0$ and some $\tau > 0$, where H is the common distribution function of random variables $\{X_n\}$ and X is a certain random variable with the same distribution function (see [336], Chap. 9).

According to Lemma 3.5, the sequence $\{q_n, n \ge 1\}$ can be chosen so slowly increasing that

$$\frac{k_n}{q_n} \to \infty, \qquad n \to \infty. \tag{3.45}$$

Lemma 3.6 *Let $d > 1$ and let the sequences $\{k_n, n \ge 1\}$ and $\{q_n, n \ge 1\}$ be such that $q_n \to \infty$ as $n \to \infty$ and satisfy condition* (3.45)*. Then there are sequences of positive integers $\{m_{1n}, n \ge 1\}, \dots, \{m_{dn}, n \ge 1\}$ for which $m_{1n} \to \infty, \dots, m_{dn} \to \infty$ as $n \to \infty$ and*

$$k_n \le m_{1n} \dots m_{dn} \le k_n + q_n. \tag{3.46}$$

Proof of Lemma 3.6 We use induction with respect to d. For $d = 2$, two sequences $\{m_{1n}, n \ge 1\}$ and $\{m_{2n}, n \ge 1\}$ can be defined as follows:

$$m_{1n} = q_n, \qquad m_{2n} = \left[\frac{k_n}{q_n}\right] + 1.$$

Assume that Lemma 3.6 holds for some $d \ge 2$ and let us prove it for $d + 1$. Put

$$k_n' = \left[\frac{k_n}{[\sqrt{q_n}]}\right] + 1, \qquad q_n' = \left[\frac{k_n + q_n}{[\sqrt{q_n}]}\right] - k_n'.$$

It is clear that $k_n' \to \infty$, $q_n' \to \infty$, and $k_n'/q_n' \to \infty$ as $n \to \infty$. By the inductive assumption applied to the sequences $\{k_n'\}$ and $\{q_n'\}$, one can choose sequences $\{m_{1n}, n \ge 1\}, \dots, \{m_{dn}, n \ge 1\}$ for which $m_{1n} \to \infty, \dots, m_{dn} \to \infty$ as $n \to \infty$ and $k_n' \le m_{1n} \dots m_{dn} \le k_n' + q_n'$. Since $k_n' \sim k_n/\sqrt{q_n}$ and $q_n' \sim \sqrt{q_n}$ as $n \to \infty$, the sequence $m_{(d+1)n} = \left[\sqrt{q_n}\right]$ is such that: $m_{(d+1)n} \to \infty$ as $n \to \infty$ and $k_n \le m_{1n} \dots m_{dn} m_{(d+1)n} \le k_n + q_n$. □

Now we turn back to the proof of Theorem 3.13. First we choose sequences $m_{1n}, \dots, m_{dn}$ satisfying conditions (3.46). Let $l_n = m_{1n} \dots m_{dn}$. Since $0 \le l_n - k_n \le q_n$, conditions (3.44) imply that

$$\begin{aligned} &(l_n - k_n)\mathsf{P}(|X| \ge \varepsilon\beta_{k_n}) \to 0, \qquad n \to \infty, \\ &\frac{l_n - k_n}{\beta_{k_n}} \int\limits_{|x|<\tau\beta_{k_n}} x\, dH(x) \to 0, \qquad n \to \infty, \\ &\frac{l_n - k_n}{\beta_{k_n}^2} \int\limits_{|x|<\tau\beta_{k_n}} x^2\, dH(x) \to 0, \qquad n \to \infty, \end{aligned} \tag{3.47}$$

for all $\varepsilon > 0$ and some $\tau > 0$. Conditions (3.47) are necessary and sufficient for the following law of large numbers in the scheme of series:

$$\frac{1}{\beta_{k_n}} \sum_{k=k_n+1}^{l_n} X_n \xrightarrow{\mathsf{P}} 0, \qquad n \to \infty.$$

The latter relation together with (3.41) and Lemma 3.7 proves that

$$\frac{1}{\beta_{k_n}} \sum_{k=1}^{l_n} X_k - \delta_{k_n} \xrightarrow{\mathscr{D}} G.$$

We need the following well-known result.

Lemma 3.7 *Let $\{X_n\}$ and $\{Y_n\}$ be two sequences of random variables defined in the same probability space. If a sequence of distribution functions* Law (X_n) *weakly converges to the distribution function F and if $Y_n \xrightarrow{\mathsf{P}} 0$, $n \to \infty$, then the sequence of distribution functions* Law $(X_n + Y_n)$ *also weakly converges to F.*

Now let $\mathbf{m}_j = (m_{1j}, \dots, m_{dj})$ and $\beta(\mathbf{m}_j) = \beta_{k_j}$, $\delta(\mathbf{m}_j) = \delta_{k_j}$. Since

$$\operatorname{Law}\left(S(\mathbf{m}_j)\right) = \operatorname{Law}\left(\sum_{k=1}^{l_j} X_k\right),$$

we get

$$F_j \stackrel{\text{def}}{=} \operatorname{Law}\left(\frac{S(\mathbf{m}_j)}{\beta(\mathbf{m}_j)} - \delta(\mathbf{m}_j)\right) \xrightarrow{\mathscr{D}} G.$$

On the other hand,

$$\frac{S(\mathbf{m}_j)}{b(\mathbf{m}_j)} - c(\mathbf{m}_j) \xrightarrow{\mathscr{D}} F.$$

Since

$$\mathsf{P}\left(\frac{S(\mathbf{m}_j)}{b(\mathbf{m}_j)} - c(\mathbf{m}_j) < x\right) = F_j(xu_j + v_j)$$

for

$$u_j = \frac{b(\mathbf{m}_j)}{\beta(\mathbf{m}_j)}, \qquad v_j = c(\mathbf{m}_j)\frac{b(\mathbf{m}_j)}{\beta(\mathbf{m}_j)} - \delta(\mathbf{m}_j),$$

we have

$$F_j(x) \xrightarrow{\mathscr{D}} G(x) \quad \text{and} \quad F_j(xu_j + v_j) \xrightarrow{\mathscr{D}} F(x). \tag{3.48}$$

Next we apply the following result due to Khintchine.

Theorem 3.15 (A.Ya. Khintchine) *Let $\{u_j\}$ and $\{v_j\}$ be two sequences of real numbers such that $u_j > 0$. Assume that a sequence of distribution functions $\{F_n\}$ weakly converges to a nondegenerate distribution function F. Then*

(A) if $F_n(u_jx + v_j) \xrightarrow{\mathscr{D}} G(x)$, where G is a nondegenerate distribution function, then $G(x) = F(ax + b)$ for some $a, b \in \mathbf{R}$ and moreover $u_j \to a$ and $v_j \to b$. In particular, if $F_n(u_jx + v_j) \xrightarrow{\mathscr{D}} F(x)$, then $u_j \to 1$ and $v_j \to 0$;

(B) if $u_j \to a$ and $v_j \to b$, then $F_n(u_jx + v_j) \xrightarrow{\mathscr{D}} F(ax + b)$.

Theorem 3.15 together with (3.48) yields $G(x) = F(ax + b)$ for some real numbers a and b. This means that only the case (iii) in Theorem 3.14 may happen.

To complete the proof we need the following observation due to Gnedenko.

Theorem 3.16 (B. V. Gnedenko) *If a distribution function belongs to a unique type, then this type is necessarily stable.*

Theorem 3.16 implies that the type of distributions in condition (iii) can only be stable, that is F is a stable distribution function. Thus the inclusion $\mathfrak{S}_d^{\min} \subseteq \mathfrak{S}$ and, as a result, Theorem 3.13, as well, is proved. □

3.6 Comments

Section 3.1 The general theory of the weak convergence for sums of infinitesimal independent random variables is developed in the monograph by Gnedenko and Kolmogorov [126]. If one drops the infinitesimal negligibility condition, then another theory is established in the monograph by Zolotarev [439].

Section 3.2 A systematic exposition of infinitely divisible distributions appeared for the first time in the monographs by Lévy [283] and Khintchine [201].

Theorem 3.1, one of the basic results of the theory of the weak convergence of distribution functions of sums of independent random variables, is proved by Khintchine [197]. Theorems 3.2 and 3.3, which coincide with Theorem 3.1 when $d = 1$, are proved in [97].

Section 3.3 Representation (3.8) is obtained by Khintchine [199] who simplified a result due to Lévy [282] (thus (3.8) is often called the *Lévy–Khintchine representation*, see [126]). Representation (3.8), as shown by Kolmogorov [256], simplifies essentially if the corresponding distribution function possesses the second moment. A particular case of representation (3.8) was obtained earlier by Finetti [108].

Theorems 3.5 and 3.6 for $d = 1$ are due to Gnedenko (see [126] and the references therein to the original Gnedenko papers). Theorems 3.5 and 3.6 for $d > 1$ are mentioned in [97].

Section 3.4 The problem of finding the limit distributions of normalized and centered cumulative sums of independent random variables is posed in 1936 by A. Khintchine. A solution to this problem was discovered by Lévy in [283]. Theorems 3.7 and 3.8 are proved by Lévy [283]; these results can also be found, for example, in [126] (Theorem 1 in §29 and Theorem 1 in §30). Theorem 3.9 is proved by Kubik in [265]. Theorem 3.15 is obtained by Khintchine in [200] (also see Theorem 13, Chap. 1, in [336]). The inclusion $\mathfrak{L}_1 \subseteq \mathfrak{L}_{d;\max} \subseteq \mathfrak{L}_{d;\min}$ was observed in [97].

Section 3.5 The basic results concerning the limit distributions of sums of independent identically distributed random variables are due to Lévy [281] and Khintchine [201]. Theorem 3.14 is proved independently by Döblin [75] and Gnedenko [125] (also see [126], §37, Proposition 1). Theorem 3.16 is obtained by Gnedenko [125] (also see [126], §37, Proposition 2).

Theorems 3.12 and 3.13 are proved in [97]. Section 3.5 above provides a corrected version of this result.

The rate of convergence in limit theorems for multi-indexed sums is studied by Gut [137].

The limit stability of the maximum of multi-indexed sums of independent random variables is considered by Li and Tomkins [293]. Katz and Stratton [193] obtained an interesting limit theorem for the convergence in probability of the maximum of multi-indexed sums. Chernoff and Teicher [52] studied the limit distributions for $\max_i \min_j X_{ij}$ and $\max_j \min_i X_{ij}$.

Chapter 4
The Law of Large Numbers for Multi-Indexed Sums of Independent Random Variables

The *Bernoulli law of large numbers* is the first limit theorem of probability theory.

Theorem 4.1 (J. Bernoulli) *Let S_n be the total number of successes in n independent and homogeneous Bernoulli trials. Then*

$$\operatorname{Law}\left(\frac{S_n}{n}\right) \xrightarrow{\mathscr{D}} \mathbb{I}_{(p,\infty)}(x), \quad n \to \infty,$$

where p denotes the probability of success in a single trial.

Theorem 4.1 is often stated in its original form in terms of convergence in probability, namely

$$\frac{S_n}{n} \xrightarrow{\mathsf{P}} p \qquad \text{or} \qquad \mathsf{P}\text{-}\lim \frac{S_n}{n} = p.$$

The efforts of many authors lead to various generalizations of Bernoulli's theorem. A natural generalization is the *Khintchine theorem* where the symbol S_n stands for a cumulative sum of independent identically distributed random variables.

Theorem 4.2 (A.Ya. Khintchine) *Let $\{X_n, n \geq 1\}$ be independent identically distributed random variables, $S_1 = X_1$, $S_n = X_1 + \cdots + X_n$, $n > 1$. If $\mathsf{E}[X_1] = a$, then*

$$\mathsf{P}\text{-}\lim \frac{S_n}{n} = a.$$

The most general form of the law of large numbers deals with sums of independent random variables in the scheme of series:

$$\begin{array}{l} \xi_{11}, \ldots, \xi_{1k_1} \\ \xi_{21}, \xi_{22}, \ldots, \xi_{2k_2} \\ \ldots\ldots\ldots\ldots\ldots\ldots \\ \xi_{n1}, \xi_{n2}, \xi_{n3}, \ldots, \xi_{nk_n} \\ \ldots\ldots\ldots\ldots\ldots\ldots\ldots\ldots \end{array} \tag{4.1}$$

O. Klesov, *Limit Theorems for Multi-Indexed Sums of Random Variables*,
Probability Theory and Stochastic Modelling 71, DOI 10.1007/978-3-662-44388-0_4

Theorem 4.3 (B.V. Gnedenko) *Let the random variables $\{\xi_{nk},\ k \le k_n\}$ be jointly independent for each $n \ge 1$. Assume that*

$$\lim_{n\to\infty} \max_{k\le k_n} \mathsf{P}(|\xi_{nk}| \ge \varepsilon) = 0 \quad \textit{for all}\ \varepsilon > 0.$$

Put $\zeta_n = \sum_{k=1}^{k_n} \xi_{nk}$. In order that a sequence of non-random constants $\{c_n, n \ge 1\}$ such that

$$\mathsf{P}\text{-}\lim(\zeta_n - c_n) = 0$$

exists it is necessary and sufficient that

$$\lim_{n\to\infty} \sum_{k=1}^{k_n} \mathsf{P}(|\xi_{nk}| \ge \varepsilon) = 0 \quad \textit{for all}\ \varepsilon > 0,$$

$$\lim_{n\to\infty} \sum_{k=1}^{k_n} \mathsf{E}\left[\xi_{nk}^c\right] = 0 \quad \textit{for some}\ c > 0,$$

$$\lim_{n\to\infty} \sum_{k=1}^{k_n} \mathsf{var}\left[\xi_{nk}^c\right] = 0 \quad \textit{for some}\ c > 0.$$

The statement above is still true if the expression "for some $c > 0$" is changed to "for all $c > 0$".

One of special cases of Theorem 4.3 which, with $\nu = 1$, implies Khintchine's theorem corresponds to the Marcinkiewicz–Zygmund normalization $b_n = n^\nu$.

Theorem 4.4 (Marcinkiewicz–Zygmund normalization) *Let $\{X_k, k \ge 1\}$ be a sequence of independent identically distributed random variables with the distribution function F. Let $\nu > \frac{1}{2}$. Then the law of large numbers*

$$\mathsf{P}\text{-}\lim_{k\to\infty} \frac{S_k}{k^\nu} = 0 \tag{4.2}$$

is equivalent to the following set of two conditions

$$k\mathsf{P}(|X| \ge k^\nu) \underset{k\to\infty}{\longrightarrow} 0 \tag{4.3}$$

and

$$k^{1-\nu} \int\limits_{|x|<k^\nu} x\, dF(x) \underset{k\to\infty}{\longrightarrow} 0. \tag{4.4}$$

In addition, if $\nu \le \frac{1}{2}$ and F is a nondegenerate distribution function, then the law of large numbers (4.2) does not hold.

In this chapter, we study the weak convergence to a degenerate limit of distribution functions of sums of random variables corresponding to a (d, r)-series. Using the results obtained in Chap. 3, we find the criteria for the law of large numbers for both lim(min)- and lim(max)-convergence. Having obtained results for the scheme of series, we turn to the special case of "cumulative" sums. Conditions for the law of large numbers simplify if the terms of "cumulative" sums have an identical distribution.

The forms of the conditions in Theorems 4.9 (for lim(max)-convergence) and in 4.10 (for lim(min)-convergence) are similar and look like the classical conditions in the Gnedenko Theorem 4.3. Hence it may seem that the conditions for the law of large numbers coincide for both convergences. In fact, the conditions are essentially different and do not coincide even for identically distributed terms and for the simplest normalization (the Marcinkiewicz–Zygmund normalization). Moreover, there are several cases where the law of large numbers holds for lim(min)-convergence but does not hold for lim(max)-convergence. Below is an appropriate example demonstrating that these modes of convergence are not equivalent.

Example 4.1 Let $d = 2$. Let $\{X(m, n)\}$ be independent identically distributed random variables and $S(m, n) = \sum_{i=1}^{m} \sum_{j=1}^{n} X(i, j)$. Consider the law of large numbers for the normalizing field $b(m, n) = m\sqrt{n}$. The relation

$$\mathsf{P}\text{-}\lim(\max) \frac{S(m, n)}{b(m, n)} = 0$$

cannot hold for any nondegenerate distribution of random variables $X(i, j)$. Otherwise,

$$\mathsf{P}\text{-}\lim(\max) \frac{S(1, n)}{b(1, n)} = 0 \qquad \Longleftrightarrow \qquad \mathsf{P}\text{-}\lim_{n\to\infty} \frac{S_n}{\sqrt{n}} = 0$$

and this contradicts the central limit theorem. On the other hand, the law of large numbers with the same normalization $b(m, n)$ holds under certain assumptions for lim(min)-convergence (see case (iii) in Theorem 4.12).

4.1 The Law of Large Numbers in the Scheme of Series

Let $\{X(\mathbf{n}, \mathbf{k}); \mathbf{k} \le \mathbf{k}(\mathbf{n})\}$, $\mathbf{n} \in \mathbf{N}^d$, be a (d, r)-sequence of series of random variables. Recall that the random variables $X(\mathbf{n}, \mathbf{k})$, $\mathbf{k} \le \mathbf{k}(\mathbf{n})$, are jointly independent for all fixed $\mathbf{n} \in \mathbf{N}^d$ (see Definition 3.2). Throughout the following we assume the infinitesimal negligibility condition (3.3).

Definition 4.1 We say that a sequence of (d, r)-series satisfies the *law of large numbers in the scheme of series for* lim(min) *-convergence* if the random variables

$S(\mathbf{n}) - c(\mathbf{n})$, $\mathbf{n} \in \mathbf{N}^d$, converge in probability to zero in the sense of lim(min)-convergence for some non-random field $\{c(\mathbf{n}), \mathbf{n} \in \mathbf{N}^d\}$, that is, if

$$\lim(\min)\mathsf{P}(|S(\mathbf{n}) - c(\mathbf{n})| \geq \varepsilon) = 0 \quad \text{for all}\, \varepsilon > 0. \tag{4.5}$$

Definition 4.2 We say that a sequence of (d, r)-series satisfies the *law of large numbers in the scheme of series for* lim(max) *-convergence* if the random variables $S(\mathbf{n}) - c(\mathbf{n})$, $\mathbf{n} \in \mathbf{N}^d$, converge in probability to zero in the sense of lim(max)-convergence for some non-random field $\{c(\mathbf{n}), \mathbf{n} \in \mathbf{N}^d\}$, that is if

$$\lim(\max)\mathsf{P}(|S(\mathbf{n}) - c(\mathbf{n})| \geq \varepsilon) = 0 \quad \text{for all}\, \varepsilon > 0. \tag{4.6}$$

Note that relations (4.5) and (4.6) are equivalent to the weak convergence of distribution functions of the random variables

$$\sum_{\mathbf{k} \leq \mathbf{n}} X(\mathbf{n}, \mathbf{k}) - c(\mathbf{n})$$

to the degenerate law

$$I_0(x) = \mathbb{I}_{(0,\infty)}(x) = \begin{cases} 0, & \text{if}\, x \leq 0, \\ 1, & \text{if}\, x > 0, \end{cases}$$

for some non-random field $\{c(\mathbf{n}), \mathbf{n} \in \mathbf{N}^d\}$.

The conditions for the weak convergence of distribution functions of sums constructed from a scheme of series to some infinitely divisible distribution function are studied in Chap. 3. Since the law $I_0(x)$ is infinitely divisible, the conditions for the law of large numbers can easily be deduced from the results of Chap. 3.

Remark 4.1 As has been remarked in Chap. 3, the case of (d, r)-fields of series does not differ from the case of $(d, 1)$-series in the study of weak convergence of sums, since the distribution of a sum $S(\mathbf{n})$ does not depend on an order which is used to sum the random variables $X(\mathbf{n}, \mathbf{k})$. Writing the random variables one by one according to a linear order in every series, we obtain a series for $r = 1$ that depends on the multi-index $\mathbf{n}$. This explains why we assume that $r = 1$ and consider $(d, 1)$-series throughout the following.

Summarizing what is said above, we consider finite collections of random variables $\{X_j(\mathbf{n}), j \leq j(\mathbf{n})\}$ for all $\mathbf{n} \in \mathbf{N}^d$, where $j(\mathbf{n})$ is a positive integer for each $\mathbf{n} \in \mathbf{N}^d$. For the sake of brevity, we refer to such a collection as a d-series. According to this language, the classical theory deals with the case of 1-series. Put

$$S(\mathbf{n}) = \sum_{j=1}^{j(\mathbf{n})} X_j(\mathbf{n}).$$

For every field of random variables $\{\xi(\mathbf{n}), \mathbf{n} \in \mathbf{N}^d\}$, we write

$$\mathsf{P}\text{-}\lim(\max)\xi(\mathbf{n}) = 0 \quad \text{or} \quad \mathsf{P}\text{-}\lim(\min)\xi(\mathbf{n}) = 0$$

if

$$\lim(\max)\mathsf{P}(|\xi(\mathbf{n})| \geq \varepsilon) = 0 \quad \text{or} \quad \lim(\min)\mathsf{P}(|\xi(\mathbf{n})| \geq \varepsilon) = 0)$$

for all $\varepsilon > 0$.

In this section, we study the conditions for the law of large numbers for both modes of convergence, namely for lim(max) convergence

$$\mathsf{P}\text{-}\lim(\max)\,[S(\mathbf{n}) - c(\mathbf{n})] = 0 \quad \text{for some field}\,\{c(\mathbf{n})\} \tag{4.7}$$

and for lim(min) convergence

$$\mathsf{P}\text{-}\lim(\min)\,[S(\mathbf{n}) - c(\mathbf{n})] = 0 \quad \text{for some field}\,\{c(\mathbf{n})\}. \tag{4.8}$$

We stress once more that the basic assumption in the classical theory, that is in the case of $d = 1$, is the infinitesimal negligibility condition for the random variables $X_j(\mathbf{n})$:

$$\lim(\max) \max_{j \leq j(\mathbf{n})} \mathsf{P}(|X_j(\mathbf{n})| \geq \varepsilon) = 0 \tag{4.9}$$

for lim(max)-convergence or

$$\lim(\min) \max_{j \leq j(\mathbf{n})} \mathsf{P}(|X_j(\mathbf{n})| \geq \varepsilon) = 0 \tag{4.10}$$

for lim(min)-convergence. In the case of $d > 1$, the infinitesimal negligibility condition also plays a crucial role (see Definitions 3.3 and 3.4).

The following criteria for the law of large numbers use the truncated versions $X_j^c(\mathbf{n})$ of the random variables $X_j(\mathbf{n})$ defined by $X_j^c(\mathbf{n}) = X_j(\mathbf{n})\,\mathbb{I}_{\{|X_j(\mathbf{n})|<c\}}$.

Theorem 4.5 *The set of conditions* (4.7), (4.9) *is equivalent to the following set of three conditions*:

$$\lim(\max) \sum_{j=1}^{j(\boldsymbol{n})} \mathsf{P}(|X_j(\boldsymbol{n})| \geq \varepsilon) = 0 \quad \textit{for all } \varepsilon > 0, \tag{4.11}$$

$$\lim(\max) \sum_{j=1}^{j(\boldsymbol{n})} \mathsf{E}\left[X_j^c(\boldsymbol{n})\right] = 0 \quad \textit{for some } c > 0, \tag{4.12}$$

$$\lim(\max) \sum_{j=1}^{j(\boldsymbol{n})} \mathsf{var}\left[X_j^c(\boldsymbol{n})\right] = 0 \quad \textit{for some } c > 0. \tag{4.13}$$

The above result is still true if the expression "for some $c > 0$" is changed to "for all $c > 0$".

Theorem 4.6 *The set of conditions* (4.8), (4.10) *is equivalent to the following set of three conditions*:

$$\lim(\min) \sum_{j=1}^{j(\boldsymbol{n})} \mathsf{P}(|X_j(\boldsymbol{n})| \geq \varepsilon) = 0 \quad \textit{for all } \varepsilon > 0, \tag{4.14}$$

$$\lim(\min) \sum_{j=1}^{j(\boldsymbol{n})} \mathsf{E}\left[X_j^c(\boldsymbol{n})\right] = 0 \quad \textit{for some } c > 0, \tag{4.15}$$

$$\lim(\min) \sum_{j=1}^{j(\boldsymbol{n})} \mathsf{var}\left[X_j^c(\boldsymbol{n})\right] = 0 \quad \textit{for some } c > 0. \tag{4.16}$$

The above result is still true if the expression "for some $c > 0$" is changed to "for all $c > 0$".

Remark 4.2 One can use the criteria for weak convergence obtained in Theorems 3.5 and 3.6 to prove Theorems 4.5 and 4.6 (this approach is usually used in the case $d = 1$). A simpler proof, given below, uses the Cauchy convergence criterion for the lim(max)- and lim(min)-convergences of non-random fields (Proposition A.3). The main idea of this approach is to pass from d-series to 1-series and then to apply the classical criteria for the law of large numbers.

Proof of Theorem 4.5 First we prove that conditions (4.11)–(4.13) imply conditions (4.7) and (4.9). Consider an increasing sequence of multi-indices $\{\mathbf{n}_k, k \geq 1\}$, $\mathbf{n}_k = (n_{1k}, \ldots, n_{dk})$, such that $\max\{n_{1k}, \ldots, n_{dk}\} \to \infty$. Recall that the monotonicity means that $\mathbf{n}_k \leq \mathbf{n}_{k+1}$, $k \geq 1$.

With $l_k = j(\mathbf{n}_k)$, consider random variables $\xi_{kj} = X_j(\mathbf{n}_k)$ for $1 \leq j \leq l_k$ and their sums $\zeta_k = \sum_{j=1}^{l_k} \xi_{kj}$. It is clear that the sequence of random variables $\{\xi_{kj}, 1 \leq j \leq l_k\}$ constitutes a 1-series. The Cauchy convergence criterion (Proposition A.3) and conditions (4.11)–(4.13) imply that

$$\lim_{k\to\infty} \sum_{j \leq j(\mathbf{n}_k)} \mathsf{P}(|X_j(\mathbf{n}_k)| \geq \varepsilon) = \lim_{k\to\infty} \sum_{j=1}^{l_k} \mathsf{P}(|\xi_{kj}| \geq \varepsilon) = 0 \tag{4.17}$$

for all $\varepsilon > 0$ and

$$\lim_{k\to\infty} \sum_{j \leq j(\mathbf{n}_k)} \mathsf{E}\left[X_j^c(\mathbf{n}_k)\right] = \lim_{k\to\infty} \sum_{j=1}^{l_k} \mathsf{E}\left[\xi_{kj}^c\right] = 0, \tag{4.18}$$

$$\lim_{k\to\infty} \sum_{j\le j(\mathbf{n}_k)} \mathsf{var}\left[X_j^c(\mathbf{n}_k)\right] = \lim_{k\to\infty} \sum_{j=1}^{l_k} \mathsf{var}\left[\xi_{kj}^c\right] = 0 \tag{4.19}$$

for some $c > 0$.

Therefore conditions (4.17)–(4.19) hold for the 1-series $\{\xi_{kj}, j \le l_k\}$ and these conditions are equivalent to the infinitesimal negligibility condition

$$\lim_{k\to\infty} \max_{j\le j_k} \mathsf{P}(|\xi_{kj}| \ge \varepsilon) = 0 \quad \text{for all } \varepsilon > 0 \tag{4.20}$$

and the law of large numbers

$$\zeta_k \xrightarrow{\mathsf{P}} 0, \quad k \to \infty \tag{4.21}$$

(see Theorem 4.3).

Conditions (4.20)–(4.21) can be rewritten in terms of the random variables $X_j(\mathbf{n})$ and their sums $S(\mathbf{n})$, namely

$$\lim_{k\to\infty} \max_{j\le j(\mathbf{n}_k)} \mathsf{P}(|X_j(\mathbf{n}_k)| \ge \varepsilon) = 0 \quad \text{for all } \varepsilon > 0 \tag{4.22}$$

and

$$S(\mathbf{n}_k) \xrightarrow{\mathsf{P}} 0, \quad k \to \infty. \tag{4.23}$$

Since the sequence of multi-indices $\{\mathbf{n}_k, k \ge 1\}$ is arbitrary, the Cauchy convergence criterion (Proposition A.3) implies that conditions (4.7) and (4.9) hold.

Now we prove the inverse statement, namely that conditions (4.7) and (4.9) imply conditions (4.11)–(4.13).

Assume that conditions (4.7) and (4.9) hold and consider an arbitrary increasing sequence of multi-indices $\{\mathbf{n}_k, k \ge 1\}$ such that $\max\{n_{1k}, \dots, n_{dk}\} \to \infty$, where $\mathbf{n}_k = (n_{1k}, \dots, n_{dk})$. The Cauchy convergence criterion (Proposition A.3) means that conditions (4.22) and (4.23) follow from conditions (4.7) and (4.9).

As above, we put $l_k = j(\mathbf{n}_k)$, $\xi_{kj} = X_j(\mathbf{n}_k)$ for $1 \le j \le l_k$, and $\zeta_k = \sum_{j=1}^{l_k} \xi_{kj}$. The sequence of random variables $\{\xi_{kj}, 1 \le j \le l_k\}$ constitutes a 1-series and conditions (4.22)–(4.23) are equivalent to conditions (4.20)–(4.21). We again use the classical law of large numbers (Theorem 4.3) to show that conditions (4.17)–(4.19) hold. By the Cauchy convergence criterion (Proposition A.3) we prove that conditions (4.11)–(4.13) hold in view of the arbitrariness of the sequence $\{\mathbf{n}_k\}$. The proof of Theorem 4.5 is complete. □

Proof of Theorem 4.6 The proof follows the lines of that of Theorem 4.5 almost literally. The only difference is the property of an arbitrary sequence of multi-indices $\{\mathbf{n}_k\}$, namely we have to assume that $\min(n_{1k}, \dots, n_{dk}) \to \infty$ instead of $\max(n_{1k}, \dots, n_{dk}) \to \infty$. □

4.2 The Law of Large Numbers for Cumulative Sums

We consider a particular, but still important, case of the scheme of series. Let $\{X(\mathbf{n}), \mathbf{n} \in \mathbf{N}^d\}$ be independent random variables and let $\{b(\mathbf{n}), \mathbf{n} \in \mathbf{N}^d\}$ be non-random positive numbers. Assume that a field $\{b(\mathbf{n}), \mathbf{n} \in \mathbf{N}^d\}$ is increasing (see Definition A.10):

$$b(\mathbf{m}) \le b(\mathbf{n}), \quad \mathbf{m} \le \mathbf{n}.$$

Put $S(\mathbf{n}) = \sum_{\mathbf{k} \le \mathbf{n}} X(\mathbf{k})$. We study the conditions for the law of large numbers for both lim(max)- and lim(min)-convergences, namely for

$$\mathsf{P}\text{-}\lim(\max) \frac{S(\mathbf{n})}{b(\mathbf{n})} = 0 \tag{4.24}$$

as well as for

$$\mathsf{P}\text{-}\lim(\min) \frac{S(\mathbf{n})}{b(\mathbf{n})} = 0. \tag{4.25}$$

It is clear that the random variables $X(\mathbf{n}, \mathbf{k}) = X(\mathbf{k})/b(\mathbf{n})$, $\mathbf{k} \le \mathbf{n}$, constitute a (d, d)-series, that is, we are indeed dealing with a particular case of the scheme of series.

The infinitesimal negligibility condition (4.9) and (4.10) for such a scheme of series is reduced to the following conditions

$$\lim(\max) \max_{\mathbf{k} \le \mathbf{n}} \mathsf{P}\big(|X(\mathbf{k})| \ge \varepsilon b(\mathbf{n})\big) = 0 \tag{4.26}$$

in the case of lim(max) convergence and

$$\lim(\min) \max_{\mathbf{k} \le \mathbf{n}} \mathsf{P}\big(|X(\mathbf{k})| \ge \varepsilon b(\mathbf{n})\big) = 0 \tag{4.27}$$

in the case of lim(min) convergence, respectively.

Remark 4.3 Conditions (4.26) and (4.27) coincide if $d = 1$ and each of them follows from the law of large numbers (4.24) (or, from (4.25)). The infinitesimal negligibility condition (4.26) also follows from the law of large numbers (4.24) if $d > 1$. Although this is not the case for lim(min)-convergence. Below is an appropriate example demonstrating, in particular, that (4.25) does not imply (4.24). It also shows that the infinitesimal negligibility condition (4.27) is not necessary for the law of large numbers (4.25) for lim(min)-convergence if $d > 1$.

Example 4.2 Let $d = 2$ and let

$$X(m,n) = \begin{cases} +n^2, & \text{for } m = 1, \\ -n^2, & \text{for } m = 2, \\ 0, & \text{for } m > 2. \end{cases}$$

The field $\{X(m,n),\ (m,n) \in \mathbf{N}^2\}$ just defined is non-random and $S(m,n) = 0$ for $m \geq 2$. This means that condition (4.25) holds for all normalizations $b(m,n)$, and in particular for $b(m,n) = mn$. In contrast, condition (4.24) does not hold for these random variables and normalizing field.

On the other hand, we show that condition (4.27) does not hold for the $\{X(m,n)\}$ and $\{b(m,n)\}$ introduced above. Indeed,

$$\max_{\substack{k\leq m \\ l \leq n}} \mathsf{P}(|X(k,l| \geq \varepsilon mn) = \max_{l\leq n} \mathsf{P}(|X(1,l)| \geq \varepsilon n) = \mathsf{P}(|X(1,n)| \geq \varepsilon n) = 1$$

for all fixed $\varepsilon > 0$, all $m \geq 1$, and for $n > \varepsilon$.

Theorem 4.7 *The law of large numbers* (4.24) *is equivalent to the following set of three conditions*:

$$\lim(\max) \sum_{\boldsymbol{k}\leq \boldsymbol{n}} \mathsf{P}\big(|X(\boldsymbol{k})| \geq b(\boldsymbol{n})\big) = 0, \tag{4.28}$$

$$\lim(\max) \frac{1}{b(\boldsymbol{n})} \sum_{\boldsymbol{k}\leq \boldsymbol{n}} \mathsf{E}\left[X^{b(\boldsymbol{n})}(\boldsymbol{k})\right] = 0, \tag{4.29}$$

$$\lim(\max) \frac{1}{b^2(\boldsymbol{n})} \sum_{\boldsymbol{k}\leq \boldsymbol{n}} \mathsf{var}\left[X^{b(\boldsymbol{n})}(\boldsymbol{k})\right] = 0. \tag{4.30}$$

Here $X^{b(\boldsymbol{n})}(\boldsymbol{k})$ is the truncation of the random variable $X(\boldsymbol{k})$ at the level $b(\boldsymbol{n})$.

Theorem 4.8 *The law of large numbers* (4.25) *together with the infinitesimal negligibility condition* (4.27) *is equivalent to the following set of three conditions*:

$$\lim(\min) \sum_{\boldsymbol{k}\leq \boldsymbol{n}} \mathsf{P}\big(|X(\boldsymbol{k})| \geq \varepsilon b(\boldsymbol{n})\big) = 0 \quad \textit{for all } \varepsilon > 0, \tag{4.31}$$

$$\lim(\min) \frac{1}{b(\boldsymbol{n})} \sum_{\boldsymbol{k}\leq \boldsymbol{n}} \mathsf{E}\left[X^{b(\boldsymbol{n})}(\boldsymbol{k})\right] = 0, \tag{4.32}$$

$$\lim(\min) \frac{1}{b^2(\boldsymbol{n})} \sum_{\boldsymbol{k}\leq \boldsymbol{n}} \mathsf{var}\left[X^{b(\boldsymbol{n})}(\boldsymbol{k})\right] = 0. \tag{4.33}$$

Here $X^{b(\boldsymbol{n})}(\boldsymbol{k})$ is the truncation of the random variable $X(\boldsymbol{k})$ at the level $b(\boldsymbol{n})$.

Theorem 4.8 is a straightforward consequence of Theorem 4.6. On the other hand, the proof of Theorem 4.7 needs some extra reasoning, since the infinitesimal negligibility condition (4.26) is not present in the statement of the theorem and, moreover, condition (4.28) (an analog of condition (4.11) used in Theorem 4.5) is assumed only for $\varepsilon = 1$ in Theorem 4.7.

Proof of Theorem 4.7 First we prove the implication (4.24) $\Longrightarrow$ (4.28)–(4.30). Consider the (d, d)-field of series $X(\mathbf{n}, \mathbf{k}) = X(\mathbf{k})/b(\mathbf{n})$, $\mathbf{k} \le \mathbf{n}$. It is clear that conditions (4.29)–(4.30) coincide with conditions (4.12)–(4.13) for $c = 1$, while condition (4.28) coincides with (4.11) for $\varepsilon = 1$. Thus, by Theorem 4.5, conditions (4.28)–(4.30) follow from (4.24) and (4.26).

To complete the proof of the implication mentioned above it is necessary to show that the infinitesimal negligibility condition (4.26) follows from the law of large numbers (4.24).

By inequality (A.7), we have

$$|X(\mathbf{k})| \le 2^d \max_{\boldsymbol{\delta} \in \mathscr{E}_d} |S(\mathbf{k} - \boldsymbol{\delta})|,$$

where the maximum is evaluated with respect to all possible vectors $\boldsymbol{\delta} = (\delta_1, \dots, \delta_d)$ each of whose coordinates is either 0 or 1. Recall that we denote the set of all such vectors $\boldsymbol{\delta}$ by $\mathscr{E}_d$. Fix $\varepsilon > 0$ and choose $\gamma = \varepsilon/2^d$. Since the field $\{b(\mathbf{n})\}$ is increasing, we conclude that

$$\left\{\omega : \left|\frac{X(\mathbf{k})}{b(\mathbf{k})}\right| \le \varepsilon\right\} \supseteq \left\{\omega : \left|\frac{S(\mathbf{k} - \boldsymbol{\delta})}{b(\mathbf{k} - \boldsymbol{\delta})}\right| \le \gamma \text{ for all } \boldsymbol{\delta} \in \mathscr{E}_d\right\}.$$

Fix $0 < p < 1$. Relation (4.24) implies that there exists an N such that

$$\mathsf{P}\left(\left|\frac{S(\mathbf{m})}{b(\mathbf{m})}\right| \le \gamma\right) \ge p \quad \text{for all } \mathbf{m} = (m_1, \dots, m_d) : \max(m_1, \dots, m_d) \ge N.$$

Now if $\max(k_1, \dots, k_d) > N$, then $\max(k_1 - \delta_1, \dots, k_d - \delta_d) \ge N$, as well, whence

$$\mathsf{P}\left(\left|\frac{X(\mathbf{k})}{b(\mathbf{k})}\right| \le \varepsilon\right) \ge 2^d(p-1) + 1.$$

Here we used the following elementary result.

Lemma 4.1 *Let $m \ge 1, p > 0$, and random events $B_1, \dots, B_m$ be such that $\mathsf{P}(B_k) \ge p$, $1 \le k \le m$. Then*

$$\mathsf{P}\left(\bigcap_{k=1}^m B_k\right) \ge m(p-1) + 1.$$

Therefore

$$\max_{\mathbf{k}\le\mathbf{n}} \mathsf{P}\left(\left|\frac{X(\mathbf{k})}{b(\mathbf{k})}\right| \le \varepsilon\right) \ge \max_{\substack{\mathbf{k}\le\mathbf{n}\\ \max(k_1,\dots,k_d)>N}} \mathsf{P}\left(\left|\frac{X(\mathbf{k})}{b(\mathbf{k})}\right| \le \varepsilon\right) \ge 2^d(p-1)+1.$$

This means that

$$\liminf(\max)\left[\max_{\mathbf{k}\le\mathbf{n}} \mathsf{P}\left(\left|\frac{X(\mathbf{k})}{b(\mathbf{k})}\right| \le \varepsilon\right)\right] \ge 2^d(p-1)+1.$$

Since p is arbitrary in the latter inequality, the proof of the infinitesimal negligibility (4.26) is complete by passing to the limit as $p \to 1$.

Now we prove the inverse implication (4.24) $\Longleftarrow$ (4.28)–(4.30). Put

$$Y(\mathbf{n},\mathbf{k}) = X^{b(\mathbf{n})}(\mathbf{k}), \quad \mathbf{k} \le \mathbf{n}.$$

By Chebyshev's inequality,

$$\mathsf{P}\left(\left|\frac{1}{b(\mathbf{n})}\sum_{\mathbf{k}\le\mathbf{n}}(Y(\mathbf{n},\mathbf{k}) - \mathsf{E}\,[Y(\mathbf{n},\mathbf{k})])\right| \ge \varepsilon\right) \le \frac{1}{b^2(\mathbf{n})\varepsilon^2}\sum_{\mathbf{k}\le\mathbf{n}} \mathsf{var}\left[X^{b(\mathbf{n})}(\mathbf{k})\right]$$

for all $\varepsilon > 0$. By condition (4.30), the lim(max)-limit of the right-hand side equals 0. Thus condition (4.29) implies that

$$\mathsf{P}\text{-}\lim(\max)\frac{1}{b(\mathbf{n})}\sum_{\mathbf{k}\le\mathbf{n}} X^{b(\mathbf{n})}(\mathbf{k}) = 0. \tag{4.34}$$

Further,

$$\mathsf{P}\left(\frac{1}{b(\mathbf{n})}\sum_{\mathbf{k}\le\mathbf{n}} X(\mathbf{k}) \ne \sum_{\mathbf{k}\le\mathbf{n}} X^{b(\mathbf{n})}(\mathbf{k})\right) \le \sum_{\mathbf{k}\le\mathbf{n}} \mathsf{P}(|X(\mathbf{k})| \ge b(\mathbf{n})).$$

By condition (4.28), the lim(max)-limit of the right-hand side equals 0, that is

$$\mathsf{P}\text{-}\lim(\max)\left[\frac{1}{b(\mathbf{n})}\sum_{\mathbf{k}\le\mathbf{n}} X(\mathbf{k}) - \sum_{\mathbf{k}\le\mathbf{n}} X^{b(\mathbf{n})}(\mathbf{k})\right] = 0,$$

which together with (4.34) proves (4.24).

The proof of Theorem 4.7 is complete. □

4.3 The Law of Large Numbers for Multi-Indexed Sums of Independent Identically Distributed Random Variables

The most interesting case of Theorems 4.7 and 4.8 concerns cumulative sums of independent identically distributed random variables. The infinitesimal negligibility conditions (4.26) and (4.27) hold automatically in this case if the normalizing constants $b(\mathbf{n})$ tend to infinity for the corresponding mode of convergence. Thus the statements of Theorems 4.7 and 4.8 simplify for the case under consideration.

Theorem 4.9 *Let $\{X(\boldsymbol{n}), \boldsymbol{n} \in \mathbf{N}^d\}$ be a field of independent identically distributed random variables with the distribution function F and let $\{b(\boldsymbol{n}), \boldsymbol{n} \in \mathbf{N}^d\}$ be increasing,*

$$b(\boldsymbol{m}) \le b(\boldsymbol{n}), \quad \boldsymbol{m} \le \boldsymbol{n},$$

and unbounded,

$$\lim(\max) b(\boldsymbol{n}) = \infty.$$

Then the law of large numbers (4.24) is equivalent to the following set of three conditions:

$$\lim(\max) |\boldsymbol{n}| \int\limits_{|x| \ge b(\boldsymbol{n})} dF(x) = 0, \tag{4.35}$$

$$\lim(\max) \frac{|\boldsymbol{n}|}{b(\boldsymbol{n})} \int\limits_{|x| < b(\boldsymbol{n})} x \, dF(x) = 0, \tag{4.36}$$

$$\lim(\max) \frac{|\boldsymbol{n}|}{b^2(\boldsymbol{n})} \left[\int\limits_{|x| < b(\boldsymbol{n})} x^2 \, dF(x) - \left(\int\limits_{|x| < b(\boldsymbol{n})} x \, dF(x) \right)^2 \right] = 0. \tag{4.37}$$

Theorem 4.10 *Let $\{X(\boldsymbol{n}), \boldsymbol{n} \in \mathbf{N}^d\}$ be a field of independent identically distributed random variables with the distribution function F and let $\{b(\boldsymbol{n}), \boldsymbol{n} \in \mathbf{N}^d\}$ be increasing,*

$$b(\boldsymbol{m}) \le b(\boldsymbol{n}), \quad \boldsymbol{m} \le \boldsymbol{n},$$

and unbounded,

$$\lim(\min) b(\boldsymbol{n}) = \infty.$$

Then the law of large numbers (4.25) *is equivalent to the following set of three conditions*:

$$\lim(\min)|\boldsymbol{n}| \int\limits_{|x|\geq b(\boldsymbol{n})} dF(x) = 0, \tag{4.38}$$

$$\lim(\min)\frac{|\boldsymbol{n}|}{b(\boldsymbol{n})} \int\limits_{|x|<b(\boldsymbol{n})} x\,dF(x) = 0, \tag{4.39}$$

$$\lim(\min)\frac{|\boldsymbol{n}|}{b^2(\boldsymbol{n})}\left[\int\limits_{|x|<b(\boldsymbol{n})} x^2\,dF(x) - \left(\int\limits_{|x|<b(\boldsymbol{n})} x\,dF(x)\right)^2\right] = 0. \tag{4.40}$$

Proof of Theorems 4.9 and 4.10 Theorem 4.9 is a straightforward consequence of Theorem 4.7. To prove Theorem 4.10 we note that the infinitesimal negligibility condition (4.27) follows from the assumption that the random variables $X(\mathbf{n})$ are identically distributed, since the field $\{b(\mathbf{n})\}$ is unbounded. Thus Theorem 4.8 yields the implication (4.25) $\Longrightarrow$ (4.38)–(4.40). The proof of the converse implication follows the lines of that used for the proof of the implication (4.24) $\Longleftarrow$ (4.28)–(4.30) in Theorem 4.7. □

4.4 The Law of Large Numbers for Multi-Indexed Sums of Independent Identically Distributed Random Variables with the Marcinkiewicz–Zygmund Normalization

Put $b(\mathbf{n}) = n_1^{\nu_1} \dots n_d^{\nu_d}$ for $\nu_1 > 0, \dots, \nu_d > 0$ and

$$\nu = \min(\nu_1, \dots, \nu_d), \quad \nu' = \max(\nu_1, \dots, \nu_d).$$

In this section, we consider a field of independent identically distributed random variables $\{X(\mathbf{n}), \mathbf{n} \in \mathbf{N}^d\}$ with the common distribution function F. The law of large numbers for this case is written as follows

$$\mathsf{P}\text{-}\lim \frac{S(\mathbf{n})}{n_1^{\nu_1} \dots n_d^{\nu_d}} = 0 \tag{4.41}$$

(see relations (4.24) and (4.25)). We study the law of large numbers (4.41) for both modes of convergence, that is for lim(max)- and lim(min)-convergence.

Theorem 4.11 *Let* $\nu > \frac{1}{2}$. *The law of large numbers* (4.41) *for* lim(max)-*convergence is equivalent to conditions* (4.3)–(4.4).

If $v \le \frac{1}{2}$ and the distribution function F is nondegenerate, then the law of large numbers (4.41) *does not hold.*

Therefore lim(max)-convergence for $d > 1$ is completely compatible with the classical case of $d = 1$. This is not the case for lim(min)-convergence.

Theorem 4.12 *Let $v \ge \frac{1}{2}$. The law of large numbers* (4.41) *for* lim(min)*-convergence is equivalent to*

(i) *conditions* (4.3)–(4.4) *in the case $v' = v > \frac{1}{2}$;*
(ii) *conditions* (4.42)–(4.43) *in the case $v' > v > \frac{1}{2}$, where*

$$\limsup_{k\to\infty} k\mathsf{P}(|X| \ge k^{v}) < \infty, \tag{4.42}$$

$$\limsup_{k\to\infty} k^{1-v}\left|\int\limits_{|x|<k^{v}} x\,dF(x)\right| < \infty; \tag{4.43}$$

(iii) *conditions* (4.44)–(4.45) *in the case $v' > v = \frac{1}{2}$, where*

$$\int\limits_{-\infty}^{+\infty} x\,dF(x) = 0, \tag{4.44}$$

$$\int\limits_{-\infty}^{+\infty} x^2\,dF(x) < \infty. \tag{4.45}$$

If $v < \frac{1}{2}$ or $v = v' = \frac{1}{2}$, then the law of large numbers (4.41) *for* lim(min)*-convergence does not hold for any nondegenerate distribution function F.*

Remark 4.4 Note that the cases (ii) and (iii) of Theorem 4.12 are meaningful for $d > 1$. This is the case where one can see the difference between lim(max)- and lim(min)-convergence. For example, if the distribution function F has the density

$$f(x) = \begin{cases} 0, & |x| < 1, \\ c|x|^{-1-\frac{1}{v}}, & |x| \ge 1, \end{cases} \qquad c = \frac{1}{2v},$$

then

$$1 - F(x) \sim cx^{-1/v}.$$

Then conditions (4.42)–(4.43) hold, that is the law of large numbers (4.41) for lim(min)-convergence holds, as well. In contrast, conditions (4.3)–(4.4) do not hold and this means that the law of large numbers for lim(max)-convergence is not valid.

Case (iii) does not have a counterpart for $d = 1$ or for lim(max)-convergence. This is explained by the central limit theorem. Indeed, the limit distribution is Gaussian if the limit is considered along one direction, while, according to the law of large numbers (4.41), the limit distribution is degenerate along another direction.

Proof of Theorem 4.11 Without loss of generality assume that $\nu = \nu_1 \le \nu_2 \le \dots \le \nu_d = \nu'$. If the weak law of large numbers (4.41) holds for lim(max)-convergence, then

$$\mathsf{P}\text{-}\lim \frac{S(n_1, 1, \dots, 1)}{n_1^{\nu_1}} = 0$$

in view of Corollary A.1. Thus Theorem 4.4 implies that conditions (4.3)–(4.4) hold.

Conversely, assume that conditions (4.3)–(4.4) hold. Then the law of large numbers (4.2) also holds, where $S_k = X_1 + \dots + X_k$ and where $\{X_k\}$ are independent random variables with the same distribution function as in the case of $\{X(\mathbf{n})\}$. Since the distribution function of the random variable $S(\mathbf{n})/|\mathbf{n}|^\nu$ depends only on $|\mathbf{n}|$, we conclude that $\mathsf{P}(|S(\mathbf{n})| \ge \varepsilon |\mathbf{n}|^\nu) = \mathsf{P}(|S_k| \ge \varepsilon k^\nu)$ for $\varepsilon > 0$, where $k = |\mathbf{n}|$. Therefore

$$\mathsf{P}\text{-} \lim_{n_1 \to \infty} \frac{S(\mathbf{n})}{|\mathbf{n}|^\nu} = 0,$$

whence

$$\mathsf{P}\text{-}\lim(\max) \frac{S(\mathbf{n})}{n_1^{\nu_1} \dots n_d^{\nu_d}} = \mathsf{P}\text{-}\lim(\max) \frac{S(\mathbf{n})}{|\mathbf{n}|^\nu} \cdot n_2^{\nu_1 - \nu_2} \dots n_d^{\nu_1 - \nu_d} = 0,$$

since $\nu_1 - \nu_2 \le 0, \dots, \nu_1 - \nu_d \le 0$. □

Proof of Theorem 4.12 : case (i) To prove case (i) of the theorem note that $\nu_1 = \dots = \nu_d$ if $\nu' = \nu$. Then

$$\mathsf{P}\text{-}\lim(\max) \frac{S(\mathbf{n})}{b(\mathbf{n})} = 0 \iff \mathsf{P}\text{-}\lim(\min) \frac{S(\mathbf{n})}{b(\mathbf{n})} = 0.$$

The proof in this case is completed in view of Theorem 4.11.

Proof of Theorem 4.12: case (ii) Theorem 4.10 implies that the law of large numbers (4.41) for lim(min)-convergence is equivalent to the following conditions

$$\lim(\min) |\mathbf{n}| \, \mathsf{P}(|X| \ge n_1^{\nu_1} \dots n_d^{\nu_d}) = 0, \tag{4.46}$$

$$\lim(\min) n_1^{1-\nu_1} \dots n_d^{1-\nu_d} \int\limits_{|x| < n_1^{\nu_1} \dots n_d^{\nu_d}} x \, dF(x) = 0, \tag{4.47}$$

$$\lim(\min) n_1^{1-2\nu_1} \dots n_d^{1-2\nu_d} \Bigg[\int\limits_{|x|<n_1^{\nu_1}\dots n_d^{\nu_d}} x^2\, dF(x) - \Bigg(\int\limits_{|x|<n_1^{\nu_1}\dots n_d^{\nu_d}} x\, dF(x) \Bigg)^2 \Bigg] = 0. \tag{4.48}$$

Lemmas 4.2 and 4.3 stated below show that conditions (4.46)–(4.47) are equivalent to conditions (4.42)–(4.43), while (4.48) follows from (4.46)–(4.47) according to Lemma 4.4. This observation completes the proof of case (ii) of Theorem 4.12.

Lemma 4.2 *Conditions* (4.42) *and* (4.46) *are equivalent.*

Proof of Lemma 4.2 It is clear that equality (4.46) with discrete parameters $n_1, \dots, n_d$ can be exchanged with the equivalent equality with continuous parameters $s_1, \dots, s_d$. More precisely, condition (4.46) is equivalent to

$$\lim_{\min(s_1,\dots,s_d)\to\infty} s_1 \dots s_d \mathsf{P}\left(|X| \ge s_1^{\nu_1} \dots s_d^{\nu_d}\right) = 0. \tag{4.49}$$

In turn, condition (4.42) is equivalent to

$$\limsup_{t\to\infty} t\mathsf{P}(|X| \ge t^{\nu}) < \infty. \tag{4.50}$$

Without loss of generality we may assume that $\nu_1 \le \dots \le \nu_d$. Let a positive integer i be such that $1 \le i < d$ and $\nu_1 = \dots = \nu_i < \nu_{i+1}$.

First we prove the implication (4.46) $\Longrightarrow$ (4.42). If $t = s_1 \cdots s_i s_{i+1}^{\nu_{i+1}/\nu} \cdots s_d^{\nu_d/\nu}$, then

$$\mathsf{P}\text{-}\lim_{\substack{s_1\to\infty\\ \dots\\ s_d\to\infty}}(\min)\, t = \infty.$$

Further, condition (4.49) is equivalent to

$$\lim_{\substack{t\to\infty\\ s_{i+1}\to\infty\\ \dots\\ s_d\to\infty}} t s_{i+1}^{1-\frac{\nu_{i+1}}{\nu}} \dots s_d^{1-\frac{\nu_d}{\nu}} \mathsf{P}(|X| \ge t^{\nu}) = 0. \tag{4.51}$$

By the Cauchy convergence criterion (see Corollary A.2), the latter result means

$$\lim_{k\to\infty} t_k s_{i+1,k}^{1-\frac{\nu_{i+1}}{\nu}} \dots s_{d,k}^{1-\frac{\nu_d}{\nu}} \mathsf{P}(|X| \ge t_k^{\nu}) = 0$$

whatever the sequences $\{t_k\}$, $\{s_{i+1,k}\}$, ..., $\{s_{d,k}\}$, all tending to infinity.

If condition (4.42) does not hold (as a consequence (4.50) also does not hold), then there exists a sequence $\{t_k\}$ increasing to ∞ and such that

$$0 < u_k \stackrel{\text{def}}{=} t_k \mathsf{P}(|X| \ge t_k^{\nu}) \to \infty, \quad k \to \infty.$$

Put

$$\alpha = \sum_{j=i+1}^{d} \left(\frac{\nu_j}{\nu} - 1 \right) = \frac{\nu_{i+1} + \cdots + \nu_d}{\nu} - (d - i)$$

and $s_{i+1,k} = \cdots = s_{d,k} = u_k^{1/\alpha}$. It is clear that $\alpha > 0$, whence $s_{i+1,k} \to \infty, \ldots, s_{d,k} \to \infty$ as $k \to \infty$. Moreover,

$$t_k s_{i+1,k}^{1-\frac{\nu_{i+1}}{\nu}} \ldots s_{d,k}^{1-\frac{\nu_d}{\nu}} \mathsf{P}(|X| \ge t_k^{\nu}) = 1$$

for all $k \ge 1$. This result contradicts (4.51), therefore condition (4.42) holds.

Now we prove the converse implication (4.42) $\Longrightarrow$ (4.46). Recall that (4.42) $\Longleftrightarrow$ (4.50) and (4.46) $\Longleftrightarrow$ (4.51). Thus we need to show that (4.50) $\Longrightarrow$ (4.51), which is trivial, since $\nu_{i+1} > \nu, \ldots, \nu_d > \nu$. □

Lemma 4.3 *The set of conditions* (4.42)–(4.43) *is equivalent to the set of conditions* (4.46)–(4.47).

Proof of Lemma 4.3 First we prove that (4.42)–(4.43) $\Longrightarrow$ (4.46)–(4.47). By Lemma 4.2, it is necessary to show that (4.47) holds. Put $k = \left[n_1^{\nu_1/\nu} \cdots n_d^{\nu_d/\nu} \right]$, where $[\cdot]$ stands for the integer part of number. Then

$$\left| n_1^{1-\nu_1} \ldots n_d^{1-\nu_d} \int\limits_{|x| < n_1^{\nu_1} \ldots n_d^{\nu_d}} x \, dF(x) \right|$$
$$\le n_1^{1-\nu_1} \ldots n_d^{1-\nu_d} \left| \int\limits_{|x| < k^{\nu}} x \, dF(x) \right| + |\mathbf{n}| \mathsf{P}\left(|X| \ge k^{\nu} \right).$$

Since

$$\lim(\min) \frac{n_1^{1-\nu_1} \ldots n_d^{1-\nu_d}}{k^{1-\nu}} = 0, \quad \lim(\min) \frac{|\mathbf{n}|}{k} = 0,$$

condition (4.47) indeed follows from (4.42)–(4.43).

Now we prove the converse implication, namely (4.46)–(4.47) $\Longrightarrow$ (4.42)–(4.43). By Lemma 4.2, only relation (4.43) needs to be proved. We follow the same method as in the proof of Lemma 4.2. First, note that (4.47) implies

$$\lim_{\min(s_1,\cdots,s_d)\to\infty} s_1^{1-v_1}\dots s_d^{1-v_d} \int\limits_{|x|<s_1^{v_1}\cdots s_d^{v_d}} x\,dF(x) = 0,$$

where $s_1, \dots, s_d$ are continuous parameters. Without loss of generality we may assume that i is an index such that

$$v = v_1 = \dots = v_i < v_{i+1} \le \dots \le v_d = v'.$$

As in the proof of Lemma 4.2, put $t = s_1 \dots s_i s_{i+1}^{v_{i+1}/v} \dots s_d^{v_d/v}$. Then (4.47) implies that

$$\lim_{\substack{t\to\infty\\ s_{i+1}\to\infty\\ \dots\\ s_d\to\infty}} t^{1-v} s_{i+1}^{1-v_{i+1}-(1-v)v_{i+1}/v} \dots s_d^{1-v_d-(1-v)v_d/v} \int\limits_{|x|<t^v} x\,dF(x) = 0. \tag{4.52}$$

By the Cauchy convergence criterion (see Corollary A.2), the latter result means

$$\lim_{k\to\infty} t_k^{1-v} s_{i+1,k}^{1-v_{i+1}-(1-v)v_{i+1}/v} \dots s_{d,k}^{1-v_d-(1-v)v_d/v} \int\limits_{|x|<t_k^v} x\,dF(x) = 0$$

whatever the sequences $\{t_k\}$, $\{s_{i+1,k}\}$, ..., $\{s_{d,k}\}$, all tending to infinity.

It is clear that $1-v_j-(1-v)v_j/v = 1-v_j/v < 0$ for $i < j \le d$. If condition (4.43) does not hold, then one can find a sequence $\{t_k, k \ge 1\}$ such that $t_k \to \infty$ and

$$0 < u_k \stackrel{\text{def}}{=} t_k^{1-v} \Bigg| \int\limits_{|x|<t_k^v} x\,dF(x) \Bigg| \to \infty, \quad k \to \infty.$$

Then we choose the sequences $\{s_{i+1,k}, k \ge 1\}, \dots, \{s_{d,k}, k \ge 1\}$ in such a way that

$$s_{i+1,k} = \dots = s_{d,k} = u_k^{1/\alpha}, \quad \text{where} \quad \alpha = \sum_{j=i+1}^{d} \left(\frac{v_j}{v} - 1\right).$$

Obviously $\alpha > 0$ and thus $s_{i+1,k} \to \infty, \dots, s_{d,k} \to \infty$ as $k \to \infty$. Moreover,

$$t_k^{1-v} s_{i+1,k}^{1-v_{i+1}-(1-v)v_{i+1}/v} \dots s_{d,k}^{1-v_d-(1-v)v_d/v} \Bigg| \int\limits_{|x|<t_k^v} x\,dF(x) \Bigg| = 1$$

for all $k \ge 1$ which contradicts (4.52).

Therefore condition (4.43) follows from (4.52) and this completes the proof of Lemma 4.3. □

Lemma 4.4 *Let* $\min\{\nu_1, \dots, \nu_d\} > \frac{1}{2}$. *Then condition* (4.48) *follows from* (4.46)–(4.47).

Proof of Lemma 4.4 Since

$$\left(\int\limits_{|x|<n_1^{\nu_1}\dots n_d^{\nu_d}} x\,dF(x) \right)^2 \le n_1^{\nu_1} \dots n_d^{\nu_d} \cdot \left| \int\limits_{|x|<n_1^{\nu_1}\dots n_d^{\nu_d}} x\,dF(x) \right|,$$

condition (4.47) implies that

$$\lim(\min) n_1^{1-2\nu_1} \dots n_d^{1-2\nu_d} \left(\int\limits_{|x|<n_1^{\nu_1}\dots n_d^{\nu_d}} x\,dF(x) \right)^2 = 0.$$

It remains to prove

$$\lim(\min) n_1^{1-2\nu_1} \dots n_d^{1-2\nu_d} \int\limits_{|x|<n_1^{\nu_1}\dots n_d^{\nu_d}} x^2\,dF(x) = 0. \tag{4.53}$$

Put $k = |\mathbf{n}| = n_1 \dots n_d$ and $s = n_1^{\nu_1} \dots n_d^{\nu_d}$. Since

$$n_1^{1-2\nu_1} \dots n_d^{1-2\nu_d} \int\limits_{|x|<n_1^{\nu_1}\dots n_d^{\nu_d}} x^2\,dF(x) \le \frac{k}{s^2} \int\limits_{|x|<k^\nu} x^2\,dF(x) + \frac{k}{s^2} \int\limits_{k^\nu \le |x| < s} x^2\,dF(x)$$

and $\lim(\min) k^\nu/s = 0$, equality (4.53) follows from

$$\limsup_{\min\{n_1,\dots,n_d\}\to\infty} k^{1-2\nu} \int\limits_{|x|<k^\nu} x^2\,dF(x) < \infty, \tag{4.54}$$

$$\lim_{\min\{n_1,\dots,n_d\}\to\infty} \frac{k}{s^2} \int\limits_{k^\nu \le |x| < s} x^2\,dF(x) = 0. \tag{4.55}$$

To prove (4.54), we introduce a random variable X with the distribution function F and put $Y = X^{1/\nu}$ and $P_j = \mathsf{P}(|Y| \ge j)$. Since $j\mathsf{P}(|X| \ge j^\nu) = O(1)$ by Lemma 4.3, we have

$$\begin{aligned}\int\limits_{|x|<k^v} x^2\, dF(x) &\le \sum_{j=1}^{k} j^{2v} \mathsf{P}\left(j-1 \le |Y| < j\right) = \sum_{j=1}^{k} j^{2v}\left(P_{j-1}-P_j\right)\\ &= P_0 + \sum_{j=1}^{k-1} P_j\left((j+1)^{2v} - j^{2v}\right) - k^{2v}P_k\\ &\le O(1) + O(1)\sum_{j=1}^{k} j^{2v-1}\mathsf{P}\left(|X| \ge j^v\right)\\ &= O(1) + O(1)\sum_{j=1}^{k} j^{2v-2} = O(1)k^{2v-1}.\end{aligned}$$

This proves relation (4.54).

It only remains to prove equality (4.55). Given $0 < \delta < 1$, we have

$$\begin{aligned}\frac{k}{s^2}\int\limits_{k^v \le |x| < s} x^2\, dF(x) &\le \frac{k}{s^2}\int\limits_{k^v \le |x| < \delta s} x^2\, dF(x) + \frac{k}{s^2}\int\limits_{\delta s \le |x| < s} x^2\, dF(x)\\ &\le \delta^2 k\mathsf{P}(|X| \ge k^v) + k\mathsf{P}(|X| \ge \delta s).\end{aligned}$$

Lemma 4.3 implies $k\mathsf{P}(|X| \ge k^v) = O(1)$ and $(\delta s)^{1/v}\mathsf{P}(|X| \ge \delta s) = O(1)$, whence

$$\begin{aligned}&\limsup_{\min\{n_1,\dots,n_d\}\to\infty} \frac{k}{s^2}\int\limits_{k^v \le |x| < s} x^2\, dF(x)\\ &\quad\le O(1)\delta^2 + O(1)\delta^{-1/v}\lim_{\min\{n_1,\dots,n_d\}\to\infty}\frac{k}{s^{1/v}}.\end{aligned} \tag{4.56}$$

It is clear that

$$\lim_{\min\{n_1,\dots,n_d\}\to\infty}\frac{k}{s^{1/v}} = \lim(\min) n_1^{1-v_1/v}\dots n_d^{1-v_d/v} = 0.$$

Since $\delta > 0$ is arbitrary, relation (4.56) yields (4.55). The proof of Lemma 4.4 is complete. □

Proof of Theorem 4.12: case (iii) If conditions (4.44)–(4.45) hold, we apply the Chebyshev inequality and obtain

$$\mathsf{P}\left(|S(\mathbf{n})| \ge \varepsilon n_1^{v_1}\dots n_d^{v_d}\right) \le \frac{\mathsf{E}\left[X^2\right]}{\varepsilon^2} n_1^{1-2v_1}\dots n_d^{1-2v_d}.$$

Thus the relation $\lim(\min) n_1^{1-2\nu_1} \dots n_d^{1-2\nu_d} = 0$ implies the law of large numbers (4.41) for lim(min)-convergence.

Next we prove that the converse is also true, namely that the law of large numbers (4.41) for lim(min)-convergence implies relations (4.44)–(4.45). As above, let $\nu_1 \le \nu_2 \le \dots \le \nu_d$. Note that $\frac{1}{2} = \nu_1 < \nu_d = \nu'$. Thus, for all $\varepsilon > 0$,

$$\lim(\min) \mathsf{P}\big(|S(\mathbf{n})| \ge \varepsilon n_1^{\nu_1} (n_2 \dots n_d)^{\nu_d}\big) = 0. \tag{4.57}$$

Put $k = n_1$, $l = n_2 \cdots n_d$, and $b(\mathbf{n}) = n_1^{\nu_1}(n_2 \dots n_d)^{\nu_d} = \sqrt{k} l^{\nu'}$. Relation (4.57) together with Theorem 4.10 implies that

$$\lim_{\substack{k \to \infty \\ l \to \infty}} kl \mathsf{P}\left(|X| \ge k^{1/2} l^{\nu'}\right) = 0, \tag{4.58}$$

$$\lim_{\substack{k \to \infty \\ l \to \infty}} \sqrt{k} l^{1-\nu'} \int\limits_{|x| < k^{1/2} l^{\nu'}} x \, dF(x) = 0, \tag{4.59}$$

$$\lim_{\substack{k \to \infty \\ l \to \infty}} l^{1-2\nu'} \left\{ \int\limits_{|x| < k^{1/2} l^{\nu'}} x^2 \, dF(x) - \left(\int\limits_{|x| < k^{1/2} l^{\nu'}} x \, dF(x) \right)^2 \right\} = 0.$$

Let $\delta > 0$. Relation (4.58) with $l = \left[k^{\delta}\right]$ implies that

$$k^{1+\delta} \mathsf{P}\left(|X| \ge k^{\delta \nu' + \frac{1}{2}}\right) \to 0, \qquad k \to \infty.$$

Next we apply the following result.

Lemma 4.5 *If*

$$k^{1+\delta} \mathsf{P}\left(|X| \ge k^{1/\gamma}\right) \to 0, \qquad k \to \infty,$$

for some $\delta > 0$ and $\gamma > 0$, then $\mathsf{E}\left[|X|^{\gamma}\right] < \infty$.

Therefore $\mathsf{E}\left[|X|^{\gamma}\right] < \infty$ for $\gamma = 1/(\delta\nu' + \frac{1}{2})$. One can choose, for example, $\delta = 1/2\nu'$ and then we obtain that $\gamma = 1$, that is $\mathsf{E}\left[|X|\right] < \infty$. Taking into account

$$\lim_{\substack{k \to \infty \\ l \to \infty}} \int\limits_{|x| < k^{1/2} l^{\nu'}} x \, dF(x) = \mathsf{E}\left[X\right],$$

we deduce from (4.59) that $\mathsf{E}\left[X\right] = 0$ (otherwise $\sqrt{k} l^{1-\nu'} \to 0$ as $k, l \to \infty$, which is impossible for any ν'). Thus condition (4.44) holds. Now relation (4.59) and (4.4) imply

$$\lim_{\substack{k\to\infty\\ l\to\infty}} l^{1-2\nu'} \int_{|x|<k^{1/2}l^{\nu'}} x^2\,dF(x) = 0. \tag{4.60}$$

Suppose condition (4.45) does not hold. Then

$$\lim_{\substack{k\to\infty\\ l\to\infty}} \int_{|x|<k^{1/2}l^{\nu'}} x^2\,dF(x) = \infty.$$

This implies that, given a fixed $l \geq 1$, there exists a large number k_l such that

$$\int_{|x|<k_l^{1/2}l^{\nu'}} x^2\,dF(x) \geq l^{2\nu'-1}.$$

This means that condition (4.60) does not hold. The contradiction just obtained proves that condition (4.45) holds, as well. Therefore the proof of Theorem 4.12 is complete. □

4.5 Comments

Jakob Bernoulli's law of large numbers, Theorem 4.1, was published in 1713 in the well known book *Ars Conjectandi*, 8 years after his death. Celebrating the 300th anniversary of the law of large numbers, an initiative, called the 2013 International Year of Statistics, involving several international associations and supported by hundreds of organizations and institutes from all over the world, has been established. 100 hundred years ago, in 1913, the Russian Academy of Sciences had a special meeting devoted to the 200th anniversary of the law of large numbers, where A. A. Markov delivered a celebrated talk about this result and its generalizations. The name *law of large numbers* was introduced by Poisson.[1]

Theorem 4.2 is proved by Khintchine in [196]. The criterion of the law of large numbers for the normalizing sequence $b_n = n$ is obtained by Kolmogoroff in [254]. Theorem 4.4 with the Marcinkiewicz–Zygmund normalizing sequence $b_n = n^\nu$ is due to Gnedenko (see the monograph [126]). This result can be proved with the help of Theorem 4.3. The case of a general normalizing sequence b_n is considered by Feller in [101]. An extra condition used in [101] can be removed with the help of Theorem 4.3.

---

[1] S. D. Poisson, *Recherches de la probabilité des jugements en matière criminelle et en matière civile, precedees des regles generales du calcul des probabilities*, 1837, Paris, Bachelier Impremiur-Libraire, p. 7.

Several results are known in the literature concerning the law of large numbers for multi-indexed sums of dependent random variables. For example, martingales with two parameters in Banach spaces are considered by Dung and Tien [81], Quang and Huan [346], and Quang and Huy [351]; $\mathscr{M}$ blockwise independent random variables are studied by Quang et al. [350]. Rosalsky and Thanh [357] consider random fields in the Rademacher spaces of type p. Some simple relationships between the law of large numbers for double sums in Banach spaces and properties of these spaces are considered by Quang and Huan in [349].

Laws of large numbers for double sums in Rademacher spaces of type p are studied by Rosalsky and Thanh [357, 359]. Adler [5] considers the law of large numbers for multi-indexed sums with random indices.

Section 4.1 The law of large numbers in the scheme of series for $d = 1$ is a partial case of general results on the weak convergence of sums of independent random variables (see Gnedenko and Kolmogorov [126]. Theorems 4.5 and 4.6 are proved in [231].

Section 4.2 The results of section for $d > 1$ are obtained in [231].

Section 4.3 Theorems 4.9–4.10 for $d > 1$ are proved in [231].

Section 4.4 Theorems 4.11–4.12 for $d > 1$ are proved in [231].

Chapter 5
Almost Sure Convergence of Multi-Indexed Series

One of the fundamental results of the classical probability theory is the so-called Kolmogorov three series theorem.

Theorem 5.1 (A.N. Kolmogorov) *Let* $\{X_n, n \ge 1\}$ *be independent random variables.*

(i) The series $\sum_{n\ge1} X_n$ *converges almost surely if, for some* $c > 0$, *the following three numerical series converge*:

$$\sum_{n=1}^{\infty} \mathsf{P}(|X_n| \ge c), \qquad \sum_{n=1}^{\infty} \mathsf{E}\left[X_n^c\right], \qquad \sum_{n=1}^{\infty} \mathsf{var}\left[X_n^c\right]. \tag{5.1}$$

Here, as usual, X_n^c *is the truncation of the random variable* X_n *at a level* $c > 0$, *that is* $X_n^c = X_n \mathbb{I}\{|X_n| < c\}$.

(ii) On the other hand, if the series $\sum_{n\ge1} X_n$ *converges almost surely, then the three series in* (5.1) *converge for all* $c > 0$.

It is natural to ask whether such a result is valid for multi-indexed sums. Recall (see Definition A.7) that a multi-indexed series $\sum X(\mathbf{n})$ is said to be convergent if the limit

$$S \stackrel{\text{def}}{=} \lim(\min)\ S(\mathbf{n}), \quad \text{where} \quad S(\mathbf{n}) \stackrel{\text{def}}{=} \sum_{\mathbf{k}\le\mathbf{n}} X(\mathbf{k})$$

exists.

The following example shows that statement (ii) of the Kolmogorov three series theorem does not hold for $d > 1$ even if the terms $X(\mathbf{n})$ are non-random. We treat the case of $d = 2$ in Example 5.1, however it can easily be generalized to other dimensions $d > 1$.

Example 5.1 Let $d = 2$ and let $X(i, j)$ be defined as follows

$$X(i,j) = \begin{cases} i(-1)^j, & \text{for } i \ge 1 \text{ and } j \le 2, \\ 0, & \text{for other } i, j. \end{cases}$$

O. Klesov, *Limit Theorems for Multi-Indexed Sums of Random Variables*,
Probability Theory and Stochastic Modelling 71, DOI 10.1007/978-3-662-44388-0_5

Consider the double sums $S(m,n) = \sum_{i=1}^{m} \sum_{j=1}^{n} X(i,j)$. It is clear that

$$S(m,n) = \begin{cases} -\dfrac{m(m+1)}{2}, & \text{for } n = 1, \\ 0, & \text{for } n \geq 2. \end{cases}$$

Therefore the limit

$$\lim_{\substack{m\to\infty \\ n\to\infty}} S(m,n)$$

exists and equals 0, that is the double series $\sum_{i,j} X(i,j)$ converges almost surely. On the other hand,

$$\sum_{i,j=1}^{\infty} \mathsf{P}(|X(i,j)| \geq c) = \sum_{i=1}^{\infty} \sum_{j=1}^{2} \mathsf{P}(|X(i,j)| \geq c) = 2 \sum_{i=1}^{\infty} \mathsf{P}(|X(i,1)| \geq c) = \infty$$

for all $c > 0$.

The reason why the Kolmogorov three series theorem fails for $d > 1$ is clear from Example 5.1, namely *all* terms X_n of a convergent series are asymptotically negligible as $n \to \infty$ if $d = 1$ but this property does not hold for $d > 1$. Example 5.1 presents a field $\{X(m,n)\}$ of non-random numbers such that the double series $\sum X(m,n)$ converges but $X(\mathbf{n}) \not\to 0$ (more precisely, $\liminf\,(\max)\, X(\mathbf{n}) = -\infty$ and $\limsup\,(\max)\, X(\mathbf{n}) = +\infty$). It turns out that such an "anomaly" never happens for a "really" random multi-indexed series and thus Kolmogorov's theorem is valid for all $d \geq 1$ if a series "really" is random. The precise statement of this result is given in the next section.

The problem of finding the necessary and sufficient conditions for the almost sure convergence of multi-indexed series of independent terms is posed by Pyke [345]. Below we solve this problem for an arbitrary dimension $d \geq 1$. The necessary and sufficient conditions in the general case are expressed in terms of the convergence of four numerical series (of three series if $d = 1$). In various particular (still important) cases for $d > 1$ only two series (sometimes only one series) are needed to guarantee the almost sure convergence of a random series. One of these cases deals with weighted independent identically distributed random variables, considered in Sect. 5.4.

5.1 Equivalence of Four Types of Convergence

We start with an assertion which is well-known for $d = 1$, that different types of convergence are equivalent for series of independent terms.

Theorem 5.2 *Let $\{X(\mathbf{n}), \mathbf{n} \in \mathbf{N}^d\}$ be a field of independent random variables. Consider a multi-indexed series*

$$\sum_{\mathbf{n}\in\mathbf{N}^d} X(\mathbf{n}) \stackrel{\text{def}}{=} \lim\,(\min)\, S(\mathbf{n}), \tag{5.2}$$

where

$$S(\mathbf{n}) = \sum_{\mathbf{k}\le\mathbf{n}} X(\mathbf{k}), \qquad \mathbf{n} \in \mathbf{N}^d.$$

The following three conditions are equivalent:

(*A*) *series* (5.2) *converges in probability*;
(*B*) *series* (5.2) *converges almost surely*;
(*C*) *series* (5.2) *converges in the sense of the weak convergence of distribution functions of sums* $S(\mathbf{n})$.
In addition, if random variables $\{X(\mathbf{n}), \mathbf{n} \in \mathbf{N}^d\}$ *have zero expectations and finite second moments, then each of the conditions* (A)–(C) *is equivalent to*
(*D*) *series* (5.2) *converges in the mean square sense.*

Proof of Theorem 5.2 The implication $(B) \Longrightarrow (A)$ in the case of $d = 1$ is a general relationship between almost sure convergence and convergence in probability. The proof of this fact for $d > 1$ is literally the same as in the case of $d = 1$ (see, for example, [257], Proposition II in Sect. 5).

Proof of implication $(A) \Longrightarrow (B)$. Let $\mathscr{D}$ be the divergence set of the series (5.2), that is

$$\mathscr{D} = \bigcup_{j=1}^{\infty} \bigcap_{\mathbf{n}\in\mathbf{N}^d} A_j(\mathbf{n}), \qquad A_j(\mathbf{n}) \stackrel{\text{def}}{=} \left\{ \sup_{\mathbf{n}',\mathbf{n}''\ge\mathbf{n}} |S(\mathbf{n}') - S(\mathbf{n}'')| > \frac{1}{j} \right\}.$$

Fix j and put $\varepsilon = \frac{2}{j(d+1)}$. Condition (A) implies that, for all $0 < \alpha < \frac{1}{d}$, there exists an $\mathbf{n}_0 \in \mathbf{N}^d$ such that

$$\mathsf{P}(|S(\mathbf{n}') - S(\mathbf{n}'')| > \varepsilon) < \alpha, \qquad \mathbf{n}', \mathbf{n}'' \ge \mathbf{n}_0. \tag{5.3}$$

Given $\mathbf{n} \ge \mathbf{n}_0$, consider the random variables $S(\mathbf{k}) - S(\mathbf{n}_0)$, $\mathbf{n}_0 \le \mathbf{k} \le \mathbf{n}$. Put $\mathbf{m} = \mathbf{n} - \mathbf{n}_0 + \mathbf{1}$ and

$$T(\mathbf{k}) = S(\mathbf{k} + \mathbf{n}_0 - \mathbf{1}) - S(\mathbf{n}_0), \qquad \mathbf{1} \le \mathbf{k} \le \mathbf{m}.$$

Finally, let $Y(\mathbf{k}) = \Delta\,[T(\mathbf{k})]$, $\mathbf{1} \le \mathbf{k} \le \mathbf{m}$. Here the backward increment operator Δ for multi-indices is defined in Sect. A.5.

Note that the random variables $\{Y(\mathbf{k}), \mathbf{k} \le \mathbf{m}\}$ are independent for all $\mathbf{m}$ and that $T(\mathbf{k}) = \sum_{\mathbf{i}\le\mathbf{k}} Y(\mathbf{i})$, $\mathbf{1} \le \mathbf{k} \le \mathbf{m}$. Moreover, inequality (5.3) implies that

$$\mathsf{P}(|T(\mathbf{k}') - T(\mathbf{k}'')| > \varepsilon) = \mathsf{P}(|S(\mathbf{k}' + \mathbf{n}_0 - \mathbf{1}) - S(\mathbf{k}'' + \mathbf{n}_0 - \mathbf{1})| > \varepsilon) < \alpha$$

for all $\mathbf{1} \leq \mathbf{k}', \mathbf{k}'' \leq \mathbf{m}$. Thus condition (2.29) holds and Ottaviani's inequality for multi-indices (Corollary 2.6) implies that

$$\mathsf{P}\left(\max_{\mathbf{k}\leq\mathbf{m}} |T(\mathbf{k})| > (d+1)\varepsilon\right) \leq \frac{\alpha}{1-d\alpha}$$

or

$$\mathsf{P}\left(\max_{\mathbf{n}_0\leq\mathbf{k}\leq\mathbf{n}} |S(\mathbf{k}) - S(\mathbf{n}_0)| > (d+1)\varepsilon\right) \leq \frac{\alpha}{1-d\alpha}.$$

Since

$$\mathsf{P}\left(\sup_{\mathbf{k}\geq\mathbf{n}_0} |S(\mathbf{k}) - S(\mathbf{n}_0)| \geq x\right) \leq \mathsf{P}\left(\bigcup_{\mathbf{k}\geq\mathbf{n}_0} \{\omega\colon |S(\mathbf{k}) - S(\mathbf{n}_0)| \geq x-\delta\}\right)$$

for all $\delta > 0$ and the random events

$$\left\{\omega\colon \max_{\mathbf{n}_0\leq\mathbf{k}\leq\mathbf{n}} |S(\mathbf{k}) - S(\mathbf{n}_0)| > x\right\}$$

are increasing with respect to $\mathbf{n}$, we get with $\delta = \varepsilon$ that

$$\begin{aligned}
&\mathsf{P}\left(\sup_{\mathbf{k}\geq\mathbf{n}_0} |S(\mathbf{k}) - S(\mathbf{n}_0)| > (d+2)\varepsilon\right) \\
&\quad\leq \mathsf{P}\left(\bigcup_{\mathbf{k}\geq\mathbf{n}_0} |S(\mathbf{k}) - S(\mathbf{n}_0)| > (d+1)\varepsilon\right) \\
&\quad= \lim\,(\min)\,\mathsf{P}\left(\max_{\mathbf{n}_0\leq\mathbf{k}\leq\mathbf{n}} |S(\mathbf{k}) - S(\mathbf{n}_0)| > (d+1)\varepsilon\right) \\
&\quad\leq \frac{\alpha}{1-d\alpha}.
\end{aligned}$$

Then we conclude that

$$\mathsf{P}\left(\sup_{\mathbf{k}',\mathbf{k}''\geq\mathbf{n}_0} |S(\mathbf{k}') - S(\mathbf{k}'')| > \frac{(d+2)\varepsilon}{2}\right) \leq \frac{2\alpha}{1-d\alpha}.$$

Thus, for sufficiently large j,

$$\mathsf{P}\left(A_j(\mathbf{n}_0)\right) \leq \frac{2\alpha}{1-d\alpha}, \quad whence\,\mathsf{P}\left(\bigcap_{\mathbf{n}\in\mathbf{N}^d} A_j(\mathbf{n})\right) \leq \frac{2\alpha}{1-d\alpha}.$$

Since $0 < \alpha < \frac{1}{d}$ is arbitrary,

$$\mathsf{P}\left(\bigcap_{\mathbf{n}\in\mathbf{N}^d} A_j(\mathbf{n})\right) = 0, \quad j \ge 1.$$

This yields $\mathsf{P}(\mathscr{D}) = 0$ and the series (5.2) converges almost surely. □

Proof of implication $(A) \Longrightarrow (C)$. This is a well-known result for $d = 1$ (see [257], a Theorem in Sect. 5). The proof for $d > 1$ is the same as that for $d = 1$.

Proof of implication $(C) \Longrightarrow (D)$. We give the proof under an additional assumption that the random variables $\{X(\mathbf{n}), \mathbf{n} \in \mathbf{N}^d\}$ are uniformly bounded, that is, there exists a non-random constant $c > 0$ such that $\mathsf{P}(|X(\mathbf{n})| \le c) = 1$.

Remark 5.1 The implication $(C) \Longrightarrow (D)$ holds in the general case (without the assumption that the random variables are uniformly bounded). The proof in the general case requires a more involved technique (which is quite similar to the classical case $d = 1$, see, for example, [183], Sect. 10, Theorem 3). An excuse for introducing such an additional restriction is that we will never use the general case.

Let condition (C) hold. Denote by $h(\mathbf{n}; t)$ the characteristic functions of the random variables $X(\mathbf{n})$ and let $h(t)$ be the characteristic function of the multi-indexed series. Then

$$\lim\,(\min) \prod_{\mathbf{k}\le\mathbf{n}} h(\mathbf{k}; t) = h(t) \quad \text{for all} \quad t \in \mathbf{R}.$$

Since h is a continuous function and $h(0) = 1$, we have $h(\mathbf{k}; t) \ne 0$ for all $\mathbf{k} \in \mathbf{N}^d$ in a neighborhood of 0. Thus

$$-\sum \log|h(\mathbf{k}; t)| = -\log|h(t)| < \infty$$

in a neighborhood of 0.

Next we use the following inequality for characteristic functions (see [299] statement 12.4.B′).

Lemma 5.1 *Let F be a distribution function and let h be its characteristic function. If $t > 0$, then*

$$\int_{|x|<1/t} x^2\, dF(x) \le \frac{3}{t^2}\left\{1 - \Re h(t)\right\}.$$

According to Lemma 5.1, for all $\mathbf{k}$ and sufficiently small $t < 1/2c$,

$$\begin{aligned}
2\,\mathsf{var}\,[X(\mathbf{k})] &= \mathsf{var}\left[X^{(s)}(\mathbf{k})\right] \le \frac{3}{t^2}\left(1 - h^{(s)}(\mathbf{k}; t)\right) \le -\frac{3}{t^2}\log h^{(s)}(\mathbf{k}; t)\\
&= -\frac{6}{t^2}\log|h(\mathbf{k}; t)|,
\end{aligned}$$

since $1 - z \le -\log z$, $0 < z \le 1$. As usual, we denote by $X^{(s)}(\mathbf{k})$ the symmetrization of the random variable $X(\mathbf{k})$ and let $h^{(s)}(\mathbf{k}; t)$ be the characteristic function of

$X^{(s)}(\mathbf{k})$. Hence

$$2\sum_{\mathbf{k}\le\mathbf{n}} \mathsf{var}\,[X(\mathbf{k})] \le -\frac{6}{t^2}\sum_{\mathbf{k}\le\mathbf{n}} \log|h(\mathbf{k};t)| \le \sum_{\mathbf{k}\in\mathbf{N}^d} \log|h(\mathbf{k};t)| < \infty.$$

Therefore the series $\sum \mathsf{var}\,[X(\mathbf{n})]$ converges.

By the Cauchy convergence criterion (Corollary A.2) the series in (5.2) converges in the mean square sense if and only if

$$\mathsf{E}\left[(S(\mathbf{m}_i) - S(\mathbf{n}_i))^2\right] = \sum_{\mathbf{k}\le\mathbf{m}_i,\mathbf{k}\not\le\mathbf{n}_i} \mathsf{var}\,[X(\mathbf{k})] \to 0, \qquad i\to\infty, \tag{5.4}$$

for all sequences of multi-indices $\{\mathbf{m}_i, i\ge 1\}$ and $\{\mathbf{n}_i, i\ge 1\}$ such that

$$\mathbf{m}_i \le \mathbf{m}_{i+1}, \quad \mathbf{n}_i \le \mathbf{n}_{i+1}, \quad \mathbf{n}_i \le \mathbf{m}_i,$$
$$\lim_{i\to\infty} \min(m_{1i},\dots,m_{di}) = \infty, \quad \lim_{i\to\infty} \min(n_{1i},\dots,n_{di}) = \infty,$$

where $\mathbf{m}_i = (m_{1i},\dots,m_{di})$ and $\mathbf{n}_i = (n_{1i},\dots,n_{di})$.

It is clear that condition (5.4) is equivalent to the convergence of the series $\sum \mathsf{var}\,[X(\mathbf{k})]$, whence we conclude that series (5.2) converges in the mean square sense.

Proof of implication $(D) \implies (A)$. This is a well-known result for $d = 1$. Its proof for $d > 1$ is the same as that for $d = 1$.

Therefore we showed that

$$(B) \iff (A) \implies (C) \implies (D) \implies (A).$$

This means that Theorem 5.2 is proved (the implication $(C) \implies (D)$ is proved under an extra assumption that the random variables $X(\mathbf{k})$ are uniformly bounded). □

5.2 The Three Series Theorem for Multi-Indexed Sums

We have already seen that the convergence of a multi-indexed series does not, in general, imply that its terms lim (max)-converge to zero. On the other hand, this property holds for lim (min)-convergence, that is, the convergence in probability of a multi-indexed series implies that

$$\forall\varepsilon > 0 \qquad \lim\,(\min)\,\mathsf{P}\,(|X(\mathbf{n})| \ge \varepsilon) = 0. \tag{5.5}$$

There are several important cases where the convergence of a multi-indexed series does imply the lim (max)-convergence to zero of its terms. One such case is considered in the following result.

Lemma 5.2 *Let the random variables $\{X(\mathbf{n}), \mathbf{n} \in \mathbf{N}^d\}$ be independent and symmetric. If series* (5.2) *converges almost surely, then*

$$\forall \varepsilon > 0 \qquad \lim\,(\max)\,\mathsf{P}\left(|X(\mathbf{n})| \geq \varepsilon\right) = 0. \tag{5.6}$$

Lemma 5.2 is a key result for the proof of the criterion for the almost sure convergence of a multi-indexed series. Condition (5.6) is a trivial consequence of the convergence of the series (5.2) if $d = 1$ (not only does almost sure convergence imply (5.6), but convergence in probability suffices, too). For $d > 1$, the result is not obvious at all.

Theorem 5.3 *Let $\{X(\mathbf{n}), \mathbf{n} \in \mathbf{N}^d\}$ be a field of independent random variables. If series* (5.2) *converges almost surely and condition* (5.6) *holds, then, for all $c > 0$, the following non-random series converge*:

$$\sum_{\mathbf{n}\in\mathbf{N}^d} \mathsf{P}\left(|X(\mathbf{n})| \geq c\right), \tag{5.7}$$

$$\sum_{\mathbf{n}\in\mathbf{N}^d} \mathsf{E}\left[X^c(\mathbf{n})\right], \tag{5.8}$$

$$\sum_{\mathbf{n}\in\mathbf{N}^d} \mathsf{var}\left[X^c(\mathbf{n})\right]. \tag{5.9}$$

On the other hand, if the three series (5.7)–(5.9) *converge for some $c > 0$, then the series* (5.2) *converges almost surely.*

Recall that $\xi^c = \xi\,\mathbb{I}_{|\xi|<c}$ for any random variable ξ.

The latter result can be rewritten as follows.

Corollary 5.1 *The almost sure convergence of series* (5.2) *together with condition* (5.6) *is equivalent to the convergence for all $c > 0$ of series* (5.7)–(5.9).

Also, the almost sure convergence of series (5.2) *together with condition* (5.6) *is equivalent to the convergence for some $c > 0$ of series* (5.7)–(5.9).

Proof of Lemma 5.2 Almost sure convergence implies convergence in probability. By the Cauchy convergence criterion (Corollary A.2), we see that, for all $\varepsilon > 0$ and all sequences of multi-indices $\{\mathbf{m}'_k, k \geq 1\}$ and $\{\mathbf{m}''_k, k \geq 1\}$ such that

(i) $\mathbf{m}'_k \leq \mathbf{m}''_k$, $k \geq 1$;
(ii) $\mathbf{m}'_k \leq \mathbf{m}'_{k+1}$ and $\mathbf{m}''_k \leq \mathbf{m}''_{k+1}$, $k \geq 1$;
(iii) $\lim \min(m'_{1k}, \ldots, m'_{dk}) = \infty$,

the following property holds:

$$\lim_{k\to\infty} \mathsf{P}\left(|S(\mathbf{m}''_k) - S(\mathbf{m}'_k)| \geq \varepsilon\right) = 0. \tag{5.10}$$

Now suppose that condition (5.6) does not hold, despite the fact that series (5.2) converges almost surely. Then there are two numbers $\varepsilon_0 > 0$ and $\delta_0 > 0$ and a sequence of multi-indices $\{\mathbf{n}_k, k \geq 1\}$ such that

$$\mathsf{P}(|X(\mathbf{n}_k)| \geq \varepsilon_0) \geq \delta_0, \qquad k \geq 1. \tag{5.11}$$

Without loss of generality we assume that the sequence $\{\mathbf{n}_k, k \geq 1\}$ is increasing and all its element are *different*. Starting from $\{\mathbf{n}_k, k \geq 1\}$ we construct two other sequences $\{\mathbf{m}'_k, k \geq 1\}$ and $\{\mathbf{m}''_k, k \geq 1\}$ satisfying conditions (i)–(iii) and such that

(iv) $\mathbf{n}_k \in \Pi_k$, where $\Pi_k = \{\mathbf{n} : \mathbf{n} \leq \mathbf{m}''_k, \mathbf{n} \not\leq \mathbf{m}'_k\}$.

For example, one can choose $\mathbf{m}'_k = \mathbf{n}_{k-1}$ and $\mathbf{m}''_k = \mathbf{n}_k$.

Fix $k \geq 1$ and introduce the random variables $\{Y(\mathbf{n}), \mathbf{n} \in \mathbf{N}^d\}$ as follows:

$$Y(\mathbf{n}) = \begin{cases} X(\mathbf{n}), & \mathbf{n} \in \Pi_k, \\ 0, & \mathbf{n} \notin \Pi_k, \end{cases} \qquad T(\mathbf{n}) = \sum_{\mathbf{m} \leq \mathbf{n}} Y(\mathbf{m}).$$

It is clear that $\{Y(\mathbf{n})\}$ are independent symmetric random variables and $T(\mathbf{m}''_k) = S(\mathbf{m}''_k) - S(\mathbf{m}'_k)$. Inequality (5.11) implies that

$$\begin{aligned}\delta_0 &\leq \mathsf{P}\left(\max_{\mathbf{n} \in \Pi_k} |X(\mathbf{n})| \geq \varepsilon_0\right) = \mathsf{P}\left(\max_{\mathbf{n} \leq \mathbf{m}''_k} |Y(\mathbf{n})| \geq \varepsilon_0\right) \\ &= \mathsf{P}\left(\max_{\mathbf{n} \leq \mathbf{m}''_k} |\Delta[T(\mathbf{n})]| \geq \varepsilon_0\right) \leq \mathsf{P}\left(\max_{\mathbf{n} \leq \mathbf{m}''_k} |T(\mathbf{n})| \geq \varepsilon_0 2^{-d}\right).\end{aligned}$$

Here we used one of the properties of the operator Δ (see (A.7)). By the Lévy inequality for multi-indexed sums of independent symmetric terms (see Corollary 2.4)

$$\begin{aligned}\delta_0 &\leq \mathsf{P}\left(\max_{\mathbf{n} \leq \mathbf{m}''_k} |T(\mathbf{n})| \geq \varepsilon_0 2^{-d}\right) \leq 2^d\, \mathsf{P}\left(|T(\mathbf{m}''_k)| \geq \varepsilon_0 2^{-d}\right) \\ &= 2^d\, \mathsf{P}\left(|S(\mathbf{m}''_k) - S(\mathbf{m}'_k)| \geq \varepsilon_0 2^{-d}\right).\end{aligned} \tag{5.12}$$

Since $k \geq 1$ is arbitrary, the latter inequality contradicts condition (5.10) and therefore condition (5.6) must hold. □

Proof of Theorem 5.3 Let condition (5.6) holds and series (5.2) converge almost surely. Choose two increasing sequences of multi-indices $\{\mathbf{n}_k, k \geq 1\}$ and $\{\mathbf{m}_k, k \geq 1\}$ such that

$$\mathbf{n}_k \geq \mathbf{m}_k, \qquad \mathbf{n}_k \leq \mathbf{n}_{k+1}, \qquad \mathbf{m}_k \leq \mathbf{m}_{k+1},$$
$$\lim_{k\to\infty} \min(m_{1k}, \ldots, m_{dk}) = \infty, \qquad \lim_{k\to\infty} \min(n_{1k}, \ldots, n_{dk}) = \infty,$$

where $\mathbf{n}_k = (n_{1k}, \dots n_{dk})$, $\mathbf{m}_k = (m_{1k}, \dots m_{dk})$. Then $S(\mathbf{n}_k) - S(\mathbf{m}_k) \to 0$ as $k \to \infty$ almost surely. Consider the (1, 1)-sequence of series of random variables $\{\Xi_k, k \ge 1\}$, where $\Xi_k = \{X(\mathbf{n}) \colon \mathbf{n} \le \mathbf{n}_k, \mathbf{n} \not\le \mathbf{m}_k\}$. The total number of random variables in the series Ξ_k equals $l_k = |\mathbf{n}_k| - |\mathbf{m}_k|$. The order used to write the random variables $X(\mathbf{n})$ in the series Ξ_k does not matter. Denote the random variables in Ξ_k by ξ_{nk}, $n \le l_k$.

Finally, put $\zeta_k = \sum_{n=1}^{l_k} \xi_{nk}$. It is clear that $\zeta_k = S(\mathbf{n}_k) - S(\mathbf{m}_k)$ and thus $\zeta_k \to 0$, $k \to \infty$, almost surely (in probability, as well). Condition (5.6) implies that

$$\forall\, \varepsilon > 0 \qquad \lim_{k\to\infty} \max_{n \le l_k} \mathsf{P}(|\xi_{nk}| \ge \varepsilon) = 0$$

which, in fact, is the classical infinitesimal negligibility condition for a scheme of series. The law of large numbers (Theorem 4.6) implies that, for all $c > 0$,

$$\begin{aligned}
\lim_{k\to\infty} \sum_{n=1}^{l_k} \mathsf{P}(|\xi_{nk}| \ge c) &= 0,\\
\lim_{k\to\infty} \sum_{n=1}^{l_k} \mathsf{E}\left[\xi_{nk}^c\right] &= 0, \qquad (5.13)\\
\lim_{k\to\infty} \sum_{n=1}^{l_k} \mathsf{var}\left[\xi_{nk}^c\right] &= 0.
\end{aligned}$$

The latter conditions can be rewritten in terms of the random variables $X(\mathbf{n})$ in an equivalent form:

$$\begin{aligned}
\lim_{k\to\infty} \sum_{\mathbf{n} \le \mathbf{n}_k, \mathbf{n} \not\le \mathbf{m}_k} \mathsf{P}(|X(\mathbf{n})| \ge c) &= 0,\\
\lim_{k\to\infty} \sum_{\mathbf{n} \le \mathbf{n}_k, \mathbf{n} \not\le \mathbf{m}_k} \mathsf{E}\left[X^c(\mathbf{n})\right] &= 0,\\
\lim_{k\to\infty} \sum_{\mathbf{n} \le \mathbf{n}_k, \mathbf{n} \not\le \mathbf{m}_k} \mathsf{var}\left[X^c(\mathbf{n})\right] &= 0.
\end{aligned}$$

Since the sequences $\{\mathbf{n}_k\}$ and $\{\mathbf{m}_k\}$ are arbitrary, the Cauchy convergence criterion (Corollary A.2) implies the convergence of the series (5.7)–(5.9).

Now our aim is to prove the second part of Theorem 5.3. Let series (5.7)–(5.9) converge for some $c > 0$. Then the convergence of series (5.8) and (5.9) yields

$$\text{the series} \quad \sum_{\mathbf{n} \in \mathbf{N}^d} X^c(\mathbf{n}) \quad \text{converges in the mean square sense.}$$

According to Theorem 5.2, this means

$$\text{the series} \quad \sum_{\mathbf{n}\in\mathbf{N}^d} X^c(\mathbf{n}) \quad \text{converges almost surely.}$$

Further, by the convergence of series (5.7) and by the Borel–Cantelli lemma we conclude that

$$\mathsf{P}(X(\mathbf{n}) \neq X^c(\mathbf{n}) \text{ i. o.}) = 0,$$

whence the almost sure convergence of series (5.2) follows. □

5.3 The Four Series Theorem

This section is devoted to finding the necessary and sufficient conditions for the almost sure convergence of multi-indexed series if condition (5.6) is dropped.

Let $\{X(\mathbf{n}), \mathbf{n} \in \mathbf{N}^d\}$ be a field of independent random variables. For $c > 0$, put

$$c(\mathbf{n}) = \mu(X(\mathbf{n})) + \mathsf{E}\left[(X(\mathbf{n}) - \mu(X(\mathbf{n})))^c\right], \tag{5.14}$$

where $\mu(X)$ is a median of the random variable X and where X^c denotes the truncation of the random variable X at a level c:

$$X^c = \begin{cases} X, & \text{if } |X| < c, \\ 0, & \text{if } |X| \geq c. \end{cases}$$

Theorem 5.4 *Let series* (5.2) *converge almost surely and let the constants* $c(\mathbf{n})$ *be defined by* (5.14). *Then, for all* $c > 0$,

(1) the series $\sum c(\mathbf{n})$ *converges;*
(2) $\lim\,(\max)\,\mathsf{P}(|X(\mathbf{n}) - c(\mathbf{n})| \geq \varepsilon) = 0$ *for all* $\varepsilon > 0$.

Remark 5.2 Theorem 5.4 asserts that if series (5.2) converges, then its terms $X(\mathbf{n})$ admit a decomposition as a sum of non-random constants $c(\mathbf{n})$ and random variables $X(\mathbf{n}) - c(\mathbf{n})$ and moreover the series $\sum c(\mathbf{n})$ converges and condition (5.6) holds with $X(\mathbf{n}) - c(\mathbf{n})$ instead of $X(\mathbf{n})$. Thus the three series theorem can be applied to random variables $X(\mathbf{n}) - c(\mathbf{n})$ which gives the convergence criterion in the general case. In contrast to the case $d = 1$, the criterion for $d > 1$ is expressed in terms of the convergence of four series of non-random terms, since, in addition to the three "classical" series, we require that the series $\sum c(\mathbf{n})$ converges.

Theorem 5.5 *The almost sure convergence of series (5.2) is equivalent to the following set of two condition:*

(1) the series $\sum c(\mathbf{n})$ *converges;*
(2) the series $\sum[X(\mathbf{n}) - c(\mathbf{n})]$ *converges almost surely.*

Theorem 5.6 *For $c > 0$, put*

$$\widehat{X}(\mathbf{n}) = (X(\mathbf{n}) - c(\mathbf{n}))\, \mathbb{I}\,(|X(\mathbf{n}) - c(\mathbf{n})| < c)\,.$$

The almost sure convergence of series (5.2) *is equivalent to the convergence for some $c > 0$ of the following four series*:

$$\sum_{\mathbf{n}\in\mathbf{N}^d} c(\mathbf{n}),$$

$$\sum_{\mathbf{n}\in\mathbf{N}^d} \mathsf{P}(|X(\mathbf{n}) - c(\mathbf{n})| \geq c), \quad \sum_{\mathbf{n}\in\mathbf{N}^d} \mathsf{E}\left[\widehat{X}(\mathbf{n})\right], \quad \sum_{\mathbf{n}\in\mathbf{N}^d} \mathsf{var}\left[\widehat{X}(\mathbf{n})\right]. \tag{5.15}$$

Moreover, the result is still true if the expression "for all $c > 0$" is substituted for "for some $c > 0$".

5.3.1 The Two Series Theorem

Theorem 5.6 can be reduced to a simpler form in many particular (still important) cases. Below we discuss some of those cases where the conditions for the almost sure convergence look identical to the case of $d = 1$.

Theorem 5.7 *Let $\{X(\mathbf{n}), \mathbf{n} \in \mathbf{N}^d\}$ be a field of independent random variables. Assume that $\mathsf{var}\,[X(\mathbf{n})] < \infty$ for all $\mathbf{n} \in \mathbf{N}^d$. Then the almost sure convergence of series* (5.2) *follows from the convergence of the following two series*

$$\sum_{\mathbf{n}\in\mathbf{N}^d} \mathsf{E}\,[X(\mathbf{n})]\,, \qquad \sum_{\mathbf{n}\in\mathbf{N}^d} \mathsf{var}\,[X(\mathbf{n})]\,.$$

Corollary 5.2 *Let $\{X(\mathbf{n}), \mathbf{n} \in \mathbf{N}^d\}$ be a field of independent symmetric random variables. Then series* (5.2) *converges almost surely if and only if, for some $c > 0$, the following two series*

$$\sum_{\mathbf{n}\in\mathbf{N}^d} \mathsf{P}\,(|X(\mathbf{n})| \geq c)\,, \qquad \sum_{\mathbf{n}\in\mathbf{N}^d} \mathsf{var}\left[X^c(\mathbf{n})\right]$$

converge. Moreover, the result is still true if the expression "for all $c > 0$" is substituted for "for some $c > 0$".

Corollary 5.3 *Let $\{X(\mathbf{n}), \mathbf{n} \in \mathbf{N}^d\}$ be a field of independent non-negative random variables. Then the series* (5.2) *converges almost surely if, for some $c > 0$, the following two series*

$$\sum_{\mathbf{n}\in\mathbf{N}^d} \mathsf{P}\left(|X(\mathbf{n})| \ge c\right), \qquad \sum_{\mathbf{n}\in\mathbf{N}^d} \mathsf{E}\left[X^c(\mathbf{n})\right]$$

converge. Moreover, the result is still true if the expression "for all $c > 0$*" is substituted for "for some* $c > 0$*".*

Corollary 5.4 *Let* $\{X(\mathbf{n}), \mathbf{n} \in \mathbf{N}^d\}$ *be a field of independent Gaussian random variables. Then the series*

$$\sum_{\mathbf{n}\in\mathbf{N}^d} X^2(\mathbf{n}) \tag{5.16}$$

converges almost surely if and only if the series

$$\sum_{\mathbf{n}\in\mathbf{N}^d} \mathsf{E}\left[X^2(\mathbf{n})\right] \tag{5.17}$$

converges.

5.3.2 Proofs

First we prove Theorem 5.7 by using Theorem 5.3. Then we pass to all other results.

Proof of Theorem 5.7 Let $c > 0$. Without loss of generality assume that $\mathsf{E}\left[X(\mathbf{n})\right] = 0$ for all $\mathbf{n} \in \mathbf{N}^d$. Then condition (5.6) holds and series (5.7) converges in view of the Chebyshev inequality, since the series $\sum \mathsf{var}\left[X(\mathbf{n})\right]$ converges. Further,

$$c\,\mathsf{E}\left[|X(\mathbf{n}) - X^c(\mathbf{n})|\right] \le \mathsf{E}\left[(X(\mathbf{n}) - X^c(\mathbf{n}))^2\right] \le \mathsf{E}\left[X^2(\mathbf{n})\right],$$

whence the convergence of the series $\sum \mathsf{E}\left[|X(\mathbf{n}) - X^c(\mathbf{n})|\right]$ follows. Since

$$\mathsf{E}\left[X^c(\mathbf{n})\right] = -\,\mathsf{E}\left[X(\mathbf{n}) - X^c(\mathbf{n})\right]$$

by the assumption $\mathsf{E}\left[X(\mathbf{n})\right] = 0$, series (5.8) converges, as well. Finally, series (5.9) converges in view of the inequality $\mathsf{var}\left[X^c(\mathbf{n})\right] \le \mathsf{var}\left[X(\mathbf{n})\right]$. It remains to use Theorem 5.3 to complete the proof of Theorem 5.7. □

Proof of Theorem 5.4 Since series (5.2) converges almost surely, the series $\sum X^{(s)}(\mathbf{n})$ of the symmetrizations also converges almost surely. Lemma 5.2 implies condition (5.6) for random variables $X^{(s)}(\mathbf{n})$. Thus we obtain from Corollary 5.1 that, for all $c > 0$,

$$\sum_{\mathbf{n}\in\mathbf{N}^d} \mathsf{P}(|X^{(s)}(\mathbf{n})| \ge c) < \infty, \qquad \sum_{\mathbf{n}\in\mathbf{N}^d} \mathsf{var}\left[\left(X^{(s)}(\mathbf{n})\right)^c\right] < \infty. \tag{5.18}$$

Further, the symmetrization inequalities (Proposition A.35) imply that

$$\mathsf{P}(|X(\mathbf{n}) - \mu(X(\mathbf{n}))| \geq \varepsilon) \leq 2\,\mathsf{P}(|X^{(s)}(\mathbf{n})| \geq \varepsilon), \tag{5.19}$$

whence

$$\forall\, \varepsilon > 0 \quad \lim\,(\max)\,\mathsf{P}(|X(\mathbf{n}) - \mu(X(\mathbf{n}))| \geq \varepsilon) = 0. \tag{5.20}$$

Since

$$\mathsf{P}(|X^c| \geq \varepsilon) = \mathsf{P}(X = X^c, |X| \geq \varepsilon) \leq \mathsf{P}(|X| \geq \varepsilon)$$

for an arbitrary random variable X and all numbers $c > 0$ and $\varepsilon > 0$, relation (5.20) implies that

$$\forall\, \varepsilon > 0 \quad \lim\,(\max)\,\mathsf{P}(|\,[X(\mathbf{n}) - \mu(X(\mathbf{n}))]^c\,| \geq \varepsilon) = 0$$

for all $c > 0$. Now the inequality $\left|[X(\mathbf{n}) - \mu(X(\mathbf{n}))]^c\right| \leq c$ and Lebesgue dominated convergence theorem (see Proposition A.35) imply that

$$\lim\,(\max)\,\mathsf{E}\left[(X(\mathbf{n}) - \mu(X(\mathbf{n})))^c\right] = 0.$$

The latter relation together with (5.20) prove statement (2) of Theorem 5.4.

To prove statement (1) we again use (5.18) and (5.19). We have

$$\sum_{\mathbf{n}\in\mathbf{N}^d} \mathsf{P}(|X(\mathbf{n}) - \mu(X(\mathbf{n}))| \geq c) \leq 2 \sum_{\mathbf{n}\in\mathbf{N}^d} \mathsf{P}(|X^{(s)}(\mathbf{n})| \geq c). \tag{5.21}$$

It is clear that

$$\left|[X(\mathbf{n}) - \mu(X(\mathbf{n}))]^c - [X'(\mathbf{n}) - \mu(X'(\mathbf{n})]^c\right| \leq \left|\left[X(\mathbf{n}) - X'(\mathbf{n})\right]^{2c}\right| + c\,\mathbb{I}_D, \tag{5.22}$$

where $X'(\mathbf{n})$ is an independent copy of the random variable $X(\mathbf{n})$ and where the symbol $\mathbb{I}_D$ stands for the indicator of the random event

$$D = \left\{\omega\colon\ \max\left[|X(\mathbf{n}) - \mu(X(\mathbf{n}))|, |X'(\mathbf{n}) - \mu(X'(\mathbf{n})|)\right] \geq c\right\}.$$

Note that $X(\mathbf{n}) - X'(\mathbf{n}) = X^{(s)}(\mathbf{n})$. Squaring both sides of inequality (5.22) and passing to the expectation we get

$$\begin{aligned} 2\,\mathsf{var}\left[(X(\mathbf{n}) - \mu(X(\mathbf{n})))^c\right] &\leq \mathsf{var}\left[\left(X^{(s)}(\mathbf{n})\right)^{2c}\right] + 2c\,\mathsf{E}\left[\mathbb{I}_D\left(X^{(s)}(\mathbf{n})\right)^{2c}\right] \\ &\quad + c^2\,\mathsf{P}(D) \\ &\leq \mathsf{var}\left[\left(X^{(s)}(\mathbf{n})\right)^{2c}\right] + 5c^2\,\mathsf{P}(D). \end{aligned}$$

We use the symmetrization inequalities (Proposition A.35) once more:

$$\mathsf{P}(D) \le 2\,\mathsf{P}(|X^{(s)}(\mathbf{n})| \ge c),$$

since the symmetrizations of the random variables $X(\mathbf{n})$ and $X'(\mathbf{n})$ are identically distributed. Thus

$$2 \sum_{\mathbf{n}\in\mathbf{N}^d} \mathsf{var}\left[(X(\mathbf{n}) - \mu(X(\mathbf{n})))^c\right]$$
$$\le \sum_{\mathbf{n}\in\mathbf{N}^d} \mathsf{var}\left[\left(X^{(s)}(\mathbf{n})\right)^{2c}\right] + 10c^2 \sum_{\mathbf{n}\in\mathbf{N}^d} \mathsf{P}(|X^{(s)}(\mathbf{n})| \ge c).$$

This result together with (5.18) and (5.21) yields

$$\sum_{\mathbf{n}\in\mathbf{N}^d} \mathsf{P}(|X(\mathbf{n}) - \mu(X(\mathbf{n}))| \ge c) < \infty, \tag{5.23}$$

$$\sum_{\mathbf{n}\in\mathbf{N}^d} \mathsf{var}\left[(X(\mathbf{n}) - \mu(X(\mathbf{n})))^c\right] < \infty. \tag{5.24}$$

Put $Y(\mathbf{n}) = X(\mathbf{n}) - \mu(X(\mathbf{n}))$. Relation (5.24) and Theorem 5.7 imply that the series $\sum_{\mathbf{n}\in\mathbf{N}^d} \left(Y^c(\mathbf{n}) - \mathsf{E}\left[Y^c(\mathbf{n})\right]\right)$ converges almost surely. Now relation (5.23) and the Borel–Cantelli lemma (see Proposition A.30) prove that $\mathsf{P}(|Y(\mathbf{n})| \ge c \text{ i. o.}) = 0$. Thus the series $\sum_{\mathbf{n}\in\mathbf{N}^d} \left(Y(\mathbf{n}) - \mathsf{E}\left[Y^c(\mathbf{n})\right]\right)$ converges almost surely. Since $Y(\mathbf{n}) - \mathsf{E}\left[Y^c(\mathbf{n})\right] = X(\mathbf{n}) - c(\mathbf{n})$, statement (1) of Theorem 5.4 holds, as well. □

Remark 5.3 Inspecting the proof of conditions (5.23)–(5.24) we see that (5.23)–(5.24) hold if the series $\sum X^{(s)}(n)$ converges almost surely. This result will be used in the proof of Theorem 5.9.

Proof of Theorem 5.5 If series (5.2) converges almost surely, then Theorem 5.4 implies the convergence of the series $\sum_{\mathbf{n}\in\mathbf{N}^d} c(\mathbf{n})$. In turn, this implies the almost sure convergence of the series $\sum_{\mathbf{n}\in\mathbf{N}^d}[X(\mathbf{n}) - c(\mathbf{n})]$.

The second part of the proof of Theorem 5.5 is obvious. □

Proof of Theorem 5.6 First we assume that series (5.2) converges almost surely. Then Theorem 5.5 implies that

(i) the series $\sum c(\mathbf{n})$ converges,
(ii) the series $\sum[X(\mathbf{n}) - c(\mathbf{n})]$ converges almost surely.
In addition, we conclude from Theorem 5.4 that
(iii) $\lim\,(\max)\, \mathsf{P}(|X(\mathbf{n}) - c(\mathbf{n})| \ge \varepsilon) = 0$ for all $\varepsilon > 0$.

Applying Theorem 5.3 to the field $\{X(\mathbf{n}) - c(\mathbf{n}),\ \mathbf{n} \in \mathbf{N}^d\}$, we prove the convergence of the three series in (5.15). Therefore the implication

$\{\text{a.s.convergence of series (5.2)}\} \Longrightarrow \{\text{convergence of series (5.15)}\}$

is proved.

For the proof of the inverse implication we assume that all four series in (5.15) converge. The convergence of the latter three series in (5.15) and Theorem 5.3 imply that the series $\sum[X(\mathbf{n}) - c(\mathbf{n})]$ converges almost surely. Since the series $\sum c(\mathbf{n})$ also converges by the assumption of the theorem, we conclude that the series $\sum X(\mathbf{n})$ converges almost surely. □

Proof of Corollary 5.2 For symmetric random variables, all constants $c(\mathbf{n})$ defined by (5.14) equal zero. Since $\mathsf{E}\left[X^c = 0\right]$ for all symmetric random variables X and constants $c > 0$, Theorem 5.6 for symmetric random variables coincides with Corollary 5.2. □

Proof of Corollary 5.3 For non-negative random variables,

$$\mathsf{var}\left[X^c(\mathbf{n})\right] \le \mathsf{E}\left[\left(X^c(\mathbf{n})\right)^2\right] \le c\,\mathsf{E}\left[X^c(\mathbf{n})\right].$$

Thus the convergence of series (5.9) follows from the convergence of series (5.8). Thus, if the series (5.7) and (5.8) converge, then we derive from Theorem 5.3 the almost sure convergence of series (5.2) and the implication

$\{$convergence of series (5.7) and (5.8)$\}$ $\Longrightarrow$ $\{$a.s. convergence of series (5.2)$\}$ is proved.

To prove the inverse implication note that condition (5.6) follows from the almost sure convergence of series (5.2) if the terms are non-negative. Now we establish the convergence of series (5.7) and (5.8) from Theorem 5.3. □

Proof of Corollary 5.4 If series (5.16) converges almost surely, then

$$\lim\,(\max)\, X(\mathbf{n}) = 0 \qquad \text{a. s.}$$

and thus condition (5.6) holds. It is clear that condition (5.6) can be rewritten with $X^2(\mathbf{n})$ instead of $X(\mathbf{n})$. Let $c > 0$ and put $Y(\mathbf{n}) = X^2(\mathbf{n})$. Then Theorem 5.3 implies that the series

$$\sum_{\mathbf{n}\in\mathbf{N}^d} \mathsf{E}\left[Y^c(\mathbf{n})\right] \quad \text{converges whatever } c > 0.$$

Since $Y^c(\mathbf{n}) = \left(X^{\sqrt{c}}(\mathbf{n})\right)^2$, the series

$$\sum_{\mathbf{n}\in\mathbf{N}^d} \mathsf{E}\left[\left(X^c(\mathbf{n})\right)^2\right] \quad \text{converges whatever } c > 0. \tag{5.25}$$

Changing the variable in the Poisson integral one can prove for every random variable $Z \in \mathcal{N}\left(a, \sigma^2\right)$ that

$$\mathsf{E}\left[\left(Z^c\right)^2\right] \sim \mathsf{E}\left[Z^2\right], \qquad a, \sigma^2 \to 0.$$

This property proves that the convergence of the series (5.17) follows from (5.25).

On the other hand, if series (5.17) converges, then

$$\sum_{\mathbf{n}\in\mathbf{N}^d} \mathsf{P}(X^2(\mathbf{n}) \ge c) \le \frac{1}{c^2} \sum_{\mathbf{n}\in\mathbf{N}^d} \mathsf{E}\left[X^2(\mathbf{n})\right] < \infty$$

for all $c > 0$ by the Markov inequality. Moreover, $\mathsf{E}\left[\left(X^2(\mathbf{n})\right)^c\right] \le \mathsf{E}\left[X^2(\mathbf{n})\right]$ and thus Corollary 5.4 follows from Corollary 5.3. □

5.4 Convergence of Series of Weighted Independent Identically Distributed Random Variables

A special type of random field is considered in this section. Any representative of such a field is constructed from a sequence of independent identically distributed random variables and a sequence of non-random numbers.

Let $Y(\mathbf{n})$ be independent identically distributed random variables and let $X(\mathbf{n}) = b(\mathbf{n})Y(\mathbf{n})$, where $\{b(\mathbf{n})\}$ are non-random numbers. We have already dealt with a representative of this kind of random field in Example 5.1, exhibiting a difference between the cases $d = 1$ and $d > 1$. The factors $b(\mathbf{n})$ can be treated as weights and $X(\mathbf{n})$ are called the random variables with weights. The series $\sum b(\mathbf{n})Y(\mathbf{n})$ appear quite often in the case $d = 1$ (as a representation for ARMA sequences, say) and the convergence of such series is quite an interesting problem.

We consider the case where the distribution function of random variables $Y(\mathbf{n})$ is not degenerate (not concentrated at a single point); otherwise random variables $Y(\mathbf{n})$ are actually non-random.

Theorem 5.8 *Let $\{Y(\mathbf{n}), \mathbf{n} \in \mathbf{N}^d\}$ be a field of independent identically distributed random variables and let $\{b(\mathbf{n}), \mathbf{n} \in \mathbf{N}^d\}$ be a field of non-random constants. Put $X(\mathbf{n}) = b(\mathbf{n})Y(\mathbf{n})$. If the distribution function of the random variable $Y(\mathbf{1})$ is not concentrated at a single point, then the almost sure convergence of series* (5.2) *is equivalent to the convergence for some $c > 0$ of series* (5.7)–(5.9).

Moreover, the result is still true if the expression "for all $c > 0$" substitutes "for some $c > 0$".

Remark 5.4 As $\{X(\mathbf{n})\}$ are independent random variables, one could say that Theorem 5.8 is a particular case of Theorem 5.3. Nevertheless, there is a difference between Theorems 5.8 and 5.3, since condition (5.6) is dropped in the statement of Theorem 5.8. This is explained by a feature of the series of random variables with weights, that its almost sure convergence implies condition (5.6).

Proof of Theorem 5.8 One step of the proof is easy: if series (5.7)–(5.9) converge, then Theorem 5.3 indeed implies that series (5.2) converges almost surely.

Now we prove the converse implication. Assume that series (5.2) converges almost surely. Since the symmetrizations are such that $\left(b(\mathbf{n})Y(\mathbf{n})\right)^{(s)} = b(\mathbf{n})Y^{(s)}(\mathbf{n})$, the series $\sum b(\mathbf{n})Y^{(s)}(\mathbf{n})$ converges almost surely, as well. Lemma 5.2 implies that, for all $\varepsilon > 0$,

$$\lim\,(\max)\,\mathsf{P}(|b(\mathbf{n})Y^{(s)}(\mathbf{n})| \geq \varepsilon) = \lim\,(\max)\,\mathsf{P}\left(b(\mathbf{n})|Y^{(s)}| \geq \varepsilon\right) = 0, \tag{5.26}$$

where $Y^{(s)}$ is a copy of the random variable $Y^{(s)}(\mathbf{n})$. Using the identical distribution of the random variables $Y(\mathbf{n})$, we conclude that $\lim\,(\max)\,b(\mathbf{n}) = 0$ if the distribution function of $Y^{(s)}$ is not concentrated at 0, that is, if the distribution function of the random variables $Y(\mathbf{n})$ is not concentrated at a single point. Relation (5.26) implies that

$$\forall \varepsilon > 0 \qquad \lim\,(\max)\,\mathsf{P}\,(b(\mathbf{n})|Y| \geq \varepsilon) = 0, \tag{5.27}$$

that is, condition (5.6) holds. Indeed, fix $\varepsilon > 0$. Then the inequality

$$\mathsf{P}(b(\mathbf{n})|Y| \geq \varepsilon) \leq \frac{1}{2}$$

may occur only for a finite number of multi-indices $\mathbf{n}$, since $\lim\,(\max)\,b(\mathbf{n}) = 0$. For the remaining multi-indices $\mathbf{n}$, we have

$$\mathsf{P}(b(\mathbf{n})|Y^{(s)}| \geq \varepsilon) \geq \mathsf{P}(b(\mathbf{n})|Y_1| \geq 2\varepsilon,\, b(\mathbf{n})|Y_2| < \varepsilon) \geq \frac{1}{2}\,\mathsf{P}(b(\mathbf{n})|Y| \geq 2\varepsilon),$$

where Y_1 and Y_2 are independent copies of the random variable Y. This together with (5.26) proves (5.27).

Now the convergence of series (5.7)–(5.9) follows from Theorem 5.3. □

Remark 5.5 What may happen if $\mathsf{P}(Y(\mathbf{n}) = y_0) = 1$ for some $y_0 \in \mathbf{R}$? This property means that the distribution function of the random variables $Y(\mathbf{n})$ is concentrated at a single point y_0. Then $\sum b(\mathbf{n})Y(\mathbf{n}) = y_0 \sum b(\mathbf{n})$ almost surely if $y_0 \neq 0$ (the same equality holds for $y_0 = 0$, as well, under the assumption that the series $\sum b(\mathbf{n})$ converges). Example 5.1 shows that the three series criterion does not always apply to non-random terms.

Consider the consequences of Theorem 5.8 for two classes of fields $\{b(\mathbf{n}), \mathbf{n} \in \mathbf{N}^d\}$ for which the conditions for almost sure convergence can be reduced to a moment condition imposed on an individual random variable $Y(\mathbf{n})$. Let $\{Y(\mathbf{n})\}$ be a field of independent identically distributed random variables. Put $Y = Y(1, \ldots, 1)$ and let F be the distribution function of the random variable Y.

Note that the almost sure convergence of the series $\sum b(\mathbf{n})Y(\mathbf{n})$ is obvious for all non-random fields $\{b(\mathbf{n}), \mathbf{n} \in \mathbf{N}^d\}$ if the distribution function F is concentrated at 0.

Corollary 5.5 *Let $\{Y(\mathbf{n}), \mathbf{n} \in \mathbf{N}^d\}$ be a field of independent identically distributed random variables. Let $v_1 > 0, \ldots, v_d > 0$; put*

$$v = \min(v_1, \dots, v_d), \qquad d_1 = \mathrm{card}(i\colon v_i = v).$$

The almost sure convergence of the series

$$\sum_{\mathbf{n}\in\mathbf{N}^d} \frac{Y(\mathbf{n})}{n_1^{v_1}\dots n_d^{v_d}} \tag{5.28}$$

is equivalent to the

(i) condition

$$\mathsf{E}\left[|Y|^{1/v}\left(\log^+|Y|\right)^{d_1-1}\right] < \infty \tag{5.29}$$

if $v > 1$;

(ii) set of conditions (5.29) *and*

$$\text{the series} \quad \sum_{\mathbf{n}\in\mathbf{N}^d} \frac{1}{n_1^{v_1}\dots n_d^{v_d}} \int_{|x|<b(\mathbf{n})} x\,dF(x) \quad \text{converges} \tag{5.30}$$

if $v = 1$ *where* $b(\mathbf{n}) = n_1^{v_1}\dots n_d^{v_d}$;

(iii) set of conditions (5.29) *and*

$$\mathsf{E}\,[Y] = 0 \tag{5.31}$$

if $\frac{1}{2} < v < 1$.

(iv) Moreover, series (5.28) *diverges almost surely for all distribution functions* F *that are not concentrated at the point* 0 *if* $v \le \frac{1}{2}$.

Corollary 5.6 *Let* $\{Y(\mathbf{n}), \mathbf{n}\in\mathbf{N}^d\}$ *be a field of independent identically distributed random variables. Further, let*

$$u > 0, \qquad v = \frac{1}{v_1} + \dots + \frac{1}{v_d}.$$

Then the almost sure convergence of the series

$$\sum_{\mathbf{n}\in\mathbf{N}^d} \frac{Y(\mathbf{n})}{(n_1^{v_1} + \dots + n_d^{v_d})^u} \tag{5.32}$$

is equivalent to the

(i) *condition*

$$\mathsf{E}\left[|Y|^{v/u}\right] < \infty \tag{5.33}$$

if $v > u$;

(ii) *set of conditions* (5.33) *and*

$$\textit{the series} \quad \sum_{\mathbf{n}\in\mathbf{N}^d} \frac{1}{(n_1^{v_1}+\cdots+n_d^{v_d})^u} \int_{|x|<b(\mathbf{n})} x\,dF(x) \quad \textit{converges} \tag{5.34}$$

if $v = u$ *where* $b(\mathbf{n}) = (n_1^{v_1}+\cdots+n_d^{v_d})^u$;

(iii) *set of conditions* (5.33) *and* (5.31) *if* $\frac{u}{2} < v < u$.

(iv) *Moreover, series* (5.32) *diverges almost surely if* $v \le \frac{u}{2}$ *for all distribution functions* F *that are not concentrated at the point* 0.

Proof of Corollary 5.5 Recall that

$$H(n) \asymp n^{1/v} (\log n)^{d_1-1}, \qquad n \to \infty$$

(see Proposition A.20).

First we consider the case of a degenerate (not necessarily at 0) distribution function of terms $Y(\mathbf{n})$. We denote the point where the whole mass of F is concentrated by y_0. Then series (5.28) almost surely is equal to

$$\sum_{\mathbf{n}\in\mathbf{N}^d} \frac{y_0}{n_1^{v_1}\dots n_d^{v_d}}.$$

Thus, whatever y_0, series (5.28) converges almost surely and condition (5.29) holds in statement (i) of Corollary 5.5 (recall that $v_1 > 1, \dots, v_d > 1$ in this case).

For statements (ii) and (iii), series (5.28) converges almost surely only if $y_0 = 0$, that is, only for a degenerate distribution function with $y_0 = 0$. In both of these two cases, conditions (5.30) and (5.31) hold only if $y_0 = 0$. Therefore Corollary 5.5 is proved for degenerate distribution functions.

For the rest of the proof, we assume that the distribution function of the random variables $\{Y(\mathbf{n})\}$ is not concentrated at a single point. We precede the proof by some general observations related to all three cases of the theorem. These observations are presented in the following three lemmas. We use the abbreviation $b(\mathbf{n}) = n_1^{v_1}\dots n_d^{v_d}$ throughout in the proof of these three results. □

Lemma 5.3 *For all* $v_1, \dots, v_d > 0$, *the almost sure convergence of series* (5.28) *is equivalent to the following set of three conditions*

$$\sum_{\mathbf{n}\in\mathbf{N}^d} \mathsf{P}(|Y(\mathbf{n})| \ge b(\mathbf{n})) < \infty, \tag{5.35}$$

$$\textit{the series} \sum_{\mathbf{n}\in\mathbf{N}^d} \frac{1}{b(\mathbf{n})} \int_{|x|<b(\mathbf{n})} x\,dF(x) \textit{ converges,} \tag{5.36}$$

$$\sum_{\mathbf{n}\in\mathbf{N}^d} \frac{1}{b^2(\mathbf{n})} \left[\int_{|x|<b(\mathbf{n})} x^2\,dF(x) - \left(\int_{|x|<b(\mathbf{n})} x\,dF(x) \right)^2 \right] < \infty. \tag{5.37}$$

This result is a direct consequence of Theorem 5.8 with $c = 1$.

Lemma 5.4 *For all* $v_1 > 0, \dots, v_d > 0$,

$$(5.29) \iff (5.35). \tag{5.38}$$

Proof of Lemma 5.4 To prove the equivalence of conditions (5.29) and (5.35) put

$$h(k) = \operatorname{card}\left\{\mathbf{n} \in \mathbf{N}^d : k-1 < b(\mathbf{n}) \le k\right\}, \qquad k \ge 1,$$

and $H(1) = h(1)$, $H(k) = h(1) + \cdots + h(k)$ for $k > 1$. It is clear that

$$\begin{aligned}
\sum_{\mathbf{n}\in\mathbf{N}^d} \mathsf{P}(|Y| \ge b(\mathbf{n})) &\le \sum_{k=1}^{\infty} h(k)\,\mathsf{P}(|Y| \ge k-1) \\
&= \sum_{k=1}^{\infty} H(k)\,\mathsf{P}(k-1 \le |Y| < k) \\
&\le \text{const} \sum_{k=1}^{\infty} k^{1/v} \left(\log^+ k\right)^{d_1-1} \mathsf{P}(k-1 \le |Y| < k).
\end{aligned}$$

We used the upper bound in Proposition A.20 for the latter inequality. Thus

$$\sum_{\mathbf{n}\in\mathbf{N}^d} \mathsf{P}(|Y| \ge b(\mathbf{n})) \le \text{const}\,\mathsf{E}\left[|Y|^{1/v}\left(\log^+ |Y|\right)^{d_1-1}\right].$$

The lower bound is obtained similarly:

$$\begin{aligned}
\sum_{\mathbf{n}\in\mathbf{N}^d} \mathsf{P}(|Y| \ge b(\mathbf{n})) &\ge \sum_{k=1}^{\infty} h(k)\,\mathsf{P}(|Y| \ge k) \\
&= \sum_{k=1}^{\infty} H(k)\,\mathsf{P}(k \le |Y| < k+1) \\
&\ge \text{const} \sum_{k=1}^{\infty} k^{1/v} \left(\log^+ k\right)^{d_1-1} \mathsf{P}(k \le |Y| < k+1).
\end{aligned}$$

Here we used the lower bound in Proposition A.20. Therefore

$$\sum_{\mathbf{n}\in\mathbf{N}^d} \mathsf{P}(|Y| \ge b(\mathbf{n})) \ge \text{const}\,\mathsf{E}\left[|Y|^{1/v}\left(\log^+ |Y|\right)^{d_1-1}\right],$$

that is, conditions (5.29) and (5.35) are indeed equivalent. □

Lemma 5.5 *For all* $v_1 > \frac{1}{2}, \dots, v_d > \frac{1}{2}$,

$$(5.35) \implies (5.37). \tag{5.39}$$

Proof of Lemma 5.5 Applying the partial summation formula for $v > \frac{1}{2}$ we get

$$\begin{aligned}\sum_{k=n}^{\infty} \frac{h(k)}{k^2} &= \sum_{k=n}^{\infty} \frac{H(k) - H(k-1)}{k^2} \le \sum_{k=n}^{\infty} H(k) \left[\frac{1}{k^2} - \frac{1}{(k+1)^2}\right] \\ &\le 2 \sum_{k=n}^{\infty} \frac{H(k)}{k^3} \le \text{const} \sum_{k=n}^{\infty} \frac{k^{1/v} \left(\log^+ k\right)^{d_1 - 1}}{k^3} \\ &= O\left(\frac{H(n)}{n^2}\right),\end{aligned}$$

where we have used Proposition A.20 and the direct half of Karamata's theorem for sums (Proposition A.14). Thus

$$\begin{aligned}\sum_{\mathbf{n} \in \mathbf{N}^d} \frac{1}{b^2(\mathbf{n})} \int_{|x|<b(\mathbf{n})} x^2 \, dF(x) &\le \text{const} + \sum_{k=2}^{\infty} \frac{h(k)}{(k-1)^2} \int_{|x|<k} x^2 \, dF(x) \\ &= O\left(1 + \sum_{k=1}^{\infty} H(k) \, \mathsf{P}(k \le |Y| < k+1)\right) \\ &= O\left(1 + \sum_{k=1}^{\infty} h(k) \, \mathsf{P}(|Y| \ge k)\right) \\ &= O\left(1 + \sum_{\mathbf{n} \in \mathbf{N}^d} \mathsf{P}(|Y| \ge b(\mathbf{n}))\right),\end{aligned}$$

that is, condition (5.37) indeed follows from (5.35). □

Now we turn back to the proof of Corollary 5.5 for the case of a nondegenerate distribution function F. First we prove statement (i). Our current goal is to prove the implication

$$(5.29) \implies (5.36). \tag{5.40}$$

It is clear that

$$\begin{aligned}\left|\sum_{\mathbf{n} \in \mathbf{N}^d} \frac{1}{b(\mathbf{n})} \int_{|x|<b(\mathbf{n})} x \, dF(x)\right| &\le \sum_{\mathbf{n} \in \mathbf{N}^d} \frac{1}{b(\mathbf{n})} \int_{|x|<b(\mathbf{n})} |x| \, dF(x) \\ &\le \int_{|x|<1} |x| \, dF(x) + \sum_{k=2}^{\infty} \frac{h(k)}{k-1} \int_{|x|<k} |x| \, dF(x) \\ &= \int_{|x|<1} |x| \, dF(x) \\ &\quad + \sum_{j=2}^{\infty} \int_{j-1 \le |x| < j} |x| \, dF(x) \sum_{k=j}^{\infty} \frac{h(k)}{k-1}.\end{aligned}$$

Recall that we are dealing with the case $v > 1$ where

$$\sum_{k=n}^{\infty} \frac{h(k)}{k} = \lim_{m\to\infty} \left[\sum_{k=n}^{m} \frac{H(k)}{k} - \sum_{k=n}^{m} \frac{H(k-1)}{k} \right] \\ \leq \sum_{k=n}^{\infty} \frac{H(k)}{k(k-1)} = O\left(\frac{H(n)}{n}\right)$$

by Proposition A.20, and Karamata's theorem for sums (Proposition A.14). Thus

$$\left| \sum_{\mathbf{n}\in\mathbf{N}^d} \frac{1}{b(\mathbf{n})} \int_{|x|<b(\mathbf{n})} x\, dF(x) \right| \leq O\left(1 + \sum_{k=2}^{\infty} H(k)\, \mathsf{P}(k-1 \leq |Y| < k)\right) \\ = O\left(|Y|^{1/v} \left(\log^+ |Y|\right)^{d_1 - 1}\right)$$

and we conclude that conditions (5.35)–(5.37) follow from (5.29) in view of (5.38)–(5.40). On the other hand if conditions (5.35)–(5.37) hold, then (5.38) implies (5.29). Part (i) is completely proved.

Next we prove statement (ii). Since condition (5.30) is just another way to write condition (5.36), the implications (5.38)–(5.39) imply that (5.29)–(5.30) $\Longleftrightarrow$ (5.35)–(5.37) by Lemma 5.3. Part (ii) is also proved.

Now we prove statement (iii). Assume that conditions (5.29) and (5.31) hold. Then (5.35) follows from (5.38), while (5.37) follows from (5.39). It remains to prove that condition (5.36) holds, too. Relation (5.31) implies that, for all $A > 0$,

$$\left| \int_{|x|<A} x\, dF(x) \right| \leq \int_{|x|\geq A} |x|\, dF(x).$$

Hence

$$\left| \sum_{\mathbf{n}\in\mathbf{N}^d} \frac{1}{b(\mathbf{n})} \int_{|x|<b(\mathbf{n})} x\, dF(x) \right| \leq \sum_{\mathbf{n}\in\mathbf{N}^d} \frac{1}{b(\mathbf{n})} \int_{|x|\geq b(\mathbf{n})} |x|\, dF(x) \\ \leq \int_{|x|>1} |x|\, dF(x) + \sum_{k=2}^{\infty} \frac{h(k)}{k-1} \int_{|x|\geq k-1} |x|\, dF(x).$$

As above, we apply Propositions A.20 and A.14:

$$\sum_{k=1}^{n} \frac{h(k)}{k} = \sum_{k=1}^{n} \frac{H(k) - H(k-1)}{k} = \frac{H(n)}{n} + \sum_{k=1}^{n-1} \frac{H(k)}{k(k-1)} = O\left(\frac{H(n)}{n}\right).$$

Thus

$$\left|\sum_{\mathbf{n}\in\mathbf{N}^d}\frac{1}{b(\mathbf{n})}\int_{|x|<b(\mathbf{n})} x\,dF(x)\right| \le O\left(1+\sum_{k=1}^{\infty} H(k)\,\mathsf{P}(k\le |Y| < k+1)\right)$$
$$\le O\left(\mathsf{E}\left[|Y|^{1/v}\left(\log^+|Y|\right)\right)^{d-1}\right).$$

Therefore, condition (5.36) follows from (5.29) and series (5.28) converges in view of Theorem 5.8.

Now let series (5.28) converge almost surely. Then Theorem 5.8 implies conditions (5.35)–(5.37). Condition (5.29) also holds by (5.38), whence we deduce that $\mathsf{E}\,[|Y|] < \infty$, since $v < 1$.

Put $a = \mathsf{E}\,[Y]$. The part of statement (iii) already proved implies the almost sure convergence of the series $\sum b(\mathbf{n})(Y(\mathbf{n}) - a)$, since $\mathsf{E}\,[Y(\mathbf{n}) - a] = 0$ and

$$\mathsf{E}\left[|Y(\mathbf{n}) - a|\left(\log^+|Y(\mathbf{n}) - a|\right)^{d-1}\right] < \infty$$

even in the case when $\mathsf{E}\,[Y] \ne 0$. Thus we conclude that the series $\sum b(\mathbf{n})a$ converges. Since $v < 1$, this may happen only if $a = 0$. Note that the proof of this result does not use the restriction $v > \frac{1}{2}$ and thus the result holds for $v \le \frac{1}{2}$, as well.

It remains to consider the case of $v \le \frac{1}{2}$ to complete the proof of Corollary 5.5. If series (5.28) converges almost surely, then we establish from Theorem 5.8 that the series of symmetrizations also converges almost surely, that is

$$\text{the series } \sum_{\mathbf{n}\in\mathbf{N}^d}\frac{Y^{(s)}(\mathbf{n})}{b(\mathbf{n})} \text{ converges almost surely.}$$

Corollary 5.2 implies that

$$\sum_{\mathbf{n}\in\mathbf{N}^d}\frac{1}{b^2(\mathbf{n})}\int_{|x|<b(\mathbf{n})} x^2\,dF^{(s)}(x) < \infty,$$

where $F^{(s)}$ is the distribution function of the random variables $Y^{(s)}(\mathbf{n})$. Thus

$$\sum_{\mathbf{n}\in\mathbf{N}^d}\frac{1}{b^2(\mathbf{n})}\int_{|x|<b(\mathbf{n})} x^2\,dF^{(s)}(x) \ge \sum_{k=1}^{\infty}\frac{h(k)}{k^2}\int_{|x|<k} x^2\,dF^{(s)}(x)$$
$$= \sum_{j=1}^{\infty}\int_{j-1\le|x|<j} x^2\,dF^{(s)}(x)\sum_{k=j}^{\infty}\frac{h(k)}{k^2},$$

that is

$$\text{the series } \sum_{k=1}^{\infty} \frac{h(k)}{k^2} \text{ converges,}$$

since the distribution function $F^{(s)}$ is not concentrated at 0 by assumption of Corollary 5.5. This implies that

$$\text{the series } \sum_{k=1}^{\infty} \frac{H(k)}{k^3} \text{ converges.} \tag{5.41}$$

On the other hand, Proposition A.20 yields $H(k) \ge \operatorname{const} k^{1/v} \ge \operatorname{const} k^2$ and series (5.29) cannot converge. This means that the assumption that series (5.28) converges almost surely is wrong.

The proof of Corollary 5.5 is complete.

Proof of Corollary 5.6 The proof is the same as that of Corollary 5.5. The difference is that now we have to refer to Proposition A.21 rather than to Proposition A.20. □

5.5 Essential Convergence

Definition 5.1 The multi-indexed series (5.2) is called *essentially convergent* if there exists a non-random field $\{d(\mathbf{n}), \mathbf{n} \in \mathbf{N}^d\}$ such that

$$\text{the series } \sum_{\mathbf{n}\in\mathbf{N}^d} [X(\mathbf{n}) - d(\mathbf{n})] \text{ converges almost surely.} \tag{5.42}$$

A criterion for essential convergence can be derived from Theorem 5.6 in terms of the convergence of some non-random series.

Theorem 5.9 *Let $\{X(\mathbf{n}), \mathbf{n} \in \mathbf{N}^d\}$ be a field of independent random variables. Then the property*

$$\textit{the series } \sum_{\mathbf{n}\in\mathbf{N}^d} X(\mathbf{n}) \textit{ converges essentially} \tag{5.43}$$

is equivalent to each of the following two conditions:

(i) *the series $\sum_{\mathbf{n}\in\mathbf{N}^d} X^{(s)}(\mathbf{n})$ converges almost surely*;
(ii) *for some $c > 0$, the series*

$$\sum_{\mathbf{n}\in\mathbf{N}^d} \mathsf{P}(|X(\mathbf{n}) - \mu(\mathbf{n})| \ge c), \qquad \sum_{\mathbf{n}\in\mathbf{N}^d} \operatorname{var}\left[(X(\mathbf{n}) - \mu(\mathbf{n}))^c\right]$$

converge where $\mu(\mathbf{n}) \stackrel{\text{def}}{=} \mu(X(\mathbf{n}))$.

Moreover, the result is still true if the expression "for all $c > 0$" substitutes "for some $c > 0$".

Proof of Theorem 5.9 Below is the diagram of the proof

$$\text{(ii)} \Longrightarrow (5.43) \Longrightarrow \text{(i)} \Longrightarrow \text{(ii)}.$$

Proof of implication (ii) $\Longrightarrow$ (5.43). Let $Y(\mathbf{n})$ denote the random variable $X(\mathbf{n})$ shifted by a median, $Y(\mathbf{n}) \stackrel{\text{def}}{=} X(\mathbf{n}) - \mu(\mathbf{n})$. Since $\sum \operatorname{var}\left[Y^c(\mathbf{n})\right] < \infty$, the series $\sum \left(Y^c(\mathbf{n}) - \mathsf{E}\left[Y^c(\mathbf{n})\right]\right)$ converges almost surely by the two series theorem (see Theorem 5.7). Moreover, $\mathsf{P}(Y(\mathbf{n}) \neq Y^c(\mathbf{n}) \text{ i. o.}) = 0$ by the Borel–Cantelli lemma in view of the convergence of the series $\sum \mathsf{P}(|Y(\mathbf{n})| \geq c)$. This means that the series $\sum \left(Y(\mathbf{n}) - \mathsf{E}\left[Y^c(\mathbf{n})\right]\right)$ converges almost surely, whence (5.42) follows for the constants $d(\mathbf{n}) = \mu(\mathbf{n}) - \mathsf{E}\left[(X(\mathbf{n}) - \mu(\mathbf{n}))^c\right]$.

Proof of implication (5.43) $\Longrightarrow$ (i). According to the definition of essential convergence, there exists a non-random field $\{d(\mathbf{n})\}$ such that the series $\sum[X(\mathbf{n}) - d(\mathbf{n})]$ converges almost surely. Thus we conclude that the series $\sum[X(\mathbf{n}) - d(\mathbf{n})]^{(s)}$ converges almost surely. Since $[X(\mathbf{n}) - d(\mathbf{n})]^{(s)} = X^{(s)}(\mathbf{n})$, statement (i) follows from (5.43).

Proof of implication (i) $\Longrightarrow$ (ii). This result has already been proved in the course of the proof of Theorem 5.4 (see Remark 5.3). □

5.6 Convergence of Multi-Indexed Series on Subsets

Definition 5.2 Let M be some subset of $\mathbf{N}^d$, $M \subseteq \mathbf{N}^d$. The restriction of a field $\{x(\mathbf{n}), \mathbf{n} \in \mathbf{N}^d\}$ to a subset $M \subseteq \mathbf{N}^d$ is called a *subfield* and is denoted by

$$x_M(\mathbf{n}) = \begin{cases} x(\mathbf{n}), & \text{if } \mathbf{n} \in M, \\ 0, & \text{if } \mathbf{n} \notin M. \end{cases}$$

We say that the series $\sum_{\mathbf{n}\in M} x(\mathbf{n})$ converges, if the series $\sum_{\mathbf{n}\in\mathbf{N}^d} x_M(\mathbf{n})$ converges. In the case of convergence, we put $\sum_{\mathbf{n}\in M} x(\mathbf{n}) = \sum_{\mathbf{n}\in\mathbf{N}^d} x_M(\mathbf{n})$.

Theorem 5.10 *Let $\{X(\mathbf{n}), \mathbf{n} \in \mathbf{N}^d\}$ be a field of independent random variables.*

(i) *If, for every subset $M \subseteq \mathbf{N}^d$,*

$$\textit{the series } \sum_{\mathbf{n}\in M} X(\mathbf{n}) \textit{ converges almost surely,} \tag{5.44}$$

then, for all $c > 0$, conditions (5.7) *and* (5.9) *hold and, for all $c > 0$,*

$$\sum_{\mathbf{n}\in\mathbf{N}^d} \left|\mathsf{E}\left[X^c(\mathbf{n})\right]\right| < \infty. \tag{5.45}$$

(ii) *If, for some* $c > 0$*, conditions* (5.7), (5.9)*, and* (5.45) *hold, then, for every subset* $M \subseteq \mathbf{N}^d$*, condition* (5.44) *holds.*

Corollary 5.7 *Let* $\{X(\mathbf{n}), \mathbf{n} \in \mathbf{N}^d\}$ *be a field of independent random variables. Condition* (5.44) *for all subsets* $M \subseteq \mathbf{N}^d$ *is equivalent to for all* $c > 0$*, conditions* (5.7), (5.9)*, and* (5.45) *hold.*

Proof of Theorem 5.10 First we prove statement (i). We start with the proof of the implication

$$(5.44) \implies (5.6). \tag{5.46}$$

Assume that this is not the case. Then there exist an infinite set of multi-indices $M = \{\mathbf{n}\}$ and numbers $\varepsilon_0 > 0$ and $\delta_0 > 0$ such that $\mathsf{P}(|X(\mathbf{n})| \geq \varepsilon_0) \geq \delta_0$ for $\mathbf{n} \in M$. The set M can be chosen to be completely ordered. Then, for all its members $\mathbf{m}$ and $\mathbf{n}$, either $\mathbf{m} \leq \mathbf{n}$ or $\mathbf{n} \leq \mathbf{m}$. We write the members of the set M in ascending order, that is, $M = \{\mathbf{n}_1, \mathbf{n}_2, \dots\}$ and $\mathbf{n}_k \leq \mathbf{n}_{k+1}$.

By the assumption of the theorem, the series $\sum_{\mathbf{n}\in M} X(\mathbf{n}) = \sum_{k=1}^{\infty} X(\mathbf{n}_k)$ converges almost surely. Thus the classical three series theorem (see Theorem 5.1) implies the convergence of the series $\sum_{k=1}^{\infty} \mathsf{P}(|X(\mathbf{n}_k)| \geq \varepsilon)$ for all $\varepsilon > 0$ which contradicts our assumption (5.46).

Thus condition (5.6) holds. Now Theorem 5.3 implies that all three conditions (5.7)–(5.9) hold. Fix $c > 0$ and consider the following two subsets of $\mathbf{N}^d$

$$M_1 = \{\mathbf{n}\colon \mathsf{E}\left[X^c(\mathbf{n})\right] > 0\} \quad \text{and} \quad M_2 = \{\mathbf{n}\colon \mathsf{E}\left[X^c(\mathbf{n})\right] \leq 0\}.$$

Then both series $\sum_{\mathbf{n}\in M_1} X(\mathbf{n})$ and $\sum_{\mathbf{n}\in M_2} X(\mathbf{n})$ converge almost surely by the assumption of the theorem. Since condition (5.6) holds, Theorem 5.3 implies that conditions (5.7) and (5.9) hold for each of the subsets M_1 and M_2. Therefore the series

$$\sum_{\mathbf{n}\in M_1} \mathsf{E}\left[X^c(\mathbf{n})\right] \quad \text{and} \quad \sum_{\mathbf{n}\in M_2} \mathsf{E}\left[X^c(\mathbf{n})\right]$$

converge, whence condition (5.45) follows.

Statement (ii) is a direct consequence of Theorem 5.3. □

5.7 Convergence of Permutations

If $\lambda : \mathbf{N}^d \to \mathbf{N}^d$ is a one-to-one mapping, then the field $\{\lambda(\mathbf{n}), \mathbf{n} \in \mathbf{N}^d\}$ is called a *permutation* in $\mathbf{N}^d$.

Theorem 5.11 *Let* $\{X(\mathbf{n}), \mathbf{n} \in \mathbf{N}^d\}$ *be a field of independent random variables and let* $\{\lambda(\mathbf{n}), \mathbf{n} \in \mathbf{N}^d\}$ *be a permutation in* $\mathbf{N}^d$. *If conditions* (5.7), (5.9) *and* (5.45) *hold for some* $c > 0$, *then the series*

$$\sum_{\mathbf{n}\in\mathbf{N}^d} X(\lambda(\mathbf{n}))$$

converges almost surely for all permutations $\{\lambda(\mathbf{n}), \mathbf{n} \in \mathbf{N}^d\}$ *in* $\mathbf{N}^d$. *Moreover,*

$$\sum_{\mathbf{n}\in\mathbf{N}^d} X(\lambda(\mathbf{n})) = \sum_{\mathbf{n}\in\mathbf{N}^d} X(\mathbf{n}). \tag{5.47}$$

Proof of Theorem 5.11 The series $\sum_{\mathbf{n}\in\mathbf{N}^d} X(\lambda(\mathbf{n}))$ converges almost surely by conditions (5.7), (5.9), (5.45) and Theorem 5.3.

It remains to prove equality (5.47). Condition (5.7) together with the Borel–Cantelli lemma (see Proposition A.30) implies that, given an arbitrary $c > 0$, only a finite number of the random variables $X(\mathbf{n}) - X^c(\mathbf{n})$ (and of the random variables $X(\lambda(\mathbf{n})) - X^c(\lambda(\mathbf{n}))$ is different from zero. Thus

$$\sum_{\mathbf{n}\in\mathbf{N}^d} [X(\mathbf{n}) - X^c(\mathbf{n})] = \sum_{\mathbf{n}\in\mathbf{N}^d} [X(\lambda(\mathbf{n})) - X^c(\lambda(\mathbf{n}))] \tag{5.48}$$

almost surely, since $X(\mathbf{n}) - X^c(\mathbf{n}) \neq 0$ if and only if $X(\lambda(\mathbf{n})) - X^c(\lambda(\mathbf{n}) \neq 0$.

Since the series $\sum_{\mathbf{n}\in\mathbf{N}^d} \mathsf{E}\left[X^c(\mathbf{n})\right]$ converges absolutely,

$$\sum_{\mathbf{n}\in\mathbf{N}^d} \mathsf{E}\left[X^c(\mathbf{n})\right] = \sum_{\mathbf{n}\in\mathbf{N}^d} \mathsf{E}\left[X^c(\lambda(\mathbf{n}))\right]. \tag{5.49}$$

Finally, we prove that

$$\sum_{\mathbf{n}\in\mathbf{N}^d} \left(X^c(\mathbf{n}) - \mathsf{E}\left[X^c(\mathbf{n})\right]\right) = \sum_{\mathbf{n}\in\mathbf{N}^d} \left(X^c(\lambda(\mathbf{n})) - \mathsf{E}\left[X^c(\lambda(\mathbf{n}))\right]\right) \tag{5.50}$$

almost surely; then (5.48)–(5.50) would imply (5.47). Put

$$\zeta(\mathbf{n}) = \sum_{\mathbf{k}\le\mathbf{n}} \left(X^c(\mathbf{k}) - \mathsf{E}\left[X^c(\mathbf{k})\right]\right), \qquad \zeta = \lim\,(\min)\,\zeta(\mathbf{n}),$$

$$\eta(\mathbf{n}) = \sum_{\mathbf{k}\le\mathbf{n}} \left(X^c(\lambda(\mathbf{k})) - \mathsf{E}\left[X^c(\lambda(\mathbf{k}))\right]\right), \qquad \eta = \lim\,(\min)\,\eta(\mathbf{n}).$$

We also put

$$A(\mathbf{n}) = \{\mathbf{k} \colon \mathbf{k} \le \mathbf{n}\}, \qquad B(\mathbf{m}) = \{\lambda(\mathbf{k}) \colon \mathbf{k} \le \mathbf{m}\}.$$

If $B(\mathbf{m}) \subseteq A(\mathbf{n})$, then

$$\mathsf{E}\left[(\zeta(\mathbf{n}) - \eta(\mathbf{m}))^2\right] = \sum_{\mathbf{k} \in A(\mathbf{n}) \setminus B(\mathbf{m})} \mathsf{var}\left[X^c(\mathbf{k})\right],$$

whence

$$\lim_{\mathbf{n}} (\min) \mathsf{E}\left[(\zeta(\mathbf{n}) - \eta(\mathbf{m}))^2\right] = \sum_{\mathbf{k} \notin B(\mathbf{m})} \mathsf{var}\left[X^c(\mathbf{k})\right],$$

$$\lim_{\mathbf{m}} (\min) \lim_{\mathbf{n}} (\min) \mathsf{E}\left[(\zeta(\mathbf{n}) - \eta(\mathbf{m}))^2\right] = 0,$$

since series (5.9) converges. On the other hand,

$$\lim_{\mathbf{n}} (\min) \mathsf{E}\left[(\zeta(\mathbf{n}) - \eta(\mathbf{m}))^2\right] = \mathsf{E}\left[(\zeta - \eta(\mathbf{m}))^2\right],$$

$$\lim_{\mathbf{m}} (\min) \lim_{\mathbf{n}} (\min) \mathsf{E}\left[(\zeta(\mathbf{n}) - \eta(\mathbf{m}))^2\right] = \mathsf{E}\left[(\zeta - \eta)^2\right]$$

by statement (D) of Theorem 5.2. Therefore $\mathsf{E}\left[(\zeta - \eta)^2\right] = 0$, whence we conclude that equality (5.50) holds almost surely. □

5.8 A Generalization of a Theorem of Chung

The following result contains a variety of sufficient conditions for the almost sure convergence of a multi-indexed series. A feature of this result is that an individual condition can be chosen for each multi-index $\mathbf{n}$ independently of others.

Theorem 5.12 *Let $\{X(\mathbf{n}), \mathbf{n} \in \mathbf{N}^d\}$ be a field of independent random variables and let $\{b(\mathbf{n}), \mathbf{n} \in \mathbf{N}^d\}$ be a field of non-random non-negative numbers. Further, let $\{g(\mathbf{n}; x), \mathbf{n} \in \mathbf{N}^d\}$ be a field of non-decreasing positive even functions. Assume that, for every $\mathbf{n} \in \mathbf{N}^d$, at least one of the following conditions hold:*

(i) *$x/g(\mathbf{n}; x)$ is non-decreasing in the domain $x > 0$;*
(ii) *$x/g(\mathbf{n}; x)$ and $g(\mathbf{n}; x)/x^2$ are non-increasing in the domain $x > 0$;*
(iii) *$x/g(\mathbf{n}; x)$ is decreasing in the domain $x > 0$ and $X(\mathbf{n})$ is a symmetric random variable.*

Let

$$\sum_{\mathbf{n} \in \mathbf{N}^d} \frac{\mathsf{E}\left[g(\mathbf{n}; X(\mathbf{n}))\right]}{g(\mathbf{n}; b(\mathbf{n}))} < \infty.$$

Then

(a) *the series*

$$\sum_{\mathbf{n}\in\mathbf{N}^d} \frac{X(\mathbf{n})}{b(\mathbf{n})}$$

converges almost surely;

(b) *the series*

$$\sum_{\mathbf{n}\in M} \frac{X(\mathbf{n})}{b(\mathbf{n})}$$

converges almost surely for all subfields $M \subseteq \mathbf{N}^d$;

(c) *the series*

$$\sum_{\mathbf{n}\in\mathbf{N}^d} \frac{X(\lambda(\mathbf{n}))}{b(\lambda(\mathbf{n}))}$$

converges almost surely for every permutation $\{\lambda(\mathbf{n}), \mathbf{n} \in \mathbf{N}^d\}$; *moreover,*

$$\sum_{\mathbf{n}\in\mathbf{N}^d} \frac{X(\mathbf{n})}{b(\mathbf{n})} = \sum_{\mathbf{n}\in\mathbf{N}^d} \frac{X(\lambda(\mathbf{n}))}{b(\lambda(\mathbf{n}))}$$

almost surely.

Corollary 5.8 *Let* $\{X(\mathbf{n}), \mathbf{n} \in \mathbf{N}^d\}$ *be a field of independent random variables and let* $\{b(\mathbf{n}), \mathbf{n} \in \mathbf{N}^d\}$ *be a field of non-random non-negative numbers. Assume that at least one of the following two conditions holds*:

(i) $0 < p < 1$;

(ii) $1 \le p \le 2$ *and* $\mathsf{E}[X(\mathbf{n})] = 0$ *for all* $\mathbf{n} \in \mathbf{N}^d$.

If

$$\sum_{\mathbf{n}\in\mathbf{N}^d} \frac{\mathsf{E}\left[|X(\mathbf{n})|^p\right]}{b^p(\mathbf{n})} < \infty,$$

then statements (a)–(c) *of Theorem 5.12 prevail.*

Proof of Theorem 5.12 Proposition A.27 implies that

$$\begin{aligned} &\max\left\{\mathsf{P}(|X(\mathbf{n})| \ge b(\mathbf{n})), \left|\frac{\mathsf{E}\left[Y^c(\mathbf{n})\right]}{b(\mathbf{n})}\right|, \left|\frac{\mathsf{E}\left[Y^c(\mathbf{n}) - \mathsf{E}\left[Y^c(\mathbf{n})\right]\right]^2}{b^2(\mathbf{n})}\right|\right\} \\ &\quad \le \frac{\mathsf{E}\left[g(\mathbf{n}; X(\mathbf{n}))\right]}{g(\mathbf{n}; b(\mathbf{n}))} \end{aligned}$$

for all $\mathbf{n} \in \mathbf{N}^d$. This means, in particular, that conditions (5.7)–(5.9) and (5.45) hold with random variables $X(\mathbf{n}) = Y(\mathbf{n})/b(\mathbf{n})$. Thus Theorems 5.10 and 5.11 imply statements (a)–(c) of Theorem 5.12. □

Proof of Corollary 5.8 The function $g(x) = |x|^p$, $0 < p \le 2$, fits the assumptions imposed in Theorem 5.12. Thus Corollary 5.8 follows from Theorem 5.12. □

5.9 Comments

Theorem 5.1 is due to Kolmogorov [254].

Section 5.1 The equivalence (A) $\iff$ (B) in Theorem 5.2 is proved by Lévy in [283] for $d = 1$; the case of $d > 1$ is considered in [210].

Section 5.2 Theorem 5.3 is proved in [210]. A similar result is obtained by Gabriel [115]. Tien and Dung [398] consider the convergence of double sums of random elements assuming values in a Banach space. Quang, Thanh, and Tien [350] study the convergence of double sums of block-wise dependent random variables.

Section 5.3 Theorems 5.4–5.7 are proved in [232]. Theorem 5.4 follows from the essential convergence lemma (see [299]) in the case $d = 1$. Another proof of Corollary 5.4 for $d = 1$ can be found in [178] (Lemma 2, Sect. 3, Chap. 1).

Section 5.4 All results of this section are obtained in [232] for $d > 1$.

Section 5.5 The essential convergence for $d = 1$ is studied in the book by Loéve [299]; the case of $d > 1$ is treated in [232].

Section 5.6 Results concerning the convergence of subsequences for $d = 1$ can be found in Skorokhod [371] and Loéve [299]; the case of $d > 1$ is studied in [232].

Section 5.7 Theorem 5.11 is proved by Skorokhod [371] in the case $d = 1$; the case of $d > 1$ in Theorem 5.11 is proved in [232]. The convergence of a series resulting from a permutation of a Fourier series for $d = 1$ appeared in the paper by Kolmogorov [252] (the modern state of the art is given by Levental, Mandrekar, and Chobanyan [280]).

Section 5.8 Cases (i) and (ii) in Theorem 5.12 are proved by Chung [63] for $d = 1$ and $g_n(x) = g(x)$; case (iii) is considered by Petrov [338]; the general case of $d > 1$ is obtained in [232].

Some results for the almost sure convergence of multi-indexed series of non-orthogonal random variables are established by Yas'kov [429].

Chapter 6
Boundedness of Multi-Indexed Series of Independent Random Variables

It is a trivial result for $d = 1$ that if a series converges, then the sequence of its partial sums is bounded. In contrast, this is not the case if $d \geq 2$ (see Example 6.1). The aim of this chapter is to find necessary and sufficient conditions for the boundedness of partial sums of a multi-indexed series and to establish a relationship between the convergence and boundedness for all $d \geq 1$.

For $d = 1$, if the terms of a series are independent random variables, then a relationship between the convergence and boundedness is described in the following assertion.

Theorem 6.1 (A.V. Skorokhod) *Let $\{X_n, n \geq 1\}$ be independent random variables. If the sequence of partial sums of the series $\sum X_n$ is bounded almost surely, then there exists a non-random sequence $\{a_n, n \geq 1\}$ such that the series $\sum [X_n - a_n]$ converges almost surely and the sums $\sum_{k \leq n} a_k$, $n \geq 1$, are bounded.*

The decomposition of a bounded series of independent random variables into two parts, a convergent random series and bounded non-random sequence, is called the *Skorokhod decomposition*.

A similar decomposition into random and non-random parts is valid for the case $d > 1$ (see Theorems 6.4 and 6.5), however the non-random part is not necessarily bounded. We try to find those cases where this is true even for $d > 1$.

In Sect. 6.1, we provide necessary definitions and auxiliary results. In Sect. 6.2, we consider a relationship between the boundedness and convergence in probability of multi-indexed series of independent random variables. We also prove that these two properties are equivalent for symmetric terms.

We find an analog of the Skorokhod decomposition for the case $d > 1$ in Sect. 6.3.

We establish the equivalence of boundedness in probability and almost sure boundedness in Sect. 6.4. Necessary and sufficient conditions for almost sure boundedness are also obtained in Sect. 6.4.

Based on the preceding results, we derive necessary and sufficient conditions for the so-called bounded convergence of multi-indexed series in Sect. 6.5.

O. Klesov, *Limit Theorems for Multi-Indexed Sums of Random Variables*,
Probability Theory and Stochastic Modelling 71, DOI 10.1007/978-3-662-44388-0_6

6.1 Definitions and Auxiliary Results

Throughout this chapter we consider a field $\{X(\mathbf{n}), \mathbf{n} \in \mathbf{N}^d\}$ of independent random variables and the field of their sums $\{S(\mathbf{n}), \mathbf{n} \in \mathbf{N}^d\}$: $S(n) = \sum_{\mathbf{k} \le \mathbf{n}} X(\mathbf{k})$.

Recall the definitions of a symmetrization (see Definition A.13) and a truncation. The field of random variables $\{X^{(s)}(\mathbf{n}), \mathbf{n} \in \mathbf{N}^d\}$ is called the *symmetrization of a field* $\{X(\mathbf{n}), \mathbf{n} \in \mathbf{N}^d\}$ if $X^{(s)}(\mathbf{n}) = X(\mathbf{n}) - X'(\mathbf{n})$ for all $\mathbf{n} \in \mathbf{N}^d$, where $\{X'(\mathbf{n}), \mathbf{n} \in \mathbf{N}^d\}$ is an independent copy of the field $\{X'(\mathbf{n}), \mathbf{n} \in \mathbf{N}^d\}$.

The random variable $X^{\varepsilon}(\mathbf{n}) = X(\mathbf{n})\mathbb{I}(|X(\mathbf{n})| < \varepsilon)$ is called the *truncation* of a random variable X at a level $\varepsilon > 0$, where $\mathbb{I}(A)$ is the indicator of a random event A.

Definition 6.1 A field of random variables $\{S(\mathbf{n}), \mathbf{n} \in \mathbf{N}^d\}$ is said to be *bounded in probability* if

$$\lim_{t\to\infty} \sup_{\mathbf{n}\in\mathbf{N}^d} \mathsf{P}(|S(\mathbf{n})| \ge t) = 0. \tag{6.1}$$

Definition 6.2 A field of random variables $\{S(\mathbf{n}), \mathbf{n} \in \mathbf{N}^d\}$ is said to be *bounded almost surely* if

$$\lim_{t\to\infty} \mathsf{P}\left(\sup_{\mathbf{n}\in\mathbf{N}^d} |S(\mathbf{n})| \ge t \right) = 0. \tag{6.2}$$

It is clear that almost sure boundedness implies boundedness in probability for every field $\{S(\mathbf{n})\}$ of random variables. For easy references we state this result below.

Lemma 6.1 *If a field of random variables $\{S(\mathbf{n}), \mathbf{n} \in \mathbf{N}^d\}$ is bounded almost surely, then it is bounded in probability.*

The next result is easy; nevertheless it helps in several proofs throughout this section.

Lemma 6.2 *Let a field $\{S(\mathbf{n})\}$ be bounded in probability (almost surely). If only a finite number of terms of the field $\{S'(\mathbf{n})\}$ differ from those of $\{S(\mathbf{n})\}$, then $\{S'(\mathbf{n})\}$ is bounded in probability (almost surely), as well.*

Finally, we recall the definition of a convergent multi-indexed series (see Definition A.7). As mentioned in Remark A.3, the convergence of a multi-indexed series is defined only in the lim (min) sense and not in the lim (max) sense.

Definition 6.3 A random field $\{S(\mathbf{n}), \mathbf{n} \in \mathbf{N}^d\}$ is said to be *almost surely convergent* if there exists a random variable S such that

$$\lim_{\substack{n_1\to\infty \\ \cdots \\ n_d\to\infty}} S(\mathbf{n}) = S$$

almost surely, where $\mathbf{n} = (n_1, \dots, n_d)$. We say that a series $\sum_{\mathbf{n}\in\mathbf{N}^d} X(\mathbf{n})$ converges almost surely if a field $\{S(\mathbf{n}), \mathbf{n} \in \mathbf{N}^d\}$ converges almost surely.

One of the nice properties of a series for $d = 1$ is that convergence implies boundedness. The following example exhibits a convergent field which is unbounded for $d = 2$. An analogous example can be constructed for all $d > 1$. We stress that the field in Example 6.1 is even non-random.

Example 6.1 Let $d = 2$ and let

$$X(m,n) = \begin{cases} (-1)^{n-1}m, & \text{for } n \le 2, \\ 0, & \text{for } n > 2. \end{cases}$$

It is clear that $S(m,n) = 0$ for $n \ge 2$ and all $m \ge 1$. Thus the limit of double sums $S(m,n)$ as $m, n \to \infty$ equals 0, that is the series converges almost surely. On the other hand,

$$\sup_{m,n\ge 1} \mathsf{P}(|S(m,n)| \ge t) \ge \sup_{m \ge 1} \mathsf{P}(|S(m,1)| \ge t) = 1$$

for all $t > 0$, whence we conclude that the series is not bounded (even in probability).

The terms $X(m,n)$ are non-random in Example 6.1. It turns out that such a phenomenon does not occur for "truly random" series. Moreover, every convergent series can be "adjusted" by non-random constants in such a way that the resulting series is bounded.

Prior to establishing the main results of this section, we consider some auxiliary assertions which are of interest in themselves.

Lemma 6.3 *If a field $\{S(\mathbf{n}), \mathbf{n} \in \mathbf{N}^d\}$ is bounded in probability, then the field $\{X(\mathbf{n}), \mathbf{n} \in \mathbf{N}^d\}$ is bounded in probability, as well, where $X(\mathbf{n}) = \Delta[S(\mathbf{n})]$.*

Proof Since every random variable $X(\mathbf{n})$, $\mathbf{n} \in \mathbf{N}^d$, is a linear combination of at most 2^d sums $S(\mathbf{m})$ (see Proposition A.5), we get

$$\sup_{\mathbf{n}\in\mathbf{N}^d} \mathsf{P}(|X(\mathbf{n})| \ge t) \le 2^d \sup_{\mathbf{n}\in\mathbf{N}^d} \mathsf{P}(|S(\mathbf{n})| \ge t2^{-d}).$$

Letting $t \to \infty$ we complete the proof of Lemma 6.3. □

Lemma 6.4 *Let a series $\sum_{\mathbf{n}\in\mathbf{N}^d} \mathsf{P}(|X(\mathbf{n})| \ge c)$ converge for some $c > 0$. Then the series $\sum_{\mathbf{n}\in\mathbf{N}^d} \left[X(\mathbf{n}) - X^c(\mathbf{n})\right]$ converges almost surely.*

Proof The random events $\{\omega\colon X(\mathbf{n}) \ne X^c(\mathbf{n})\}$, $\mathbf{n} \in \mathbf{N}^d$, are jointly independent. Thus the Borel–Cantelli lemma implies that $\mathsf{P}(X(\mathbf{n}) \ne X^c(\mathbf{n}) \text{ i. o.}) = 0$. This means that only a finite number of terms of the series $\sum_{\mathbf{n}\in\mathbf{N}^d} \left[X(\mathbf{n}) - X^c(\mathbf{n})\right]$ are non-zero and this proves Lemma 6.4. □

Lemma 6.5 *Let*

$$\mathsf{P}(|X(\mathbf{n})| < c) = 1$$

for some $c > 0$ and all $\mathbf{n} \in \mathbf{N}^d$. Assume that the field $\{S(\mathbf{n}), \mathbf{n} \in \mathbf{N}^d\}$ is bounded in probability. Then

(i) *the field of expectations* $\{\mathsf{E}[S(\mathbf{n})], \mathbf{n} \in \mathbf{N}^d\}$ *is bounded*;
(ii) *the series of variances* $\sum_{\mathbf{n}\in\mathbf{N}^d} \mathsf{E}\left[(X(\mathbf{n}) - \mathsf{E}[X(\mathbf{n})])^2\right]$ *converges*;
(iii) *if the series* $\sum X(\mathbf{n})$ *converges almost surely, then the series* $\sum \mathsf{E}[X(\mathbf{n})]$ *converges.*

Proof Since the field $\{S(\mathbf{n}), \mathbf{n} \in \mathbf{N}^d\}$ is bounded in probability, one can find a number $t_0 > 0$ such that $\mathsf{P}(|S(\mathbf{n})| \ge t_0) \le 1/(8e)$ for all $\mathbf{n} \in \mathbf{N}^d$. Applying Skorokhod's inequality for $r = 1$ (Theorem 2.13) we find a number $q_1 > 0$ that depends on t_0 and c such that

$$\mathsf{E}[|S(\mathbf{n})|] \le q_1 \quad \text{for all } \mathbf{n} \in \mathbf{N}^d.$$

This proves statement (i).

Now we use the same Skorokhod inequality with $r = 2$, and find another number $q_2 > 0$ that depends on t_0 and c such that

$$\mathsf{E}\left[S^2(\mathbf{n})\right] \le q_2 \quad \text{for all } \mathbf{n} \in \mathbf{N}^d,$$

whence $\mathsf{E}\left[(S(\mathbf{n}) - \mathsf{E}[S(\mathbf{n})])^2\right] \le q_2$. Therefore statement (ii) is also proved.

Finally, using statement (ii) and the two series theorem for multi-indexed series (see Theorem 5.7), we obtain that the series $\sum_{\mathbf{n}\in\mathbf{N}^d} (X(\mathbf{n}) - \mathsf{E}[X(\mathbf{n})])$ converges almost surely, that is, statement (iii) follows. □

Lemma 6.6 *Let* $\{X(\mathbf{n}), \mathbf{n} \in \mathbf{N}^d\}$ *be a field of independent random variables. If*

$$\sum_{\mathbf{n}\in\mathbf{N}^d} \mathsf{P}(|X(\mathbf{n})| \ge c) < \infty \tag{6.3}$$

for some $c = c_0 > 0$, *then*

$$\inf_{c>0}\left[\sum_{\mathbf{n}\in\mathbf{N}^d} \mathsf{P}(|X(\mathbf{n})| \ge c)\right] = \lim_{c\to\infty}\left[\sum_{\mathbf{n}\in\mathbf{N}^d} \mathsf{P}(|X(\mathbf{n})| \ge c)\right] = 0. \tag{6.4}$$

Proof Denote the sum of the series on the right-hand side of (6.3) by $f(c)$. Since the terms of this series are non-negative, one can write them in an arbitrary order and the sum does not change. As usual, we write the terms of series (6.3) according to the lexicographical order $\preccurlyeq$ (see A.1.2). Denote the terms of the resulting series by $f_n(c)$. Since the series $f(c) = \sum f_n(c)$ converges uniformly, we conclude

$$\lim_{c\to\infty} f(c) = \sum_{n=1}^{\infty} \lim_{c\to\infty} f_n(c) = 0$$

and (6.4) follows. □

Lemma 6.7 *Let* $\{X(\mathbf{n}), \mathbf{n} \in \mathbf{N}^d\}$ *be a field of independent random variables. Assume that, for all* $\varepsilon > 0$,

(i) *both series* $\sum \mathsf{P}(|X(\mathbf{n})| \geq \varepsilon)$ *and* $\sum \mathsf{var}\,[X^\varepsilon(\mathbf{n})]$ *converge,*
(ii) *the series* $\sum \mathsf{E}\,[X^\varepsilon(\mathbf{n})]$ *is bounded.*

Then the series $\sum X(\mathbf{n})$ *is bounded in probability.*

Proof Put $E_0 = \sup_{\mathbf{n}\in\mathbf{N}^d} \left|\sum_{\mathbf{k}\leq\mathbf{n}} \mathsf{E}\,[X^\varepsilon(\mathbf{k})]\right|$. Using the Chebyshev inequality we get

$$\mathsf{P}(|S(\mathbf{n})| \geq t) \leq \sum_{\mathbf{m}\leq\mathbf{n}} \mathsf{P}(|X(\mathbf{m})| \geq \varepsilon) + \mathsf{P}\left\{\left|\sum_{\mathbf{m}\leq\mathbf{n}} X^\varepsilon(\mathbf{m})\right| \geq t\right\} \tag{6.5}$$

$$\leq \sum_{\mathbf{m}\leq\mathbf{n}} \mathsf{P}(|X(\mathbf{m})| \geq \varepsilon) + \frac{1}{(t - E_0)^2} \sum_{\mathbf{m}\leq\mathbf{n}} \mathsf{var}\left[X^\varepsilon(\mathbf{m})\right] \tag{6.6}$$

for all $t > E_0$ and $\mathbf{n} \in \mathbf{N}^d$. Evaluating sup with respect to $\mathbf{n} \in \mathbf{N}^d$ and then passing to the limit as $t \to \infty$ we get

$$\lim_{t\to\infty} \sup_{\mathbf{n}\in\mathbf{N}^d} \mathsf{P}(|S(\mathbf{n})| \geq t) \leq \sum_{\mathbf{m}\in\mathbf{N}^d} \mathsf{P}(|X(\mathbf{m})| \geq \varepsilon).$$

Now evaluating inf with respect to $\varepsilon > 0$, we derive Lemma 6.7 from Lemma 6.6. □

Lemma 6.8 *Let a field* $\{S(\mathbf{n}), \mathbf{n} \in \mathbf{N}^d\}$ *be bounded in probability and let*

$$\sum_{\mathbf{n}\in\mathbf{N}^d} \mathsf{P}(|X^{(s)}(\mathbf{n})| \geq c) < \infty$$

for all $c > 0$. *Then, for sufficiently large* $c > 0$,

(i) *the series* $\sum \mathsf{P}(|X(\mathbf{n})| \geq c)$ *converges,*
(ii) *the series* $\sum \mathsf{E}\left[X^c(\mathbf{n})\right]$ *is bounded,*
(iii) *the series* $\sum \mathsf{var}\left[X^c(\mathbf{n})\right]$ *converges.*

Proof For all $c > 0$,

$$\infty > \sum_{\mathbf{n}\in\mathbf{N}^d} \mathsf{P}(|X^{(s)}(\mathbf{n})| \geq c/2) \geq \sum_{\mathbf{n}\in\mathbf{N}^d} \mathsf{P}(|X(\mathbf{n})| \geq c)\mathsf{P}(|X(\mathbf{n})| < c/2)$$

$$\geq \left[\inf_{\mathbf{n}\in\mathbf{N}^d} \mathsf{P}(|X(\mathbf{n})| < c/2)\right] \sum_{\mathbf{n}\in\mathbf{N}^d} \mathsf{P}(|X(\mathbf{n})| \geq c).$$

According to Lemma 6.3, the field $\{X(\mathbf{n}), \mathbf{n} \in \mathbf{N}^d\}$ is bounded in probability, whence

$$\inf_{\mathbf{n}\in\mathbf{N}^d} \mathsf{P}(|X(\mathbf{n})| < c/2) > 0$$

for sufficiently large c. Thus condition (6.3) holds for sufficiently large c and this proves statement (i).

Further,

$$\begin{aligned}\mathsf{P}\left\{\left|\sum_{\mathbf{m}\le\mathbf{n}} X^c(\mathbf{m})\right| \ge t\right\} &\le \sum_{\mathbf{m}\le\mathbf{n}} \mathsf{P}(|X(\mathbf{m})| \ge c) + \mathsf{P}(|S(\mathbf{n})| \ge t)\\ &\le \sum_{\mathbf{m}\in\mathbf{N}^d} \mathsf{P}(|X(\mathbf{m})| \ge c) + \sup_{\mathbf{m}\in\mathbf{N}^d} \mathsf{P}(|S(\mathbf{m})| \ge t) \qquad (6.7)\end{aligned}$$

for all $c > 0$ and $t > 0$. Statement (i) together with Lemma 6.6 proves that

$$\inf_{c>0} \sum_{\mathbf{n}\in\mathbf{N}^d} \mathsf{P}(|X(\mathbf{n})| \ge c) = 0.$$

Thus the first term on the right-hand side of (6.7) is less than $1/(16e)$ if the constant c is chosen appropriately. Since the field $\{S(\mathbf{n}), \mathbf{n} \in \mathbf{N}^d\}$ is bounded, the second term in (6.7) can also be made less than $1/(16e)$ if t is sufficiently large. Therefore

$$\mathsf{P}\left(\left|\sum_{\mathbf{m}\le\mathbf{n}} X^c(\mathbf{m})\right| \ge t\right) \le \frac{1}{8e}$$

for sufficiently large c and t.

Fix a pair of numbers c and t for which the preceding inequality holds. Using the Skorokhod inequality for $r = 1$ (see Theorem 2.13) we prove that there exists a number $q_1 > 0$ such that

$$\mathsf{E}\left[\left|\sum_{\mathbf{m}\le\mathbf{n}} X^c(\mathbf{m})\right|\right] \le q_1$$

for all $\mathbf{n} \in \mathbf{N}^d$, whence statement (ii) follows.

Using the Skorokhod inequality once more, now for $r = 2$, we prove that there exists a number $q_2 > 0$ such that

$$\mathsf{E}\left[\left(\sum_{\mathbf{m}\le\mathbf{n}} X^c(\mathbf{m})\right)^2\right] \le q_2$$

for all $\mathbf{n} \in \mathbf{N}^d$, whence statement (iii) follows. □

Lemma 6.9 *Let $c > 0$ and $t > 0$. If the random variables $\{X(\mathbf{k}), \mathbf{k} \le \mathbf{n}\}$ are independent and symmetric, then*

$$\mathsf{P}(|S(\mathbf{n}) - S'(\mathbf{n})| \ge t) \le 2\mathsf{P}(|S(\mathbf{n})| \ge t), \tag{6.8}$$

$$\mathsf{P}(|S'(\mathbf{n})| \ge t) \le 2\mathsf{P}(|S(\mathbf{n})| \ge t), \tag{6.9}$$

where $S'(\mathbf{n}) = \sum_{\mathbf{k}\le\mathbf{n}} X^c(\mathbf{k})$.

Proof of both inequalities (6.8) and (6.9) is the same We choose to prove inequality (6.8). It is clear that

$$\mathsf{P}(S(\mathbf{n}) - S'(\mathbf{n}) \ge t) \le \mathsf{P}(S(\mathbf{n}) - S'(\mathbf{n}) \ge t, S'(\mathbf{n}) \ge 0) \\ + \mathsf{P}(S(\mathbf{n}) - S'(\mathbf{n}) \ge t, S'(\mathbf{n}) \le 0).$$

We are going to prove that the two terms on the right-hand side are equal each to other.

Put $\Pi = \{\mathbf{k} : \mathbf{k} \le \mathbf{n}\}$. Let $\mathscr{I} = 2^{\Pi}$ be the set of all subsets of Π. Then

$$\mathsf{P}(S(\mathbf{n}) - S'(\mathbf{n}) \ge t, S'(\mathbf{n}) \le 0) = \sum_{I\in\mathscr{I}} \mathsf{P}(S(\mathbf{n}) - S'(\mathbf{n}) \ge t, S'(\mathbf{n}) \le 0, A_I, B_I),$$

where

$$A_I = \{\omega\colon |X(\mathbf{k})| \ge c, \mathbf{k} \in I\}, \quad B_I = \{\omega\colon |X(\mathbf{k})| < c, \mathbf{k} \notin I\}.$$

Since

$$\{\omega\colon S(\mathbf{n}) - S'(\mathbf{n}) \ge t\} \cap A_I \cap B_I = \left\{\omega\colon \sum_{\mathbf{k}\in I} X(\mathbf{k}) \ge t\right\} \cap A_I \cap B_I,$$

$$\{\omega\colon S'(\mathbf{n}) \le 0\} \cap A_I \cap B_I = \left\{\omega\colon \sum_{\mathbf{k}\notin I} X(\mathbf{k}) \le 0\right\} \cap A_I \cap B_I$$

for all $I \in \mathscr{I}$, we conclude that

$$\begin{aligned} &\mathsf{P}(S(\mathbf{n}) - S'(\mathbf{n}) \ge t, S'(\mathbf{n}) \le 0, A_I, B_I) \\ &\qquad = \mathsf{P}(S(\mathbf{n}) - S'(\mathbf{n}) \ge t, A_I)\mathsf{P}(S'(\mathbf{n}) \le 0, B_I) \\ &\qquad = \mathsf{P}(S(\mathbf{n}) - S'(\mathbf{n}) \ge t, A_I)\mathsf{P}(S'(\mathbf{n}) \ge 0, B_I) \\ &\qquad = \mathsf{P}(S(\mathbf{n}) - S'(\mathbf{n}) \ge t, S'(\mathbf{n}) \ge 0, A_I, B_I) \end{aligned}$$

for all $I \in \mathscr{I}$. Adding together these inequalities with respect to all $I \in \mathscr{I}$, we prove the result desired.

To complete the proof of (6.8), note that the random variables $S(\mathbf{n}) - S'(\mathbf{n})$ and $S(\mathbf{n})$ are symmetric and thus

$$\begin{aligned}\mathsf{P}(|S(\mathbf{n}) - S'(\mathbf{n})| \geq t) &= 2\mathsf{P}(S(\mathbf{n}) - S'(\mathbf{n}) \geq t)\\ &\leq 4\mathsf{P}(S(\mathbf{n}) - S'(\mathbf{n}) \geq t,\ S'(\mathbf{n}) \geq 0)\\ &\leq 4\mathsf{P}(S(\mathbf{n}) \geq t) = 2\mathsf{P}(|S(\mathbf{n})| \geq t).\end{aligned}$$

The proof of Lemma 6.9 is complete. □

6.2 Equivalence of Convergence and Boundedness in Probability of Series of Symmetric Terms

6.2.1 Sums of Independent Symmetric Random Variables

Theorem 6.2 *Let $\{X(\mathbf{n}), \mathbf{n} \in \mathbf{N}^d\}$ be a field of independent symmetric random variables. The convergence in probability of a field $\{S(\mathbf{n}), \mathbf{n} \in \mathbf{N}^d\}$ is equivalent to its boundedness in probability.*

Proof of Theorem 6.2 If a series $\{S(\mathbf{n}), \mathbf{n} \in \mathbf{N}^d\}$ converges in probability, then it converges almost surely by Theorem 5.2. In the case of symmetric terms $X(\mathbf{n})$, this means that the following two series

$$\sum \mathsf{P}(|X(\mathbf{n})| \geq \varepsilon), \quad \sum \mathsf{var}\left[X^{\varepsilon}(\mathbf{n})\right]$$

converge for all $\varepsilon > 0$ (see Corollary 5.2). Note that $\mathsf{E}\,[X^{\varepsilon}(\mathbf{n})] = 0$. Thus Lemma 6.7 implies that the field $\{S(\mathbf{n}), \mathbf{n} \in \mathbf{N}^d\}$ is bounded in probability.

Now we prove the converse statement. The Lévy inequality for multi-indexed sums (Corollary 2.4) implies that

$$\mathsf{P}\left(\sup_{\mathbf{n}\in\mathbf{N}^d} |S(\mathbf{n})| \geq t\right) \leq 2^d \sup_{\mathbf{n}\in\mathbf{N}^d} \mathsf{P}\left(|S(\mathbf{n})| \geq t\right)$$

for all $t > 0$. Relation (A.7) yields

$$\mathsf{P}\left(\sup_{\mathbf{n}\in\mathbf{N}^d} |X(\mathbf{n})| \geq t2^d\right) \leq \mathsf{P}\left(\sup_{\mathbf{n}\in\mathbf{N}^d} |S(\mathbf{n})| \geq t\right) \leq 2^d \sup_{\mathbf{n}\in\mathbf{N}^d} \mathsf{P}\left(|S(\mathbf{n})| \geq t\right).$$

Since the field $\{S(\mathbf{n}), \mathbf{n} \in \mathbf{N}^d\}$ is bounded in probability, one can choose a number $t = t_0$ such that

$$\sup_{\mathbf{n}\in\mathbf{N}^d} \mathsf{P}\left(|S(\mathbf{n})| \geq t_0\right) \leq \frac{1}{2^{d+1}}.$$

Then

$$\mathsf{P}\left(\sup_{\mathbf{n}\in\mathbf{N}^d} |X(\mathbf{n})| \geq c\right) \leq \frac{1}{2}$$

for $c \geq c_0 \stackrel{\text{def}}{=} t_0 2^d$. On the other hand, $1 - x \leq e^{-x}$, $x \geq 0$, whence

$$\begin{aligned}\frac{1}{2} &\leq \mathsf{P}\left(\sup_{\mathbf{n}\in\mathbf{N}^d} |X(\mathbf{n})| < c\right) \leq \prod_{\mathbf{n}\in\mathbf{N}^d} (1 - \mathsf{P}(|X(\mathbf{n})| \geq c_0)) \\ &\leq \exp\left\{-\sum_{\mathbf{n}\in\mathbf{N}^d} \mathsf{P}(|X(\mathbf{n})| \geq c_0)\right\}\end{aligned}$$

for $c > c_0$ and condition (6.3) follows.

Further, the random variables $X^c(\mathbf{n})$ are independent and symmetric. Thus Lemma 6.9 yields

$$\mathsf{P}(|S(\mathbf{n}) - S'(\mathbf{n})| \geq t) \leq 2\mathsf{P}(|S(\mathbf{n})| \geq t),$$

that is, the field $S(\mathbf{n}) - S'(\mathbf{n})$, as well as the field $S'(\mathbf{n})$, is bounded in probability. The series $\sum_{\mathbf{n}\in\mathbf{N}^d} \mathsf{var}\left[X^c(\mathbf{n})\right]$ converges by Lemma 6.5. This together with (6.3) implies that the series $\sum_{\mathbf{n}\in\mathbf{N}^d} X(\mathbf{n})$ converges almost surely in view of Corollary 5.2. Therefore the series converges in probability. □

Corollary 6.1 *Let $\{X(\mathbf{n}), \mathbf{n} \in \mathbf{N}^d\}$ be a field of independent symmetric random variables with finite second moments. If*

$$\sum_{\mathbf{n}\in\mathbf{N}^d} \mathsf{E}\left[X^2(\mathbf{n})\right] < \infty, \tag{6.10}$$

then the series $\sum X(\mathbf{n})$ converges almost surely.

Proof Corollary 6.1 is a particular case of the two series theorem for symmetric random variables (see Theorem 5.7). Using Theorem 6.2 one can provide another proof of this result. Indeed, it follows from the Chebyshev inequality that

$$\mathsf{P}(|S(\mathbf{n})| \geq t) \leq \frac{1}{t^2} \sum_{\mathbf{n}\in\mathbf{N}^d} \mathsf{E}\left[X^2(\mathbf{n})\right],$$

whence we derive equality (6.1). By Theorem 6.2, this means that the series $\sum X(\mathbf{n})$ converges in probability. Now we conclude that the series $\sum X(\mathbf{n})$ converges almost surely (Theorem 5.2). □

What happens if (6.10) fails? Next we study the case where

$$\operatorname{limsup}\,(\max) \sum_{\mathbf{k}\leq\mathbf{n}} \mathsf{E}\left[X^2(\mathbf{k})\right] = \infty. \tag{6.11}$$

Corollary 6.2 *Let* $\{X(\mathbf{n}), \mathbf{n} \in \mathbf{N}^d\}$ *be a field of independent symmetric random variables that are bounded almost surely:* $|X(\mathbf{n})| \le c(\mathbf{n})$ *for some non-random constant* $c(\mathbf{n})$ *and all* $\mathbf{n} \in \mathbf{N}^d$. *Put*

$$C(\mathbf{n}) = \max_{\mathbf{k} \le \mathbf{n}} c(\mathbf{n}).$$

If condition (6.11) *holds and*

$$\limsup\,(\max) \frac{C^2(\mathbf{n})}{\mathsf{E}\left[S^2(\mathbf{n})\right]} < \infty, \tag{6.12}$$

then the series $\sum X(\mathbf{n})$ *is not bounded in probability.*

Proof It is clear that

$$\begin{aligned}\mathsf{E}\left[S^2(\mathbf{n})\right] &= \mathsf{E}\left[S^2(\mathbf{n})\,\mathbb{I}_{\{|S(\mathbf{n})| \ge \lambda\}}\right] + \mathsf{E}\left[S^2(\mathbf{n})\,\mathbb{I}_{\{|S(\mathbf{n})| < \lambda\}}\right] \\ &\le \sqrt{\mathsf{E}\left[S^4(\mathbf{n})\right] \mathsf{P}(|S(\mathbf{n})| > \lambda)} + \lambda^2\end{aligned}$$

for all $\lambda > 0$ by the Cauchy–Bunyakovskiĭ inequality. We choose $\lambda = B(\mathbf{n})/\sqrt{2}$, where $B(\mathbf{n}) = \mathsf{E}\left[S^2(\mathbf{n})\right]$. The fourth moment of $S(\mathbf{n})$ can be estimated as follows:

$$\begin{aligned}\mathsf{E}\left[S^4(\mathbf{n})\right] &= \sum_{\mathbf{k} \le \mathbf{n}} \mathsf{E}\left[X^4(\mathbf{k})\right] + 3 \sum_{\mathbf{k} \ne \mathbf{l}} \mathsf{E}\left[X^2(\mathbf{k})\right] \mathsf{E}\left[X^2(\mathbf{l})\right] \\ &\le C^2(\mathbf{n}) B(\mathbf{n}) + 3B^2(\mathbf{n}).\end{aligned}$$

Let $t > 0$ be a fixed number. If $\mathbf{n}$ is such that $B(\mathbf{n}) \ge 2t^2$, then

$$\begin{aligned}\mathsf{P}(|S(\mathbf{n})| \ge t) \ge \mathsf{P}(|S(\mathbf{n})| \ge B(\mathbf{n})/2) &\ge \frac{B^2(\mathbf{n})}{4\mathsf{E}\left[S^4(\mathbf{n})\right]} \\ &\ge \frac{B^2(\mathbf{n})}{4\left(C^2(\mathbf{n})B(\mathbf{n}) + 3B^2(\mathbf{n})\right)}.\end{aligned}$$

Thus (6.12) implies that condition (6.1) does not hold, that is, the series is not bounded in probability. □

Corollary 6.3 *Let* $\{X(\mathbf{n}), \mathbf{n} \in \mathbf{N}^d\}$ *be a field of independent symmetric random variables with finite moments of order* $p > 0$. *If the numerical field* $\{\mathsf{E}\left[|S^p(\mathbf{n})|\right]\}$ *is bounded, then the field* $\{S(\mathbf{n})\}$ *is bounded in probability.*

Proof According to the Markov inequality,

$$\mathsf{P}(|S(\mathbf{n})| \geq t) \leq \frac{1}{t^p} \sup_{\mathbf{n}\in\mathbf{N}^d} \mathsf{E}\left[|S^p(\mathbf{n})|\right],$$

whence condition (6.1) follows. □

With the help of Theorem 6.2 one can provide another proof of the implications $(C) \Rightarrow (A)$ and $(C) \Rightarrow (B)$ in Theorem 5.2.

Corollary 6.4 *Let $\{X(\mathbf{n}), \mathbf{n} \in \mathbf{N}^d\}$ be a field of independent symmetric random variables. If the series $\sum X(\mathbf{n})$ converges in the sense of the weak convergence of distribution functions of their rectangular sums, then the series converges in probability and almost surely.*

Proof Any sequence of random variables convergent in the sense of the weak convergence of their distribution functions is dense, that is, condition (6.1) holds. By Theorem 6.2, this means that the series $\sum X(\mathbf{n})$ converges in probability. Now, Theorem 5.2 implies that the series $\sum X(\mathbf{n})$ converges almost surely, as well. □

6.2.2 Sums of Weighted Independent Identically Distributed Random Variables

Yet another class of multi-indexed series of independent random variables for which convergence implies boundedness is represented by multi-indexed series of weighted identically distributed terms. If the distribution function of the random variables is degenerate (is concentrated at a single point), then the series is "non-random" and Example 6.1 shows that boundedness is not necessarily a consequence of convergence. Otherwise, if the distribution function is nondegenerate, then boundedness does indeed follow from convergence.

Theorem 6.3 *Let $\{Y(\mathbf{n}), \mathbf{n} \in \mathbf{N}^d\}$ be independent identically distributed random variables whose common distribution function is not concentrated at a single point. Further let $\{w(\mathbf{n}), \mathbf{n} \in \mathbf{N}^d\}$ be a field of non-random numbers. If the series $\sum w(\mathbf{n})Y(\mathbf{n})$ converges in probability, then the field*

$$\left\{\sum_{\mathbf{m}\leq\mathbf{n}} w(\mathbf{m})Y(\mathbf{m}), \quad \mathbf{n} \in \mathbf{N}^d\right\} \tag{6.13}$$

is bounded in probability.

Proof of Theorem 6.3 Put $X(\mathbf{n}) = w(\mathbf{n})Y(\mathbf{n})$, $S(\mathbf{n}) = \sum_{\mathbf{m}\leq\mathbf{n}} X(\mathbf{m})$. The convergence in probability of a series of independent terms is equivalent to its almost sure convergence (see Theorem 5.2). Thus the convergence in probability of the series $\sum X(\mathbf{n})$ together with Theorem 5.8 implies that $\sum \mathsf{P}(|X(\mathbf{n})| \ge \varepsilon) < \infty$ and $\sum \mathsf{var}\,[X^{\varepsilon}(\mathbf{n})] < \infty$ for all $\varepsilon > 0$. Then

$$\sup_{\mathbf{n}\in\mathbf{N}^d} \left| \sum_{\mathbf{k}\le\mathbf{n}} \mathsf{E}\left[X^{\varepsilon}(\mathbf{k})\right] \right| < \infty$$

for all $\varepsilon > 0$ by Lemma 6.5. Therefore Theorem 6.3 follows from Lemma 6.7. □

6.3 Skorokhod's Decomposition of a Bounded Series

Now we are ready to establish some results for $d > 1$ similar to Theorem 6.1 proved only for $d = 1$.

Theorem 6.4 *Let $\{X(\mathbf{n}), \mathbf{n} \in \mathbf{N}^d\}$ be independent random variables. If a field $\{S(\mathbf{n}), \mathbf{n} \in \mathbf{N}^d\}$ is bounded in probability, then there exists a field of non-random numbers $\{a(\mathbf{n}), \mathbf{n} \in \mathbf{N}^d\}$ such that*

$$\text{the sums} \quad \sum_{\mathbf{k}\le\mathbf{n}} a(\mathbf{k}) \quad \text{are uniformly bounded}$$

and the series $\sum[X(\mathbf{n}) - a(\mathbf{n})]$ converges almost surely. Moreover, one can choose $a(\mathbf{n}) = \mathsf{E}\left[X^c(\mathbf{n})\right]$ if $c > 0$ is sufficiently large.

Theorem 6.5 *If a series $\sum X(\mathbf{n})$ converges almost surely, then there exists a non-random field $\{b(\mathbf{n}), \mathbf{n} \in \mathbf{N}^d\}$ such that $\{S(\mathbf{n}) - b(\mathbf{n}), \mathbf{n} \in \mathbf{N}^d\}$ is bounded in probability. Moreover, one can choose $b(\mathbf{n}) = \mu(S(\mathbf{n}))$.*

Remark 6.1 The field $\{b(\mathbf{n})\}$ involved in Theorem 6.5 is not necessarily bounded if $d > 1$. Indeed, the non-random field $\{X(\mathbf{n})\}$ constructed in Example 6.1 is such that the series $\sum X(\mathbf{n})$ converges and the sums $\{S(\mathbf{n})\}$ are unbounded. Thus there is no bounded field $\{b(\mathbf{n})\}$ for which Theorem 6.5 is valid.

Proof of Theorem 6.4 It is easy to see that the field

$$\left\{ \sum_{\mathbf{m}\le\mathbf{n}} X^{(s)}(\mathbf{m}), \ \mathbf{n} \in \mathbf{N}^d \right\} \tag{6.14}$$

is bounded in probability and thus it converges in probability (see Theorem 6.2). Theorem 5.2 implies that the latter field converges almost surely. Further, Corollary 5.2 proves that the series $\sum \mathsf{P}(|X^{(s)}(\mathbf{n})| \ge c)$ converges for all $c > 0$. Using Lemma 6.8 we obtain

$$\sum \mathsf{var}\left[X^c(\mathbf{n})\right] < \infty$$

for sufficiently large $c > 0$. The two series theorem (Theorem 5.7) implies that the series $\sum \left(X^c(\mathbf{n}) - \mathsf{E}\left[X^c(\mathbf{n})\right]\right)$ converges almost surely for such numbers $c > 0$ and thus Theorem 6.4 holds with $a(\mathbf{n}) = \mathsf{E}\left[X^c(\mathbf{n})\right]$ if $c > 0$ is sufficiently large. The boundedness of the field

$$\sum_{\mathbf{k} \leq \mathbf{n}} a(\mathbf{k})$$

also follows from Lemma 6.8. □

Proof of Theorem 6.5 By the assumptions of the theorem, the series $\sum X^{(s)}(\mathbf{n})$ converges almost surely and in probability, as well. Theorem 6.2 implies that the field (6.14) is bounded in probability. Since $\mathsf{P}(|X - \mu(X)| \geq t) \leq 2\mathsf{P}(|X^{(s)}| \geq t)$ for any random variable X and its symmetrization $X^{(s)}$, we get

$$\mathsf{P}(|S(\mathbf{n}) - \mu(S(\mathbf{n}))| \geq t) \leq 2\mathsf{P}(|S^{(s)}(\mathbf{n})| \geq t)$$

(see Proposition A.35). Evaluating $\sup_{\mathbf{n} \in \mathbf{N}^d}$ on both sides of this inequality and then passing to the limit as $t \to \infty$ we complete the proof for $b(\mathbf{n}) = \mu(S(\mathbf{n}))$. □

6.4 Conditions for the Almost Sure Boundedness of Multi-Indexed Sums

We have seen in Chap. 5 that the almost sure convergence of a series is equivalent to its convergence in probability. We can also prove the equivalence for boundedness.

Theorem 6.6 *If $\{X(\mathbf{n}), \mathbf{n} \in \mathbf{N}^d\}$ are independent random variables, then the boundedness in probability of the field $\{S(\mathbf{n}), \mathbf{n} \in \mathbf{N}^d\}$ is equivalent to the almost sure boundedness of the series $\sum X(\mathbf{n})$.*

The conditions in the following criterion look similar to those in the three series theorem (Theorem 5.3) but now the three series are assumed to be bounded rather than convergent.

Theorem 6.7 *The field $\{S(\mathbf{n}), \mathbf{n} \in \mathbf{N}^d\}$ is bounded almost surely if and only if the following three numerical series*

$$\sum_{\mathbf{n} \in \mathbf{N}^d} \mathsf{P}(|X(\mathbf{n})| \geq c), \quad \sum_{\mathbf{n} \in \mathbf{N}^d} \mathsf{E}\left[X^c(\mathbf{n})\right], \quad \sum_{\mathbf{n} \in \mathbf{N}^d} \mathsf{var}\left[X^c(\mathbf{n})\right] \tag{6.15}$$

are bounded for some $c > 0$.

Remark 6.2 Since the terms of the first and third series in (6.15) are non-negative, the boundedness of these series is equivalent to their convergence.

Proof of Theorem 6.6 One step of the proof is easy: almost sure boundedness implies boundedness in probability (see Lemma 6.1). It remains to prove the inverse implication.

If the field $\{S(\mathbf{n})\}$ is bounded in probability, then the field (6.14) is bounded in probability, as well, and thus it converges in probability (see Theorem 6.2). This implies its almost sure convergence (Theorem 5.2). Now Corollary 5.2 yields the convergence of the series $\sum \mathsf{P}(|X^{(s)}(\mathbf{n})| \geq c)$ for all $c > 0$. Let

$$E_c = \sup_{\mathbf{n}\in\mathbf{N}^d} \left| \sum_{\mathbf{m}\leq\mathbf{n}} \mathsf{E}\left[X^c(\mathbf{m})\right] \right|.$$

By Lemma 6.8, $E_c < \infty$ for sufficiently large $c > 0$. For such c and for $t > E_c$, we have

$$\mathsf{P}\left\{ \sup_{\mathbf{n}\in\mathbf{N}^d} |S(\mathbf{n})| \geq t \right\} \leq \sum_{\mathbf{n}\in\mathbf{N}^d} \mathsf{P}(|X(\mathbf{n})| \geq c) + \mathsf{P}\left\{ \sup_{\mathbf{n}\in\mathbf{N}^d} \left| \sum_{\mathbf{m}\leq\mathbf{n}} \left(X^c(\mathbf{m}) - \mathsf{E}\left[X^c(\mathbf{m})\right]\right) \right| \geq t - E_c \right\}.$$

Since

$$\mathsf{P}\left(\sup_{\mathbf{n}\in\mathbf{N}^d} |S(\mathbf{n})| \geq t + \delta \right) \leq \mathsf{P}\left(\bigcup_{\mathbf{n}\in\mathbf{N}^d} \{\omega\colon |S(\mathbf{n})| \geq t\} \right)$$

for all $\delta > 0$, we apply the multi-index Kolmogorov inequality to the second term (Corollary 2.11) and obtain

$$\mathsf{P}\left\{ \sup_{\mathbf{n}\in\mathbf{N}^d} |S(\mathbf{n})| \geq 2t \right\} \leq \sum_{\mathbf{n}\in\mathbf{N}^d} \mathsf{P}(|X(\mathbf{n})| \geq c) + \frac{2^{d+2}}{(t - E_c)^2} \sum_{\mathbf{n}\in\mathbf{N}^d} \mathsf{E}\left[\left(X^c(\mathbf{n}) - \mathsf{E}\left[X^c(\mathbf{n})\right]\right)^2\right].$$

The series on the right-hand side of the latter inequality converge by Lemma 6.8. Passing to the limit in this inequality, first as $t \to \infty$ and then as $c \to \infty$, we prove the almost sure boundedness by Lemma 6.6. □

Proof of Theorem 6.7 For $c > 0$, put

$$P(\mathbf{n}; c) = \sum_{\mathbf{k}\leq\mathbf{n}} \mathsf{P}(|X(\mathbf{k})| \geq c), \tag{6.16}$$

$$E(\mathbf{n}; c) = \sum_{\mathbf{k}\leq\mathbf{n}} \mathsf{E}\left[X^c(\mathbf{k})\right], \quad V(\mathbf{n}; c) = \sum_{\mathbf{k}\leq\mathbf{n}} \mathsf{var}\left[X^c(\mathbf{k})\right].$$

Clearly, the fields $\{P(\mathbf{n}; c), \mathbf{n} \in \mathbf{N}^d\}$ and $\{V(\mathbf{n}; c), \mathbf{n} \in \mathbf{N}^d\}$ are bounded if and only if the series $\sum \mathsf{P}(|X(\mathbf{n})| \geq c)$ and $\sum \mathsf{var}\left[X^c(\mathbf{n})\right]$ converge.

Let the field $\{S(\mathbf{n}), \mathbf{n} \in \mathbf{N}^d\}$ be bounded almost surely. Our aim is to prove that the fields $\{P(\mathbf{n}; c)\}$, $\{E(\mathbf{n}; c)\}$, and $\{V(\mathbf{n}; c)\}$ are bounded for some $c > 0$. Since the field $\{S^{(s)}(\mathbf{n})\}$ is bounded in probability, Theorem 6.2 implies that the field $\{S^{(s)}(\mathbf{n})\}$ converges in probability. Then we conclude that the field $\{S^{(s)}(\mathbf{n})\}$ converges almost surely (Theorem 5.2). According to Corollary 5.2, the latter implies that the series $\sum \mathsf{P}(|X^{(s)}(\mathbf{n})| \geq c)$ converges for all $c > 0$. Hence

$$\begin{aligned}\infty > \sum_{\mathbf{n}\in\mathbf{N}^d} \mathsf{P}(|X^{(s)}(\mathbf{n})| \geq c/2) &\geq \sum_{\mathbf{n}\in\mathbf{N}^d} \mathsf{P}(|X(\mathbf{n})| > c, |X'(\mathbf{n})| \leq c/2) \\ &\geq \left[\inf_{\mathbf{n}\in\mathbf{N}^d} \mathsf{P}(|X(\mathbf{n})| \leq c/2)\right] \sum_{\mathbf{n}\in\mathbf{N}^d} \mathsf{P}(|X(\mathbf{n})| \geq c), \end{aligned} \tag{6.17}$$

where the independent random variables $\{X'(\mathbf{n}), \mathbf{n} \in \mathbf{N}^d\}$ do not depend on the field $\{X(\mathbf{n}), \mathbf{n} \in \mathbf{N}^d\}$ and have the same distributions as the random variables $X(\mathbf{n})$.

Now we deduce from Lemma 6.3 that the field $\{X(\mathbf{n}), \mathbf{n} \in \mathbf{N}^d\}$ is bounded in probability. Therefore there exists a $c > 0$ such that $\mathsf{P}(|X(\mathbf{n})| \leq c) > 1/2$ for all $\mathbf{n} \in \mathbf{N}^d$. This property together with relation (6.17) implies that, for such a number c,

$$\sup_{\mathbf{n}\in\mathbf{N}^d} P(\mathbf{n}; c) < \infty.$$

Moreover, the Borel–Cantelli lemma for independent random events (see Proposition A.30) yields $\mathsf{P}(X(\mathbf{k}) \neq X^c(\mathbf{k}) \text{ i. o.}) = 0$ and thus the field

$$\left\{\sum_{\mathbf{k}\leq\mathbf{n}} X^c(\mathbf{k}), \quad \mathbf{n} \in \mathbf{N}^d\right\}$$

is bounded almost surely and in probability, as well. By Lemma 6.5, this means that

$$\sup_{\mathbf{n}\in\mathbf{N}^d} |E(\mathbf{n}; c)| < \infty \quad \text{and} \quad \sup_{\mathbf{n}\in\mathbf{N}^d} V(\mathbf{n}; c) < \infty.$$

It remains to prove the inverse implication. Assume that all three numerical fields $\{P(\mathbf{n}; c)\}$, $\{E(\mathbf{n}; c)\}$ and $\{V(\mathbf{n}; c)\}$ are bounded for some $c > 0$. Our aim is to prove that the field $\{S(\mathbf{n})\}$ is bounded almost surely. Since the field $\{V(\mathbf{n}; c)\}$ converges, the two series theorem (Theorem 5.7) implies that

$$\sum_{\mathbf{k}\in\mathbf{N}^d} \left(X^c(\mathbf{k}) - \mathsf{E}\left[X^c(\mathbf{k})\right]\right)$$

converges almost surely and in probability, as well.

Now let $M(\mathbf{n}; c)$ denote a median of the random variable

$$\sum_{\mathbf{k}\le\mathbf{n}} \left(X^c(\mathbf{k}) - \mathsf{E}\left[X^c(\mathbf{k})\right]\right).$$

Then Theorem 6.5 implies that the field

$$\left\{\sum_{\mathbf{k}\le\mathbf{n}} \left(X^c(\mathbf{k}) - \mathsf{E}\left[X^c(\mathbf{k})\right]\right) - M(\mathbf{n}; c), \quad \mathbf{n} \in \mathbf{N}^d\right\}$$

is bounded in probability. Since the field $\{E(\mathbf{n}; c)\}$ is bounded, the latter result means that the field $\sum_{\mathbf{k}\le\mathbf{n}} X^c(\mathbf{k}) - M(\mathbf{n}; c)$ is bounded in probability. According to the Mallows bound, $|M(\mathbf{n}; c)| \le \sqrt{V(\mathbf{n}; c)}$ (see (A.80)) and thus the field $\{M(\mathbf{n}; c)\}$ is bounded, whence we conclude that the field

$$\left\{\sum_{\mathbf{k}\le\mathbf{n}} X^c(\mathbf{k}), \quad \mathbf{n} \in \mathbf{N}^d\right\}$$

is bounded in probability. Now Theorem 6.6 implies that the latter field is bounded almost surely. The convergence of the field $\{P(\mathbf{n}; c)\}$ together with the Borel–Cantelli lemma implies that $\mathsf{P}(X(\mathbf{k}) \ne X^c(\mathbf{k}) \text{ i. o.}) = 0$. Thus the field

$$\left\{\sum_{\mathbf{k}\le\mathbf{n}} X(\mathbf{k}), \mathbf{n} \in \mathbf{N}^d, \quad \mathbf{n} \in \mathbf{N}^d\right\}$$

is bounded almost surely. □

6.5 Bounded Convergence of Multi-Indexed Series

The unpleasant phenomenon exhibited in Example 6.1, where a convergent series is not bounded, suggests separating the cases with this phenomenon.

The theory of multi-indexed series studies various types of convergence that eliminate in one way or another the phenomenon exhibited in Example 6.1. One such type is called *bounded convergence* (see, for example, [426]).

Definition 6.4 A multi-indexed series $\sum_{\mathbf{n}\in\mathbf{N}^d} X(\mathbf{n})$ is called *boundedly convergent* if it converges almost surely and if, additionally, the field $\{S(\mathbf{n}), \mathbf{n} \in \mathbf{N}^d\}$ is bounded.

This definition is equivalent to the usual definition of convergent series if $d = 1$. However, the situation changes for $d > 1$, since in that case the class of boundedly convergent series is narrower than the class of convergent series. The conditions for the boundedness of random multi-indexed series considered in the preceding sections allow one to easily obtain the criterion for the bounded convergence of a series of independent random variables.

Theorem 6.8 *Let* $\{X(\mathbf{n}), \mathbf{n} \in \mathbf{N}^d\}$ *be independent random variables. If the following three series*

$$\sum_{\mathbf{n}\in\mathbf{N}^d} \mathsf{P}(|X(\mathbf{n})| \geq c), \tag{6.18}$$

$$\sum_{\mathbf{n}\in\mathbf{N}^d} \mathsf{E}\left[X^c(\mathbf{n})\right], \tag{6.19}$$

$$\sum_{\mathbf{n}\in\mathbf{N}^d} \mathsf{var}\left[X^c(\mathbf{n})\right] \tag{6.20}$$

are boundedly convergent for all $c > 0$, *then*

$$\text{the series } \sum_{\mathbf{n}\in\mathbf{N}^d} X(\mathbf{n}) \text{ almost surely converges boundedly.} \tag{6.21}$$

If condition (6.21) *holds, then series* (6.18)–(6.20) *are boundedly convergent for some* $c > 0$.

Proof Let the series $\sum X(\mathbf{n})$ almost surely converge boundedly. Since the field $\{S(\mathbf{n})\}$ is bounded almost surely, Theorem 6.7 implies that the three series (6.18)–(6.20) are bounded for some $c > 0$. Since the terms of series (6.18) and (6.20) are non-negative, the boundedness is equivalent to their convergence even in the case where $d > 1$. The boundedness of series (6.18) together with the Borel–Cantelli lemma implies that $\mathsf{P}(X(\mathbf{n}) \neq X^c(\mathbf{n}) \text{ i. o.}) = 0$ and thus the series $\sum X^c(\mathbf{k})$ converges almost surely. Then the two series theorem (Theorem 5.7) yields the almost sure convergence of the series $\sum(X(\mathbf{n}) - \mathsf{E}\left[X^c(\mathbf{n})\right])$, since series (6.20) converges. This proves the convergence of series (6.19).

The inverse implication is a consequence of Lemma 6.7 and the three series theorem (Theorem 5.3).

The proof of Theorem 6.8 is complete. □

6.6 Moments of the Supremum of Multi-Indexed Sums

Let φ be an unbounded function defined for non-negative arguments and let ξ be a non-negative random variable. The "boundedness" of the random variable ξ can be "measured" in terms of its moments. The finiteness of the moment $\mathsf{E}[\varphi(\xi)]$ is one of the indicators for the "boundedness" of ξ. In this section, we study the conditions for the finiteness of moments for the random variable

$$\mathfrak{M} = \sup_{\mathbf{n}\in\mathbf{N}^d} |S(\mathbf{n})|.$$

It turns out that the conditions for the existence of the moment $\mathsf{E}[\varphi(\mathfrak{M})]$ are related to the boundedness in probability of the field $\{S(\mathbf{n})\}$.

Theorem 6.9 *Let $\{X(\mathbf{n}), \mathbf{n} \in \mathbf{N}^d\}$ be independent random variables and let a function φ be non-negative, increasing, and such that*

$$\varphi(2x) \le C\varphi(x), \quad x \ge 0, \tag{6.22}$$

for some constant $C > 0$. If the field $\{S(\mathbf{n}), \mathbf{n} \in \mathbf{N}^d\}$ is bounded in probability, then the condition

$$\mathsf{E}\left[\varphi\left(\sup_{\mathbf{n}\in\mathbf{N}^d} |S(\mathbf{n})|\right)\right] < \infty \tag{6.23}$$

holds if and only if

$$\mathsf{E}\left[\varphi\left(\sup_{\mathbf{n}\in\mathbf{N}^d} |X(\mathbf{n})|\right)\right] < \infty. \tag{6.24}$$

Remark 6.3 Condition (6.22) is satisfied by all regularly varying functions including $\varphi(x) = |x|^q$, $q > 0$. On the other hand, the function $\varphi(x) = e^x$ does not fit this condition.

Considering only increasing functions, condition (6.22) describes the class of the so-called ORV functions defined by

$$\forall\lambda > 0 \qquad \limsup_{x\to\infty} \frac{\varphi(\lambda x)}{\varphi(x)} < \infty. \tag{6.25}$$

Condition (6.22) seems less restrictive than (6.25). Actually, they are equivalent in the class of non-decreasing functions. Indeed, if $\lambda \le 2$, then (6.25) follows from (6.22) in view of the monotonicity of φ. To prove that (6.25) follows from (6.22) in the case of $\lambda > 2$, as well, let a positive integer $m \ge 1$ be such that $2^m \ge \lambda$. Then

$$\limsup_{x\to\infty} \frac{\varphi(\lambda x)}{\varphi(x)} \le \limsup_{x\to\infty} \frac{\varphi(2^m x)}{\varphi(x)} \le \left(\limsup_{x\to\infty} \frac{\varphi(2x)}{\varphi(x)}\right)^m \le C^m.$$

Remark 6.4 Theorem 6.9 fails if the assumption on the boundedness of the field $\{S(\mathbf{n})\}$ is dropped. Indeed, if a field $\{S(\mathbf{n})\}$ is not bounded in probability, then it is not bounded almost surely (see Theorem 6.6) and thus (6.23) does not hold. On the other hand, an example of a field $\{X(\mathbf{n})\}$ that fits condition (6.24) is easy to construct even in the case of an unbounded field $\{S(\mathbf{n})\}$.

The following auxiliary result (which is of interest in itself) is helpful in the course of proof of Theorem 6.9.

Theorem 6.10 (V. V. Buldygin, J. Hoffman-Jørgensen) *Let $\{X_k\}$ be a sequence of independent symmetric random elements in a separable Frechet space and let $\|\cdot\|$ be a measurable seminorm in this space. Put $S_n = X_1 + \cdots + X_n$. Then*

$$\mathsf{P}\left(\max_{1\le m\le n} \|S_m\| \ge 2t + s\right) \le 8\mathsf{P}^2(\|S_n\| \ge t) + 2\mathsf{P}\left(\max_{1\le k\le n} \|X_k\| \ge s\right),$$
$$\mathsf{P}\left(\sup_{m\ge 1} \|S_m\| \ge 2t + s\right) \le 8\sup_{n\ge 1} \mathsf{P}^2(\|S_n\| \ge t) + 2\mathsf{P}\left(\sup_{k\ge 1} \|X_k\| \ge s\right)$$

for all $n \ge 1$, $t > 0$, and $s > 0$.

Below is yet another auxiliary result that follows from equality (A.79).

Proposition 6.1 *Let X be a random element in a measurable linear space. Assume that $\|\cdot\|$ is a measurable seminorm and $\varphi : [0, \infty) \to \mathbf{R}$ is a continuous increasing function. Then $\mathsf{E}\,[\varphi(\|X\|)] < \infty$ if and only if*

$$\int_0^\infty \mathsf{P}(\|X\| \ge u)\, d\varphi(u) < \infty$$

(*the integral is understood in the Lebesgue–Stieltjes sense*). *Moreover,*

$$\int_a^\infty \varphi(x)\, d\mathsf{P}(\|X\| < x) = \varphi(a)\mathsf{P}(\|X\| \ge a) + \int_a^\infty \mathsf{P}(\|X\| \ge x)\, d\varphi(x) \tag{6.26}$$

for all $a > 0$.

Proof of Theorem 6.9 Let

$$\mathfrak{X} = \sup_{\mathbf{n}\in\mathbf{N}^d} |X(\mathbf{n})|, \quad \mathfrak{M} = \sup_{\mathbf{n}\in\mathbf{N}^d} |S(\mathbf{n})|.$$

If condition (6.23) holds, then inequality (A.7) implies that $\mathfrak{X} \le 2^d\mathfrak{M}$, whence

$$\mathsf{E}\,[\varphi(\mathfrak{X})] \le C^d\mathsf{E}\,[\varphi(\mathfrak{M})],$$

that is, condition (6.24) holds.

Next we prove the converse implication (6.24)⇒(6.23). Let

$$\mathfrak{X}(\mathbf{n}) = \max_{\mathbf{m}\le\mathbf{n}} |X(\mathbf{m})|, \quad \mathfrak{M}(\mathbf{n}) = \max_{\mathbf{m}\le\mathbf{n}} |S(\mathbf{m})|.$$

Step 1. First we treat the case of symmetric random variables and a function φ such that $\varphi(0) = 0$. Using the Lévy inequality for multi-indexed sums (Corollary 2.4), we get

$$\mathsf{P}(\mathfrak{M}(\mathbf{n}) \geq t) \leq 2^d \mathsf{P}(|S(\mathbf{n})| \geq t) \tag{6.27}$$

for all positive t.

We apply Theorem 6.10 with $s = t$ to the right-hand side of (6.27):

$$\mathsf{P}(|S(\mathbf{n})| \geq 3t) \leq 8\mathsf{P}^2(|S(\mathbf{n})| \geq t) + \mathsf{P}\left(\mathfrak{X}(\mathbf{n}) \geq t\right).$$

Thus

$$\mathsf{P}(\mathfrak{M}(\mathbf{n}) \geq 3t) \leq 2^{d+3}\mathsf{P}^2(|S(\mathbf{n})| \geq t) + 2^d \mathsf{P}(\mathfrak{X} \geq x),$$

whence

$$\int_0^\infty \mathsf{P}(\mathfrak{M}(\mathbf{n}) \geq 3t)\, d\varphi(t) \leq 2^{d+3} \int_0^\infty \mathsf{P}^2(|S(\mathbf{n})| \geq t)\, d\varphi(t) + 2^d \int_0^\infty \mathsf{P}(\mathfrak{X} \geq t)\, d\varphi(t). \tag{6.28}$$

The second term on the right-hand side is finite in view of condition (6.24) and Proposition 6.1. Since $\varphi(0) = 0$, relation (6.26) implies that

$$\int_0^\infty \mathsf{P}(\mathfrak{M}(\mathbf{n}) \geq 3t)\, d\varphi(t) = \int_0^\infty \varphi(t/3)\, d\mathsf{P}(\mathfrak{M}(\mathbf{n}) < t).$$

It is easily seen that condition (6.22) imposed on the function φ yields

$$\varphi(3t) \leq \varphi(4t) \leq C^2 \varphi(t).$$

Thus relation (6.26) implies that

$$\int_0^\infty \mathsf{P}(\mathfrak{M}(\mathbf{n}) \geq 3t)\, d\varphi(t) \geq C^{-2} \int_0^\infty \mathsf{P}(\mathfrak{M}(\mathbf{n}) \geq t)\, d\varphi(t),$$

whence

$$\int_0^\infty \left[C^{-2}\mathsf{P}(\mathfrak{M}(\mathbf{n}) \geq t) - 2^{d+3}\mathsf{P}^2(|S(\mathbf{n})| \geq t)\right] d\varphi(t) \leq 2^d \mathsf{E}\left[\varphi(\mathfrak{X})\right]. \tag{6.29}$$

Since the field $\{S(\mathbf{n}), \mathbf{n} \in \mathbf{N}^d\}$ is bounded in probability, there exists a number t_0 such that

$$\frac{1}{C^2} - 2^{d+3} \sup_{\mathbf{n}\in\mathbf{N}^d} \mathsf{P}(|S(\mathbf{n})| \ge t) \ge \frac{1}{2C^2}, \quad t \ge t_0.$$

Since $\mathsf{P}(|S(\mathbf{n})| \ge t) \le \mathsf{P}(\mathfrak{M}(\mathbf{n}) \ge t)$, we have

$$\begin{aligned}
&\int_0^\infty \left[C^{-2}\mathsf{P}(\mathfrak{M}(\mathbf{n}) \ge t) - 2^{d+3}\mathsf{P}^2(|S(\mathbf{n})| \ge t)\right] d\varphi(t) \\
&\ge \int_0^\infty \mathsf{P}(\mathfrak{M}(\mathbf{n}) \ge t)\left[C^{-2} - 2^{d+3}\mathsf{P}(|S(\mathbf{n})| \ge t)\right] d\varphi(t) \\
&\ge \int_{t_0}^\infty \mathsf{P}(\mathfrak{M}(\mathbf{n}) \ge t)\left[C^{-2} - 2^{d+3}\sup_{\mathbf{n}\in\mathbf{N}^d} \mathsf{P}(|S(\mathbf{n})| \ge t)\right] d\varphi(t) \\
&\ge \frac{1}{2C^2}\int_{t_0}^\infty \mathsf{P}(\mathfrak{M}(\mathbf{n}) \ge t)\, d\varphi(t).
\end{aligned}$$

Therefore (6.29) implies that

$$\int_{t_0}^\infty \mathsf{P}(\mathfrak{M}(\mathbf{n}) \ge t)\, d\varphi(t) \le 2^{d+1}C^2\mathsf{E}\left[\varphi(\mathfrak{X})\right]$$

for all $\mathbf{n} \in \mathbf{N}^d$. As the function φ is increasing and $\mathfrak{M}(\mathbf{m}) \le \mathfrak{M}(\mathbf{n})$ for all $\mathbf{m} \le \mathbf{n}$, the Beppo Levi lemma (Proposition A.31) implies that

$$\int_{t_0}^\infty \mathsf{P}(\mathfrak{M} \ge t)\, d\varphi(t) \le 2^{d+1}C^2\mathsf{E}\left[\varphi(\mathfrak{X})\right],$$

that is, condition (6.23) holds. The theorem is completely proved in the case of symmetric random variables and a function φ such that $\varphi(0) = 0$.

Step 2. Now we drop the assumption that random variables are symmetric. For the general case of not necessarily symmetric random variables, we consider an independent copy $\{X'(\mathbf{n})\}$ of $\{X(\mathbf{n})\}$. Put

$$\mathfrak{X}' = \sup_{\mathbf{n}\in\mathbf{N}^d} |X'(\mathbf{n})|, \quad S'(\mathbf{n}) = \sum_{\mathbf{m}\le\mathbf{n}} X'(\mathbf{m}), \quad \mathfrak{M}' = \sup_{\mathbf{n}\in\mathbf{N}^d} |S'(\mathbf{n})|.$$

It is clear that the field $\{S(\mathbf{n}) - S'(\mathbf{n}), \mathbf{n} \in \mathbf{N}^d\}$ is bounded in probability and that

$$\mathsf{E}\left[\varphi\left(\sup_{\mathbf{n}\in\mathbf{N}^d}|X(\mathbf{n})-X'(\mathbf{n})|\right)\right] \le \mathsf{E}\left[\varphi(2\max\{\mathfrak{X},\mathfrak{X}'\})\right]$$
$$\le C\left(\mathsf{E}\left[\varphi(\mathfrak{X})\right]+\mathsf{E}\left[\varphi(\mathfrak{X}')\right]\right)<\infty,$$

since $\mathsf{E}\left[\varphi(\mathfrak{X})\right]=\mathsf{E}\left[\varphi(\mathfrak{X}')\right]$. According to Step 1,

$$\mathsf{E}\left[\varphi\left(\sup_{\mathbf{n}\in\mathbf{N}^d}|S(\mathbf{n})-S'(\mathbf{n})|\right)\right]<\infty,$$

whence $\mathsf{E}\left[\varphi(|\mathfrak{M}-\mathfrak{M}'|)\right]<\infty$. By Fubini's theorem, there exists a number $x_0\in\mathbf{R}$ such that $\mathsf{E}\left[\varphi(|\mathfrak{M}-x_0|)\right]<\infty$. Then

$$\mathsf{E}\left[\varphi\left(\mathfrak{M}/2\right)\right]\le\int\limits_{\{\mathfrak{M}\ge 2x_0\}}\varphi\left(\mathfrak{M}/2\right)\,d\mathsf{P}+\varphi(x_0)\le\int\limits_{\Omega}\varphi(|\mathfrak{M}-x_0|)\,d\mathsf{P}+\varphi(x_0).$$

Observing that $\mathsf{E}\left[\varphi\left(\mathfrak{M}\right)\right]\le C\mathsf{E}\left[\varphi\left(\mathfrak{M}/2\right)\right]$, we complete the proof of Theorem 6.9 for nonsymmetric random variables.

Step 3. It remains to avoid the assumption that $\varphi(0)=0$. Since $\mathsf{E}\left[\varphi(\mathfrak{M})\right]=\varphi(0)+\mathsf{E}\left[\psi(\mathfrak{M})\right]$, where $\psi(t)=\varphi(t)-\varphi(0)$, and $d\psi(t)=d\varphi(t)$, the finiteness of $\mathsf{E}\left[\varphi(\mathfrak{M})\right]$ is equivalent to the finiteness of $\mathsf{E}\left[\psi(\mathfrak{M})\right]$. Applying the result of Step 2 to the function ψ we complete the proof of the theorem, since $\psi(0)=0$. □

Remark 6.5 Examining the proof of Theorem 6.9 we see that it also holds for multi-indexed sums of independent Banach space-valued elements. This is the result we will use in Chap. 12. The only point that needs to be explained for multi-indexed sums of Banach space-valued elements is the multi-index Lévy inequality (6.27). This result can be obtained by induction starting with the Lévy inequality for Banach spaces in the case $d=1$.

Theorem 6.11 *Let $X(\mathbf{k})$, $\mathbf{k}\le\mathbf{n}$, be independent symmetric random elements in some Frechet space. Put*

$$S(\mathbf{m})=\sum_{\mathbf{k}\le\mathbf{m}}X(\mathbf{k}).$$

If $\|\cdot\|$ is a measurable norm in this space, then

$$\mathsf{P}\left(\max_{\mathbf{m}\le\mathbf{n}}\|S(\mathbf{m})\|\ge x\right)\le 2^d\mathsf{P}(\|S(\mathbf{n})\|\ge x) \tag{6.30}$$

for all $x>0$.

Proof Assume that inequality (6.30) holds for all indices whose dimension is less than d and let us prove it for indices of dimension d. Denote the space where the random elements $X(\mathbf{k})$ are living by $\mathscr{B}$. Put

$$\xi_k = \left(\sum_{k_2=1}^{m_2} \cdots \sum_{k_d=1}^{m_d} X(k, k_2, \ldots, k_d), 1 \le m_2 \le n_2, \ldots, 1 \le m_d \le n_d \right),$$

$$\zeta_k = \sum_{i=1}^{k} \xi_i \in \mathscr{B}^N = \underbrace{\mathscr{B} \times \cdots \times \mathscr{B}}_{N \text{ times}},$$

where $N = n_2 \ldots n_d$. As the norm $\|\cdot\|_1$ in the space $\mathscr{B}^N$, we take the max-norm, that is, $\|\mathbf{x}\|_1 = \max_{k \le N} \|x_k\|$ for $\mathbf{x} = (x_1, \ldots, x_N)$, where $x_1, \ldots, x_N \in \mathscr{B}$. The coordinates of the random element ζ_k are the sums $S(k, m_2, \ldots, m_d)$ and

$$\|\zeta_k\|_1 = \max_{\substack{m_2 \le n_2 \\ \cdots \\ m_d \le n_d}} \|S(k, m_2, \ldots, m_d)\|.$$

Thus

$$\mathsf{P}\left(\max_{k \le n_1} \|\zeta_k\|_1 \ge t\right) = \mathsf{P}\left(\max_{\mathbf{k} \le \mathbf{n}} \|S(\mathbf{k})\| \ge t\right).$$

Therefore the Lévy inequality in Frechet spaces for $d = 1$ implies that

$$\mathsf{P}\left(\max_{\mathbf{k} \le \mathbf{n}} \|S(\mathbf{k})\| \ge t\right) = \mathsf{P}\left(\max_{k \le n_1} \|\zeta_k\|_1 \ge t\right) \le 2\mathsf{P}\left(\|\zeta_{n_1}\|_1 \ge t\right)$$

$$= 2\mathsf{P}\left(\max_{\substack{m_2 \le n_2 \\ \cdots \\ m_d \le n_d}} \|S(n_1, m_2, \ldots, m_d)\| \ge t \right).$$

Using the inductive assumption we complete the proof. □

6.7 Comments

The first to study the boundedness of usual series of independent random variables was Doob [77]. More details concerning the boundedness of series of independent terms for the case $d = 1$ is discussed in the book by Skorokhod [371]. A different method for studying the boundedness of series of independent terms for $d = 1$ is developed by Velez in [402]. Hoffman-Jørgensen [164] considered bounded random series in Banach spaces for $d = 1$.

The case of $d > 1$ is studied in [238].

Section 6.1 The approach we are following is due to Skorokhod [371] for the case $d = 1$. Lemma 6.9 is new. Other results for $d > 1$ are obtained in [238].

Section 6.2 Corollary 6.4 for $d = 1$ is due to Lévy and is called the *Lévy equivalence theorem* for independent random variables.

Section 6.3 Theorem 6.4 for $d = 1$ is proved by Skorokhod [371].

Section 6.4 Theorem 6.6 for $d = 1$ is actually proved by Doob [77]. Some other variants of this result are considered by Loéve [299].

Section 6.5 The notion of bounded convergence is well known in the theory of multi-indexed series (see, for example, Yanushauskas [426]). Theorem 6.8 is proved in [239].

Section 6.6 Theorem 6.9 for $d = 1$ is due to Jain and Marcus [186]. Hoffman-Jørgensen [163] and independently Buldygin [33] (see also Corollary 3.7.1 in the book by Buldygin [34]) proved Theorem 6.10. Lévy's inequality (6.27) for Banach spaces can be found in Kahane's book [188] (also see Chap. 3 in Buldygin's book [34]).

The functions satisfying condition (6.25) are called *ORV* or *Avakumović–Karamata* functions after the Yugoslavian mathematicians Avakumović (1910–1990) and Karamata (1902–1967). More detail on the Avakumović–Karamata functions are given in the monograph [37].

Chapter 7
Rate of Convergence of Multi-Indexed Series

The results on convergence of multi-indexed series obtained in Chap. 5 also allow us to study their rate of convergence. The case $d > 1$ is more challenging than the case $d = 1$, since even the setting of the problem for $d > 1$ requires a careful consideration of peculiarities arising for multi-indexed sums.

7.1 The Case $d = 1$

Let $\{X_n, n \geq 1\}$ be independent random variables. If a series $\sum X_n$ converges almost surely, then the tail

$$\zeta_n = \sum_{k=n}^{\infty} X_k$$

is well-defined almost surely for all $n \geq 1$. For the sake of definiteness, we put $\zeta_n(\omega) = 0$ for those elementary random events $\omega \in \Omega$ for which the series $\sum X_n(\omega)$ diverges. The probability of this event equals 0 if the series converges almost surely.

One of the results concerning the rate of convergence of the series $\sum X_n$ resembles the Kolmogorov law of the iterated logarithm for cumulative sums (see Theorem 10.2).

Theorem 7.1 (Y. S. Chow, H. Teicher) *Let $\{X_n, n \geq 1\}$ be independent random variables such that the series $\sum X_n$ converges almost surely. Assume that* $\operatorname{var}[\zeta_n] = B_n > 0$ *for all $n \geq 1$. If there exists a decreasing sequence of real numbers $\{M_n, n \geq 1\}$ such that $\lim_{n\to\infty} M_n = 0$ and*

$$|X_n| \leq M_n \qquad a.\ s.,$$

$$M_n = o\left(\sqrt{\frac{B_n}{\log\log B_n^{-1}}}\right), \qquad n \to \infty,$$

O. Klesov, *Limit Theorems for Multi-Indexed Sums of Random Variables*,
Probability Theory and Stochastic Modelling 71, DOI 10.1007/978-3-662-44388-0_7

then

$$\limsup_{n\to\infty} \frac{\zeta_n}{\sqrt{2B_n \log\log B_n^{-1}}} = 1 \tag{7.1}$$

almost surely.

We consider similar results in this chapter for the general case of $d \geq 1$.

7.2 The Case $d \geq 1$

Let $\{X(\mathbf{n}), \mathbf{n} \in \mathbf{N}^d\}$ be a field of independent random variables such that *all* its tails

$$\zeta(\mathbf{n}) = \sum_{\mathbf{k}\geq\mathbf{n}} X(\mathbf{k}), \quad \mathbf{n} \in \mathbf{N}^d \tag{7.2}$$

converge almost surely. Sometimes we write $\zeta_X(\mathbf{n})$ instead of $\zeta(\mathbf{n})$ to indicate that the tails are constructed from the random variables $\{X(\mathbf{n})\}$.

Consider a field $\{b(\mathbf{n}), \mathbf{n} \in \mathbf{N}^d\}$ of non-random unbounded and increasing (see Definition A.10) real constants. The relation

$$b(\mathbf{n})\zeta(\mathbf{n}) \to 0 \qquad \text{a. s.} \tag{7.3}$$

means that the speed of convergence of the series $\sum_{\mathbf{n}\in\mathbf{N}^d} X(\mathbf{n})$ is higher than the speed of divergence of $b(\mathbf{n})$. Below we find some sufficient conditions for relations like (7.3).

Relation (7.3) resembles the strong law of large numbers for cumulative sums and thus we call it the *strong law of large numbers for tails of a multi-indexed series*.

It is worth mentioning that we assume more than the convergence of a multi-indexed series if $d > 1$, namely we assume that *all* its tails converge. This property need not be mentioned if $d = 1$, since it is a trivial consequence of the convergence of a series. The situation changes drastically if $d > 1$, where the tails may diverge even if a series converges.

Below is an example of a convergent double series with a divergent tail. Note that the terms of the series constructed in Example 7.1 are non-random.

Example 7.1 We consider the case of $d = 2$, however the idea of the construction can also be applied to other dimensions $d > 1$. Consider the field of real numbers $\{X(m, n); m \geq 1, n \geq 1\}$ defined by

$$X(k, l) = \begin{cases} (-1)^l k, & k \geq 1,\ l \leq 2, \\ 0, & k \geq 1,\ l > 2. \end{cases}$$

The rectangular sums $S(m,n)$ are easy to evaluate: for example, $S(m,n) = 0$ if $m \geq 1$ and $n \geq 2$. Thus lim (min) $S(m,n) = 0$, that is the series $\sum_{m,n\geq 1} X(m,n)$ converges (see Definition A.7). Despite the convergence of the series, the tail $\zeta(1,2)$ diverges, since

$$\zeta(1,2) \stackrel{\text{def}}{=} \lim_{\substack{m\to\infty \\ n\to\infty}} \sum_{k=1}^{m} \sum_{l=2}^{n} X(k,l) = \infty.$$

This example exhibits yet another anomaly of multi-indexed sums: if all the tails $\zeta(\mathbf{m})$, $\mathbf{m} \geq \mathbf{n}_1$, converge for some $\mathbf{n}_1 \in \mathbf{N}^d$, then it need not be the case that all tails $\zeta(\mathbf{m})$, $\mathbf{m} \geq \mathbf{n}_2$, converge for $\mathbf{n}_2 \neq \mathbf{n}_1$. Indeed, the field defined above is such that the tails $\zeta(m,n)$, $m,n \geq 2$, equal 0 (they converge) but the tail $\zeta(1,2)$ diverges.

Note also that if a multi-indexed series converges absolutely, then all its tails converge.

7.3 Almost Sure Convergence of All Tails of a Multi-Indexed Series

The criterion for the almost sure convergence of *all* tails of a random multi-indexed series is given in the following assertion.

Theorem 7.2 *Let $\{X(\mathbf{n}), \mathbf{n} \in \mathbf{N}^d\}$ be a field of independent random variables. All its tails $\zeta(\mathbf{n}) = \sum_{\mathbf{k}\geq\mathbf{n}} X(\mathbf{k})$, $\mathbf{n} \in \mathbf{N}^d$, converge almost surely if and only if, for some $c > 0$,*

$$\text{the series } \sum_{\mathbf{n}\in\mathbf{N}^d} \mathsf{P}(|X(\mathbf{n})| \geq c) \text{ and } \sum_{\mathbf{n}\in\mathbf{N}^d} \mathsf{var}\left[X^c(\mathbf{n})\right] \text{ converge,} \tag{7.4}$$

$$\text{the tail } \sum_{\mathbf{k}\geq\mathbf{n}} \mathsf{E}\left[X^c(\mathbf{k})\right] \text{ converges for every } \mathbf{n} \in \mathbf{N}^d. \tag{7.5}$$

Moreover, the result is still true if the expression "for all $c > 0$" substitutes "for some $c > 0$" in the statement above.

Remark 7.1 Since the terms of two series in (7.4) are non-negative, the convergence of these series is equivalent to the convergence of all their tails

$$\sum_{\mathbf{k}\geq\mathbf{n}} \mathsf{P}(|X(\mathbf{k})| \geq c), \quad \sum_{\mathbf{k}\geq\mathbf{n}} \mathsf{var}\left[X^c(\mathbf{k})\right], \quad \mathbf{n} \in \mathbf{N}^d.$$

A simpler sufficient condition is presented in the following corollary in terms of the convergence of the series of some absolute moments of random variables.

Corollary 7.1 *Let* $\{X(\mathbf{n}), \mathbf{n} \in \mathbf{N}^d\}$ *be a field of independent random variables. Consider a field of real numbers* $\{r(\mathbf{n}), \mathbf{n} \in \mathbf{N}^d\}$ *such that at least one of the two conditions holds for every* $\mathbf{n} \in \mathbf{N}^d$: *either* $0 < r(\mathbf{n}) < 1$, *or* $1 \le r(\mathbf{n}) \le 2$ *and* $\mathsf{E}[X(\mathbf{n})] = 0$. *Assume that*

$$\sum_{\mathbf{n}\in\mathbf{N}^d} \mathsf{E}\left[|X(\mathbf{n})|^{r(\mathbf{n})}\right] < \infty.$$

Then all the tails $\zeta(\mathbf{n})$ *of the random series* $\sum X(\mathbf{n})$ *converge almost surely.*

Proof of Theorem 7.2 First we assume that conditions (7.4)–(7.5) hold for some $c > 0$; then our aim is to prove that the tail $\zeta(\mathbf{n})$ converges almost surely for every $\mathbf{n} \in \mathbf{N}^d$. Fix $\mathbf{n} \in \mathbf{N}^d$ and put

$$Y(\mathbf{k}) = \begin{cases} X(\mathbf{k}), & \mathbf{k} \ge \mathbf{n}, \\ 0, & \mathbf{k} \not\ge \mathbf{n}. \end{cases}$$

The three series

$$\sum_{\mathbf{k}\in\mathbf{N}^d} \mathsf{P}(|Y(\mathbf{k})| \ge c), \qquad \sum_{\mathbf{k}\in\mathbf{N}^d} \mathsf{E}\left[Y^c(\mathbf{k})\right], \qquad \sum_{\mathbf{k}\in\mathbf{N}^d} \mathsf{var}\left[Y^c(\mathbf{k})\right]$$

converge by the assumption of the theorem. Applying the three series theorem (Theorem 5.3) to the random variables $Y(\mathbf{k})$ we prove that the series

$$\sum_{\mathbf{k}\in\mathbf{N}^d} Y(\mathbf{k}) = \zeta_X(\mathbf{n})$$

converges almost surely.

To prove the converse implication we assume that the tail $\zeta_X(\mathbf{n})$ converges almost surely for every $\mathbf{n} \in \mathbf{N}^d$. If it so happens that for all $\varepsilon > 0$

$$\lim\,(\max)\, \mathsf{P}(|X(\mathbf{n})| \ge \varepsilon) = 0, \tag{7.6}$$

then the three series theorem (Theorem 5.3) implies that the following three numerical series

$$\sum \mathsf{P}(|Y(\mathbf{k})| \ge c), \quad \sum \mathsf{E}\left[Y^c(\mathbf{k})\right], \quad \sum \mathsf{var}\left[Y^c(\mathbf{k})\right]$$

converge. Since $\mathbf{n}$ is an arbitrary element of $\mathbf{N}^d$, conditions (7.4)–(7.5) hold.

Generally speaking, condition (7.6) does not follow from the convergence of the series $\sum_{\mathbf{n}\in\mathbf{N}^d} X(\mathbf{n})$ (see Example 5.1); however, one can assert in this case that

$$\lim\,(\min)\, \mathsf{P}(|X(\mathbf{n})| \ge \varepsilon) = 0. \tag{7.7}$$

To prove relation (7.6) under the assumption of the theorem, we will use Proposition A.2. Let $1 \le u < d$ and $v = d - u$. Fix coordinates of multi-indices of $\mathbf{N}^d$ at positions $i_1, \dots, i_u$ and let $\{j_1, \dots, j_v\} = \{1, \dots, d\} \setminus \{i_1, \dots, i_u\}$ be the set of other coordinates. Our current goal is to prove that, for all $\varepsilon > 0$,

$$\lim_{\substack{n_{j_1} \to \infty \\ \dots \\ n_{j_v} \to \infty}} \mathsf{P}(|X(\mathbf{n})| \ge \varepsilon) = 0,$$

where the indices $n_{i_1}, \dots, n_{i_u}$ are fixed. Together with (7.7) the latter results implies condition (7.6), since the coordinates $i_1, \dots, i_u$ are arbitrary.

Without loss of generality we may restrict the consideration to the case of $j_1 = 1, \dots, j_v = v$, that is $i_1 = v + 1, \dots, i_u = d$ (the same proof is valid for other sets of fixed coordinates). We fix the coordinates $n_{v+1}, \dots, n_d$ in the element $\mathbf{n} = (n_1, \dots, n_d)$.

We are going to prove that the v-tuple series

$$\sum_{(k_1,\dots,k_v) \in \mathbf{N}^v} X(k_1, \dots, k_v, n_{v+1}, \dots, n_d) \tag{7.8}$$

converges almost surely for arbitrary fixed $n_{v+1}, \dots, n_d$ (we call series (7.8) a v-tuple series, since the number of indices used to form the sum equals v). First we consider the case of $v = d - 1$. Then

$$\sum_{(k_1,\dots,k_{d-1}) \ge (n_1,\dots,n_{d-1})} X\left(k_1, \dots, k_{d-1}, n_d\right) = \zeta\left(n_1, \dots, n_d\right) - \zeta\left(n_1, \dots, n_{d-1}, n_d + 1\right)$$

and thus the multi-indexed series (7.8) with $v = d - 1$ converges almost surely, since the tails $\zeta\left(n_1, \dots, n_{d-1}, n_d\right)$ and $\zeta\left(n_1, \dots, n_{d-1}, n_d + 1\right)$ converge almost surely by the assumption of the theorem.

Next we use induction for decreasing values of v. Put

$$\zeta^{(v)}\left(n_1, \dots, n_v, n_{v+1}, \dots, n_d\right) = \sum_{k_1=n_1}^{\infty} \cdots \sum_{k_v=n_v}^{\infty} X\left(k_1, \dots, k_v, n_{v+1}, \dots, n_d\right).$$

Then

$$\begin{aligned} \zeta^{(v-1)}(n_1, \dots, n_{v-1}, n_v, n_{v+1}, \dots, n_d) &= \zeta^{(v)}(n_1, \dots, n_{v-1}, n_v, n_{v+1}, \dots, n_d) \\ &\quad - \zeta^{(v)}(n_1, \dots, n_{v-1}, n_v + 1, n_{v+1}, \dots, n_d). \end{aligned}$$

By the inductive assumption, the v-tuple series $\zeta^{(v)}(n_1, \dots, n_{v-1}, m, n_{v+1}, \dots, n_d)$ converges almost surely for all $m, n_{v+1}, \dots, n_d \ge 1$, whence we obtain that the $(v-1)$-tuple series

$$\zeta^{(v-1)}(n_1, \dots, n_{v-1}, n_v, n_{v+1}, \dots, n_d)$$

converges almost surely for all $n_v, n_{v+1}, \dots, n_d \ge 1$.

Therefore the almost sure convergence of series (7.8) is proved for all $1 \le v < d$ and all fixed $n_{v+1}, \dots, n_d$. Now Proposition A.4, implies that

$$\lim_{\substack{n_{j_1}\to\infty \\ \dots \\ n_{j_v}\to\infty}} X(\mathbf{n}) = 0 \qquad \text{a. s.,}$$

for all fixed $n_{v+1}, \dots, n_d$. This yields, for all $\varepsilon > 0$, that

$$\lim_{\substack{n_{j_1}\to\infty \\ \dots \\ n_{j_v}\to\infty}} \mathsf{P}(|X(\mathbf{n})| \ge \varepsilon) = 0,$$

for all $n_{v+1}, \dots, n_d$. Relation (7.7) together with Proposition A.2 implies property (7.6). Recalling the three series theorem (Theorem 5.3) we complete the proof of Theorem 7.2. □

Proof of Corollary 7.1 Applying bounds (A.73)–(A.75) to random variables $X(\mathbf{n})$ we obtain, for all $c > 0$,

$$\mathsf{P}(|X(\mathbf{n})| \ge c) \le \frac{\mathsf{E}\left[|X(\mathbf{n})|^{r(\mathbf{n})}\right]}{c^{r(\mathbf{n})}},$$

$$|\mathsf{E}\left[X^c(\mathbf{n})\right]| \le c^{1-r(\mathbf{n})}\mathsf{E}\left[|X(\mathbf{n})|^{r(\mathbf{n})}\right], \qquad \mathsf{E}\left[\left(X^c(\mathbf{n})\right)^2\right] \le c^{2-r(\mathbf{n})}\mathsf{E}\left[|X(\mathbf{n})|^{r(\mathbf{n})}\right].$$

Recalling Theorem 7.2 we complete the proof. □

Fix $\mathbf{n} \in \mathbf{N}^d$ and consider the tails $\zeta(\mathbf{m})$ for $\mathbf{m} \ge \mathbf{n}$. The convergence of these tails is not equivalent to the convergence of *all* tails if $d > 1$ (see Example 7.1). A criterion for the convergence of the tails $\zeta(\mathbf{m})$, $\mathbf{m} \ge \mathbf{n}$, can easily be obtained from Theorem 7.2.

Corollary 7.2 *Let $\{X(\mathbf{n}), \mathbf{n} \in \mathbf{N}^d\}$ be a field of independent random variables. Fix $\mathbf{n} \in \mathbf{N}^d$. All the tails $\zeta(\mathbf{m}) = \sum_{\mathbf{k}\ge\mathbf{m}} X(\mathbf{k})$, $\mathbf{m} \ge \mathbf{n}$, converge almost surely if and only if, for some $c > 0$, the following conditions hold*

$$\text{the series } \sum_{\mathbf{k}\ge\mathbf{n}} \mathsf{P}(|X(\mathbf{k})| \ge c), \qquad \sum_{\mathbf{k}\ge\mathbf{n}} \operatorname{var}\left[X^c(\mathbf{k})\right] \text{ converge,}$$

$$\text{the tails } \sum_{\mathbf{k}\ge\mathbf{m}} \mathsf{E}\left[X^c(\mathbf{k})\right] \text{ converge for all } \mathbf{m} \ge \mathbf{n}.$$

Moreover, the result remains valid if the expression "for all $c > 0$" substitutes "for some $c > 0$" in the statement above.

Proof Put

$$Y(\mathbf{k}) = \begin{cases} X(\mathbf{k}), & \mathbf{k} \ge \mathbf{n}, \\ 0, & \mathbf{k} \not\ge \mathbf{n}. \end{cases}$$

Since $\zeta_Y(m_1,\ldots,m_d)=\zeta_X(m_1\vee n_1,\ldots,m_d\vee n_d)$, *all* tails $\zeta_Y(\mathbf{m})$ of the series of terms $Y(\mathbf{k})$ converge almost surely if and only if all tails $\zeta_X(\mathbf{m})$, $\mathbf{m}\ge\mathbf{n}$, of the series of the terms $X(\mathbf{k})$ converge almost surely. Now applying Theorem 7.2 we complete the proof. □

7.4 The Doob–Bahr–Esseen Inequality for Tails

The Doob inequality for multi-indexed sums allows one to estimate the moments of the maximal rectangular sum via the moments of the sum with the "largest" index (see Corollary 2.9). Further, the Bahr–Esseen inequality provides a bound for the latter sum in terms of the sum of moments of individual random variables (see Theorem 2.6). We combine these two results and obtain a bound for the moments of the supremum of tails of a convergent multi-indexed series in terms of the sum of moments of the individual random variables.

Theorem 7.3 *Let $0<r\le 2$ and let $\{X(\mathbf{n}),\mathbf{n}\in\mathbf{N}^d\}$ be independent random variables such that $\mathsf{E}\left[|X(\mathbf{n})|^r\right]<\infty$, $\mathbf{n}\in\mathbf{N}^d$. If $r\ge 1$, then we additionally assume that $\mathsf{E}\left[X(\mathbf{n})\right]=0$ for all $\mathbf{n}\in\mathbf{N}^d$.*

Assume that

$$\sum_{\mathbf{n}\in\mathbf{N}^d}\mathsf{E}\left[|X(\mathbf{n})|^r\right]<\infty. \tag{7.9}$$

Then there exists a universal constant $\mathscr{D}_{r,d}$ that depends on r and d such that

$$\mathsf{E}\left[\left(\sup_{\mathbf{n}\in\mathbf{N}^d}|\zeta(\mathbf{n})|\right)^r\right]\le\mathscr{D}_{r,d}\sum_{\mathbf{n}\in\mathbf{N}^d}\mathsf{E}\left[|X(\mathbf{n})|^r\right], \tag{7.10}$$

where $\zeta(\mathbf{n})$ is the tail of the multi-indexed series defined by equality (7.2).

Proof Condition (7.9) together with the bounds obtained in the proof of Corollary A.3 implies the convergence of the series

$$\sum_{\mathbf{n}\in\mathbf{N}^d}\mathsf{P}(|X(\mathbf{n})|\ge c),\qquad\sum_{\mathbf{k}\ge\mathbf{n}}\left|\mathsf{E}\left[X^c(\mathbf{k})\right]\right|,\qquad\sum_{\mathbf{n}\in\mathbf{N}^d}\mathsf{var}\left[X^c(\mathbf{n})\right] \tag{7.11}$$

for all $c>0$ and $\mathbf{n}\in\mathbf{N}^d$. This result combined with Theorem 7.2 yields the almost sure convergence of all tails $\zeta(\mathbf{n})$.

Now we fix $\mathbf{m}\in\mathbf{N}^d$. We prove that there exists a universal constant $\mathscr{D}_{r,d}$ ($\mathscr{D}_{r,d}$ does not depend on $\mathbf{m}$) such that

$$\mathsf{E}\left[\left(\max_{\mathbf{n}\le\mathbf{m}}|\zeta(\mathbf{n})|\right)^r\right]\le\mathscr{D}_{r,d}\sum_{\mathbf{k}\in\mathbf{N}^d}\mathsf{E}\left[|X(\mathbf{k})|^r\right]. \tag{7.12}$$

The tails of a multi-indexed series admit the following representation

$$\zeta(\mathbf{n}) = \sum_{I \in \mathscr{I}} \sum_{\mathbf{k} \in M(I)} X(\mathbf{k}), \tag{7.13}$$

where $\mathscr{I} = 2^{\{1,\dots,d\}}$ is the set of all subsets of $\{1, 2, \dots, d\}$ and where $M(I), I \in \mathscr{I}$, is the subfield constituted by two multi-indices $\mathbf{n}$ and $\mathbf{m}$, $\mathbf{n} \le \mathbf{m}$, as follows:

$$M(I) = \left\{\mathbf{k} = (k_1, \dots, k_d) \in \mathbf{N}^d : k_i \ge m_i, i \in I;\; n_i \le k_i < m_i, i \in I^c\right\}$$

(here the symbol I^c denotes the complement of the set I). The sets $M(I)$ are also well-defined for $I = \varnothing$ and $I^c = \varnothing$ if we put

$$M(\varnothing) = \left\{\mathbf{k} \in \mathbf{N}^d : n_i \le k_i < m_i,\ 1 \le i \le d\right\},$$
$$M(\{1, \dots, d\}) = \left\{\mathbf{k} \in \mathbf{N}^d : k_i \ge m_i,\ 1 \le i \le d\right\}.$$

Yet another undefined case occurs if $n_i = m_i$ for at least one index $i \in I^c$. We put $M(I) = \varnothing$ in such a case.

The almost sure convergence of random series corresponding to the summation over a subfield is considered in Sect. 5.6. Since series (7.11) converges for $\mathbf{n} = \mathbf{1}$, statement (ii) of Theorem 5.10 implies that every term $\sum_{\mathbf{k}\in M(I)} X(\mathbf{k})$ in representation (7.13) converges almost surely.

The set $\mathscr{I}$ consists of 2^d members and the total number of terms of the form $\sum_{\mathbf{k}\in M(I)} X(\mathbf{k})$ in (7.13) does not exceed 2^d for all $\mathbf{n} \in \mathbf{N}^d$. The triangle inequality in the case $0 < r \le 1$ or Hölder inequality in the case $r > 1$ applied to (7.13) yields

$$\mathsf{E}\left[\left(\max_{\mathbf{n}\le\mathbf{m}} |\zeta(\mathbf{n})|\right)^r\right] \le A_{r,d} \sum_{I\in\mathscr{I}} \mathsf{E}\left[\left(\max_{\mathbf{n}\le\mathbf{m}} \left|\sum_{\mathbf{k}\in M(I)} X(\mathbf{k})\right|\right)^r\right], \tag{7.14}$$

where

$$A_{r,d} = \begin{cases} 1, & \text{if } r \le 1, \\ 2^{d(r-1)}, & \text{if } r > 1. \end{cases} \tag{7.15}$$

Each of the terms on the right-hand side of (7.14) can be estimated by the multi-index Doob inequality (see inequality (2.59)) and the Bahr–Esseen inequality (see inequality (2.6)) as follows:

$$\mathsf{E}\left[\left(\max_{\mathbf{n}\le\mathbf{m}} \left|\sum_{\mathbf{k}\in M(I)} X(\mathbf{k})\right|\right)^r\right] \le B \sum_{\mathbf{k}\in\mathbf{N}^d} \mathsf{E}\left[|X(\mathbf{k})|^r\right]$$

for some universal constant $B > 0$.

Just for the sake of demonstration, we exhibit the above estimation procedure for the set $I = \{\nu + 1, \dots, d\}$, where ν is a positive integer less than d. All other cases are considered similarly.

For $\mathbf{k} = (k_1, \dots, k_\nu, k_{\nu+1}, \dots, k_d)$, put

$$Y(k_1, \dots, k_\nu) \stackrel{\text{def}}{=} \sum_{\substack{k_{\nu+1} \geq m_{\nu+1} \\ \dots \\ k_d \geq m_d}} X(\mathbf{k}).$$

As shown in the proof of Theorem 7.2, the series defining the random variables $Y(k_1, \dots, k_\nu)$ converge almost surely. Moreover, the random variables $Y(k_1, \dots, k_\nu)$, $k_1 \leq m_1, \dots, k_\nu \leq m_\nu$, are jointly independent. We have for $I = \{\nu + 1, \dots, d\}$

$$\sum_{\mathbf{k} \in M(I)} X(\mathbf{k}) = \sum_{\substack{n_1 \leq k_1 < m_1 \\ \dots \\ n_\nu \leq k_\nu < m_\nu}} Y(k_1, \dots, k_\nu) \tag{7.16}$$

(if the summation set is empty, then the corresponding sum is assumed to be equal to zero). Thus

$$\begin{aligned} &\mathsf{E}\left[\left(\max_{\substack{1 \leq n_1 \leq m_1 \\ \dots \\ 1 \leq n_\nu \leq m_\nu}} \left| \sum_{\substack{n_1 \leq k_1 < m_1 \\ \dots \\ n_\nu \leq k_\nu < m_\nu}} Y(k_1, \dots, k_\nu) \right| \right)^r\right] \\ &\leq D_{r,\nu} \sum_{\substack{1 \leq k_1 \leq m_1 \\ \dots \\ 1 \leq k_\nu \leq m_\nu}} \mathsf{E}\left[|Y(k_1, \dots, k_\nu)|^r\right] \end{aligned} \tag{7.17}$$

by the Doob inequality for multi-indexed sums (Corollary 2.10 and inequality (2.41) if $0 < r \leq 1$ or inequality (2.42) if $r > 1$).

Now we use the Bahr–Esseen inequality (2.8) for multi-indexed sums to estimate the moments $\mathsf{E}\left[|Y(k_1, \dots, k_\nu)|^r\right]$:

$$\mathsf{E}\left[|Y(k_1, \dots, k_\nu)|^r\right] \leq C_r \sum_{\substack{k_{\nu+1} \geq m_{\nu+1} \\ \dots \\ k_d \geq m_d}} \mathsf{E}\left[|X(\mathbf{k})|^r\right]. \tag{7.18}$$

The Bahr–Esseen inequality (2.8) is valid for finite sums of independent random variables; nevertheless it can also be used for sums of an infinite number of terms under the assumption that the series $\sum \mathsf{E}\left[|X(\mathbf{n})|^r\right]$ converges (this is the result we need in order to prove inequality (7.18)). To justify the Bahr–Esseen inequality for infinite sums under the convergence of the series $\sum \mathsf{E}\left[|X(\mathbf{n})|^r\right]$, let $1 \leq \nu < d$. For all fixed $k_1, \dots, k_\nu \in \mathbf{N}^\nu$ and $m_{\nu+1}, \dots, m_d \in \mathbf{N}^{d-\nu}$, we have

$$|Y(k_1,\dots,k_\nu)|^r = \lim_{\substack{n_{\nu+1}\to\infty\\ \cdots\\ n_d\to\infty}} \left|\sum_{k_{\nu+1}=m_{\nu+1}}^{n_{\nu+1}} \cdots \sum_{k_d=m_d}^{n_d} X(\mathbf{k})\right|^r \qquad \text{a. s.}$$

by the definition of a convergent multi-indexed series (see Definition A.7). The Bahr–Esseen inequality (2.8) implies that

$$\begin{aligned}\mathsf{E}\left[\left|\sum_{k_{\nu+1}=m_{\nu+1}}^{n_{\nu+1}} \cdots \sum_{k_d=m_d}^{n_d} X(\mathbf{k})\right|^r\right] &\le C_r \sum_{k_{\nu+1}=m_{\nu+1}}^{n_{\nu+1}} \cdots \sum_{k_d=m_d}^{n_d} \mathsf{E}\left[|X(\mathbf{k})|^r\right]\\ &\le C_r \sum_{k_{\nu+1}=m_{\nu+1}}^{\infty} \cdots \sum_{k_d=m_d}^{\infty} \mathsf{E}\left[|X(\mathbf{k})|^r\right].\end{aligned}$$

This result combined with the Fatou lemma (Lemma A.33) implies bound (7.18).

Since any of the terms $\sum_{\mathbf{k}\in M(I)} X(\mathbf{k})$ does not depend on $n_{\nu+1},\dots,n_d$, inequalities (7.16)–(7.18) yield

$$\begin{aligned}\mathsf{E}\left[\max_{\mathbf{n}\le\mathbf{m}}\left|\sum_{\mathbf{k}\in M(I)} X(\mathbf{k})\right|^r\right] &\le C_r D_{r,\nu} \sum_{\substack{k_1\le m_1\\ \cdots\\ k_\nu\le m_\nu}} \sum_{\substack{k_{\nu+1}\ge m_{\nu+1}\\ \cdots\\ k_d\ge m_d}} \mathsf{E}\left[|X(\mathbf{k})|^r\right]\\ &\le B_{r,d} \sum_{\mathbf{k}\in\mathbf{N}^d} \mathsf{E}\left[|X(\mathbf{k})|^r\right],\end{aligned}$$

where $B_{r,d} = C_r \max_{1\le\nu<d} D_{r,\nu}$. Substituting this inequality into (7.14) we prove bound (7.12) with the constant

$$\mathscr{D}_{r,d} \stackrel{\text{def}}{=} 2^d A_{r,d} C_r \max_{1\le\nu<d} D_{r,\nu}. \tag{7.19}$$

The random variables $\max_{\mathbf{n}\le\mathbf{m}} |\zeta(\mathbf{n})|^r$ increase with respect to $\mathbf{m}$. Thus Beppo Levi's lemma (Proposition A.31) together with bound (7.12) completes the proof of Theorem 7.3.

7.5 The Hájek–Rényi Inequality for Tails

Now we are going to prove an inequality for a moment of the supremum of weighted tails of a multi-indexed series. Such an inequality for $d = 1$ is due to J. Hájek and A. Rényi in the case of cumulative sums (see Corollary 2.13).

Let $\{\zeta(\mathbf{n}), \mathbf{n}\in\mathbf{N}^d\}$ be the field of tails of a multi-indexed series of independent random variables $\sum_{\mathbf{n}\in\mathbf{N}^d} X(\mathbf{n})$ (see Definition (7.2)).

Theorem 7.4 *Let $\{b(\mathbf{n}), \mathbf{n} \in \mathbf{N}^d\}$ be an increasing field of non-negative non-random constants and let $\{X(\mathbf{n}), \mathbf{n} \in \mathbf{N}^d\}$ be independent random variables. Assume that either $0 < r < 1$ or $1 \le r \le 2$ and $\mathsf{E}[X(\mathbf{n})] = 0$ for all $\mathbf{n} \in \mathbf{N}^d$. Then*

$$\mathsf{E}\left[\left(\sup_{\mathbf{n}\in\mathbf{N}^d} |b(\mathbf{n})\zeta(\mathbf{n})|\right)^r\right] \le H_{r,d} \sum_{\mathbf{n}\in\mathbf{N}^d} b^r(\mathbf{n})\mathsf{E}\left[|X(\mathbf{n})|^r\right], \tag{7.20}$$

where

$$H_{r,d} = 4\mathscr{D}_{r,d} \tag{7.21}$$

and where the constant $\mathscr{D}_{r,d}$ is defined by (7.19).

Remark 7.2 If $b(\mathbf{n}) = 1$ for all $\mathbf{n} \in \mathbf{N}^d$, then Theorem 7.4 coincides with Theorem 7.3.

Proof of Theorem 7.4 Clearly the only case requiring a proof is that where the series on the right-hand of (7.20) converges. First we consider the case of $b(\mathbf{1}) = 1$. Fix $\rho > 1$ and introduce the subsets of $\mathbf{N}^d$ as follows

$$A_i = \{\mathbf{n} \in \mathbf{N}^d : \rho^i \le b(\mathbf{n}) < \rho^{i+1}\}, \qquad i \ge 0.$$

It is clear that $A_i \cap A_j = \varnothing$, $i \ne j$, and $\bigcup_{i\ge 0} A_i = \mathbf{N}^d$. Since

$$\sup_{\mathbf{n}\in\mathbf{N}^d} \xi(\mathbf{n}) \le \sum_{i=0}^{\infty} \sup_{\mathbf{n}\in A_i} \xi(\mathbf{n})$$

for all non-negative numbers $\{\xi(\mathbf{n}), \mathbf{n} \in \mathbf{N}^d\}$, we have

$$\begin{aligned}\mathsf{E}\left[\left(\sup_{\mathbf{n}\in\mathbf{N}^d} |b(\mathbf{n})\zeta(\mathbf{n})|\right)^r\right] &\le \sum_{i=0}^{\infty} \mathsf{E}\left[\left(\sup_{\mathbf{n}\in A_i} |b(\mathbf{n})\zeta(\mathbf{n})|\right)^r\right] \\ &\le \sum_{i=0}^{\infty} \rho^{r(i+1)} \mathsf{E}\left[\left(\sup_{\mathbf{n}\in A_i} |\zeta(\mathbf{n})|\right)^r\right]. \end{aligned} \tag{7.22}$$

Our current aim is to prove that, for all $i \ge 0$,

$$\mathsf{E}\left[\left(\sup_{\mathbf{n}\in A_i} |\zeta(\mathbf{n})|\right)^r\right] \le \mathscr{D}_{r,d} \sum_{j=i}^{\infty} \sum_{\mathbf{n}\in A_j} \mathsf{E}\left[|X(\mathbf{n})|^r\right], \tag{7.23}$$

where the constant $\mathscr{D}_{r,d}$ is defined in Theorem 7.3 (see (7.19)). Indeed, put

$$\mathfrak{A}_i = \bigcup_{j=i}^{\infty} A_j, \qquad Y(\mathbf{n}) = \begin{cases} X(\mathbf{n}), & \mathbf{n} \in \mathfrak{A}_i, \\ 0, & \mathbf{n} \notin \mathfrak{A}_i. \end{cases}$$

First we show that all tails of the series $\sum_{\mathbf{n}\in\mathbf{N}^d} Y(\mathbf{n})$ converge almost surely. Put $\zeta_Y(\mathbf{n}) = \sum_{\mathbf{k}\geq\mathbf{n}} Y(\mathbf{k})$. In view of $b(\mathbf{1}) = 1$,

$$\sum_{\mathbf{n}\in\mathbf{N}^d} \mathsf{E}\left[|X(\mathbf{n})|^r\right] < \infty, \tag{7.24}$$

since the field $\{b(\mathbf{n})\}$ increases. Thus Corollary 7.1 implies that all tails of the series $\sum X(\mathbf{n})$ converge almost surely. Obviously $\mathsf{E}\left[|Y(\mathbf{n})|^r\right] \leq \mathsf{E}\left[|X(\mathbf{n})|^r\right]$, thus (7.24) together with Corollary 7.1 implies that all tails of the series $\sum Y(\mathbf{n})$ converge almost surely, too.

Now we apply Theorem 7.3 to the random variables $Y(\mathbf{n})$:

$$\mathsf{E}\left[\left(\sup_{\mathbf{n}\in\mathbf{N}^d} |\zeta_Y(\mathbf{n})|\right)^r\right] \leq \mathscr{D}_{r,d} \sum_{\mathbf{n}\in\mathbf{N}^d} \mathsf{E}\left[|Y(\mathbf{n})|^r\right].$$

The latter inequality can be rewritten in the following form:

$$\mathsf{E}\left[\left(\sup_{\mathbf{n}\in\mathbf{N}^d} |\zeta_Y(\mathbf{n})|\right)^r\right] \leq \mathscr{D}_{r,d} \sum_{j=0}^{\infty} \sum_{\mathbf{n}\in A_j} \mathsf{E}\left[|Y(\mathbf{n})|^r\right] = \mathscr{D}_{r,d} \sum_{j=i}^{\infty} \sum_{\mathbf{n}\in A_j} \mathsf{E}\left[|X(\mathbf{n})|^r\right].$$

Finally, note that $\zeta_X(\mathbf{n}) = \zeta_Y(\mathbf{n})$ for all $\mathbf{n} \in A_i$, whence

$$\mathsf{E}\left[\sup_{\mathbf{n}\in A_i} |\zeta_X(\mathbf{n})|^r\right] \leq \mathsf{E}\left[\sup_{\mathbf{n}\in\mathbf{N}^d} |\zeta_Y(\mathbf{n})|^r\right].$$

Therefore inequality (7.23) holds and thus the estimate in (7.22) can be continued as follows

$$\begin{aligned}
\mathsf{E}\left[\left(\sup_{\mathbf{n}\in\mathbf{N}^d} |b(\mathbf{n})\zeta(\mathbf{n})|\right)^r\right] &\leq \mathscr{D}_{r,d} \sum_{i=0}^{\infty} \rho^{r(i+1)} \sum_{j=i}^{\infty} \sum_{\mathbf{n}\in A_j} \mathsf{E}\left[|X(\mathbf{n})|^r\right] \\
&= \mathscr{D}_{r,d} \sum_{j=0}^{\infty} \left(\sum_{\mathbf{n}\in A_j} \mathsf{E}\left[|X(\mathbf{n})|^r\right]\right) \sum_{i=0}^{j} \rho^{r(i+1)} \\
&\leq \mathscr{D}_{r,d} \frac{\rho^{2r}}{\rho^r - 1} \sum_{j=0}^{\infty} \rho^{rj} \left(\sum_{\mathbf{n}\in A_j} \mathsf{E}\left[|X(\mathbf{n})|^r\right]\right) \qquad (7.25) \\
&\leq \mathscr{D}_{r,d} \frac{\rho^{2r}}{\rho^r - 1} \sum_{\mathbf{n}\in\mathbf{N}^d} b^r(\mathbf{n}) \mathsf{E}\left[|X(\mathbf{n})|^r\right].
\end{aligned}$$

The minimum of the expression $\rho^{2r}/(\rho^r - 1)$ in the domain $\rho > 1$ is attained at $\rho = 2^{1/r}$ and is equal to 4; this means that the constant (7.21) fits the case under consideration. Theorem 7.4 is proved for the particular case $b(\mathbf{1}) = 1$.

We will derive all other cases from the part of the theorem just proved. Indeed, if $b(\mathbf{1}) > 0$, then consider the field $b'(\mathbf{n}) = b(\mathbf{n})/b(\mathbf{1})$. Obviously $b'(\mathbf{1}) = 1$ and the part of Theorem 7.4 just proved can be applied to the field $\{b'(\mathbf{n})\}$, which implies Theorem 7.4 in the case of $b(\mathbf{1}) > 0$.

Finally, we consider the case of $b(\mathbf{1}) = 0$. If $b(\mathbf{n}) = 0$ for all $\mathbf{n} \in \mathbf{N}^d$, then the result is obvious. Otherwise, that is, if $b(\mathbf{n})$ is not 0 for all $\mathbf{n} \in \mathbf{N}^d$, then consider the set of non-zero weights $\mathfrak{M} = \{\mathbf{n} \in \mathbf{N}^d : b(\mathbf{n}) > 0\}$. Put $\mu = \inf_{\mathbf{n} \in \mathfrak{M}} b(\mathbf{n})$. Since the field $\{b(\mathbf{n}), \mathbf{n} \in \mathbf{N}^d\}$ increases, we conclude that $\mu > 0$. Introduce two auxiliary fields of random variables $\{X'(\mathbf{n}), \mathbf{n} \in \mathbf{N}^d\}$ and non-random numbers $\{b'(\mathbf{n}), \mathbf{n} \in \mathbf{N}^d\}$ such that

$$X'(\mathbf{n}) = \begin{cases} X(\mathbf{n}), & \text{if } \mathbf{n} \in \mathfrak{M}, \\ 0, & \text{if } \mathbf{n} \notin \mathfrak{M}, \end{cases} \qquad b'(\mathbf{n}) = \begin{cases} b(\mathbf{n}), & \text{if } \mathbf{n} \in \mathfrak{M}, \\ \mu, & \text{if } \mathbf{n} \notin \mathfrak{M}. \end{cases}$$

It is clear that the field $\{b'(\mathbf{n}), \mathbf{n} \in \mathbf{N}^d\}$ increases, $b'(1) > 0$, and

$$\sum_{\mathbf{n} \in \mathbf{N}^d} (b'(\mathbf{n}))^r \, \mathsf{E}\left[|X'(\mathbf{n})|^r\right] < \infty.$$

Moreover, if $\zeta'(\mathbf{n}) = \zeta_{X'}(\mathbf{n})$ is a tail of the series constructed from the random variables $X'(\mathbf{k})$, then

$$b(\mathbf{n})\zeta(\mathbf{n}) = b'(\mathbf{n})\zeta'(\mathbf{n}) \qquad \text{for all } \mathbf{n} \in \mathbf{N}^d.$$

We apply the part of Theorem 7.4 already proved to the fields $\{X'(\mathbf{n}), \mathbf{n} \in \mathbf{N}^d\}$ and $\{b'(\mathbf{n}), \mathbf{n} \in \mathbf{N}^d\}$:

$$\begin{aligned} \mathsf{E}\left[\sup_{\mathbf{n} \in \mathbf{N}^d} |b(\mathbf{n})\zeta(\mathbf{n})|^r\right] &= \mathsf{E}\left[\left(\sup_{\mathbf{n} \in \mathbf{N}^d} |b'(\mathbf{n})\zeta'(\mathbf{n})|\right)^r\right] \le H_{r,d} \sum_{\mathbf{n} \in \mathbf{N}^d} (b'(\mathbf{n}))^r \, \mathsf{E}\left[|X'(\mathbf{n})|^r\right] \\ &= H_{r,d} \sum_{\mathbf{n} \in \mathbf{N}^d} b^r(\mathbf{n}) \mathsf{E}\left[|X(\mathbf{n})|^r\right]. \end{aligned}$$

The proof of Theorem 7.4 is complete. □

A partial case of Theorem 7.4 corresponding to $r = 2$ can be called the *Kolmogorov moment inequality for tails of multi-indexed series.*

Corollary 7.3 *Let $\{b(\mathbf{n}), \mathbf{n} \in \mathbf{N}^d\}$ be an increasing field of non-random constants and let $\{X(\mathbf{n}), \mathbf{n} \in \mathbf{N}^d\}$ be a field of independent random variables. Assume that all tails $\zeta(\mathbf{n})$ of the multi-indexed series $\sum_{\mathbf{n} \in \mathbf{N}^d} X(\mathbf{n})$ converge almost surely and that $\mathsf{E}[X(\mathbf{n})] = 0$ and $\mathsf{E}\left[X^2(\mathbf{n})\right] < \infty$. Then*

$$\mathsf{E}\left[\left(\sup_{\mathbf{n} \in \mathbf{N}^d} |b(\mathbf{n})\zeta(\mathbf{n})|\right)^2\right] \le 2^{3d+2} \sum_{\mathbf{n} \in \mathbf{N}^d} b^2(\mathbf{n}) \mathsf{E}\left[X^2(\mathbf{n})\right]. \tag{7.26}$$

Only the value of the constant 2^{3d+2} on the right-hand side of (7.26) needs clarification. Theorem 7.4 with $r = 2$ implies that the constant on the right-hand side of the inequality should be equal to

$$H_{2,d} = 4 \cdot 2^d A_{2,d} C_2 \max_{1 \le \nu < d} D_{2,\nu}$$

(see equalities (7.21) and (7.19)) where $A_{2,d} = 2^d$ (see (7.15)), $C_2 = 1$ is the Bahr–Esseen constant (2.10) (see (7.18)), and $D_{2,\nu} = 2^{\nu+1}$ is the constant involved in the Doob–Kolmogorov inequality (2.42) (see (7.17)). This value of the constant $D_{2,\nu}$ is explained in Remark 2.5.

7.6 The Strong Law of Large Numbers for Tails of a Multi-Indexed Series

This section is devoted to finding sufficient conditions for

$$\lim\,(\max)\, b(\mathbf{n})\zeta(\mathbf{n}) = 0 \qquad \text{almost surely.} \tag{7.27}$$

This relation is called the *strong law of large numbers for tails of a multi-indexed series*.

Remark 7.3 If the field $\{b(\mathbf{n})\}$ increases and $\limsup\,(\max)\, b(\mathbf{n}) < \infty$, then relation (7.27) means that $\lim\,(\max)\, \zeta(\mathbf{n}) = 0$ almost surely. Otherwise, that is, if $\lim\,(\max)\, b(\mathbf{n}) = \infty$, then (7.27) is an assertion about the rate of convergence of the series $\sum X(\mathbf{n})$.

Theorem 7.5 *Let* $\{b(\mathbf{n}), \mathbf{n} \in \mathbf{N}^d\}$ *be an increasing field of positive non-random constants and let* $\{X(\mathbf{n}), \mathbf{n} \in \mathbf{N}^d\}$ *be a field of independent random variables. Assume that* $\lim\,(\max)\, b(\mathbf{n}) = +\infty$. *Let either* $0 < r < 1$ *or* $1 \le r \le 2$ *and* $\mathsf{E}\,[X(\mathbf{n})] = 0$ *for all* $\mathbf{n} \in \mathbf{N}^d$. *If*

$$\sum_{\mathbf{n}\in\mathbf{N}^d} b^r(\mathbf{n})\mathsf{E}\left[|X(\mathbf{n})|^r\right] < \infty,$$

then the strong law of large numbers (7.27) holds.

Proof Note that all tails $\zeta(\mathbf{n})$ of the multi-indexed series $\sum_{\mathbf{n}\in\mathbf{N}^d} X(\mathbf{n})$ converge almost surely by Theorem 7.2.

Let $a(\mathbf{n}) = \mathsf{E}\left[|X(\mathbf{n})|^r\right]$ and choose an increasing field $\{\beta(\mathbf{n}), \mathbf{n} \in \mathbf{N}^d\}$ such that

$$\begin{aligned} &\lim\,(\max)\, \beta(\mathbf{n}) = +\infty, \\ &\lim\,(\max)\, b(\mathbf{n})/\beta(\mathbf{n}) = 0, \\ &\sum_{\mathbf{n}\in\mathbf{N}^d} \beta(\mathbf{n})a(\mathbf{n}) < \infty. \end{aligned} \tag{7.28}$$

We show that such a field $\{\beta(\mathbf{n})\}$ exists in the case where $\sum_{\mathbf{n}\in\mathbf{N}^d} b(\mathbf{n})a(\mathbf{n}) \neq 0$. Otherwise, everything is easy, since one can choose $\beta(\mathbf{n}) = b^2(\mathbf{n})$. It is also convenient to agree that the sum of the series $\sum_{\mathbf{n}\in\mathbf{N}^d} b(\mathbf{n})a(\mathbf{n})$ equals $1/2$. If this is not the case, then we pass to the field $\tilde{a}(\mathbf{n}) = a(\mathbf{n})/2s$, where $s = \sum_{\mathbf{n}\in\mathbf{N}^d} b(\mathbf{n})a(\mathbf{n})$, and the proof below runs for the field $\{\tilde{a}(\mathbf{n})\}$.

Given an arbitrary increasing sequence of positive integers $\{m_k, k \geq 0\}$ we introduce a family of sets $\{M_i, i \geq 1\}$ by

$$M_i = \{\mathbf{n} = (n_1, \dots, n_d) \in \mathbf{N}^d : n_1 + \dots + n_d \geq m_i\}.$$

It is clear that $M_{i+1} \subseteq M_i$ for all $i \geq 0$, whatever the sequence $\{m_k\}$. Now we choose a special subsequence $\{m_k, k \geq 1\}$ in such a way that $m_1 = d$ and

$$\sum_{\mathbf{n}\in M_i} b(\mathbf{n})a(\mathbf{n}) \leq \frac{1}{2^i}, \qquad i \geq 1.$$

Note that $M_1 = \mathbf{N}^d$. Next we define the field $\{c(\mathbf{n})\}$ by putting $c(\mathbf{n}) = 2^{i/2}$ for $\mathbf{n} \in M_i \setminus M_{i+1}$. The field $\{c(\mathbf{n})\}$ increases and $\lim\,(\max)\, c(\mathbf{n}) = \infty$, that is, the first condition in (7.28) holds. Now we let $\beta(\mathbf{n}) = b(\mathbf{n})c(\mathbf{n})$. Then

$$\begin{aligned}\sum_{\mathbf{n}\in\mathbf{N}^d} \beta(\mathbf{n})a(\mathbf{n}) &= \sum_{i=1}^{\infty} 2^{i/2} \sum_{\mathbf{n}\in M_i\setminus M_{i+1}} b(\mathbf{n})a(\mathbf{n}) \leq \sum_{i=1}^{\infty} 2^{i/2} \sum_{\mathbf{n}\in M_i} b(\mathbf{n})a(\mathbf{n}) \\ &\leq \sum_{i=1}^{\infty} \frac{1}{2^{i/2}} < \infty,\end{aligned}$$

that is the third condition in (7.28) also holds. Moreover, the field $\{\beta(\mathbf{n})\}$ increases, is unbounded, and the second property in (7.28) holds, as well.

Now we are in position to apply Theorem 7.4:

$$\mathsf{E}\left[\left(\sup_{\mathbf{n}\in\mathbf{N}^d} \beta(\mathbf{n})|\zeta(\mathbf{n})|\right)^r\right] \leq H_{r,d} \sum_{\mathbf{n}\in\mathbf{N}^d} \beta^r(\mathbf{n})\mathsf{E}\left[|X(\mathbf{n})|^r\right]. \tag{7.29}$$

The latter inequality implies that

$$\sup_{\mathbf{n}\in\mathbf{N}^d} \beta(\mathbf{n})|\zeta(\mathbf{n})| < +\infty \qquad \text{almost surely,}$$

whence

$$\limsup\,(\max)\, b(\mathbf{n})|\zeta(\mathbf{n})| \leq \limsup\,(\max)\, \frac{b(\mathbf{n})}{\beta(\mathbf{n})} \cdot \limsup\,(\max)\, \beta(\mathbf{n})|\zeta(\mathbf{n})| = 0$$

almost surely. Theorem 7.5 is proved.

A partial case of Theorem 7.5 with $r = 2$ can be called the *Kolmogorov strong law of large numbers for tails of convergent series.*

Corollary 7.4 *Let* $\{b(\mathbf{n}), \mathbf{n} \in \mathbf{N}^d\}$ *be an increasing field of non-random constants and let* $\{X(\mathbf{n}), \mathbf{n} \in \mathbf{N}^d\}$ *be a field of independent random variables such that* $\mathsf{E}[X(\mathbf{n})] = 0, \mathbf{n} \in \mathbf{N}^d$. *Assume that* $\lim(\max) b(\mathbf{n}) = +\infty$. *If*

$$\sum_{\mathbf{n}\in\mathbf{N}^d} b^2(\mathbf{n})\mathsf{E}\left[X^2(\mathbf{n})\right] < \infty,$$

then the strong law of large numbers (7.27) *holds*

7.6.1 Rate of Convergence for d = 1

The problems for usual series (in the case of $d = 1$) are technically simpler than their counterparts for multi-indexed series. Below we provide an example of an assertion about the rate of convergence for series of independent weighted identically distributed random variables (in other words, about the strong law of large numbers for tails of a convergent series).

Corollary 7.5 *Let* $d = 1$ *and let* $\{X_n, n \geq 1\}$ *be independent identically distributed random variables. Let* $t > 1/2$ *and* $s > 0$. *Then the strong law of large numbers*

$$\lim_{n\to\infty} n^s \sum_{k=n}^{\infty} \frac{X_k}{k^t} = 0 \quad a.\ s. \tag{7.30}$$

holds if and only if

$$\begin{aligned} &\mathsf{E}\left[|X|^{1/(t-s)}\right] < \infty, \quad \textit{if} \ \ t - s > 1, \\ &\mathsf{E}\left[|X|^{1/(t-s)}\right] < \infty \ \ \textit{and} \ \ \mathsf{E}[X] = 0, \quad \textit{if} \ \ \frac{1}{2} < t - s \leq 1. \end{aligned} \tag{7.31}$$

Moreover, if $t - s \leq \frac{1}{2}$, *then the strong law of large numbers* (7.30) *does not hold.*

Proof We restrict the consideration to the case of symmetric random variables X_n to demonstrate the main steps of the proof and to avoid technicalities (the assumption about the symmetric distribution function of X_n could be removed by a standard desymmetrization procedure).

We start with the proof of the implication (7.30) $\Longrightarrow$ (7.31). Put

$$\zeta_n = \sum_{k\geq n} \frac{X_k}{k^t}.$$

Then (7.30) means that $n^s \zeta_n \to 0$ almost surely. Since

$$\frac{X_n}{n^{t-s}} = n^s(\zeta_n - \zeta_{n+1}) \to 0 \qquad \text{a. s.,}$$

the Borel–Cantelli lemma implies that the series

$$\sum_{n=1}^{\infty} \mathsf{P}(|X_n| \ge n^{t-s})$$

converges, which in turn is equivalent to the existence of the moment $\mathsf{E}\left[|X|^{1/(t-s)}\right]$.

Now we prove the converse implication (7.31) $\Longrightarrow$ (7.30). Condition (7.31) implies, in particular, that $\mathsf{E}\left[|X|^{1/t}\right] < \infty$. Thus series $\sum X_k/k^t$ converges almost surely in view of Corollary 5.5. Consider the truncated random variables

$$Y_k = \begin{cases} X_k, & |X_k| < k^{t-s}, \\ 0, & |X_k| \ge k^{t-s}. \end{cases}$$

Clearly condition (7.31) yields

$$\mathsf{P}(X_k \ne Y_k \text{ i. o.}) = 0,$$
$$\sum_{k=1}^{\infty} \frac{\mathsf{E}\left[Y_k^2\right]}{k^{2(t-s)}} < \infty.$$

The latter result together with Corollary 7.4 proves that $n^s(Y_n + Y_{n+1} + \cdots) \to 0$ almost surely, whence we conclude that $n^s \zeta_n \to 0$ almost surely. □

7.7 The Law of the Iterated Logarithm for Tails of a Series

Relation (7.27) in the case of $d = 1$ is written as follows

$$b_n \zeta_n \xrightarrow{\text{a.s.}} 0$$

which looks similar to the strong law of large numbers for cumulative sums

$$\frac{S_n}{b_n} \xrightarrow{\text{a.s.}} 0.$$

This could lead to a conclusion that other results for cumulative sums may also have their counterparts for tails. In this section, we consider the law of the iterated logarithm for tails in the case of $d = 1$.

Let $\{X_n, n \geq 1\}$ be independent random variables with finite first and second moments, that is, $\mathsf{E}[X_n] = 0$ and $\mathsf{E}\left[X_n^2\right] \stackrel{\text{def}}{=} \sigma_n^2 < \infty$. Assume that

$$\sum_{k=1}^{\infty} \sigma_k^2 < \infty.$$

The three series theorem (Theorem 5.3) implies that the series $\sum X_k$ converges almost surely. If we put

$$\Omega_1 \stackrel{\text{def}}{=} \left\{\omega \in \Omega : \text{ series } \sum X_k(\omega) \text{ converges}\right\},$$

then the latter result can be written as follows $\mathsf{P}(\Omega_1) = 1$. Put

$$\zeta_n = \mathbb{I}_{\Omega_1} \sum_{k \geq n} X_k \quad \text{and} \quad B_n = \mathsf{E}\left[\zeta_n^2\right], \qquad n \geq 1.$$

It is clear that $B_n = \sum_{k \geq n} \sigma_k^2$.

Theorem 7.6 *Assume that* $B_n > 0$ *for all* $n \geq 1$. *If there exists a decreasing sequence of real numbers* $\{M_n, n \geq 1\}$ *such that* $\lim_{n \to \infty} M_n = 0$ *and*

$$|X_n| \leq M_n \qquad \text{a. s.}, \tag{7.32}$$

$$M_n = o\left(\sqrt{\frac{B_n}{\log \log B_n^{-1}}}\right), \qquad n \to \infty, \tag{7.33}$$

then the set of limit points of the sequence

$$\left\{\frac{\zeta_n}{\sqrt{2 B_n \log \log B_n^{-1}}}, \quad n \geq 1\right\}, \tag{7.34}$$

almost surely coincides with the interval $[-1, 1]$.

In particular, relation (7.1) *prevails.*

Proof Applying relation (7.1) to the sequence of random variables $\{-X_n\}$ we get

$$\liminf_{n \to \infty} \frac{\zeta_n}{\sqrt{2 B_n \log \log B_n^{-1}}} = -1 \qquad \text{a. s.}$$

This together with (7.1) proves that $x = \pm 1$ are limit points and that $x \notin [-1, 1]$ are not limit points for the sequence (7.34).

Now we show that all $x \in (0, 1)$ are limit points for the sequence (7.34). Then we deduce from this that $x = 0$ is also a limit point. Applying the result about $x \in (0, 1)$ to the random variables $\{-X_n, n \geq 1\}$ we prove that all $x \in (-1, 0)$ are limit points. Summarizing, the proof of the theorem will be complete if we can show that all $x \in (0, 1)$ almost surely are limit points of the sequence (7.34).

Without loss of generality we assume that $B_1 = 1$. Assumptions (7.32)–(7.33) imply that $\sigma_n^2 = o(B_n / \log\log B_n^{-1})$, whence we deduce that

$$\frac{B_{n+1}}{B_n} = 1 + o\left(\frac{B_{n+1}}{B_n \log\log B_{n+1}^{-1}}\right).$$

This means that $B_{n+1} \sim B_n$ and thus

$$\frac{B_{n+1}}{B_n} = 1 + o\left(\frac{1}{\log\log B_n^{-1}}\right). \tag{7.35}$$

Now let $x \in (0, 1)$ and $\delta = x^{-1/2}$ (note that $\delta > 1$). Further let τ be an arbitrary number of the interval $(0, 1)$. Introduce the sequence of integers $\{m_l, l \geq 0\}$ as follows:

$$m_l = \max\{m : B_m \geq \tau^l\}, \qquad l \geq 0.$$

It is clear that $B_{m_l} \geq \tau^l$ and $B_{m_l - 1} < \tau^l$, whence

$$\frac{B_{m_l-1}}{B_{m_l}} \cdot \frac{B_{m_l}}{\tau^l} < 1.$$

Taking into account (7.35), we conclude

$$B_{m_l} \sim \tau^l, \qquad l \to \infty. \tag{7.36}$$

Finally, let $n_k = m_{[k^\delta]}$, $k \geq 1$, where $[\cdot]$ denotes the integer part of a real number.

Note that $B_{m_{l+1}} < B_{m_l}$ for sufficiently large l. Indeed, if $B_{m_{l+1}} = B_{m_l}$ for some l, then

$$\frac{B_{m_{l+1}}}{\tau^{l+1}} = \frac{1}{\tau} \cdot \frac{B_{m_l}}{\tau^l}.$$

If the inequality $B_{m_{l+1}} = B_{m_l}$ is true for an infinite number of indices l, then the latter equality contradicts (7.36). Therefore $B_{m_{l+1}} < B_{m_l}$ for sufficiently large l.

To shorten the notation, we write $\chi_n^2 = 2B_n \log\log B_n^{-1}$. Our current aim is to show that

$$\limsup_{k\to\infty} \frac{\zeta_{n_k}}{\chi_{n_k}} = x \qquad \text{a. s.}$$

Of course, this means that x is almost surely a limit point. The remainder of the proof uses the principal idea of the Kolmogorov law of the iterated logarithm for cumulative sums. Namely, we prove the following two relations:

$$\limsup_{k\to\infty} \frac{\zeta_{n_k}}{\chi_{n_k}} \le x \qquad \text{a. s.,} \tag{7.37}$$

$$\limsup_{k\to\infty} \frac{\zeta_{n_k}}{\chi_{n_k}} \ge x \qquad \text{a. s.} \tag{7.38}$$

We start with the proof of inequality (7.37). In view of the Borel–Cantelli lemma, inequality (7.37) follows if

$$\sum_{k=1}^{\infty} \mathsf{P}(\zeta_{n_k} \ge (1+\varepsilon)x\chi_{n_k}) < \infty \tag{7.39}$$

for all $\varepsilon > 0$. The probabilities involved in series (7.39) are estimated with the help of Kolmogorov's upper exponential inequality.

Theorem 7.7 (Kolmogorov's upper exponential bound) *Let $X_1, \dots, X_n$ be independent random variables such that $\mathsf{E}[X_k] = 0$, $1 \le k \le n$. Assume that there exist non-random numbers $M_1, \dots, M_n$ such that*

$$|X_k| \le M_k, \qquad 1 \le k \le n, \tag{7.40}$$

almost surely. Put $B_n = \mathsf{var}[S_n]$, $M = \max\{M_1, \dots, M_n\}$. If $0 \le xM \le B_n$, then

$$\mathsf{P}(S_n \ge x) \le \exp\left\{-\frac{x^2}{2B_n}\left(1 - x\frac{M}{2B_n}\right)\right\}. \tag{7.41}$$

Below is an analog of the latter result for an infinite series.

Lemma 7.1 *Let $\{X_k, k \le n\}$ be independent random variables such that $\mathsf{E}[X_k] = 0$, $1 \le k \le n$. Assume that the series $\sum X_k$ converges almost surely and denote its sum by S (we agree that $S(\omega) = 0$ for elementary random events $\omega \in \Omega$, where the series diverges). Assume that there exists a sequence of real numbers $\{M_k\}$ for which*

$$|X_k| \le M_k, \qquad k \ge 1,$$

almost surely. Put $M = \sup\{M_1, M_2, \dots\}$ and $B = \mathsf{var}[S]$. If $0 < xM \le B$, then

$$\mathsf{P}(S \ge x) \le \exp\left\{-\frac{x^2}{2B}\left(1 - x\frac{M}{2B}\right)\right\}. \tag{7.42}$$

Proof of Lemma 7.1 If the distribution of the random variable S is continuous, then (7.42) follows from (7.41), since the almost sure convergence of a series of independent random variables implies its convergence in distribution. If the distribution function of S is not continuous, then we repeat the reasoning used in the proof of Theorem 7.7. Put

$$S_n = \sum_{k=1}^{n} X_k.$$

As in the proof of the classical Kolmogorov upper exponential bound,

$$\mathsf{E}\left[\exp\{tX_n\}\right] \le \exp\left\{\frac{1}{2}t^2\sigma_n^2\left(1+\frac{tM_n}{2}\right)\right\} \le \exp\left\{\frac{1}{2}t^2\sigma_n^2\left(1+\frac{tM}{2}\right)\right\}$$

for all $n \ge 1$ and $0 < t \le 1/M_n$, where $\sigma_n^2 = \mathsf{var}\,[X_n]$. Thus

$$\mathsf{E}\left[\exp\{tS_n\}\right] \le \exp\left\{\frac{1}{2}t^2(\sigma_1^2+\cdots+\sigma_n^2)\left(1+\frac{tM}{2}\right)\right\} \le \exp\left\{\frac{1}{2}t^2B\left(1+\frac{tM}{2}\right)\right\}.$$

Passing to the limit as $n \to \infty$, we get

$$\mathsf{E}\left[\exp\{tS\}\right] \le \exp\left\{\frac{1}{2}t^2B\left(1+\frac{tM}{2}\right)\right\}.$$

Choosing $t = x/B$ and using the Markov–Chebyshev inequality we prove that

$$\begin{aligned}\mathsf{P}(S \ge x) &\le \frac{\mathsf{E}\left[e^{xS/B}\right]}{e^{x^2/B}} \le \exp\left\{\frac{x^2}{2B}\left(1+\frac{xM}{2B}\right)-\frac{x^2}{B}\right\}\\ &= \exp\left\{-\frac{x^2}{2B}\left(1-\frac{xM}{2B}\right)\right\}.\end{aligned}$$

Now we are in position to provide a bound needed for the proof of Theorem 7.6.

Lemma 7.2 *If all the assumptions of Theorem 7.6 hold, then, for all $y > 0$ and $0 < \mu < 1$, there is a number $n_0 = n_0(y, \mu)$ such that*

$$\mathsf{P}(\zeta_n \ge y\chi_n) \le \left(\frac{1}{\log B_n^{-1}}\right)^{-y^2(1-\mu)}, \tag{7.43}$$

$$\mathsf{P}(|\zeta_n| \ge y\chi_n) \le 2\left(\frac{1}{\log B_n^{-1}}\right)^{-y^2(1-\mu)} \tag{7.44}$$

for all $n \ge n_0$.

Applying Lemma 7.2 to the terms of series (7.39) with sufficiently large indices we conclude that its convergence follows from that of the series

$$\sum_{k=1}^{\infty}\left(\frac{1}{\log B_{n_k}^{-1}}\right)^{-(1+\varepsilon)^2x^2(1-\mu)},$$

whose general term is bounded from above by const $k^{-(1+\varepsilon)^2(1-\mu)}$. Choosing μ such that $(1+\varepsilon)^2(1-\mu) > 1$, we prove (7.37).

Next we prove (7.38). Put $\varkappa_k = \zeta_{n_k} - \zeta_{n_{k+1}}$, $V_k = \mathsf{var}[\varkappa_k]$, $\varphi_k^2 = 2V_k \log\log V_k^{-1}$. Note that $V_k = B_{n_k} - B_{n_{k+1}} > 0$ for sufficiently large k and thus φ_k is well-defined for such numbers k. The proof of inequality (7.38) consists of two steps

$$\limsup_{k\to\infty} \frac{\varkappa_k}{\chi_{n_k}} \ge x \qquad \text{a. s.,} \tag{7.45}$$

$$\lim_{k\to\infty} \frac{\zeta_{n_{k+1}}}{\chi_{n_k}} = 0 \qquad \text{a. s.} \tag{7.46}$$

First we prove (7.46). As shown above, the sequence

$$\frac{\zeta_{n_k}}{\chi_{n_k}}, \qquad k \ge 1,$$

is bounded almost surely. Moreover, the asymptotic equivalence (7.36) implies that

$$\frac{\chi_{n_{k+1}}^2}{\chi_{n_k}^2} \le \frac{B_{n_{k+1}}}{B_{n_k}} \asymp \tau^{(k+1)^\delta - k^\delta} \le \tau^{\delta k^{\delta-1}} \to 0, \qquad k \to \infty,$$

since $\tau \in (0, 1)$. Finally,

$$\frac{\zeta_{n_{k+1}}}{\chi_{n_k}} = \frac{\zeta_{n_{k+1}}}{\chi_{n_{k+1}}} \cdot \frac{\chi_{n_{k+1}}}{\chi_{n_k}}$$

and relation (7.46) is proved.

The proof of relation (7.45) relies on an analog of Kolmogorov's lower exponential inequality for tails of a convergent series of independent random variables. First we provide the original Kolmogorov inequality for cumulative sums in a somewhat different form, being convenient for our purposes.

Theorem 7.8 (Kolmogorov's lower exponential inequality) *Let $X_1, \dots, X_n$ be independent random variables such that $\mathsf{E}[X_k] = 0$, $1 \le k \le n$. Assume that condition* (7.40) *holds; put*

$$M = \max\{M_1, \dots, M_n\}$$

and let $S = X_1 + \dots + X_n$ and $B = \mathsf{var}[S]$. Then, given $\mu > 0$, there exists a number $\rho > 0$ such that

$$\mathsf{P}(S \ge y) \ge \exp\left\{-\frac{y^2(1+\mu)}{2B}\right\}$$

for all numbers y such that

$$0 < y < \rho, \qquad \frac{yM}{B} \le \rho, \qquad \frac{y^2}{B} \le \rho.$$

It is important that the number ρ does not depend on the distributions of the random variables $X_1, \dots, X_n$. This number is determined by a relation between M and B (in this sense, the number ρ is a universal constant). Therefore, the result of Theorem 7.8 holds for $\varkappa_n$, too.

Lemma 7.3 *If all the assumptions of Theorem 7.6 hold, then, given $y > 0$ and $\mu > 0$, there exists a number $n_0 = n_0(y, \mu)$ such that*

$$\mathsf{P}(\varkappa_n \ge y\varphi_n) \ge \left(\frac{1}{\log V_n^{-1}}\right)^{-y^2(1+\mu)}$$

for all $n \ge n_0$.

Lemma 7.3 implies that

$$\mathsf{P}(\varkappa_k \ge (1-\varepsilon)x\varphi_k) \ge \left(\frac{1}{\log V_k^{-1}}\right)^{(1-\varepsilon)^2x^2(1+\mu)}$$

for all $\varepsilon > 0$, $\mu > 0$, and $n \ge n_0(\varepsilon, \mu)$. Since

$$\frac{B_{n_k}}{V_k} \to 1, \qquad k \to \infty,$$

we obtain

$$\mathsf{P}(\varkappa_k \ge (1-\varepsilon)x\varphi_k) \ge \text{const}\,\frac{1}{k^{(1-\varepsilon)^2(1+\mu)}}.$$

Choosing $\mu > 0$ in such a way that $(1-\varepsilon)^2(1+\mu) < 1$, we establish

$$\sum_{k=1}^{\infty} \mathsf{P}(\varkappa_k \ge (1-\varepsilon)x\varphi_k) = \infty.$$

Since random variables $\{\varkappa_k, k \ge 1\}$ are jointly independent, the Borel–Cantelli lemma implies that

$$\mathsf{P}(\varkappa_k \ge (1-\varepsilon)x\varphi_k \text{ i. o.}) = 1$$

for all $\varepsilon > 0$. Since $\varepsilon > 0$ is arbitrary, relation (7.45) follows.

Theorem 7.6 is proved. □

The case of independent weighted identically distributed random variables is considered in the following result.

Corollary 7.6 *Let $\{X_n, n \ge 1\}$ be independent identically distributed random variables with zero expectations and unit variances:* $\mathsf{E}[X_n] = 0$ *and* $\mathsf{E}\left[X_n^2\right] = 1$. *Let* $t > 1/2$. *Then the set of limit points of the sequence*

$$\sqrt{\frac{n^{2t-1}}{\log\log n}} \sum_{k=n}^{\infty} \frac{X_k}{k^t}, \qquad n \ge 3, \tag{7.47}$$

coincides with the interval $[-\theta, \theta]$, *where* $\theta = 2\sqrt{2t-1}$.

It is instructive to compare this result with Corollary 7.5. With $s = t - \frac{1}{2}$, sequence (7.47) is given by

$$\frac{n^s}{\sqrt{\log\log n}} \sum_{k=n}^{\infty} \frac{X_k}{k^t}.$$

Corollary 7.5 asserts that relation (7.30) does not hold if $2(t-s) = 1$. At the same time, Corollary 7.6 implies that

$$\limsup_{n\to\infty} \left| n^s \sum_{k=n}^{\infty} \frac{X_k}{k^t} \right| = \infty \qquad \text{a. s.}$$

7.8 Comments

Section 7.1 Theorem 7.1 is proved by Chow and Teicher [56]. Other laws of the iterated logarithm for tails of a convergent series in the case $d = 1$ are considered by Rosalsky [356] and Budianu [32]. Barbour [18] studies the weak convergence of tails $\{\zeta_n\}$. The law of the iterated logarithm for tails of convergent series in Banach spaces is studied by Deng in [69, 72]. Related results for Banach spaces are obtained by Mikosch in [316]. The strong law of large numbers for tails of usual convergent series is studied by Tomkins [401], Sung and Volodin [389]. The Kolmogorov strong law of large numbers for tails is obtained in [216] (this research is continued by Nam and Rosalsky [328] and Nam [327]).

Sections 7.3–7.5 The main results are obtained in [219].

Section 7.6 Theorem 7.5 for $d = 1$ is proved in [216].

Section 7.7 These results are from [216].

Chapter 8
The Strong Law of Large Numbers for Independent Random Variables

The celebrated *Kolmogorov theorem* is a masterpiece on the strong law of large numbers for cumulative sums of independent random variables.

Theorem 8.1 (Kolmogorov SLLN) *Let $\{X_n, n \ge 1\}$ be independent random variables such that $\mathsf{E}[X_n] = 0$ and $\mathsf{E}\left[X_n^2\right] = \sigma_n^2 < \infty$ for all $n \ge 1$. Let $\{b_n, n \ge 1\}$ be an increasing and unbounded numeric sequence. If*

$$\sum_{n=1}^{\infty} \frac{\sigma_n^2}{b_n^2} < \infty, \tag{8.1}$$

then the following strong law of large numbers holds

$$\lim_{n\to\infty} \frac{S_n}{b_n} = 0 \quad \text{a. s.} \tag{8.2}$$

The Kolmogorov condition (8.1) cannot be improved in the following sense: *if condition* (8.1) *does not hold, then one can find a sequence of independent random variables $\{X_n, n \ge 1\}$ such that $\mathsf{E}[X_n] = 0$ and $\mathsf{E}\left[X_n^2\right] = \sigma_n^2$ and such that the strong law of large numbers* (8.2) *is not valid* (see Example 8.2).

There are many generalizations of Kolmogorov's theorem for cumulative sums. It is also true that most of the proofs of these generalizations are based on Kolmogorov's theorem and thus these results cannot, in fact, be treated as real generalizations of Theorem 8.1.

One of the modern methods to prove the strong law of large numbers (8.2) consists of the following two steps.

(a) Using the two series theorem (see Theorem 5.7) we deduce from condition (8.1) that the series $\sum_{n=1}^{\infty} X_n/b_n$ converges almost surely;
(b) the almost sure convergence of the latter series and Kronecker's lemma imply the strong law of large numbers (8.2).

O. Klesov, *Limit Theorems for Multi-Indexed Sums of Random Variables*, Probability Theory and Stochastic Modelling 71, DOI 10.1007/978-3-662-44388-0_8

This method does not apply in the case of multi-indexed sums, since Kronecker's lemma fails for $d > 1$ (see Example A.2). Moreover, if this method were applicable for $d > 1$ it would prove the strong law of large numbers only for fields $\{b(\mathbf{n}), \mathbf{n} \in \mathbf{N}^d\}$ with non-negative increments but not for monotone fields $\{b(\mathbf{n})\}$ (see Definition A.11 and A.10).

In this chapter, we discuss another method to prove the strong law of large numbers that combines two features, generality and simplicity. For multi-indexed sums, it allows us to obtain the strong law of large numbers for monotone fields $\{b(\mathbf{n}), \mathbf{n} \in \mathbf{N}^d\}$, too. Even for $d = 1$, this method leads to some new results.

First we obtain the most general result and then derive from it several corollaries. It is worth mentioning that the general result is proved for random variables with an arbitrary dependence structure which leads to many applications and not only for independent random variables.

8.1 A Necessary Condition for the Strong Law of Large Numbers

Prior to studying sufficient conditions, we provide a necessary condition for the strong law of large numbers.

Theorem 8.2 *Let $\{X(\mathbf{n}), \mathbf{n} \in \mathbf{N}^d\}$ be independent random variables and let $\{b(\mathbf{n})\}$ be an increasing unbounded field of non-random numbers. Denote by $S(\mathbf{n})$ the rectangular sums of random variables $X(\mathbf{k})$ (see Sect. A.4). If*

$$\lim(\max) \frac{S(\mathbf{n})}{b(\mathbf{n})} = 0 \quad \text{a. s.,} \tag{8.3}$$

then, for all $\varepsilon > 0$,

$$\sum_{\mathbf{n} \in \mathbf{N}^d} \mathsf{P}(|X(\mathbf{n})| \ge \varepsilon b(\mathbf{n})) < \infty. \tag{8.4}$$

Proof Since $X(\mathbf{n}) = \Delta[S(\mathbf{n})]$ (see Definition A.8), we get

$$\left| \frac{X(\mathbf{n})}{b(\mathbf{n})} \right| \le \sum_{\boldsymbol{\varepsilon} \in \mathscr{E}_d} \left| \frac{S(\mathbf{n} - \boldsymbol{\varepsilon})}{b(\mathbf{n} - \boldsymbol{\varepsilon})} \right| \cdot \frac{b(\mathbf{n} - \boldsymbol{\varepsilon})}{b(\mathbf{n})},$$

where, as usual, the symbol $\mathscr{E}_d$ denotes the set of all multi-indices of dimension d each of whose coordinates equals either 0 or 1 (the sum on the right-hand side means the sum with respect to all multi-indices $\boldsymbol{\varepsilon}$ belonging to the set $\mathscr{E}_d$). To avoid a problem with multi-indices $\mathbf{n}$ containing unit coordinates, we agree that $S(\mathbf{n} - \boldsymbol{\varepsilon}) = 0$ if at least one of the coordinates equals 0. Thus

$$\left|\frac{X(\mathbf{n})}{b(\mathbf{n})}\right| \le 2^d \max_{\boldsymbol{\varepsilon}\in\mathscr{E}_d} \left|\frac{S(\mathbf{n}-\boldsymbol{\varepsilon})}{b(\mathbf{n}-\boldsymbol{\varepsilon})}\right|.$$

According to the assumption of the theorem, this means that

$$\lim(\max)\left|\frac{X(\mathbf{n})}{b(\mathbf{n})}\right| = 0 \qquad \text{a. s.}$$

Now the Borel–Cantelli lemma (Proposition A.30) implies (8.4). □

8.2 Generalizations of Kolmogorov's Theorem

Let $\{X(\mathbf{n}), \mathbf{n} \in \mathbf{N}^d\}$ be a field of random variables. We would like to stress that no assumption is imposed on the dependence structure of these random variables.

8.2.1 The Field $\{b(\mathbf{n})\}$

Consider an increasing field $\{b(\mathbf{n}), \mathbf{n} \in \mathbf{N}^d\}$ of positive numbers, that is

$$b(\mathbf{m}) \le b(\mathbf{n}) \quad \text{for all} \quad \mathbf{m} \le \mathbf{n}. \tag{8.5}$$

Since the field $\{b(\mathbf{n}), \mathbf{n} \in \mathbf{N}^d\}$ is used as a normalization, a natural assumption is that $b(\mathbf{1}) = \min_{\mathbf{n}\in\mathbf{N}^d} b(\mathbf{n}) > 0$ (of course, this does not restrict the generality). Put

$$b'(\mathbf{n}) = b(\mathbf{n})/b(\mathbf{1}).$$

In what follows we also assume that the field $\{b(\mathbf{n}), \mathbf{n} \in \mathbf{N}^d\}$ is unbounded, that is

$$\sup_{\mathbf{n}\in\mathbf{N}^d} b(\mathbf{n}) = +\infty. \tag{8.6}$$

8.2.2 The Sets A_t

Let $c > 1$ be an arbitrary constant and let the sets $\{A_t, t \ge 0\}$ be defined as follows

$$A_t = \{\mathbf{n} \colon b'(\mathbf{n}) \le c^t\}. \tag{8.7}$$

It is clear that $A_0 = \{\mathbf{n} : b'(\mathbf{n}) = 1\}$. The main properties of these sets to be used below are

$$A_t \subseteq A_{t+1}, \qquad \bigcup_{t \geq 0} A_t = \mathbf{N}^d.$$

8.2.3 The Majorizing Field $\{\lambda(\mathbf{n})\}$

Let $p > 0$. Consider a field $\{\lambda(\mathbf{n}), \mathbf{n} \in \mathbf{N}^d\}$ of positive numbers such that

$$\mathsf{E}\left[\left(\max_{\mathbf{m} \in A_t} |S(\mathbf{m})|\right)^p\right] \leq \sum_{\mathbf{m} \in A_t} \lambda(\mathbf{m}) \quad \text{for all} \quad t \geq 0. \tag{8.8}$$

Remark 8.1 A numeric field $\{\lambda(\mathbf{n})\}$ for which condition (8.8) holds exists for all random variables $\{X(\mathbf{n})\}$ with finite moments of order p. Moreover, there are infinitely many such fields and the question is how to find an "optimal" majorizing field $\{\lambda(\mathbf{n})\}$.

One possible majorizing field $\{\lambda(\mathbf{n})\}$ can be constructed as follows. Let $g(t) = \mathsf{E}\left[\left(\max_{\mathbf{m} \in A_t} |S(\mathbf{m})|\right)^p\right]$. It is clear that the sequence $\{g(t)\}$ is increasing with respect to t. Put $B_0 = A_0$ and $B_t = A_t \setminus A_{t-1}$, $t \geq 1$. If a set B_t, $t \geq 0$, is non-empty, then we choose an arbitrary $\mathbf{m}_t \in B_t$. Otherwise, if $B_t = \varnothing$ for some $t \geq 0$, then we put $\mathbf{m}_t = \mathbf{0}$. We also set $\lambda(\mathbf{0}) = 0$. Now we introduce the field $\{\lambda(\mathbf{n})\}$ as follows: for all $\mathbf{k} \in B_t$, let

if $t = 0$ if $t \geq 1$

$$\lambda(\mathbf{k}) = \begin{cases} g(0), & \text{if } \mathbf{k} = \mathbf{m}_t, \\ 0, & \text{if } \mathbf{k} \neq \mathbf{m}_t, \end{cases} \qquad \lambda(\mathbf{k}) = \begin{cases} g(t) - g(t-1), & \text{if } \mathbf{k} = \mathbf{m}_t, \\ 0, & \text{if } \mathbf{k} \neq \mathbf{m}_t. \end{cases}$$

For this field $\{\lambda(\mathbf{n})\}$, condition (8.8) holds (as usual, we agree that $\sum_{\mathbf{m} \in \varnothing} \lambda(\mathbf{m})$ equals 0):

$$\begin{aligned} \mathsf{E}\left[\left(\max_{\mathbf{m} \in A_t} |S(\mathbf{m})|\right)^p\right] &= g(t) = g(0) + \sum_{s=1}^{t} \big(g(s) - g(s-1)\big) \\ &= \sum_{s=0}^{t} \sum_{\mathbf{m} \in B_s} \lambda(\mathbf{m}) = \sum_{\mathbf{m} \in A_t} \lambda(\mathbf{m}). \end{aligned}$$

Remark 8.2 In the case of $d = 1$, another construction of the majorizing field $\{\lambda_n\}$ is available and it often leads to "quasi optimal' sequences $\{\lambda_n, n \geq 1\}$. Put $\lambda_1 = \mathsf{E}\left[|S_1|^p\right]$ and

$$\lambda_n = \mathsf{E}\left[\left(\max_{k \leq n} |S_k|\right)^p\right] - \mathsf{E}\left[\left(\max_{k \leq n-1} |S_k|\right)^p\right], \qquad n > 1. \tag{8.9}$$

Remark 8.3 The sequence $\{\lambda_n, n \geq 1\}$ introduced in Remark 8.2 in the case of $d = 1$ may lead to an improvement of Kolmogorov's theorem not only for independent random variables but also for martingales as well as for orthogonal and stationary sequences. Unfortunately, an analog of the majorizing sequence (8.9) for $d > 1$ does not exist for all families of random variables $\{X(\mathbf{n})\}$. The field

$$\lambda(\mathbf{n}) = \Delta\left[\left(\mathsf{E}\left[\left(\max_{\mathbf{k}\leq\mathbf{n}} |S(\mathbf{k})|\right)^p\right]\right)\right], \tag{8.10}$$

seems to be a natural generalization of Definition (8.9) for $d > 1$, since bound (8.8) holds for this field. Nevertheless, it is possible for $d > 1$ that $\lambda(\mathbf{n}) < 0$ for some $\mathbf{n} \in \mathbf{N}^d$ for the majorizing field defined by (8.10) (we demonstrate this property in Example 8.1). This circumstance means we cannot always use definition (8.10) for the problem discussed in this chapter.

Example 8.1 Indeed, let $d = 2$. Consider the following random (non-random, in fact) variables

$$X(1,1) = 0, \quad X(1,2) = 2, \quad X(2,1) = 1, \quad X(2,2) = -3.$$

Put $M(\mathbf{n}) = \max_{\mathbf{k}\leq\mathbf{n}} |S(\mathbf{k})|$. The values of the fields $S(\mathbf{n})$ and $M(\mathbf{n})$ for $\mathbf{n} \leq \mathbf{2}$ are given in the following table:

$\mathbf{n}$	$S(\mathbf{n})$	$M(\mathbf{n})$
(1, 1)	0	0
(1, 2)	2	2
(2, 1)	1	1
(2, 2)	0	2

If $p = 1$ and the field λ is defined by (8.10), then $\Delta[\lambda(2, 2)] = -1 < 0$, while $\Delta[\lambda(1, 1)] \geq 0$, $\Delta[\lambda(1, 2)] \geq 0$, and $\Delta[\lambda(2, 1)] \geq 0$.

Remark 8.4 The choice of the field $\{\lambda(\mathbf{n})\}$ according to the rule (8.10) is not available in general. However, it can be done in various particular (and important) cases. For example, this can be done for independent, orthogonal, and stationary random fields $\{X(\mathbf{n})\}$.

8.2.4 The General Form of the Strong Law of Large Numbers

Theorem 8.3 *Let $p > 0$. Let $c > 1$ and let the sets $\{A_t, t \geq 0\}$ be defined according to the rule* (8.7). *Further, assume that a field of random variables $\{X(\mathbf{n}), \mathbf{n} \in \mathbf{N}^d\}$ and an increasing field of positive numbers $\{b(\mathbf{n}), \mathbf{n} \in \mathbf{N}^d\}$ satisfy condition* (8.8) *for some field of non-negative numbers $\{\lambda(\mathbf{n}), \mathbf{n} \in \mathbf{N}^d\}$. Assume that*

$$\lim(\max) b(\mathbf{n}) = \infty,$$
$$\sum_{\mathbf{n}\in\mathbf{N}^d} \frac{\lambda(\mathbf{n})}{b^p(\mathbf{n})} < \infty. \tag{8.11}$$

Then the strong law of large numbers (8.3) *holds.*

Remark 8.5 If $\lim(\min) b(\mathbf{n}) = \infty$ is assumed instead of $\lim(\max) b(\mathbf{n}) = \infty$ in Theorem 8.3, then Theorem 8.3 remains true but for the convergence $\lim(\min)$.

Proof First we prove the following auxiliary result.

Lemma 8.1 *Let* $a(\mathbf{n}) \geq 0$ *for all* $\mathbf{n} \in \mathbf{N}^d$. *Assume that a field of real numbers* $\{b(\mathbf{n}), \mathbf{n} \in \mathbf{N}^d\}$ *satisfies the following conditions:*

(a_1) $b(\mathbf{m}) \leq b(\mathbf{n})$ *for all* $\mathbf{m} \leq \mathbf{n}$; *(monotonicity)*

(a_2) $\lim(\max) b(\mathbf{n}) = \infty$; *(unboundedness)*

(a_3) $\displaystyle\sum_{\mathbf{n}\in\mathbf{N}^d} \frac{a(\mathbf{n})}{b(\mathbf{n})} < \infty.$

Then there exists a field $\{\beta(\mathbf{n}), \mathbf{n} \in \mathbf{N}^d\}$ *such that*

(b_1) $\beta(\mathbf{m}) \leq \beta(\mathbf{n})$ *for all* $\mathbf{m} \leq \mathbf{n}$;

(b_2) $\lim(\max) \beta(\mathbf{n}) = \infty$;

(b_3) $\displaystyle\sum_{\mathbf{n}\in\mathbf{N}^d} \frac{a(\mathbf{n})}{\beta(\mathbf{n})} < \infty$;

(b_4) $\displaystyle\lim(\max) \frac{b(\mathbf{n})}{\beta(\mathbf{n})} = \infty.$

Proof of Lemma 8.1 Without loss of generality, we assume that $b(\mathbf{n}) \geq 1$ for all $\mathbf{n} \in \mathbf{N}^d$. For $k \geq 0$, put

$$V_k = \left\{\mathbf{n}\colon 2^k \leq b(\mathbf{n}) < 2^{k+1}\right\}, \qquad \nu_k = \sum_{\mathbf{n}\in V_k} a(\mathbf{n}).$$

It is convenient to agree that $\nu_k = 0$, if $V_k = \varnothing$ for some k. By assumption (a_2), each of the sets V_k is finite, that is $\nu_k < \infty$ for all $k \geq 0$. Assumption (a_3) implies that

$$\sum_{k=1}^{\infty} \frac{\nu_k}{2^k} < \infty.$$

Now we are going to prove that there exists an increasing and unbounded sequence $\{\gamma_k, k \geq 1\}$ such that

$$\sum_{k=1}^{\infty} \frac{\nu_k}{\gamma_k} < \infty, \tag{8.12}$$

$$\lim_{k\to\infty} \frac{\gamma_k}{2^k} = 0. \tag{8.13}$$

The existence of such a sequence is obvious for the case where only a finite number of members of the sequence $\{\nu_k, k \ge 1\}$ is positive, while all other numbers are zero. If there are infinitely many members in this sequence, then we put $r_n = \sum_{k\ge n} \nu_k 2^{-k}$, $n \ge 1$. It is clear that $r_n > 0$ for all $n \ge 1$. To continue the proof, we need the following analog of Dini's theorem.

Theorem 8.4 (U. Dini) *If $c_n > 0$ for all $n \ge 1$ and if the series $\sum c_n$ converges, then the series*

$$\sum_{n=1}^{\infty} \frac{c_n}{t_n^{1-\sigma}}, \quad \textit{where} \quad t_n \stackrel{\text{def}}{=} \sum_{k=n}^{\infty} c_k,$$

converges for all $0 < \sigma < 1$.

Applying Dini's theorem for $\sigma = \frac{1}{2}$, we get

$$\sum_{k=1}^{\infty} \frac{\nu_k}{2^k \sqrt{r_k}} < \infty.$$

If the series $\sum \nu_k$ converges, then we put $\gamma_n = 2^{n/2}$. Conditions (8.12) and (8.13) are obvious in this case.

If the series $\sum \nu_k$ diverges, then we set $\gamma_n = \max_{k\le n} 2^k \sqrt{r_k}$. It is clear that the sequence $\{\gamma_n\}$ is increasing and satisfies condition (8.12). Moreover, since $r_k \to 0$ as $k \to \infty$, condition (8.13) also holds. We prove the unboundedness of the sequence $\{\gamma_n\}$ by contradiction. Indeed, if $2^k \sqrt{r_k} \le C, k \ge 1$, then, for some constant $C < \infty$,

$$\infty > \sum \frac{\nu_k}{2^k \sqrt{r_k}} \ge \frac{1}{C} \sum \nu_k$$

and this contradicts the divergence of the series $\sum \nu_k$.

Using the sequence $\{\gamma_k, k \ge 1\}$ constructed above put

$$\beta(\mathbf{n}) = \gamma_k \qquad \text{for} \qquad \mathbf{n} \in V_k.$$

Now we check that the field $\{\beta(\mathbf{n}), \mathbf{n} \in \mathbf{N}^d\}$ satisfies conditions (b_1)–(b_4). Properties (b_1) and (b_2) follow directly from the monotonicity and unboundedness of the sequence $\{\gamma_k, k \ge 1\}$. Property (b_3) is a straightforward corollary of (8.12), while (b_4) follows from property (8.13). The lemma is proved. □

We turn back to the proof of Theorem 8.3. Choose the field $\{\beta(\mathbf{n}), \mathbf{n} \in \mathbf{N}^d\}$ according to Lemma 8.1. For this field, we apply the Hájek–Rényi inequality for multi-indexed sums (Theorem 2.19):

$$\mathsf{E}\left[\left(\max_{\mathbf{m}\le\mathbf{n}}\left|\frac{S(\mathbf{m})}{\beta(\mathbf{m})}\right|\right)^p\right] \le \frac{c^{2p}}{c^p-1}\sum_{\mathbf{m}\le\mathbf{n}}\frac{\lambda(\mathbf{m})}{\beta^p(\mathbf{m})} \le \frac{c^{2p}}{c^p-1}\sum_{\mathbf{m}\in\mathbf{N}^d}\frac{\lambda(\mathbf{m})}{\beta^p(\mathbf{m})}.$$

Then we pass to the limit in this inequality as $n_1 \to \infty, \dots, n_d \to \infty$. Thus

$$\mathsf{E}\left[\left(\sup_{\mathbf{m}\in\mathbf{N}^d}\left|\frac{S(\mathbf{m})}{\beta(\mathbf{m})}\right|\right)^p\right] \le \frac{c^{2p}}{c^p-1}\sum_{\mathbf{m}\in\mathbf{N}^d}\frac{\lambda(\mathbf{m})}{\beta^p(\mathbf{m})} < \infty$$

by the Beppo Levi theorem (Proposition A.31), that is

$$\sup_{\mathbf{m}\in\mathbf{N}^d}\left|\frac{S(\mathbf{m})}{\beta(\mathbf{m})}\right| < \infty \quad \text{a. s.}$$

Now we use property b_4) established in the proof of Lemma 8.1:

$$\begin{aligned}\lim(\max)\left|\frac{S(\mathbf{n})}{b(\mathbf{n})}\right| &= \lim(\max)\left|\frac{S(\mathbf{n})}{\beta(\mathbf{n})}\cdot\frac{\beta(\mathbf{n})}{b(\mathbf{n})}\right| \\ &\le \sup_{\mathbf{n}\in\mathbf{N}^d}\left|\frac{S(\mathbf{n})}{\beta(\mathbf{n})}\right|\cdot\lim(\max)\left|\frac{\beta(\mathbf{n})}{b(\mathbf{n})}\right| = 0 \quad \text{a. s.}\end{aligned}$$

□

8.3 Kolmogorov's Strong Law of Large Numbers

Below is a direct analog of Theorem 8.1 for multi-indexed sums. All other corollaries of Theorem 8.3 are given in the next section.

Corollary 8.1 *Let $\{X(\mathbf{n}), \mathbf{n} \in \mathbf{N}^d\}$ be a field of independent random variables such that $\mathsf{E}[X(\mathbf{n})] = 0$ and $\mathsf{E}\left[X^2(\mathbf{n})\right] < \infty$. Assume that a field $\{b(\mathbf{n}), \mathbf{n} \in \mathbf{N}^d\}$ is increasing and* $\lim(\max) b(\mathbf{n}) = \infty$. *If*

$$\sum_{\mathbf{n}\in\mathbf{N}^d}\frac{\mathsf{E}\left[X^2(\mathbf{n})\right]}{b^2(\mathbf{n})} < \infty, \tag{8.14}$$

then the strong law of large numbers (8.3) *holds.*

Remark 8.6 If one assumes that $\lim(\min) b(\mathbf{n}) = \infty$ instead of $\lim(\max) b(\mathbf{n}) = \infty$ in Corollary 8.1, then relation (8.3) remains true but for the convergence $\lim(\min)$.

Proof First we consider the case of $b(\mathbf{1}) = 1$. Define the sets $\{A_t\}$ according to the rule (8.7) for $c = 2^{1/2}$. We are going to check condition (8.8) with $p = 2$ for this field. Let $t \geq 0$ and let $\mathbf{n}_t$ be an arbitrary multi-index chosen such that $\Pi(\mathbf{n}_t) \supseteq A_t$, where $\Pi(\mathbf{n}_t) = \{\mathbf{n} \colon \mathbf{n} \leq \mathbf{n}_t\}$. We also put

$$Y(\mathbf{n}) = \begin{cases} X(\mathbf{n}), & \mathbf{n} \in A_t, \\ 0, & \mathbf{n} \in \Pi(\mathbf{n}_t) \setminus A_t, \end{cases} \qquad T(\mathbf{n}) = \sum_{\mathbf{k} \leq \mathbf{n}} Y(\mathbf{k}).$$

It is clear that if $\mathbf{m} \in \Pi(\mathbf{n}_t)$, then $S(\mathbf{m}) = T(\mathbf{m})$. Thus

$$\mathsf{E}\left[\left(\max_{\mathbf{m} \in A_t} |S(\mathbf{m})|\right)^2\right] \leq \mathsf{E}\left[\left(\max_{\mathbf{n} \leq \mathbf{n}_t} |T(\mathbf{n})|\right)^2\right].$$

Now we use Kolmogorov's inequality for multi-indexed sums (Theorem 2.19):

$$\begin{aligned} \mathsf{E}\left[\left(\max_{\mathbf{m} \in A_t} |S(\mathbf{m})|\right)^2\right] &\leq \mathsf{E}\left[\left(\max_{\mathbf{n} \leq \mathbf{n}_t} |T(\mathbf{n})|\right)^2\right] \leq D_2 \mathsf{E}\left[|T(\mathbf{n}_t)|^2\right] \\ &= D_2 \sum_{\mathbf{k} \leq \mathbf{n}_t} \mathsf{E}\left[Y^2(\mathbf{k})\right] = D_2 \sum_{\mathbf{k} \in A_t} \mathsf{E}\left[X^2(\mathbf{k})\right]. \end{aligned}$$

Let $\lambda(\mathbf{n}) = D_2 \mathsf{E}\left[X^2(\mathbf{n})\right]$. With this notation, the latter inequality is equivalent to (8.8), hence Corollary 8.1 follows from Theorem 8.3 in the case of $b(\mathbf{1}) = 1$.

In the general case of $0 < b(\mathbf{1}) \neq 1$, consider the field $b'(n) = b(\mathbf{n})/b(\mathbf{1})$. Applying the part of the corollary already proved, we complete the proof in the general case. □

8.3.1 Optimality of Kolmogorov's Conditions

Let us show that the assumption of Corollary 8.1 cannot be improved in general.

Example 8.2 We show that, given an arbitrary numerical field $\{b(\mathbf{n}), \mathbf{n} \in \mathbf{N}^d\}$, there exists a field of independent random variables $\{X(\mathbf{n}), \mathbf{n} \in \mathbf{N}^d\}$ such that the strong law of large numbers (8.3) does not hold.

Indeed, let a field $\{b(\mathbf{n}), \mathbf{n} \in \mathbf{N}^d\}$ of positive numbers satisfy the assumptions of Corollary 8.1 and let a field $\{p(\mathbf{n}), \mathbf{n} \in \mathbf{N}^d\}$ be such that $0 < p(\mathbf{n}) < 1, \mathbf{n} \in \mathbf{N}^d$, and $\sum p(\mathbf{n}) = \infty$. Further, let the independent random variables $\{X(\mathbf{n}), \mathbf{n} \in \mathbf{N}^d\}$ have the following discrete distributions

$$X(\mathbf{n}) = \begin{cases} \pm b(\mathbf{n}), & \text{with probability } \frac{1}{2} p(\mathbf{n}), \\ 0, & \text{with probability } 1 - p(\mathbf{n}). \end{cases}$$

Then $\mathsf{E}[X(\mathbf{n})] = 0$ and $\mathsf{E}\left[X^2(\mathbf{n})\right] = b^2(\mathbf{n})p(\mathbf{n})$. Thus the series on the left-hand side of (8.14) diverges. If the strong law of large numbers (8.3) is nevertheless valid, then condition (8.4) holds for an arbitrary $\varepsilon > 0$ (see Theorem 8.2). This conclusion is obviously wrong for $\varepsilon = 1$, since $\mathsf{P}(|X(\mathbf{n})| \geq b(\mathbf{n})) = p(\mathbf{n})$ and since the series $\sum p(\mathbf{n})$ diverges and therefore the strong law of large numbers does not hold for this field $\{X(\mathbf{n})\}$.

Example 8.3 Below is yet another example supporting the point of view that the Kolmogorov condition is optimal. Let $\{\sigma^2(\mathbf{n}), \mathbf{n} \in \mathbf{N}^d\}$ be an arbitrary field of non-negative numbers and let $\{b(\mathbf{n}), \mathbf{n} \in \mathbf{N}^d\}$ be an arbitrary field of increasing positive numbers such that

$$\sum_{\mathbf{n}\in\mathbf{N}^d} \frac{\sigma^2(\mathbf{n})}{b(\mathbf{n})} = \infty. \tag{8.15}$$

We construct a field of independent random variables $\{X(\mathbf{n}), \mathbf{n} \in \mathbf{N}^d\}$ such that $\mathsf{E}[X(\mathbf{n})] = 0$, $\mathsf{var}[X(\mathbf{n})] = \sigma^2(\mathbf{n})$, and such that strong law of large numbers (8.3) does not hold. Clearly the latter property follows from condition (8.15). Introduce independent random variables $X(\mathbf{n})$ with the following distributions:

$$\begin{array}{ll} \text{If } \sigma^2(\mathbf{n}) < b^2(\mathbf{n}) & \text{If } \sigma^2(\mathbf{n}) \geq b^2(\mathbf{n}) \\[2mm] \mathsf{P}(X(\mathbf{n}) = \pm b(\mathbf{n})) = \dfrac{\sigma^2(\mathbf{n})}{2b^2(\mathbf{n})}, & \mathsf{P}(X(\mathbf{n}) = \pm\sigma(\mathbf{n})) = \dfrac{1}{2}. \\ \mathsf{P}(X(\mathbf{n}) = 0) = 1 - \dfrac{\sigma^2(\mathbf{n})}{b^2(\mathbf{n})}. & \end{array}$$

It is clear that $\mathsf{E}[X(\mathbf{n})] = 0$ and $\mathsf{var}[X(\mathbf{n})] = \sigma^2(\mathbf{n})$. Moreover,

$$\begin{array}{ll} \text{If } \sigma^2(\mathbf{n}) \leq b^2(\mathbf{n}) & \text{If } \sigma^2(\mathbf{n}) > b^2(\mathbf{n}) \\ \mathsf{P}(|X(\mathbf{n})| \geq b(\mathbf{n})) = \dfrac{\sigma^2(\mathbf{n})}{b^2(\mathbf{n})}, & \mathsf{P}(|X(\mathbf{n})| \geq b(\mathbf{n})) = 1. \end{array}$$

This means that

$$\sum_{\mathbf{n}\in\mathbf{N}^d} \mathsf{P}(|X(\mathbf{n})| \geq b(\mathbf{n})) = \infty,$$

that is, condition (8.4) does not hold for $\varepsilon = 1$. Now we deduce from Theorem 8.2 that the strong law of large numbers (8.3) does not hold for this field $\{X(\mathbf{n})\}$.

8.3.2 Kolmogorov's Condition is not Optimal

At the same time, Kolmogorov's condition (8.14) is not optimal in the sense that it may happen that the strong law of large numbers (8.3) holds but condition (8.14) does not hold.

Example 8.4 We restrict ourselves to the case $d = 1$. Consider an arbitrary increasing unbounded sequence $\{b_n, n \geq 1\}$. Without loss of generality we may assume that $b_1 = 1$. Introduce the sequence of positive integers $\{n_k, k \geq 0\}$ as follows

$$n_k \stackrel{\text{def}}{=} \max\{n : b_n \leq 2^k\}, \qquad k \geq 0.$$

It is clear that if $\limsup \frac{b_{k+1}}{b_k} < 2$, then $n_{k-1} < n_k$ for sufficiently large k. To avoid technicalities suppose that $n_{k-1} < n_k$ for $k \geq k_0$, where k_0 is a fixed number.

Consider independent random variables $X_i = 0, i \leq n_{k_0-1}$, and $X_i \in \mathcal{N}\left(0, \sigma_i^2\right)$, $i > n_{k_0-1}$, where

$$\sigma_i^2 = \frac{2^{2k}}{k(n_k - n_{k-1})}, \qquad n_{k-1} < i \leq n_k, \quad k \geq k_0.$$

Since $b_i \leq 2^k$, $n_{k-1} < i \leq n_k$, we obtain

$$\begin{aligned}\sum_{i>n_{k_0-1}} \frac{\mathsf{E}\left[X_i^2\right]}{b_i^2} &= \sum_{k=k_0}^{\infty} \sum_{n_{k-1}<i\leq n_k} \frac{\mathsf{E}\left[X_i^2\right]}{b_i^2} = \sum_{k=k_0}^{\infty} \frac{2^{2k}}{k(n_k - n_{k-1})} \sum_{n_{k-1}<i\leq n_k} \frac{1}{b_i^2}\\ &\geq \sum_{k=k_0}^{\infty} \frac{1}{k} = \infty,\end{aligned}$$

that is, condition (8.14) does not hold. On the other hand, since $b_{n_k} > 2^{k-1}$, the well-known bounds for tails of the standard Gaussian distribution imply that

$$\begin{aligned}\sum_{k\geq k_0} \mathsf{P}(|S_{n_k} - S_{n_{k-1}}| \geq \varepsilon b_{n_k}) &\asymp \sum_{k\geq k_0} \frac{1}{b_{n_k}}\\ e^{-\varepsilon^2 b_{n_k}^2/2\,\mathsf{var}\left[S_{n_k}-S_{n_{k-1}}\right]} &\leq \sum_{k\geq k_0} e^{-\varepsilon^2 2^{2(k-1)}/(2^{2k+1}/k)} \\ &= \sum_{k\geq k_0} e^{-\varepsilon^2 k/8} < \infty\end{aligned} \tag{8.16}$$

for all $\varepsilon > 0$. Since $b_{n_k} \leq 2^k$, $b_{n_{k-1}+1} \geq 2^{k-1}$, we conclude that $b_{n_k} \leq 2b_{n_{k-1}+1}$, whence

$$\sum_{k\ge k_0} \mathsf{P}(|S_{n_k} - S_{n_{k-1}}| \ge \varepsilon b_{n_{k-1}+1}) < \infty$$

for all $\varepsilon > 0$. The Lévy inequality for independent symmetric random variables (see inequality (2.4)) yields

$$\mathsf{P}\Big(\max_{n_{k-1}<n\le n_k} |S_n - S_{n_{k-1}}| \ge x\Big) \le 2\mathsf{P}(|S_{n_k} - S_{n_{k-1}}| \ge x), \qquad x > 0,$$

whence

$$\sum_{k\ge k_0} \mathsf{P}\Big(\max_{n_{k-1}<n\le n_k} |S_n - S_{n_{k-1}}| \ge \varepsilon b_{n_{k-1}+1}\Big) < \infty, \qquad \varepsilon > 0.$$

Recalling the Borel–Cantelli lemma, we see that

$$\frac{1}{b_{n_{k-1}+1}} \max_{n_{k-1}<n\le n_k} |S_n - S_{n_{k-1}}| \to 0 \quad \text{a. s.,} \tag{8.17}$$

whence we conclude that

$$\frac{1}{b_{n_{k-1}+1}} |S_{n_k} - S_{n_{k-1}}| \to 0 \quad \text{a. s.} \tag{8.18}$$

Finally,

$$\frac{S_{n_k} - S_{n_{k_0-1}}}{b_{n_k}} = \frac{1}{b_{n_k}} \sum_{i=k_0}^{k} \frac{S_{n_i} - S_{n_{i-1}}}{b_{n_{i-1}+1}} \cdot b_{n_{i-1}+1}, \quad \limsup_{k\to\infty} \frac{1}{b_{n_k}} \sum_{i=k_0}^{k} b_{n_{i-1}+1} < \infty.$$

Now we derive from (8.18) that

$$\frac{S_{n_k}}{b_{n_k}} \to 0 \quad \text{a. s.}$$

This together with (8.17) implies the strong law of large numbers

$$\frac{S_n}{b_n} \to 0 \quad \text{a. s.}$$

8.4 The Marcinkiewicz–Zygmund Strong Law of Large Numbers

An analog of the strong law of large numbers for the Marcinkiewicz–Zygmund normalization is another corollary of Theorem 8.3.

Corollary 8.2 *Let $0 < p \le 2$. Assume that the random variables $\{X(\mathbf{n}), \mathbf{n} \in \mathbf{N}^d\}$ are jointly independent and $\mathsf{E}\left[|X(\mathbf{n})|^p\right] < \infty$. In the case of $1 \le p \le 2$, we additionally assume that $\mathsf{E}\,[X(\mathbf{n})] = 0$. If*

$$\sum_{\mathbf{n}\in\mathbf{N}^d} \frac{\mathsf{E}\left[|X(\mathbf{n})|^p\right]}{b^p(\mathbf{n})} < \infty \tag{8.19}$$

for an increasing field $\{b(\mathbf{n}), \mathbf{n} \in \mathbf{N}^d\}$ such that $\lim(\max)\, b(\mathbf{n}) = \infty$, then the strong law of large numbers (8.3) *holds.*

Proof of Corollary 8.2 The idea of the proof is similar to that used to prove Corollary 8.1. Given $t > 0$, introduce the random variables $Y(\mathbf{n})$, $T(\mathbf{n})$, and multi-index $\mathbf{n}_t$ in the same way as in the proof of Corollary 8.1. Using Doob's inequality for multi-indexed sums (Corollary 2.9) and Bahr–Esseen's inequality (see (2.5) and (2.6)), we establish

$$\begin{aligned}\mathsf{E}\left[\left(\max_{\mathbf{n}\in A_t}|S(\mathbf{n})|\right)^p\right] &\le D_p\mathsf{E}\left[|T(\mathbf{n}_t)|^p\right] \le B_pD_p\sum_{\mathbf{k}\le\mathbf{n}_t}\mathsf{E}\left[|Y(\mathbf{k})|^p\right]\\ &= B_pD_p\sum_{\mathbf{n}\in A_t}\mathsf{E}\left[|X(\mathbf{n})|^p\right].\end{aligned}$$

Putting $\lambda(\mathbf{n}) = B_pD_p\mathsf{E}\left[|X(\mathbf{n})|^p\right]$ we prove inequality (8.8). Thus Corollary 8.2 follows from Theorem 8.3. □

8.4.1 Remarks and Examples

Condition (8.19) is reduced to the Kolmogorov condition (8.14) if $p = 2$. Condition (8.19) can be used even if the second moments are infinite. Note, however, that the choice of a number p is not always obvious. Below we show that condition (8.19) becomes "better" if p increases. In other words, one should choose the number p as large as possible in order to obtain the "best" normalization.

Example 8.5 We restrict our attention to the case $d = 1$. Fix $q > 0$ and a sequence $\{b_n\}$. We construct a sequence of independent random variables $\{X_n\}$ such that

(z_1) $\sum \frac{\mathsf{E}[|X_n|^p]}{b_n^p} = \infty$ for all $p < q$,

$(z_2)\quad \sum \frac{\mathsf{E}[|X_n|^q]}{b_n^q} < \infty.$

Let $\{Y_n\}$ be a sequence of independent identically distributed random variables for which the moment $\mathsf{E}\left[|Y_n|^q\right]$ exists. Put $\mu_p = \mathsf{E}\left[|Y_1|^p\right]$, $p \le q$, and introduce the following random variables

$$X_n \stackrel{\text{def}}{=} \frac{b_n Y_n}{n^{1/q}\,(\ln n)^{2/q}}.$$

Then

$$\frac{\mathsf{E}\left[|X_n|^p\right]}{b_n^p} = \frac{\mu_p}{n^{p/q}\,(\ln n)^{2p/q}}, \qquad p \le q.$$

This proves conditions z_1) and z_2).

Remark 8.7 At first glance, Example 8.5 clearly indicates that the Marcinkiewicz–Zygmund condition (8.19) indeed becomes better if p grows. Nevertheless, if the distribution of the random variables Y_n in Example 8.5 is chosen such that $\mathsf{E}\left[|Y_n|^p\right] = \infty$, $p > q$, then condition (8.19) does not hold for $p > q$. This means that only one condition in the set of conditions (8.19) holds in Example 8.5, namely the one corresponding to the case of $p = q$.

Remark 8.8 If one chooses $q = 2$ in Example 8.5, then we see that Kolmogorov's condition (8.14) is better than any other Marcinkiewicz–Zygmund condition (8.19) with $p < 2$.

In contrast to what we have said above, now we show that condition (8.19) becomes "worse" if p increases. In other words, one should choose p as small as possible in order to find the "best" normalization.

Example 8.6 We restrict the consideration to the case of $d = 1$. Fix $q > 0$ and a sequence $\{b_n\}$. We construct a sequence of independent random variables $\{X_n\}$ such that

$z_3)\quad \sum \frac{\mathsf{E}[|X_n|^p]}{b_n^p} < \infty$ for all $p < q$,

$z_4)\quad \sum \frac{\mathsf{E}[|X_n|^q]}{b_n^q} = \infty.$

Let $\{Y_n\}$ be a sequence of independent discrete random variables whose distributions are given by

$$p_k \stackrel{\text{def}}{=} \mathsf{P}(Y_n = \pm k) = \frac{1}{4k^{2+q}}, \quad k \ge n; \qquad \mathsf{P}(Y_n = 0) = 1 - \frac{1}{2}\sum_{k=n}^{\infty} \frac{1}{k^{2+q}}.$$

Put $X_n = b_n Y_n$. Then

$$\mathsf{E}\left[|X_n|^p\right] = \frac{b_n^p}{2}\sum_{k=n}^{\infty} \frac{1}{k^{2+q-p}} \asymp \frac{b_n^p}{n^{1+q-p}}.$$

This implies condition z_3). Moreover,

$$\mathsf{E}\left[|X_n|^q\right] = \frac{b_n^q}{2}\sum_{k=n}^{\infty}\frac{1}{k^2} \asymp \frac{b_n^q}{n}.$$

This implies condition z_4).

Remark 8.9 It is easy to see that $\sum \mathsf{E}\left[|X_n|^p\right]/b_n^p = \infty$ for all $p > q$ in Example 8.6.

Remark 8.10 If one chooses $q = 2$ in Example 8.6, then we see that Marcinkiewicz–Zygmund's condition (8.19) with any $p < 2$ is "better" than Kolmogorov's condition (8.14).

8.5 Chung's Strong Law of Large Numbers

It is also possible to obtain an analog of a theorem of Chung for the case of $d > 1$.

Corollary 8.3 *Let $\{g(\mathbf{n}; x), \mathbf{n} \in \mathbf{N}^d\}$ be a field of even positive functions that are non-decreasing for $x > 0$. Assume that, for every $\mathbf{n} \in \mathbf{N}^d$, at least one of the following conditions holds*:

(i) the function $\frac{x}{g(\mathbf{n};x)}$ is non-decreasing for $x > 0$;
(ii) the functions $\frac{x}{g(\mathbf{n};x)}$ and $\frac{g(\mathbf{n};x)}{x^2}$ are non-increasing for $x > 0$ and, moreover, $\mathsf{E}[X(\mathbf{n})] = 0$;
(iii) the function $g(\mathbf{n}; x))/x^2$ is non-increasing for $x > 0$ and $X(\mathbf{n})$ is a symmetric random variable.

Assume that a non-random field $\{b(\mathbf{n}), \mathbf{n} \in \mathbf{N}^d\}$ is increasing and $\lim(\max) b(\mathbf{n}) = \infty$. *If*

$$\sum_{\mathbf{n}\in\mathbf{N}^d} \frac{\mathsf{E}[g(\mathbf{n}; X(\mathbf{n}))]}{g(\mathbf{n}; b(\mathbf{n}))} < \infty, \tag{8.20}$$

then the strong law of large numbers (8.3) *prevails.*

Proof of Corollary 8.3 Put $Y(\mathbf{n}) = X(\mathbf{n})I(|X(\mathbf{n})| < b(\mathbf{n}))$. Then Proposition A.27 implies that

$$\sum_{\mathbf{n}\in\mathbf{N}^d} \left|\frac{\mathsf{E}[Y(\mathbf{n})]}{b(\mathbf{n})}\right| < \infty \tag{8.21}$$

and

$$\sum_{\mathbf{n}\in\mathbf{N}^d} \frac{\mathsf{E}\left[Y^2(\mathbf{n})\right]}{b^2(\mathbf{n})} < \infty.$$

Combining these two results we get

$$\sum_{\mathbf{n}\in\mathbf{N}^d} \frac{\mathsf{var}\,[Y(\mathbf{n})]}{b^2(\mathbf{n})} < \infty.$$

Now we derive from Corollary 8.1 that

$$\lim(\max) \frac{1}{b(\mathbf{n})} \sum_{\mathbf{k}\le\mathbf{n}} (Y(\mathbf{k}) - \mathsf{E}\,[Y(\mathbf{k})]) = 0 \qquad \text{a. s.} \tag{8.22}$$

The counterpart of Kronecker's lemma (Proposition A.9) together with (8.21) yields

$$\lim(\max) \frac{1}{b(\mathbf{n})} \sum_{\mathbf{k}\le\mathbf{n}} \mathsf{E}\,[Y(\mathbf{k})] = 0. \tag{8.23}$$

Then we derive from Chebyshev's inequality that

$$\sum_{\mathbf{n}\in\mathbf{N}^d} \mathsf{P}(|X(\mathbf{n})| \ge b(\mathbf{n})) \le \sum_{\mathbf{n}\in\mathbf{N}^d} \frac{\mathsf{E}\,[g(\mathbf{n};\, X(\mathbf{n}))]}{g(\mathbf{n};\, b(\mathbf{n}))} < \infty$$

and thus the Borel–Cantelli lemma (Proposition A.30) proves that

$$\mathsf{P}(X(\mathbf{n}) \ne Y(\mathbf{n}) \text{ i. o.}) = 0. \tag{8.24}$$

Using (8.22)–(8.24) we complete the proof of the strong law of large numbers (8.3). □

8.5.1 Remarks and Examples

It is clear that Corollary 8.2 follows from Corollary 8.3 if $g(\mathbf{n}; x) = |x|^p, 0 < p \le 2$, for all $\mathbf{n} \in \mathbf{N}^d$. In addition, Corollary 8.3 allows us to prove the strong laws of large numbers for random variables for which there is no common p with finite moments $\mathsf{E}\left[|X(\mathbf{n})|^p\right]$ but, for each $\mathbf{n}$, there exists an individual $p(\mathbf{n})$ with finite moments $\mathsf{E}\left[|X(\mathbf{n})|^{p(\mathbf{n})}\right] < \infty$.

Example 8.7 We show that condition (8.20) is "better" than condition (8.19) in statement (i) of Corollary 8.3. As usual, we restrict the consideration to the case of $d = 1$.

More precisely, we find some functions $g_n(x)$, that satisfy the assumption of statement (i), a sequence of real numbers $\{b_n\}$, and a sequence of random variables $\{X_n\}$ for which condition (8.20) holds, while condition (8.19) does not hold for any $0 < p \le 2$.

Let

$$g(x) = 1 + \mathbb{I}_{[1,\infty)}(x) \cdot \frac{x}{\ln^+ x}, \qquad g_n(x) = g(x).$$

Let $\{X_n, n \ge 1\}$ be a sequence of independent identically distributed random variables such that $\mathsf{E}[g(|X_n|)] < \infty$ and $\mathsf{E}[|X_n|] = \infty$. Also let $b_n = n(\ln n)^3$. Then $\mathsf{E}[g_n(|X_n|)] = \text{const}$ and $\mathsf{E}\left[|X_n|^p\right] = \text{const}$, $p < 1$. Thus, for $0 < p < 1$,

$$g_n(b_n) = 1 + \frac{n(\ln n)^3}{\ln\left[n(\ln n)^3\right]} \asymp n(\ln n)^2,$$

that is, condition (8.20) holds. On the other hand, $b_n^p = n^p(\ln n)^{3p}$, whence

$$\sum \frac{\mathsf{E}\left[|X_n|^p\right]}{b_n^p} = \infty,$$

that is condition (8.19) does not hold for $0 < p < 1$. Condition (8.19) also does not hold for $1 \le p \le 2$ since the corresponding moments of random variables are infinite.

A similar example can be constructed for statement (ii) of Corollary 8.3.

Example 8.8 We again consider the case of $d = 1$ only. We are going to find some functions $g_n(x)$ that satisfy the assumption of statement (ii), a sequence of real numbers $\{b_n\}$, and a sequence of random variables $\{X_n\}$ for which condition (8.20) holds but the Marcinkiewicz–Zygmund condition (8.19) does not hold for any $0 < p \le 2$.

Let

$$g(x) = 1 + \mathbb{I}_{[1,\infty)}(x) \cdot \frac{x^2}{\ln^+ x}, \qquad g_n(x) = g(x).$$

Let $\{X_n\}$ be a sequence of independent identically distributed random variables such that $\mathsf{E}[g(|X_n|)] < \infty$ and $\mathsf{E}\left[X_n^2\right] = \infty$. Also let $b_n = \sqrt{n(\ln n)^3}$. Then $\mathsf{E}[g_n(|X_n|)] = \text{const}$, $\mathsf{E}\left[|X_n|^p\right] = \text{const}$, $p < 2$. Thus, for $0 < p < 2$,

$$g_n(b_n) = 1 + \frac{n(\ln n)^3}{\frac{1}{2}\ln\left[n(\ln n)^3\right]} \asymp n(\ln n)^2,$$

that is, condition (8.20) holds. On the other hand, $b_n^p = \left(n(\ln n)^3\right)^{p/2}$, whence

$$\sum \frac{\mathsf{E}\left[|X_n|^p\right]}{b_n^p} = \infty,$$

that is, condition (8.19) does not hold for $0 < p < 2$. Condition (8.19) also does not hold for $p = 2$ since the second moment is infinite.

Remark 8.11 Another feature of Example 8.8 is that condition (8.20) does not hold not only for functions $|x|^p, 0 < p \le 2$, but also for an arbitrary sequence of functions $\{g_n(x)\}$ that satisfy the assumption of statement (i) of Corollary 8.3. Indeed, without loss of generality, we suppose that $E_* \stackrel{\text{def}}{=} \inf_{n\ge1} \mathsf{E}[g_n(|X_1|)] > 0$ and $G^* \stackrel{\text{def}}{=} \sup_{n\ge1} g_n(b_1) < \infty$. The assumption of statement (i) implies that $g_n(b_n)/b_n \le g_n(b_1)/b_1$ whence

$$\sum \frac{\mathsf{E}[g_n(|X_n|)]}{g_n(b_n)} \ge E_* \sum \frac{1}{g_n(b_n)} \ge \frac{E_* b_1}{G^*} \sum \frac{1}{b_n} = \infty.$$

Now we are going to compare Corollary 8.3 for the cases (i) and (ii).

Example 8.9 We construct a sequence of independent random variables $\{X_n\}$ such that

(a) condition (8.20) does not hold for any sequence of functions $\{g_n\}$ that satisfy the assumption of statement (ii) and for any sequence $\{b_n\}$;
(b) there exists a sequence of functions $\{g_n\}$ satisfying assumption of statement (i) and such that condition (8.20) holds for all increasing sequences $\{b_n\}$.

Property (b) means that $S_n/b_n \to 0$ as $n \to \infty$ almost surely for any increasing and unbounded sequence $\{b_n\}$. In turn, this means that $\limsup |S_n| < \infty$ almost surely. Actually, $\lim S_n$ exists with probability one for the sequence $\{X_n\}$ constructed below.

The sequence $\{X_n\}$ is constructed by using the following result.

Lemma 8.2 *Let $p > 0$, $q > p$, and $\mu > 0$. Then there exists a random variable X such that*

$$\mathsf{E}\left[|X|^p\right] = \mu, \qquad \mathsf{E}\left[|X|^q\right] = \infty.$$

To prove Lemma 8.2 we consider the probability density $f(x) = cx^{-1-q} \mathbb{I}_{[a,\infty)}(x)$, $x \in \mathbf{R}$, where the constants a and c satisfy the equations

$$c = qa^q, \qquad c = \mu(q-p)a^{q-p}.$$

Turning back to Example 8.9, we choose a sequence of positive numbers $\{p_n\}$ increasing to 1. Using Lemma 8.2, we choose a sequence of independent random variables $\{X_n\}$ such that

$$\mathsf{E}\left[|X_n|^{p_n}\right] = \frac{1}{n^2}, \qquad \mathsf{E}[|X_n|] = \infty.$$

Then $\mathsf{E}[g_n(|X_n|)] = \infty$ for all $n \ge 1$ and any function g_n satisfying condition (ii), since $x/g_n(x)$ is a non-increasing function. Thus condition (8.20) does not hold for any sequence of functions $\{g_n\}$ satisfying condition (ii) and any sequence $\{b_n\}$.

On the other hand, all functions $g_n(x) = |x|^{p_n}$ satisfy condition (i). Moreover,

$$\sum_{n=1}^{\infty} \frac{\mathsf{E}\left[g_n(X_n)\right]}{g_n(b_n)} = \sum_{n=1}^{\infty} \frac{1}{n^2} \cdot \frac{1}{b_n^{p_n}} \le \frac{1}{b_1^{p_1}} \sum_{n=1}^{\infty} \frac{1}{n^2} < \infty$$

for an arbitrary increasing sequence $\{b_n\}$.

8.6 The Brunk–Prokhorov Strong Law of Large Numbers

In all preceding results, the sufficient conditions for the strong law of large numbers are expressed in terms of $\mathsf{E}\left[|X(\mathbf{n})|^p\right]$ for $0 < p \le 2$. Now we study the case of $p > 2$ and hope to obtain better results by using the moments of higher orders. Applying Doob's inequality for multi-indexed sums (Corollary 2.9), we get for $p \ge 1$ that

$$\mathsf{E}\left[\left(\max_{\mathbf{k}\le\mathbf{n}} |S(\mathbf{k})|\right)^p\right] \le D_p \mathsf{E}\left[|S(\mathbf{n})|^p\right].$$

It is easy to show that the field $\Lambda(\mathbf{n}) \stackrel{\text{def}}{=} \mathsf{E}\left[|S(\mathbf{n})|^p\right]$ increases: $\Lambda(\mathbf{m}) \le \Lambda(\mathbf{n})$ for all $\mathbf{m} \le \mathbf{n}$ (this result follows from the fact that $|S(\mathbf{n})|^p$ is a submartingale with respect to every coordinate of $\mathbf{n}$ and the expectations of a martingale increases).

8.6.1 Cumulative Sums

Theorem 8.3 implies the following generalization of the strong law for a monotone field $\{\Lambda(\mathbf{n})\}$. The result is stated for $d = 1$.

Corollary 8.4 *Let $p \ge 1$ and let $\{X_n, n \ge 1\}$ be a sequence of independent random variables such that $\mathsf{E}[X_n] = 0$ and $\mathsf{E}\left[|X_n|^p\right] < \infty$ for all $n \ge 1$. If $\{b_n, n \ge 1\}$ is a non-decreasing unbounded sequence such that*

$$\sum_{k=1}^{\infty} \frac{\mathsf{E}\left[|S_k|^p\right] - \mathsf{E}\left[|S_{k-1}|^p\right]}{b_k^p} < \infty, \tag{8.25}$$

then the strong law of large numbers

$$\lim_{n\to\infty} \frac{S_n}{b_n} = 0 \quad \text{a. s.} \tag{8.26}$$

holds.

Corollary 8.4 improves several results known for $d = 1$. Moreover, it holds for martingales, too.

The assumptions of Corollary 8.4 can be simplified for a wide class of normalizing sequences $\{b_n, n \ge 1\}$. Using the summation by parts formula one can prove that (8.25) follows from the following set of two conditions:

$$\limsup_{n\to\infty} \mathsf{E}\left[\left|\frac{S_n}{b_n}\right|^p\right] < \infty, \tag{8.27}$$

$$\sum_{n=1}^{\infty} \mathsf{E}\left[|S_n|^p\right] \nabla\left[\frac{1}{b_n^p}\right] < \infty, \tag{8.28}$$

where $\nabla\left[\frac{1}{b_n^p}\right] = \frac{1}{b_n^p} - \frac{1}{b_{n+1}^p}$. Conditions (8.27)–(8.28) are sometimes easier to check than (8.25).

Corollary 8.5 *Let $\{X_n, n \ge 1\}$ be independent random variables with zero expectations and finite moments of order $p \ge 2$, that is $\mathsf{E}[X_n] = 0$ and $\mathsf{E}\left[|X_n|^p\right] < \infty$ for all $n \ge 1$. Assume that $b_n = n^r L(n)$, where $r > 0$ and where $L(\cdot)$ is a measurable slowly varying function. If*

$$\sum_{n=1}^{\infty} \frac{\mathsf{E}\left[|S_n|^p\right]}{nb_n^p} < \infty, \tag{8.29}$$

then the strong law of large numbers (8.26) *holds.*

Proof of Corollary 8.5 Let $Z_0 = 0$, $Z_n = \mathsf{E}\left[|S_n|^p\right]$, $n \ge 1$, and $c_n = nb_n^p$. Since the sequence $\{Z_n\}$ is increasing,

$$\sum_{n=1}^{\infty} \frac{Z_n}{c_n} = \sum_{n=1}^{\infty} \frac{1}{c_n} \sum_{i=1}^{n} (Z_i - Z_{i-1}) = \sum_{i=1}^{\infty} (Z_i - Z_{i-1}) \sum_{n=i}^{\infty} \frac{1}{c_n} \asymp \sum_{i=1}^{\infty} (Z_i - Z_{i-1}) \frac{i}{c_i}. \tag{8.30}$$

The proof of the latter relation is as follows. Choose an arbitrary $0 < \theta < rp$. Then

$$\sum_{n=i}^{\infty} \frac{1}{c_n} = \sum_{n=i}^{\infty} \frac{1}{n^{1+rp} L^p(n)} \le \frac{1}{A_i} \sum_{n=i}^{\infty} \frac{1}{n^{1+rp-\theta}} \asymp \frac{1}{i^{rp-\theta} A_i},$$

where

$$A_i \stackrel{\text{def}}{=} \inf_{n\ge i} \left[n^\theta L^p(n)\right].$$

Since $A_i \sim i^\theta L^p(i)$ (see, for example, [368]), we have

$$\sum_{n=i}^{\infty} \frac{1}{c_n} \le \text{const} \cdot \frac{i}{c_i}.$$

A bound from below is proved similarly by choosing $\theta < 0$. Note also that sup should be used instead of inf in the definition of A_i.

Condition (8.30) implies that the series $\sum (Z_i - Z_{i-1}) i / c_i = \sum (Z_i - Z_{i-1}) / b_i^p$ converges. Therefore condition (8.25) holds and Corollary 8.5 follows from Corollary 8.4.

Remark 8.12 The function L in Corollary 8.5 is not assumed to be increasing. Moreover, Corollary 8.5 remains true even if the sequence $\{b_n\}$ is not monotone. Indeed, suppose $\{b_n\}$ is not monotone but condition (8.29) holds. Put

$$b_n' \stackrel{\text{def}}{=} \inf_{i \ge n} i^r L(i), \qquad n \ge 1.$$

Then $b_n' \sim b_n$, $n \to \infty$, and thus condition (8.29) holds for $\{b_n'\}$, as well. Since $\{b_n'\}$ is non-decreasing, Corollary 8.5 implies that $S_n / b_n' \to 0$ almost surely, whence we get (8.26).

The simplest case of Corollary 8.5 corresponds to the case where $b_n = n$.

Corollary 8.6 *If* $\{X_n, n \ge 1\}$ *are independent random variables with zero expectations and finite moments of order* $p \ge 2$, *that is* $\mathsf{E}[X_n] = 0$, $\mathsf{E}\left[|X_n|^p\right] < \infty$, *then the strong law of large numbers*

$$\frac{S_n}{n} \to 0 \quad \text{a. s.} \tag{8.31}$$

follows from the following condition

$$\sum_{n=1}^{\infty} \frac{\mathsf{E}\left[|S_n|^p\right]}{n^{p+1}} < \infty. \tag{8.32}$$

The Brunk–Prokhorov theorem, well-known for $d = 1$, follows from Corollary 8.6.

Theorem 8.5 (H. D. Brunk, Yu. V. Prokhorov) *Let* $d = 1$ *and* $p \ge 2$. *Let* $\{X_n\}$ *be a sequence of independent random variables such that* $\mathsf{E}[X_n] = 0$ *and* $\mathsf{E}\left[|X_n|^p\right] < \infty$ *for all* $n \ge 1$. *If*

$$\sum_{n=1}^{\infty} \frac{\mathsf{E}\left[|X_n|^p\right]}{n^{1+p/2}} < \infty, \tag{8.33}$$

then the strong law of large numbers (8.31) *holds.*

Indeed, the Dharmadhikari–Jogdeo inequality (see inequality (2.9)) implies that

$$\mathsf{E}\left[|S_n|^p\right] \le C_p n^{\frac{p}{2}-1} \sum_{i=1}^{n} \mathsf{E}\left[|X_i|^p\right],$$

whence

$$\sum_{n=1}^{\infty} \frac{\mathsf{E}\left[|S_n|^p\right]}{n^{p+1}} \le C_p \sum_{n=1}^{\infty} \frac{1}{n^{\frac{p}{2}+2}} \sum_{i=1}^{n} \mathsf{E}\left[|X_i|^p\right] \le \text{const} \sum_{i=1}^{\infty} \frac{\mathsf{E}\left[|X_i|^p\right]}{i^{\frac{p}{2}+1}}.$$

Therefore condition (8.32) follows from the Brunk–Prokhorov condition (8.33) and thus strong law of large numbers (8.31) holds by Corollary 8.6.

Remark 8.13 Below we provide an example of random variables $\{X_n\}$ for which the Brunk–Prokhorov condition (8.33) holds but the Kolmogorov condition (8.1) does not.

Example 8.10 Let $p > 2$ and let independent random variables $\{\beta_n, n \ge 1\}$ have the identical Rademacher distribution function, that is $\mathsf{P}(\beta_n = +1) = \mathsf{P}(\beta_n = -1) = \frac{1}{2}$. Let $0 < \gamma \le \frac{1}{2} - \frac{1}{p}$. Put $w_n = \sqrt{n}/(\log n)^{\frac{1}{p}+\gamma}$. Finally, let $X_n = w_n \beta_n$. It is clear that $\mathsf{E}[X_n] = 0$, $\mathsf{E}\left[X_n^2\right] = n/(\log n)^{\frac{2}{p}+2\gamma}$, $\mathsf{E}\left[|X_n|^p\right] = n^{\frac{p}{2}}/(\log n)^{1+p\gamma}$. Thus

$$\sum_{n=1}^{\infty} \frac{\operatorname{var}\left[X_n^2\right]}{n^2} = \sum_{n=1}^{\infty} \frac{1}{n(\log n)^{\frac{2}{p}+2\gamma}} = \infty,$$

that is, Kolmogorov's condition is not valid, since $\frac{2}{p} + 2\gamma \le 1$. On the other hand,

$$\sum_{n=1}^{\infty} \frac{\mathsf{E}\left[|X_n|^p\right]}{n^{\frac{p}{2}+1}} = \sum_{n=1}^{\infty} \frac{1}{n(\log n)^{1+p\gamma}} < \infty,$$

that is the Brunk–Prokhorov condition is valid and thus the strong law of large numbers holds.

The Brunk–Prokhorov condition (8.33) is sometimes more restrictive than condition (8.32). Below we provide an example of random variables for which the assumption of Corollary 8.6 holds for all $p > 2$ but the Brunk–Prokhorov condition does not hold for any $p \ge 2$.

Example 8.11 Let $\{X_n, n \ge 1\}$ be a sequence of independent Gaussian random variables with zero expectations and variances σ_n^2, that is, $\mathsf{E}[X_n] = 0$ and $\mathsf{E}\left[X_n^2\right] = \sigma_n^2$. We know that for any random variable $\gamma \in \mathscr{N}\left(0, \sigma^2\right)$ and for $p > 0$,

$$\mathsf{E}\left[|\gamma|^p\right] = \kappa_p \sigma^p, \quad \text{where} \quad \kappa_p \stackrel{\text{def}}{=} \frac{1}{\sqrt{2\pi}} \int_{-\infty}^{\infty} |x|^p e^{-x^2/2}\, dx.$$

Thus $\mathsf{E}\left[|S_n|^p\right] = \kappa_p \left(\sigma_1^2 + \cdots + \sigma_n^2\right)^{p/2}$. In what follows we specify the variances:

$$\sigma_n^2 = \begin{cases} 2^{m\theta_m}, & n = 2^m, \\ 0, & n \notin \{2^m\}, \end{cases} \quad \text{where} \quad \theta_m = 2 - \frac{\log_2 m}{m}.$$

Let $p \geq 2$. Then

$$\sum_{n=1}^{\infty} \frac{\mathsf{E}\left[|X_n|^p\right]}{n^{\frac{p}{2}+1}} = \kappa_p \sum_{m=0}^{\infty} \frac{2^{m\theta_m p/2}}{2^{-m(1+p/2)}}.$$

Since

$$-1 - \frac{p}{2} + p - \frac{p \log_2 m}{2m} \geq -\log_2 m^{p/2m},$$

we obtain

$$\sum_{n=1}^{\infty} \frac{\mathsf{E}\left[|X_n|^p\right]}{n^{\frac{p}{2}+1}} \geq \kappa_p \sum_{m=1}^{\infty} \frac{1}{m^{p/2m}} = \infty,$$

that is, the Brunk–Prokhorov condition (8.33) is not valid.

On the other hand, for all $p > 2$,

$$\sum_{n=1}^{\infty} \frac{\mathsf{E}\left[|S_n|^p\right]}{n^{p+1}} = \kappa_p \sum_{n=1}^{\infty} \frac{\left(\sigma_1^2 + \cdots + \sigma_n^2\right)^{p/2}}{n^{p+1}} = \kappa_p \sum_{m=0}^{\infty} \sum_{2^m \leq n < 2^{m+1}} \frac{\left(\sigma_1^2 + \cdots + \sigma_n^2\right)^{p/2}}{n^{p+1}}$$
$$\leq \kappa_p \sum_{m=0}^{\infty} 2^{-mp} \left(\sigma_1^2 + \cdots + \sigma_{2^m}^2\right)^{p/2}.$$

Since

$$\sigma_1^2 + \cdots + \sigma_{2^m}^2 = \sum_{i=1}^{m} 2^{i\theta_i} = \sum_{i=1}^{m} \frac{2^{2i}}{i} \asymp \frac{2^{2m}}{m},$$

we conclude that

$$\sum_{n=2}^{\infty} \frac{\mathsf{E}\left[|S_n|^p\right]}{n^{p+1}} \leq \text{const} \sum_{m=1}^{\infty} 2^{-mp} \left(\frac{2^{2m}}{m}\right)^{p/2} = \sum_{m=1}^{\infty} \frac{1}{m^{p/2}} < \infty.$$

Thus condition (8.32) holds.

8.6.2 Multi-Indexed Sums

We turn to the general case of dimension $d \ge 1$. If a field $\Lambda(\mathbf{n})$ increases, then its increments are not necessarily non-negative if $d > 1$. In other words, the property $\Delta[\Lambda(\mathbf{n})] \ge 0$ may fail even if $\Lambda(\mathbf{m}) \le \Lambda(\mathbf{n})$ for all $\mathbf{m} \le \mathbf{n}$. This phenomenon does not allow us to directly use the method of Corollary 8.4 to prove the strong law of large numbers for $d > 1$ and $p > 2$, since the property $\Delta[\Lambda(\mathbf{n})] \ge 0$ is crucial for this method. Nevertheless, there exists another approach to establishing the strong law of large numbers as in Corollary 8.5. First we prove an auxiliary result.

Lemma 8.3 *Assume that the fields $\{A(\mathbf{n}), \mathbf{n} \in \mathbf{N}^d\}$ and $\{B(\mathbf{n}), \mathbf{n} \in \mathbf{N}^d\}$ have non-negative increments:*

$$\Delta[A(\mathbf{n})] \ge 0, \qquad \Delta[B(\mathbf{n})] \ge 0.$$

Then the field $C(\mathbf{n}) = A(\mathbf{n})B(\mathbf{n})$ also has non-negative increments.

By induction, Lemma 8.3 implies the following assertion.

Lemma 8.4 *If a field $\{A(\mathbf{n}), \mathbf{n} \in \mathbf{N}^d\}$ has non-negative increments and p is a positive integer, then the field $\{A^p(\mathbf{n}), \mathbf{n} \in \mathbf{N}^d\}$ also has non-negative increments.*

Proof of Lemma 8.3 Since both fields $\{A(\mathbf{n})\}$ and $\{B(\mathbf{n})\}$ have non-negative increments, there exist non-negative numbers $\{a(\mathbf{n})\}$ and $\{b(\mathbf{n})\}$ such that

$$A(\mathbf{n}) = \sum_{\mathbf{k} \le \mathbf{n}} a(\mathbf{k}), \qquad B(\mathbf{n}) = \sum_{\mathbf{k} \le \mathbf{n}} b(\mathbf{k})$$

for all $\mathbf{n} \in \mathbf{N}^d$ (see Definition A.8). We are going to establish a similar representation for the field $C(\mathbf{n}) = A(\mathbf{n})B(\mathbf{n})$. Introduce the set $M(c; I) = M(\mathbf{n}; c; I)$ of pairs of multi-indices $(\mathbf{k}, \mathbf{l})$ for all $\mathbf{n} \in \mathbf{N}^d$, all $0 \le c \le d$, and all subsets $I \subseteq \{1, \dots, d\}$ with $\operatorname{card}(I) = c$. The construction is as follows. First we define the sets $M(c; I)$ for $0 < c < d$. Let $I = \{i_1, \dots, i_c\}$. Put $J \stackrel{\text{def}}{=} \{1, \dots, d\} \setminus I = \{j_1, \dots, j_{d-c}\}$ and

$$\begin{aligned} M(c; I) = \{(\mathbf{k}, \mathbf{l}) : &\mathbf{k} \le \mathbf{n}, \mathbf{l} \le \mathbf{n}, \mathbf{k} \ne \mathbf{l}, \text{ and} \\ &k_i \le l_i \text{ for } i \in I, \text{ but } k_j > l_j \text{ for } j \in J\}. \end{aligned}$$

For these sets $M(c; I)$, we need to add two others that correspond to the cases $c = 0$ and $c = d$, respectively, that is $M(0; \varnothing)$ and $M(d; \{1, \dots, d\})$. The definition of the latter two sets differs from the definition of the previous ones, since now a group of specific conditions disappears, namely these groups correspond to the sign "$\le$" for $c = 0$ or "$>$"for $c = d$. The family of such sets $M = M(c; I)$, $0 \le c \le d$, $I \in \{1, \dots, d\}$, is denoted by $\mathscr{M}$. Clearly the members of $\mathscr{M}$ are disjoint and their union equals the hypercube without the diagonal $\{(\mathbf{k}, \mathbf{l}) : \mathbf{k} \le \mathbf{n}, \mathbf{l} \le \mathbf{n}, \mathbf{k} \ne \mathbf{l}\}$. Note also that $\operatorname{card}(\mathscr{M}) = 2^d$.

Then

$$C(\mathbf{n}) = \sum_{\mathbf{k}\le\mathbf{n}}\sum_{\mathbf{l}\le\mathbf{n}} a(\mathbf{k})b(\mathbf{l}) = \sum_{\mathbf{k}\le\mathbf{n}} a(\mathbf{k})b(\mathbf{k}) + \sum_{M\in\mathscr{M}}\ \sum_{(\mathbf{k},\mathbf{l})\in M} a(\mathbf{k})b(\mathbf{l}). \tag{8.34}$$

The first term on the right-hand side of (8.34) is such that

$$\Delta\left[\sum_{\mathbf{k}\le\mathbf{n}} a(\mathbf{k})b(\mathbf{k})\right] = a(\mathbf{n})b(\mathbf{n}) \ge 0, \qquad \mathbf{n}\in\mathbf{N}^d,$$

that is, its increments are non-negative (see Definition A.8).

We prove that the second term has the same property. For this, we show that every sum $\sum_{(\mathbf{k},\mathbf{l})\in M} a(\mathbf{k})b(\mathbf{l})$ corresponding to a specific set $M = M(c; I) \in \mathscr{M}$ possesses the same property. For the sake of simplicity but without loss of generality, we treat the case of $0 < c < d$ and $I = \{1, \dots, c\}$. In this case,

$$\sum_{(\mathbf{k},\mathbf{l})\in M} a(\mathbf{k})b(\mathbf{l}) = \sum_{\mathbf{l}\le\mathbf{n}}\left[\sum_{k_1=1}^{l_1}\cdots\sum_{k_c=1}^{l_c}\ \sum_{k_{c+1}=l_{c+1}+1}^{n_{c+1}}\cdots\sum_{k_d=l_d+1}^{n_d} a(\mathbf{k})b(\mathbf{l})\right]. \tag{8.35}$$

The expression in the brackets is not defined if the index $\mathbf{l} = (l_1, \dots, l_d)$ is such that $l_\nu = n_\nu$ for at least one $\nu : c+1 \le \nu \le d$. In such a case, we agree that the expression in brackets is equal to 0.

For each $\nu : c+1 \le \nu \le d$, we change the order of summation on the right-hand side of (8.35) as follows:

$$\sum_{l_\nu=1}^{n_\nu}\left[\sum_{k_\nu=l_\nu+1}^{n_\nu}\right] = \sum_{l_\nu=1}^{n_\nu-1}\ \sum_{k_\nu=l_\nu+1}^{n_\nu} = \sum_{k_\nu=2}^{n_\nu}\ \sum_{l_\nu=1}^{k_\nu-1} = \sum_{k_\nu=1}^{n_\nu}\left[\sum_{l_\nu=1}^{k_\nu-1}\right].$$

The expression in the latter brackets is supposed to be equal to zero if $k_\nu = 1$. Therefore

$$\sum_{(\mathbf{k},\mathbf{l})\in M} a(\mathbf{k})b(\mathbf{l}) = \sum_{l_1=1}^{n_1}\cdots\sum_{l_c=1}^{n_c}\ \sum_{k_{c+1}=1}^{n_{c+1}}\cdots\sum_{k_d=1}^{n_d}\left[\sum_{k_1=1}^{l_1}\cdots\sum_{k_c=1}^{l_c}\ \sum_{l_{c+1}=1}^{k_{c+1}-1}\cdots\sum_{l_d=1}^{k_d-1} a(\mathbf{k})b(\mathbf{l})\right].$$

The expression in the latter brackets is supposed to be equal to zero if at least one of the equalities $k_{c+1} = 1, \dots, k_d = 1$ holds. Denoting $m_1 = l_1, \dots, m_c = l_c, m_{c+1} = k_{c+1}, \dots, m_d = k_d$, we obtain the following representation

$$\sum_{(\mathbf{k},\mathbf{l})\in M} a(\mathbf{k})b(\mathbf{l}) = \sum_{\mathbf{m}\le\mathbf{n}} c(\mathbf{m}),$$

where

$$c(\mathbf{m}) \stackrel{\text{def}}{=} \sum_{k_1=1}^{l_1} \cdots \sum_{k_c=1}^{l_c} \sum_{l_{c+1}=1}^{k_{c+1}-1} \cdots \sum_{l_d=1}^{k_d-1} a(\mathbf{k})b(\mathbf{l}).$$

This representation proves that the increments of $\sum_{(\mathbf{k},\mathbf{l})\in M} a(\mathbf{k})b(\mathbf{l})$ with respect to the multi-index $\mathbf{n}$ are non-negative for all $M \in \mathscr{M}$ (recall that every $M \in \mathscr{M}$ depends on $\mathbf{n}$). □

Corollary 8.7 *Let p be an even number and let $\{X(\mathbf{n}), \mathbf{n} \in \mathbf{N}^d\}$ be a field of independent random variables with zero expectations and finite moments of order p, that is* $\mathsf{E}[X(\mathbf{n})] = 0$ *and* $\mathsf{E}\left[|X(\mathbf{n})|^p\right] < \infty$. *Assume that $\{b(\mathbf{n}), \mathbf{n} \in \mathbf{N}^d\}$ is an unbounded field of non-random increasing numbers, that is*

$$b(\mathbf{m}) \le b(\mathbf{n}) \ \ \textit{for} \ \ \mathbf{m} \le \mathbf{n} \ \ \textit{and} \ \ \lim(\max) b(\mathbf{n}) = +\infty.$$

Put

$$\mathfrak{B}(\mathbf{n}) = \sum_{\mathbf{k}\le\mathbf{n}} X^2(\mathbf{k}).$$

If

$$\sum_{\mathbf{n}\in\mathbf{N}^d} \frac{\mathsf{E}\left[\Delta\left[\mathfrak{B}^{p/2}(\mathbf{n})\right]\right]}{b^p(\mathbf{n})} < \infty, \tag{8.36}$$

then the strong law of large numbers (8.3) *holds. Here* $\mathfrak{B}^{p/2}(\mathbf{n}) = (\mathfrak{B}(\mathbf{n}))^{p/2}$.

Proof As in the proof of Corollary 8.1 one can treat only the case of $b(\mathbf{1}) = 1$. Let $c = 2^{1/p}$ and let the sets $\{A_t, t \ge 0\}$ be defined according to the rule (8.7). Fix $t \ge 0$ and choose $\mathbf{n}_t$ such that $\mathbf{k} \le \mathbf{n}_t$ for all $\mathbf{k} \in A_t$. Now we define a field of independent random variables $\{Y(\mathbf{n})\}$ and their sums $\{T(\mathbf{n})\}$:

$$Y(\mathbf{n}) = \begin{cases} X(\mathbf{n}), & \mathbf{n} \in A_t, \\ 0, & \mathbf{n} \notin A_t, \end{cases} \qquad T(\mathbf{n}) = \sum_{\mathbf{k}\le\mathbf{n}} Y(\mathbf{k}), \qquad \mathbf{n} \le \mathbf{n}_t.$$

We apply Doob's inequality for multi-indexed sums (Corollary 2.9) and then we use the upper Marcinkiewicz–Zygmund inequality (see (2.14)) for the random variables $\{Y(\mathbf{n})\}$:

$$\mathsf{E}\left[\left(\max_{\mathbf{m}\le\mathbf{n}_t} |T(\mathbf{m})|\right)^p\right] \le D_p \mathsf{E}\left[|T(\mathbf{n}_t)|^p\right] \le D_p Z_p'' \sum_{\mathbf{k}\le\mathbf{n}_t} \Delta_Y(\mathbf{k}),$$

where

$$\Delta_Y(\mathbf{k}) = \Delta\left[\mathsf{E}\left[\left(\sum_{\mathbf{m}\le\mathbf{k}} Y^2(\mathbf{m})\right)^{p/2}\right]\right].$$

Lemma 8.4 implies that $\Delta_Y(\mathbf{k}) \ge 0$ for all $\mathbf{k} \le \mathbf{n}_t$, since p is an even number. If $\mathbf{k} \in A_t$ and $\mathbf{m} \le \mathbf{k}$, then $\mathbf{m} \in A_t$. Thus $\Delta_Y(\mathbf{k}) = \Delta_X(\mathbf{k})$ for all $\mathbf{k} \in A_t$, where

$$\Delta_X(\mathbf{k}) = \Delta\left[\mathsf{E}\left[\left(\sum_{\mathbf{m}\le\mathbf{k}} X^2(\mathbf{m})\right)^{p/2}\right]\right] = \mathsf{E}\left[\Delta\left[\mathfrak{B}^{p/2}(\mathbf{k})\right]\right].$$

Moreover, $\max_{\mathbf{k}\in A_t} |S(\mathbf{k})| \le \max_{\mathbf{m}\le\mathbf{n}_t} |T(\mathbf{m})|$. Therefore

$$\mathsf{E}\left[\left(\max_{\mathbf{k}\in A_t} |S(\mathbf{k})|\right)^p\right] \le \mathsf{E}\left[\left(\max_{\mathbf{m}\le\mathbf{n}_t} |T(\mathbf{m})|\right)^p\right] \le D_p Z_p'' \sum_{\mathbf{k}\in A_t} \mathsf{E}\left[\Delta\left[\mathfrak{B}^{p/2}(\mathbf{k})\right]\right].$$

Now Corollary 8.7 follows from Theorem 8.3. □

8.6.3 Remarks and Examples

Conditions similar to (8.36) are studied in Sect. 8.7, where we consider the case of an arbitrary (in general, non-integer) number p.

Remark 8.14 Since p is an even number, $\Delta\left[\mathfrak{B}^{p/2}(\mathbf{n})\right] \ge 0$ by Lemma 8.4.

Remark 8.15 Corollary 8.7 for $p = 2$ coincides with Corollary 8.1. Indeed,

$$\Delta\left[\mathfrak{B}^{p/2}(\mathbf{n})\right] = X^2(\mathbf{n})$$

in this case. The increments of the field $\{\mathfrak{B}^{p/2}(\mathbf{n}), \mathbf{n} \in \mathbf{N}^d\}$ can be evaluated similarly for some other numbers p and thus one can obtain a number of new results on the strong law of large numbers. Note that we do indeed obtain new results if p is large.

Example 8.12 We show that condition (8.36) with a parameter p does not follow from the same condition but with a smaller parameter. This is the explanation of the statement that "the sequence of Corollaries 8.7 improves if p grows".

Consider the case $d = 1$ (a similar construction can be carried out for $d > 1$, too). Let independent random variables X_n assume only two values ± 1 with equal probabilities $\frac{1}{2}$. Then $B_n \stackrel{\text{def}}{=} X_1^2 + \cdots + X_n^2 = n$, $\mathsf{E}\left[B_n^{p/2}\right] - \mathsf{E}\left[B_{n-1}^{p/2}\right] \asymp n^{-1+p/2}$ for all $p \ge 2$. Let $\theta > 0$, $b_n = \sqrt{n}(\ln n)^\theta$. Condition (8.36) for $d = 1$ reduces to the requirement that the series

$$\sum_{n=1}^{\infty} \frac{\mathsf{E}\left[B_n^{p/2}\right] - \mathsf{E}\left[B_{n-1}^{p/2}\right]}{b_n^p} \tag{8.37}$$

converges. Series (8.37) diverges if $p \le 1/\theta$, and converges if $p > 1/\theta$. As a result, the strong law of large numbers (8.26) does not follow from Corollary 8.7 if $p \le 1/\theta$ but it does follow if $p > 1/\theta$.

A similar result holds for every $d \ge 1$.

Corollary 8.8 *Let* $\{X(\mathbf{n}), \mathbf{n} \in \mathbf{N}^d\}$ *be a random walk with d-dimensional time, that is,* $\{X(\mathbf{n})\}$ *are jointly independent random variables and* $X(\mathbf{n}) = \pm 1$ *with probability* $\frac{1}{2}$. *Then, for all* $\theta > 0$,

$$\lim(\max) \frac{S(\mathbf{n})}{\sqrt{|\mathbf{n}|}(\ln |\mathbf{n}|)^{\theta}} = 0 \quad \text{a. s.} \tag{8.38}$$

Proof It is clear that $\mathsf{E}[\mathfrak{B}(\mathbf{n})] = |\mathbf{n}|$, $\mathsf{E}\left[\Delta\left[\mathfrak{B}^{p/2}(\mathbf{n})\right]\right] \asymp |\mathbf{n}|^{-1+p/2}$ for all $p > 2$. If one chooses $p > \frac{1}{\theta}$ and $b(\mathbf{n}) = \sqrt{|\mathbf{n}|}(\ln |\mathbf{n}|)^{\theta}$, then series (8.36) converges. Thus Corollary 8.8 follows from Corollary 8.7. □

Condition (8.36) simplifies if a field $\{b(\mathbf{n}), \mathbf{n} \in \mathbf{N}^d\}$ satisfies the following property:

$$\nabla\left[\frac{1}{b^p(\mathbf{n})}\right] \ge 0, \qquad \mathbf{n} \in \mathbf{N}^d, \tag{8.39}$$

(see Definition A.9).

Corollary 8.9 *Let* $p \ge 2$ *be a positive integer and let* $\{X(\mathbf{n}), \mathbf{n} \in \mathbf{N}^d\}$ *be a field of independent random variables with finite expectations and finite moments of order* p, *that is*

$$\mathsf{E}[X(\mathbf{n})] = 0 \quad \text{and} \quad \mathsf{E}\left[|X(\mathbf{n})|^p\right] < \infty.$$

Assume that $\{b(\mathbf{n}), \mathbf{n} \in \mathbf{N}^d\}$ *is an unbounded field of non-random increasing numbers, that is*

$$b(\mathbf{m}) \le b(\mathbf{n}) \ \text{ for } \ \mathbf{m} \le \mathbf{n} \ \text{ and } \ \lim(\max) b(\mathbf{n}) = +\infty.$$

Assume further that condition (8.39) *holds for* $\{b(\mathbf{n})\}$. *Then the strong law of large numbers* (8.3) *follows from the condition*

$$\sum_{\mathbf{n} \in \mathbf{N}^d} \mathsf{E}\left[|S(\mathbf{n})|^p\right] \nabla\left[\frac{1}{b^p(\mathbf{n})}\right] < \infty \tag{8.40}$$

(*cf. condition* (8.28) *for the case* $d = 1$).

Proof Condition (8.36) can be rewritten in this case in the following equivalent form

$$\sum_{\mathbf{n}\in\mathbf{N}^d} \mathsf{E}\left[\mathfrak{B}^{p/2}(\mathbf{n})\right] \nabla\left[\frac{1}{b^p(\mathbf{n})}\right] < \infty. \tag{8.41}$$

Using the lower bound in the Marcinkiewicz–Zygmund inequality (see inequality (2.14)) we prove that condition (8.41) follows from (8.40). □

Condition (8.40) further simplifies for the field $\{b(\mathbf{n}), \mathbf{n} \in \mathbf{N}^d\}$ of some special structures. For example, let $b(\mathbf{n}) = R_1(n_1)\cdots R_d(n_d)$, where $R_i(\cdot)$ is a regularly varying functions of indices $\rho_i > 0$, that is $R_i(x) = x^{\rho_i} L_i(x)$ and $L_i(\cdot)$ are slowly varying functions. The following result is a generalization of Corollary 8.5 to the case $d > 1$.

Corollary 8.10 *Let $p \geq 2$ be a positive integer and let $\{X(\mathbf{n}), \mathbf{n} \in \mathbf{N}^d\}$ be a field of independent random variables with finite expectations and finite moments of order p, that is, $\mathsf{E}[X(\mathbf{n})] = 0$ and $\mathsf{E}\left[|X(\mathbf{n})|^p\right] < \infty$. Assume that*

$$b(\mathbf{n}) = R_1(n_1)\cdots R_d(n_d),$$

where $R_i(x) = x^{\rho_i} L_i(x)$, $\rho_i > 0$, and $L_i(\cdot)$ are measurable slowly varying functions. Then the strong law of large numbers (8.3) *follows from the condition*

$$\sum_{\mathbf{n}\in\mathbf{N}^d} \frac{\mathsf{E}\left[|S(\mathbf{n})|^p\right]}{|\mathbf{n}| b^p(\mathbf{n})} < \infty.$$

An advantage of Corollary 8.10 is that one does not require the monotonicity of the field $\{b(\mathbf{n})\}$ itself. This assumption is dropped, since one can construct an asymptotically equivalent increasing field $\{b'(\mathbf{n})\}$, $b'(\mathbf{n}) \sim b(\mathbf{n})$, for which all the assumptions of Corollary 8.10 hold. The asymptotically equivalent field $\{b'(\mathbf{n})\}$ is constructed with the help of asymptotically equivalent functions $R_i'(x)$ for which $R_i'(x)/R_i(x) \to 1$ as $x \to \infty$ (see, Sect. A.9.1).

Proof For the sake of simplicity, we assume that the functions R_i themselves are non-decreasing. Then we use the integral representations for slowly varying functions $L_i^p(x)$, being true if x is sufficiently large, say for $x \geq A_i$:

$$L_i^p(x) = \exp\left\{\eta_i(x) + \int_{A_i}^{x} \frac{\varepsilon_i(t)}{t}\, dt\right\},$$

where the functions η_i and ε_i are such that $\eta_i(x) \to c_i$ and $\varepsilon_i(x) \to 0$ as $x \to \infty$. The functions ε_i, $1 \leq i \leq d$, in the latter representation can be chosen to be continuous (see [368], see (1.10)). Then $L_i^p(x) = e^{\eta_i(x)} L_i^*(x)$, where

$$L_i^*(x) = \exp\left\{\int\limits_{A_i}^{x} \frac{\varepsilon_i(t)}{t}\,dt\right\}, \qquad x \ge A_i.$$

Note that the functions L_i^* are differentiable. The final simplification is that, without loss of generality, we exclude the factors $e^{\eta_i(x)}$ from our consideration.

From criterion (1.11) [368] for slowly varying functions, we conclude that $L_i^*(n+1) - L_i^*(n) \le L_i^*(n)/n$ for sufficiently large n. Using the Lagrange mean value theorem one can prove that

$$\frac{1}{n_i^{p\rho_i} L_i^*(n_i)} - \frac{1}{(n_i+1)^{p\rho_i} L_i^*(n_i+1)} \asymp \frac{1}{n_i^{p\rho_i+1} L_i^*(n_i)}.$$

Putting $R_i^*(x) = x^{p\rho_i} L_i^*(x)$ and $b^*(\mathbf{n}) = R_1^*(n_1)\cdots R_d^*(n_d)$, we obtain for sufficiently large n that

$$\frac{1}{R_i^*(n+1)} - \frac{1}{R_i^*(n)} \asymp \frac{1}{nR_i^*(n)},$$

whence

$$\nabla\left[\frac{1}{b^*(\mathbf{n})}\right] = \prod_{i=1}^{d} \nabla\left[\frac{1}{R_i^*(n_i)}\right] \asymp \frac{1}{|\mathbf{n}|b^*(\mathbf{n})}. \asymp \frac{1}{|\mathbf{n}|b^p(\mathbf{n})}.$$

This property yields

$$\sum_{\mathbf{n}\in\mathbf{N}^d} \mathsf{E}\left[|S(\mathbf{n})|^p\right] \nabla\left[\frac{1}{b^*(\mathbf{n})}\right] \le \sum_{\mathbf{n}\in\mathbf{N}^d} \mathsf{E}\left[|S(\mathbf{n})|^p\right] \frac{1}{|\mathbf{n}|b^p(\mathbf{n})}.$$

Then we apply Corollary 8.9 to the field $\beta(\mathbf{n}) = (b^*(\mathbf{n}))^{1/p}$. The strong law of large numbers (8.3) gives us $\lim S(\mathbf{n})/\beta(\mathbf{n}) = 0$ almost surely. Since $\beta(\mathbf{n}) \le b(\mathbf{n})$, the proof of Corollary 8.10 is complete. □

All the assumptions of Corollary 8.10 concerning the field $\{b(\mathbf{n}), \mathbf{n} \in \mathbf{N}^d\}$ hold if $b(\mathbf{n}) = |\mathbf{n}|^\theta$, $\theta > 0$.

Corollary 8.11 *Let $p \ge 2$ be a positive integer and let $\{X(\mathbf{n}), \mathbf{n} \in \mathbf{N}^d\}$ be a field of independent random variables with zero expectations and finite moments of order p, that is*

$$\mathsf{E}[X(\mathbf{n})] = 0 \quad \textit{and} \quad \mathsf{E}\left[|X(\mathbf{n})|^p\right] < \infty.$$

Assume that $\theta > 0$. Then the strong law of large numbers

$$\lim(\max)\frac{S(\mathbf{n})}{|\mathbf{n}|^{\theta}}=0 \quad \text{a. s.}$$

follows from the following condition

$$\sum_{\mathbf{n}\in\mathbf{N}^d}\frac{\mathsf{E}\left[|S(\mathbf{n})|^p\right]}{|\mathbf{n}|^{1+p\theta}}<\infty. \tag{8.42}$$

Corollary 8.12 *Let $p \geq 2$ be a positive integer and let $\{X(\mathbf{n}), \mathbf{n}\in\mathbf{N}^d\}$ be a field of independent random variables with finite expectations and finite moments of order p, that is*

$$\mathsf{E}[X(\mathbf{n})]=0 \quad \textit{and} \quad \mathsf{E}\left[|X(\mathbf{n})|^p\right]<\infty.$$

Assume that $\theta > 0$. Then the strong law of large numbers

$$\lim(\max)\frac{S(\mathbf{n})}{|\mathbf{n}|^{\theta}}=0 \quad \text{a. s.}$$

follows from the condition

$$\sum_{\mathbf{n}\in\mathbf{N}^d}\frac{\mathsf{E}\left[|X(\mathbf{n})|^p\right]}{|\mathbf{n}|^{1+p\theta/2}}<\infty. \tag{8.43}$$

Corollary 8.13 *Let $p \geq 2$ be a positive integer and let $\{X(\mathbf{n}), \mathbf{n}\in\mathbf{N}^d\}$ be a field of independent random variables with finite expectations and finite moments of order p, that is*

$$\mathsf{E}[X(\mathbf{n})]=0 \quad \textit{and} \quad \mathsf{E}\left[|X(\mathbf{n})|^p\right]<\infty.$$

Then the strong law of large numbers

$$\lim(\max)\frac{S(\mathbf{n})}{|\mathbf{n}|}=0 \quad \text{a. s.}$$

follows from the condition

$$\sum_{\mathbf{n}\in\mathbf{N}^d}\frac{\mathsf{E}\left[|X(\mathbf{n})|^p\right]}{|\mathbf{n}|^{1+p/2}}<\infty. \tag{8.44}$$

8.7 The Teicher–Egorov Strong Law of Large Numbers

One of the generalizations of Theorem 8.1 in the case $d = 1$ is called the *Teicher–Egorov strong law of large numbers.*

Theorem 8.6 (H. Teicher, V. Egorov) *Let $\{X_n\}$ be independent random variables. Assume that $\mathsf{E}[X]_n = 0$ and $\sigma_n^2 \overset{\text{def}}{=} \mathsf{var}[X_n] < \infty$. If*

$$\sum_{n=1}^{\infty} \mathsf{P}(|X_n| \ge n\varepsilon) < \infty, \qquad \varepsilon > 0, \tag{8.45}$$

$$\sum_{n=2}^{\infty} \frac{\sigma_n^2}{n^4} \sum_{k=1}^{n-1} \sigma_k^2 < \infty, \tag{8.46}$$

then the strong law of large numbers (8.31) *holds.*

Note that (8.45) is a necessary condition for (8.31). Condition (8.46) seems to be more complicated. Its role becomes clear if one assumes that Kolmogorov's condition (8.1) holds for $b_n = n$. Then

$$\frac{1}{n^2} \sum_{k=1}^{n-1} \sigma_k^2 \to 0, \qquad n \to \infty, \tag{8.47}$$

and condition (8.46) obviously follows from (8.1). Otherwise, if condition (8.1) does not hold, then nevertheless condition (8.46) can hold if (8.47) is valid.

Condition (8.46) for $d > 1$ is written in a different form as compared to the case $d = 1$.

Theorem 8.7 *Let $p \ge 2$ be an even number and let $\{X(\mathbf{n}), \mathbf{n} \in \mathbf{N}^d\}$ be a field of independent random variables with finite expectations and finite moments of order p, that is*

$$\mathsf{E}[X(\mathbf{n})] = 0 \quad \text{and} \quad \mathsf{E}\left[|X(\mathbf{n})|^p\right] < \infty.$$

Assume that $\{b(\mathbf{n}), \mathbf{n} \in \mathbf{N}^d\}$ is an unbounded non-random increasing field of real numbers, that is

$$b(\mathbf{m}) \le b(\mathbf{n}) \ \text{ for } \ \mathbf{m} \le \mathbf{n} \ \text{ and } \ \lim(\max) b(\mathbf{n}) = +\infty.$$

Let $c > 1$ and let the sets $\{A_t\}$ be defined according to the rule (8.7). *Set $\sigma^2(\mathbf{n}) = \mathsf{var}[X(\mathbf{n})]$ and*

$$\Lambda_t = \left(\sum_{\mathbf{n} \in A_t} \sigma^2(\mathbf{n}) \right)^{p/2}, \qquad t \ge 0.$$

If

$$\sum_{\mathbf{n} \in \mathbf{N}^d} \frac{\mathsf{E}\left[|X(\mathbf{n})|^p\right]}{b^p(\mathbf{n})} < \infty \quad \text{and} \quad \sum_{t=1}^{\infty} \frac{\Lambda_t - \Lambda_{t-1}}{c^{pt}} < \infty, \tag{8.48}$$

then the strong law of large numbers (8.3) *holds.*

Remark 8.16 If $p = 2$, then both assumptions in (8.48) coincide with the Kolmogorov condition (8.14) and, in turn, Theorem 8.7 itself is equivalent to Corollary 8.1. If $p = 4$, then

$$\Lambda_t - \Lambda_{t-1} \le 2 \Bigg(\sum_{\mathbf{n} \in A_t \setminus A_{t-1}} \sigma^2(\mathbf{n}) \Bigg) \Lambda_t$$

and the second condition in (8.48) follows from

$$\sum_{t=1}^{\infty} \Bigg(\sum_{\mathbf{n} \in A_t \setminus A_{t-1}} \frac{\sigma^2(\mathbf{n})}{b^2(\mathbf{n})} \Bigg) \frac{\Lambda_t}{c^{2t}} < \infty. \tag{8.49}$$

It is also clear that the Kolmogorov condition (8.14) yields

$$\frac{\Lambda_t}{c^{2t}} \to 0, \qquad t \to \infty,$$

whence we derive (8.49), that is, Kolmogorov's condition (8.14) is more restrictive than (8.49). Therefore (8.14) is more restrictive than (8.48).

Example 8.13 Let $d = 1$, $p = 4$, $b_n = n$, $\sigma_1^2 = 1$, and

$$\sigma_n^2 = \frac{n}{\log_2 n}, \qquad n \ge 2.$$

Then Kolmogorov's condition (8.1) does not hold and

$$\sum_{k=1}^{n} \sigma_k^2 \asymp \frac{n^2}{\log_2 n}.$$

If we choose $c = 2$, then $A_t = \{1, 2, \ldots, 2^t\}$. Thus

$$\Lambda_t = \left(\sum_{k=1}^{2^t} \sigma_k^2\right)^2 \asymp \frac{2^{4t}}{t^2},$$

whence we conclude that $\sum 2^{-4t}\Lambda_t < \infty$ and this proves the second condition in (8.48). If one chooses Gaussian random variables $\{X_n\}$ with the distributions $\mathscr{N}\left(0, \frac{n}{\log_2 n}\right)$, then the first condition in (8.48) also holds.

Proof of Theorem 8.7 As in the proof of Corollary 8.1, one can restrict the consideration to the case of $b(\mathbf{1}) = 1$. For any $t \geq 0$, we choose a multi-index $\mathbf{n}_t$ such that $\mathbf{k} \leq \mathbf{n}_t$ for all $\mathbf{k} \in A_t$. Then we introduce the independent random variables $\{Y(\mathbf{n})\}$ and their rectangular sums $\{T(\mathbf{n})\}$:

$$Y(\mathbf{n}) = \begin{cases} X(\mathbf{n}), & \mathbf{n} \in A_t, \\ 0, & \mathbf{n} \notin A_t, \end{cases} \qquad T(\mathbf{n}) = \sum_{\mathbf{k} \leq \mathbf{n}} Y(\mathbf{k}).$$

Now we use Doob's inequality for multi-indexed sums (Corollary 2.9) and then Rosenthal's inequality (inequality (2.12)) for sums $\{T(\mathbf{n})\}$:

$$\mathsf{E}\left[\left(\max_{\mathbf{m} \leq \mathbf{n}_t} |T(\mathbf{m})|\right)^p\right] \leq D_p \mathsf{E}\left[|T(\mathbf{n}_t)|^p\right] \leq D_p R_p \left(\sum_{\mathbf{k} \leq \mathbf{n}_t} \mathsf{E}\left[|Y(\mathbf{k})|^p\right] + V_Y^{p/2}(\mathbf{n}_t)\right),$$

where

$$V_Y(\mathbf{n}_t) = \sum_{\mathbf{n} \leq \mathbf{n}_t} \mathsf{E}\left[Y^2(\mathbf{n})\right] = \sum_{\mathbf{n} \in A_t} \sigma^2(\mathbf{n}).$$

Now we choose the field $\{\lambda(\mathbf{n})\}$ such that $\lambda(\mathbf{n}) \geq 0$ and

$$\Lambda_t = \sum_{\mathbf{n} \in A_t} \lambda(\mathbf{n}) \qquad \text{for every } t \geq 0. \tag{8.50}$$

Since

$$\mathsf{E}\left[\left(\max_{\mathbf{k} \in A_t} |S(\mathbf{k})|\right)^p\right] = \mathsf{E}\left[\left(\max_{\mathbf{m} \leq \mathbf{n}_t} |T(\mathbf{m})|\right)^p\right] \leq 2 D_p R_p \sum_{\mathbf{k} \in A_t} \left(\mathsf{E}\left[|X(\mathbf{k})|^p\right] + \lambda(\mathbf{k})\right),$$

we conclude from Theorem 8.3 that the strong law of large numbers (8.3) follows from the first condition in (8.48) if

$$\sum_{\mathbf{n} \in \mathbf{N}^d} \frac{\lambda(\mathbf{n})}{b^p(\mathbf{n})} < \infty.$$

To complete the proof of Theorem 8.7 note that

$$\sum_{\mathbf{n}\in\mathbf{N}^d} \frac{\lambda(\mathbf{n})}{b^p(\mathbf{n})} = \sum_{t=1}^{\infty} \sum_{\mathbf{n}\in A_t\setminus A_{t-1}} \frac{\lambda(\mathbf{n})}{b^p(\mathbf{n})} \le c^p \sum_{t=1}^{\infty} \frac{1}{c^{pt}} \sum_{\mathbf{n}\in A_t\setminus A_{t-1}} \lambda(\mathbf{n})$$

$$= c^p \sum_{t=1}^{\infty} \frac{\Lambda_t - \Lambda_{t-1}}{c^{pt}} < \infty. \qquad \square$$

Example 8.14 We are going to compare condition (8.48) with Egorov's condition that generalizes (8.46) for an arbitrary sequence $\{b_n\}$. For the sake of simplicity, let $d = 1$ and $p = 4$. Egorov's condition in this case is written as follows

$$\sum_{n=2}^{\infty} \frac{\sigma_n^2}{b_n^4} \sum_{k=1}^{n-1} \sigma_k^2 < \infty, \tag{8.51}$$

where $\sigma_k^2 = \mathsf{var}\,[X_k]$.

Looking back at the proof of Theorem 8.7, we see that one can put $\lambda_n = B_n^2 - B_{n-1}^2$. Thus

$$\lambda_n = \sigma_n^2(B_n + B_{n-1}) = \sigma_n^4 + 2\sigma_n^2 \sum_{k=1}^{n-1} \sigma_k^2.$$

Therefore condition (8.48) is in fact constituted by (8.51) and two other conditions

$$\sum_{n=1}^{\infty} \frac{\mathsf{E}\left[X_n^4\right]}{b_n^4} < \infty, \tag{8.52}$$

$$\sum_{n=1}^{\infty} \left(\frac{\sigma_n^2}{b_n^2}\right)^2 < \infty.$$

The convergence of the second series above follows from that of the first one, since $\left(\mathsf{E}\left[X_k^2\right]\right)^2 \le \mathsf{E}\left[X_k^4\right]$ by the Cauchy inequality. Hence,

$$(8.51),\ (8.52) \quad \Longrightarrow \quad \frac{S_n}{b_n} \to 0 \ \text{ a. s.}$$

The following condition

$$\sum_{n=1}^{\infty} \mathsf{P}(|X_n| \ge \varepsilon b_n) < \infty \qquad \text{for all } \varepsilon > 0 \tag{8.53}$$

is an analog of (8.45) for a general sequence $\{b_n\}$. It is worth mentioning that condition (8.53) is in fact necessary for the strong law of large numbers. As shown by Egorov [82],

$$(8.51), (8.53) \quad \Longrightarrow \quad \frac{S_n}{b_n} \to 0 \;\text{ a. s.}$$

Condition (8.53) follows from (8.52) by the Chebyshev–Markov inequality and hence Theorem 8.7 for $d = 1$ follows from Egorov' theorem. In turn, Theorem 8.7 has some advantages if compared to the Teicher–Egorov theorem (even in the case $d = 1$), since it is valid not only for integer p. We restrict ourselves to the case of even p.

8.8 Sets Determining the Strong Law of Large Numbers

In the classical case of $d = 1$, one can provide a criterion for the strong law of large numbers for the simplest normalization $b_n = n$ that avoids an assumption about the existence of moments.

Theorem 8.8 (Yu. V. Prokhorov) *Let $\{X_n\}$ be independent random variables. Then the strong law of large numbers* (8.31) *holds if and only if*

$$\sum_{n=1}^{\infty} \mathsf{P}(|S_{2^n} - S_{2^{n-1}}| \geq \varepsilon 2^n) < \infty \quad \textit{for all } \varepsilon > 0; \tag{8.54}$$

$$\operatorname{med}\left(\frac{S_n}{n}\right) \to 0, \qquad n \to \infty. \tag{8.55}$$

Note that (8.55) is a necessary condition for (8.31). Put $\mathcal{I}_n = \{2^n + 1, \dots, 2^{n+1}\}$ and rewrite condition (8.54) as follows

$$\sum_{n=1}^{\infty} \mathsf{P}(|S_{\mathcal{I}_n}| \geq \varepsilon 2^n) < \infty \quad \text{for all } \varepsilon > 0, \tag{8.56}$$

where

$$S_{\mathcal{I}_n} \stackrel{\text{def}}{=} \sum_{k \in \mathcal{I}_n} X_k.$$

Definition 8.1 Let $\{b_n\}$ be a non-random sequence. We say that a family of sets $\{\mathcal{I}_n\}$ of positive integers determine the strong law of large numbers for the sequence $\{b_n\}$

if condition (8.2) is equivalent to (8.56) for any sequence of independent random variables $\{X_n\}$ for which condition (8.55) holds.

In the sense of Definition 8.1, the sets $\mathscr{I}_n = \{2^n + 1, \dots, 2^{n+1}\}$ determine the strong law of large numbers for the sequence $b_n = n$.

8.8.1 Sets Determining the Strong Law of Large Numbers for Multi-Indexed Sums

A similar definition can be given for multi-indexed sums, too.

Definition 8.2 Let $\{b(\mathbf{n})\}$ be a non-random field. We say that a family of sets $\{\mathscr{I}(\mathbf{n})\}$ determine the strong law of large numbers for the field $\{b(\mathbf{n})\}$ if condition (8.3) is equivalent to

$$\sum_{\mathbf{n}\in\mathbf{N}^d} \mathsf{P}(|S(\mathscr{I}(\mathbf{n}))| \geq \varepsilon 2^{n_1+\cdots+n_d}) < \infty \qquad \text{for all} \varepsilon > 0 \tag{8.57}$$

for all sequences of independent random variables $\{X_n\}$ for which

$$\lim(\max) \operatorname{med}\left(\frac{S(\mathbf{n})}{b(\mathbf{n})}\right) = 0, \tag{8.58}$$

where

$$S(\mathscr{I}(\mathbf{n})) \stackrel{\text{def}}{=} \sum_{\mathbf{k}\in\mathscr{I}(\mathbf{n})} X(\mathbf{k}).$$

Finding a family of sets $\{\mathscr{I}(\mathbf{n})\}$ determining the strong law of large numbers is a rather complicated problem for $d > 1$. Below we consider the case where the field $\{b(\mathbf{n})\}$ is "factorized", that is

$$b(\mathbf{n}) = b_{n_1}^{(1)} \dots b_{n_d}^{(d)}. \tag{8.59}$$

Theorem 8.9 *Assume that a field $\{b(\mathbf{n})\}$ satisfies condition (8.59) and the sequences $\{b_n^{(1)}\}, \dots, \{b_n^{(d)}\}$ are increasing to infinity. Without loss of generality, we assume that $b_1^{(1)} = \cdots = b_1^{(d)} = 1$. For $n \geq 0$, put*

$$\mathscr{I}_n^{(j)} = \{k : 2^n \leq b_k^{(j)} < 2^{n+1}\}, \qquad 1 \leq j \leq d,$$

and, for $\mathbf{n} = (n_1, \dots, n_d)$, set $\mathscr{I}(\mathbf{n}) = \mathscr{I}_{n_1}^{(1)} \times \cdots \times \mathscr{I}_{n_d}^{(d)}$, that is

$$\mathscr{I}(\mathbf{n}) = \{\mathbf{k} \in \mathbf{N}^d : 2^{n_1} \le b_{k_1}^{(1)} < 2^{n_1+1}, \dots, 2^{n_d} \le b_{k_d}^{(d)} < 2^{n_d+1}\}.$$

Then the sets $\{I(\mathbf{n})\}$ determine the strong law of large numbers for the field $\{b(\mathbf{n})\}$.

Proof First we assume that the strong law of large numbers (8.3) holds for some field of independent random variables $\{X(\mathbf{n})\}$. Let

$$m_n^{(j)} = \min \mathscr{I}_n^{(j)}, \qquad M_n^{(j)} = \max \mathscr{I}_n^{(j)}.$$

There are 2^d multi-indices $\mathbf{k}$ each of whose coordinates equals either $m_n^{(j)}$ or $M_n^{(j)}$. The random variable $S(\mathscr{I}(\mathbf{n}))$ can be represented as a linear combination of at most 2^d sums $S(\mathbf{k})$ with the multi-indices $\mathbf{k}$. Moreover, for each of these multi-indices $\mathbf{k}$,

$$2^{n_1+\dots+n_d} \le b(\mathbf{k}) < 2^d \cdot 2^{n_1+\dots+n_d}.$$

Thus we establish from (8.3) that

$$\lim(\max) \frac{S(\mathscr{I}(\mathbf{n}))}{2^{n_1+\dots+n_d}} = 0 \qquad \text{a. s.} \tag{8.60}$$

The random events $\{|S(\mathscr{I}(\mathbf{n}))| \ge \varepsilon 2^{n_1+\dots+n_d}\}$ are independent for all $\varepsilon > 0$ and thus (8.57) follows from the Borel–Cantelli lemma.

Now we prove the converse, that is, we prove that condition (8.57) implies the strong law of large numbers (8.3). First we note that (8.60) follows from (8.57) in view of the Borel–Cantelli lemma. Thus

$$\lim(\max) \frac{1}{2^{n_1+\dots+n_d}} \sum_{\mathbf{k}\le\mathbf{n}} S(\mathscr{I}(\mathbf{k})) = 0 \qquad \text{a. s.}$$

This result is a corollary of an elementary lemma.

Lemma 8.5 *Let $\{\beta(\mathbf{n})\}$ be a non-random field. If*

$$\lim(\max) \beta(\mathbf{n}) = 0$$

then

$$\lim(\max) \frac{1}{2^{n_1+\dots+n_d}} \sum_{\mathbf{k}\le\mathbf{n}} 2^{k_1+\dots+k_d} \beta(\mathbf{k}).$$

Now we choose a positive integer $1 \le r < d$ and introduce the random variables

$$\zeta(\mathbf{n}) \stackrel{\text{def}}{=} \frac{1}{b(\mathscr{I}(\mathbf{n}))} \max_{\substack{m_1 \in \mathscr{I}_{n_1}^{(1)} \\ \dots \\ m_r \in \mathscr{I}_{n_r}^{(r)}}} \left| \sum_{k_1=m_{n_1}^{(1)}}^{m_1} \cdots \sum_{k_r=m_{n_r}^{(r)}}^{m_r} \sum_{k_{r+1}\in\mathscr{I}_{n_{r+1}}^{(r+1)}} \cdots \sum_{k_d\in\mathscr{I}_{n_d}^{(d)}} X(\mathbf{k}) \right|,$$

where

$$b(\mathscr{I}(\mathbf{n})) = b(M_{n_1}^{(1)}, \dots, M_{n_d}^{(d)}).$$

The definition extends to the case $r = d$ if we discard the second group of sums corresponding to $r = d$ in the definition of the random variables $\zeta(\mathbf{n})$, i.e.

$$\zeta(\mathbf{n}) \stackrel{\text{def}}{=} \frac{1}{b(\mathscr{I}(\mathbf{n}))} \max_{\substack{m_1 \in \mathscr{I}_{n_1}^{(1)} \\ \dots \\ m_d \in \mathscr{I}_{n_d}^{(d)}}} \left| \sum_{k_1=m_{n_1}^{(1)}}^{m_1} \cdots \sum_{k_d=m_{n_d}^{(d)}}^{m_d} X(\mathbf{k}) \right|.$$

Now our aim is to prove that

$$\lim(\max) \frac{\zeta(\mathbf{n})}{2^{n_1+\dots+n_d}} = 0 \quad \text{a. s.} \tag{8.61}$$

First we consider the case of symmetric random variables $\{X(\mathbf{n})\}$. The Lévy inequality (see relation (2.27)) implies that, for all $\varepsilon > 0$,

$$\mathsf{P}(\zeta(\mathbf{n}) \ge \varepsilon) \le 2^r \mathsf{P}(|S(\mathscr{I}(\mathbf{n}))| \ge \varepsilon 2^{n_1+\dots+n_d}).$$

Thus (8.61) follows from (8.57) in view of the Borel–Cantelli lemma. Similarly to the proof of Lemma 8.5,

$$\lim(\max) \frac{1}{2^{n_1+\dots+n_d}} \sum_{k_{r+1}=0}^{n_{r+1}} \cdots \sum_{k_d=0}^{n_d} \zeta(n_1, \dots, n_r, k_{r+1}, \dots, k_d) = 0 \quad \text{a. s.,}$$

whence we establish that almost surely

$$\lim(\max) \frac{1}{2^{n_1+\dots+n_r} b_{n_{r+1}}^{(r+1)} \dots b_{n_d}^{(d)}} \sum_{k_1=m_{n_1}^{(1)}}^{M_{n_1}^{(1)}} \cdots \sum_{k_1=m_{n_r}^{(r)}}^{M_{n_r}^{(r)}} \sum_{k_{r+1}=1}^{n_{r+1}} \cdots \sum_{k_d=1}^{n_d} X(\mathbf{k}) = 0.$$

To make the reasoning more clear, we write $\zeta_{\{1,\dots,r\}}(\mathbf{n})$ instead of $\zeta(\mathbf{n})$ to highlight the set of indices $\{1, \dots, r\}$ used in the first group of sums in the definition of the random variable $\zeta(\mathbf{n})$.

Choose an arbitrary subset $\mathscr{J}$ of $\{1, \dots, d\}$ (in the reasoning above, $\mathscr{J} = \{1, \dots, r\}$). Now we construct the random variables $\zeta_{\mathscr{J}}(\mathbf{n})$ by using the same rule as in the case of $\zeta(\mathbf{n})$ but now we take $\mathscr{J}$ instead of $\{1, \dots, r\}$. Following the same method as in the case of $\{1, \dots, r\}$ we prove that

$$\lim(\max) \frac{1}{2^{n_1+\dots+n_d}} \sum_{j \in \mathscr{J}} \sum_{k_j=0}^{n_j} \zeta_{\mathscr{J}}(\langle \mathbf{n}, \mathbf{k} \rangle) = 0 \quad \text{a. s.,} \tag{8.62}$$

where the vector $\langle \mathbf{n}, \mathbf{k} \rangle$ is constructed from the vectors $\mathbf{n}$ and $\mathbf{k}$ as explained below: the ith coordinate of the vector $\langle \mathbf{n}, \mathbf{k} \rangle$ is denoted by $\langle \mathbf{n}, \mathbf{k} \rangle_i$ and evaluated as follows

$$\langle \mathbf{n}, \mathbf{k} \rangle_i = \begin{cases} n_i, & i \in \mathscr{J}, \\ k_i, & i \notin \mathscr{J}. \end{cases}$$

As in the case of $\{1, \ldots, r\}$, we obtain

$$\lim(\max) \frac{1}{b_{\mathscr{J}}(\mathbf{n})} \sum_{i \in \mathscr{J}} \sum_{k_i = m_{n_i}^{(i)}}^{M_{n_i}^{(i)}} \sum_{j \notin \mathscr{J}} \sum_{k_j = 1}^{n_j} X(\mathbf{k}) = 0 \qquad \text{a. s.},$$

where the field $\{b_{\mathscr{J}}(\mathbf{n})\}$ is given by

$$b_{\mathscr{J}}(\mathbf{n}) = \beta_{n_1}^{(1)} \ldots \beta_{n_d}^{(d)}, \qquad \beta_n^{(i)} = \begin{cases} 2^n, & i \in \mathscr{J}, \\ n, & i \notin \mathscr{J}. \end{cases}$$

Finally, denote the sum on the left-hand side of (8.62) by $S_{\mathscr{J}}(\mathbf{n})$. Since every random variable $S(\mathbf{n})$ is a linear combination of at most 2^d sums $S_{\mathscr{J}}(\mathbf{n})$ for different sets $\mathscr{J}$, we conclude from (8.62) that (8.3) holds and thus Theorem 8.9 is proved for random variables with symmetric distributions.

To complete the proof in the general case, consider the symmetrizations of random variables $X(\mathbf{n})$ and denote them by $X^{(s)}(\mathbf{n})$. Relation (8.57) implies that

$$\sum_{\mathbf{n} \in \mathbf{N}^d} \mathsf{P}(|S^{(s)}(\mathscr{I}(\mathbf{n}))| \geq \varepsilon 2^{n_1 + \cdots + n_d}) < \infty \qquad \text{for all } \varepsilon > 0,$$

where $S^{(s)}(\mathscr{I}(\mathbf{n}))$ are the sums of random variables $X^{(s)}(\mathbf{n})$ over the sets $\mathscr{I}(\mathbf{n})$. The part of the theorem for symmetric random variables yields

$$\lim(\max) \frac{S^{(s)}(\mathbf{n})}{b(\mathbf{n})} = 0 \qquad \text{a. s.}$$

This relation is equivalent to

$$\lim(\max) \left[\frac{S(\mathbf{n})}{b(\mathbf{n})} - \operatorname{med}\left(\frac{S(\mathbf{n})}{b(\mathbf{n})} \right) \right] = 0 \qquad \text{a. s.}$$

which together with (8.58) completes the proof. □

8.8.2 Some Applications

Here we show how Theorem 8.9 can be used to prove other results on the strong law of large numbers. This method has the disadvantage that the field $\{b(\mathbf{n})\}$ should satisfy condition (8.59).

Proof of Corollary 8.1 We estimate the probabilities on the right-hand side of (8.57) with the help of Chebyshev's inequality; then we prove that the series of those probabilities converges:

$$\begin{aligned}\sum_{\mathbf{n}\in\mathbf{N}^d} \mathsf{P}(|S(\mathscr{I}(\mathbf{n}))| \ge \varepsilon 2^{n_1+\dots+n_d}) &\le \frac{1}{\varepsilon^2} \sum_{\mathbf{n}\in\mathbf{N}^d} \frac{1}{2^{2(n_1+\dots+n_d)}} \sum_{\mathbf{k}\in\mathscr{I}(\mathbf{n})} \mathsf{E}\left[X^2(\mathbf{k})\right] \\ &\le \frac{1}{\varepsilon^2} \sum_{\mathbf{n}\in\mathbf{N}^d} \sum_{\mathbf{k}\in\mathscr{I}(\mathbf{n})} \frac{\mathsf{E}\left[X^2(\mathbf{k})\right]}{b^2(\mathbf{k})} \\ &= \frac{1}{\varepsilon^2} \sum_{\mathbf{n}\in\mathbf{N}^d} \frac{\mathsf{E}\left[X^2(\mathbf{n})\right]}{b^2(\mathbf{n})} < \infty. \end{aligned}$$

□

Proof of Corollary 8.2 We use the same idea as in the preceding case:

$$\mathsf{P}(|S(\mathscr{I}(\mathbf{n}))| \ge \varepsilon 2^{n_1+\dots+n_d}) \le \frac{\mathsf{E}\left[|S(\mathscr{I}(\mathbf{n}))|^p\right]}{\varepsilon^p} \le \frac{1}{\varepsilon^p} \sum_{\mathbf{k}\in\mathscr{I}(\mathbf{n})} \mathsf{E}\left[|X(\mathbf{k})|^p\right]$$

according to Berry–Esseen's inequality (see inequality (2.6)). Combining these bounds, we complete the proof. □

8.9 Comments

Theorem 8.1 for $b_n = n$ was proved by Kolmogorov in 1930 in [255] (see also [336]). In Loève's book [299] the strong law of large numbers is given for a general sequence $\{b_n\}$. Kolmogorov's original proof differs from that described above on p. 165. In the same paper [255], Kolmogorov constructs an example showing that the sufficient condition of this theorem cannot be improved.

Feller[1] believes that

> *[The weak law of large numbers] is of very limited interest and should be replaced by the more precise and more useful strong law of large numbers*

while the van der Waerden[2] point of view is different:

> *[The strong law of large numbers] scarcely plays a role in mathematical statistics.*

---

[1] W. Feller, *An Introduction to Probability Theory and Its Applications*, vol. I, Wiley, third edition, 1968.

[2] B.L. Waerden, van der, *Mathematische Statistik*, Springer-Verlag, Berlin, third edition, 1971.

Section 8.1 Theorem 8.2 is a well-known result if $d = 1$.

Section 8.2 Theorem 8.3 is proved in [234]. The idea of the proof is developed in [211]; the case $d = 1$ is considered in [99]. This method turned out to be quite general (the case $d = 1$ is treated by Fazekas and Chuprunov [96], Hu [169], Kuczmaszewska [266], Shuhe and Ming [367], Hu, Chen, and Wang [168], Wang, Hu, Shen, and Ling [408], Kuczmaszewska [267], Yaskov [428], Crujeiras and Keilegom [64], Wang, Hu, and Yang [412], Quang, Thanh, and Tien [350], Stoica [382]). The method for $d > 1$ is applied in Noszaly and Tomacs [332], Hu [167], Yang, Su, and Yu [425], Ryu [363], Huan and Quang [172], Ndiye and Lo [329], Sung, Hu, and Volodin [388], Chen, Antonini, Hu, and Volodin [50], Li, Rao, and Wang [289]).

Section 8.3 Theorem 8.4 is called the Dini theorem (see [107], §375.5). Corollary 8.1, a direct generalization of Kolmogorov's theorem 8.1, is proved for $d > 1$ in [211] for an even more general case of a martingale field $\{S(\mathbf{n})\}$. Smythe [374] considers the strong law of large numbers for sums of random variables whose multi-indices belong to a partially ordered set.

Section 8.4 Corollary 8.2 is proved by Marcinkiewicz and Zygmund [301] in the case $d = 1$ (also see [336]). The case $d > 1$ is considered in [211] if $\{S(\mathbf{n})\}$ is a martingale. A similar result was later obtained by Dung, Ngamkham, Tien, and Volodin [80].

Section 8.5 Corollary 8.3 is proved by Chung [63] for $d = 1$ and $g_n(x) = g(x)$ for all $n \geq 1$. Case (iii) is not treated by Chung [63]. Loève [298] notes for the case $d = 1$ that the sequence $g_n(x) = |x|^{r_n}$ also fits this result. Case (iii) for $d = 1$ is considered by [338]. A particular case of Corollary 8.3 for $d > 1$ is obtained by Smythe [375].

Section 8.6 Theorem 8.5 is proved by Brunk [31] for integers p and by Prokhorov [343] in the general case. A generalization of Theorem 8.5 for $d > 1$ is obtained in [209]. A particular case of Corollary 8.10 is proved in [211] for a more general case of a martingale $\{S(\mathbf{n})\}$ (also see a paper by Lagodowski [273]). The case of martingales with multi-indices is considered by Son and Thang [376].

Section 8.7 Theorem 8.6 is proved by Teicher [393] for $p = 4$, $b_n = n$ and by Egorov [82] for even p and a general sequence $\{b_n\}$. Egorov [82] uses some p-forms instead of quadratic forms of variances in condition (8.51) if $p > 4$. Theorem 8.7 can be improved to obtain Egorov's theorem [82] by using the Nagaev–Pinelis inequality [326] instead of the Rosenthal inequality in the proof.

Section 8.8 Theorem 8.9 for $d = 1$ and $b_n = n$ is proved by Prokhorov [343]. More general normalizing sequences $\{b_n\}$ are considered by Loève [299] in the case $d = 1$. A criterion for the strong law of large numbers for a general sequence $\{b_n\}$ is obtained by Petrov and Martikainen [309] in the case $d = 1$. Theorem 8.9 for $d > 1$ is proved in [218]. Related results for multi-indexed sums are established by Mikosch [315], and Mikosch and Norvaisa [317].

Sufficient conditions for the strong law of large numbers for second order stationary random fields are obtained in [213]. Other dependence schemes for multi-indexed sums are considered by Ko, Kim, and Kim [250] (asymptotically independent random fields), Móricz, Stadtmüller, and Thalmaier [322] and Stadtmüller and Thanh [379] ($\mathcal{M}$-dependent random fields), Gan and Chen [121] (random variables with weights),

Móricz [319] and [320] (quasi-orthogonal random variables), Thanh [397] (blockwise orthogonal random variables), Fazekas and Tómács [100] and Hong and Hwang [166] (pairwise independent random variables), Huan and Quang [171] (blockwise independent and blockwise orthogonal random variables), Quang and Huan [347] and [348] (Banach space-valued random variables).

Chapter 9
The Strong Law of Large Numbers for Independent Identically Distributed Random Variables

It is commonly accepted that *Borel's theorem* for Bernoulli trials was the first strong law of large numbers.

Theorem 9.1 (A. Borel (1909)) *Let $\{X_n, n \geq 1\}$ be a sequence of independent identically distributed Bernoulli random variables with success probability p. Let $S_n = \sum_{k=1}^{n} X_k$ be the number of successes in the first n trials. Then*

$$\frac{S_n}{n} \xrightarrow{a.s.} p, \quad n \to \infty.$$

Kolmogorov's theorem (Corollary 8.1), proved for the general case of independent random variables (not necessarily identically distributed random variables), can be used to obtain necessary and sufficient conditions in the case of identically distributed random variables. This method leads to the following results in the classical case of $d = 1$.

Theorem 9.2 (A.N. Kolmogorov (1933)) *Let X and $\{X_n\}$ be independent identically distributed random variables; $S_n = X_1 + \cdots + X_n, n > 1, S_1 = X_1$. Then the strong law of large numbers*

$$\lim_{n\to\infty} \frac{S_n}{n} = \mu \quad a.\ s. \tag{9.1}$$

holds for some constant $\mu \in \mathbf{R}$ if and only if

$$\mathsf{E}[X] = \mu. \tag{9.2}$$

The rate of convergence in (9.1) can be obtained under higher moment conditions. In the following result, this corresponds to the case of $r > 1$.

Theorem 9.3 (J. Marcinkiewicz and A. Zygmund (1937)) *Let X and $\{X_n\}$ be independent identically distributed random variables; $S_n = X_1 + \cdots + X_n$, $n > 1$, $S_1 = X_1$. Then*

O. Klesov, *Limit Theorems for Multi-Indexed Sums of Random Variables*,
Probability Theory and Stochastic Modelling 71, DOI 10.1007/978-3-662-44388-0_9

(i) *if* $0 < r < 1$, *then*

$$\lim_{n\to\infty} \frac{S_n}{n^{1/r}} = 0 \quad \text{a. s.} \quad \Longleftrightarrow \quad \mathsf{E}\left[|X_1|^r\right] < \infty;$$

(ii) *if* $1 \le r < 2$, *then*

$$\lim_{n\to\infty} \frac{S_n - n\mu}{n^{1/r}} = 0 \quad \text{a. s.} \tag{9.3}$$

holds for some constant $\mu \in \mathbf{R}$ *if and only if*

$$\mathsf{E}[X_1] = \mu, \quad \mathsf{E}\left[|X_1|^r\right] < \infty.$$

To see why (9.3) describes the rate of convergence in (9.1), we rewrite the expression in (9.3) as follows

$$\frac{S_n - n\mu}{n^{1/r}} = n^{1-\frac{1}{r}}\left(\frac{S_n}{n} - \mu\right)$$

and note that $1 - \frac{1}{r} > 0$ if $r > 1$. The proof of Theorem 9.3 as well as that of Theorem 9.2 uses Kolmogorov's strong law of large numbers for non-identically distributed random variables (Corollary 8.1).

The strong law of large numbers for multi-indexed sums of independent identically distributed random variables in the case of $d > 1$ is proved by Smythe [373].

Theorem 9.4 (R. Smythe (1973)) *Let* X, $\{X(\mathbf{n})\}$ *be independent identically distributed random variables. The relation*

$$\mathsf{P}\left(\left|\frac{S(\mathbf{n})}{|\mathbf{n}|} - \mu\right| \ge \varepsilon \ \text{i. o.}\right) = 0 \qquad \text{for all } \varepsilon > 0 \tag{9.4}$$

holds for some non-random constant $\mu \in \mathbf{R}$ *if and only if*

$$\mathsf{E}[X] = \mu, \quad \mathsf{E}\left[|X|\left(\log^+|X|\right)^{d-1}\right] < \infty. \tag{9.5}$$

This result coincides with Theorem 9.2 in the case $d = 1$, while it means that $\lim(\max) S(\mathbf{n})/|\mathbf{n}| = 0$ almost surely if $d > 1$ (see Corollary 9.4).

The class of normalizing fields $\{b(\mathbf{n}), \mathbf{n} \in \mathbf{N}^d\}$ for which the strong law of large numbers has been studied in the case $d > 1$ includes

$$b(\mathbf{n}) = |\mathbf{n}|, \tag{9.6}$$

$$b(\mathbf{n}) = |\mathbf{n}|^t, \quad t > \frac{1}{2}, \tag{9.7}$$

$$b(\mathbf{n}) = n_1^{t_1} \times \cdots \times n_d^{t_d}, \quad \min(t_1, \ldots, t_d) > \frac{1}{2}. \tag{9.8}$$

Below we significantly extend the class of normalizing fields $\{b(\mathbf{n}), \mathbf{n} \in \mathbf{N}^d\}$ in the d-parameter strong law of large numbers for multi-indexed sums of independent identically distributed random variables.

9.1 A Generalization of Feller's Theorem

Let $\{X(\mathbf{n}), \mathbf{n} \in \mathbf{N}^d\}$ be independent identically distributed random variables. Their common distribution function is denoted by F. Further, let a numerical field $\{b(\mathbf{n})\}$ be unbounded and increasing:

$$\lim(\max)b(\mathbf{n}) = \infty,$$
$$b(\mathbf{m}) \le b(\mathbf{n}) \quad \text{for all } \mathbf{m} \le \mathbf{n}.$$

Since the field $\{b(\mathbf{n}), \mathbf{n} \in \mathbf{N}^d\}$ is used to normalize the sums of random variables, we assume without loss of generality that $b(\mathbf{1}) = \min b(\mathbf{n}) > 0$.

We introduce two sequences related to the field $\{b(\mathbf{n}), \mathbf{n} \in \mathbf{N}^d\}$:

$$\begin{aligned} a_k &= \operatorname{card}\left(\mathbf{n} \in \mathbf{N}^d : k \le b(\mathbf{n}) < k+1\right), \quad k \ge 1, \\ A_1 &= a_1, \quad A_k = a_1 + \dots + a_k, \quad k \ge 2. \end{aligned} \tag{9.9}$$

It is clear that $A_k = \operatorname{card}\left(\mathbf{n} \in \mathbf{N}^d : 1 \le b(\mathbf{n}) < k+1\right)$.

The following is the main assumption imposed below on the field $\{b(\mathbf{n})\}$:

$$\sum_{k=m}^{\infty} \frac{A_k}{k^3} = O\left(\frac{A_m}{m^2}\right). \tag{9.10}$$

Note that the inverse bound holds for all monotone sequences $\{A_k\}$, since

$$\sum_{k=m}^{\infty} \frac{A_k}{k^3} \ge A_m \sum_{k=m}^{\infty} \frac{1}{k^3} \ge \frac{A_m}{2m^2}. \tag{9.11}$$

Remark 9.1 Another form of condition (9.10) is usually used for $d = 1$. This condition is written in terms of the sequence $\{b_n\}$ explicitly, namely *there exists a constant* $C > 0$ *such that*

$$\sum_{k=m}^{\infty} \frac{1}{b_k^2} \le C \frac{m}{b_m^2} \tag{9.12}$$

for all $m \ge 1$.

For $d = 1$, condition (9.12), also called the *Feller condition*, is equivalent to (9.10).

Proposition 9.1 *If* $d = 1$, *then conditions* (9.10) *and* (9.12) *are equivalent.*

The proof of Proposition 9.1 is given at the end of this chapter.

A direct analog of condition (9.12) for $d > 1$ is naturally written as follows

$$\sum_{\mathbf{k}\geq\mathbf{m}}^{\infty} \frac{1}{b(\mathbf{k})^2} = O\left(\frac{|\mathbf{m}|}{b(\mathbf{m})^2}\right). \tag{9.13}$$

Nevertheless, condition (9.10) for $d > 1$ is much more convenient to check than condition (9.13).

Feller's condition (9.12) for $d = 1$ holds for every sequence $b_n = b(n)$ if the function $b(\cdot)$ is regularly varying and if the series $\sum \frac{1}{b_n^2}$ converges. A similar property also holds in the case $d > 1$ (see Sect. 9.3).

The importance of Feller's condition is explained by the fact that one can also obtain a criterion of the strong law of large numbers in the class of normalizing fields satisfying condition (9.10).

Theorem 9.5 *Let* $\{X(\mathbf{n}), \mathbf{n} \in \mathbf{N}^d\}$ *be a field of independent identically distributed random variables with a common distribution function* F *and let a field* $\{b(\mathbf{n}), \mathbf{n} \in \mathbf{N}^d\}$ *be unbounded and increasing and satisfy condition* (9.10). *Then the strong law of large numbers*

$$\lim(\max)\frac{S(\mathbf{n})}{b(\mathbf{n})} = 0 \quad a.\ s. \tag{9.14}$$

is equivalent to the following set of two conditions:

$$\sum_{\mathbf{n}\in\mathbf{N}^d} \mathsf{P}(|X| \geq b(\mathbf{n})) < \infty, \tag{9.15}$$

$$\lim(\max)\frac{|\mathbf{n}|}{b(\mathbf{n})} \int\limits_{|x|<b(\mathbf{n})} x\, dF(x) = 0, \tag{9.16}$$

where the random variable X *has the same distribution function as all other* $X(\mathbf{n})$.

The main corollaries of Theorem 9.5 are listed in Sect. 9.3. Note also that Theorems 9.1–9.3 follow from Theorem 9.5 if $d = 1$.

When studying the strong laws of large numbers, one also uses another condition

$$\sum_{k=m}^{\infty} \frac{A_k}{k^2} = O\left(\frac{A_m}{m}\right). \tag{9.17}$$

As compared to (9.10), condition (9.17) is more restrictive. As in the case of condition (9.12), its analog written in terms of the sequence $\{b_n\}$

$$\sum_{k=m}^{\infty} \frac{1}{b_k^2} = O\left(\frac{m}{b_m^2}\right) \tag{9.18}$$

is more convenient than (9.17) if $d = 1$.

The normalizing sequences $b(\mathbf{n}) = n_1^{1/r_1} \dots n_d^{1/r_d}$ with $\max(r_1, \dots, r_d) < 1$ satisfy condition (9.17). Since

$$\sum_{k=m}^{\infty} \frac{A_k}{k^3} \le \frac{1}{m} \sum_{k=m}^{\infty} \frac{A_k}{k^2}, \tag{9.19}$$

condition (9.17) does indeed follow from condition (9.10). An advantage of condition (9.17) is that it can also be used for random variables (still identically distributed) which are not jointly independent. Here, however, we restrict our attention to the case of independent random variables.

Theorem 9.6 *Let* $\{X(\mathbf{n}), \mathbf{n} \in \mathbf{N}^d\}$ *be a field of independent identically distributed random variables. Further, let the field* $\{b(\mathbf{n}), \mathbf{n} \in \mathbf{N}^d\}$ *be unbounded and increasing and satisfy condition* (9.17). *Then the strong law of large numbers* (9.14) *is equivalent to condition* (9.15), *where the random variable* X *has the same distribution function as all other* $X(\mathbf{n})$.

9.2 Proof of the Generalized Feller Theorem

Proof of Theorem 9.5 Necessity. Assume that the strong law of large numbers (9.14) holds. Proposition A.5 implies that

$$X(\mathbf{n}) = \sum_{\boldsymbol{\varepsilon} \in \mathscr{E}_d} (-1)^{\varepsilon_1 + \dots + \varepsilon_d} S(n_1 - \varepsilon_1, \dots, n_d - \varepsilon_d), \tag{9.20}$$

where the sum is considered with respect to all multi-indices $\boldsymbol{\varepsilon} = (\varepsilon_1, \dots, \varepsilon_d)$ whose coordinates are either 0 or 1. Note also that the number of terms in this sum does not exceed 2^d. Thus

$$\lim(\max) \left| \frac{X(\mathbf{n})}{b(\mathbf{n})} \right| \le \sum_{\boldsymbol{\varepsilon} \in \mathscr{E}_d} \lim(\max) \left| \frac{S(n_1 - \varepsilon_1, \dots, n_d - \varepsilon_d)}{b(n_1 - \varepsilon_1, \dots, n_d - \varepsilon_d)} \right| = 0 \quad \text{a. s.},$$

since the field $\{b(\mathbf{n}), \mathbf{n} \in \mathbf{N}^d\}$ is increasing. Hence we conclude that $\mathsf{P}(|X(\mathbf{n})| \ge b(\mathbf{n}) \text{ i. o.}) = 0$. Now the Borel–Cantelli lemma for independent random events (Proposition A.30) implies condition (9.15).

Finally, the strong law of large numbers is a stronger statement than the weak law of large numbers. Thus condition (9.16) holds in view of Theorem 4.9. □

Proof of Theorem 9.5 *Sufficiency* First we consider the case where all the numbers $b(\mathbf{n})$ are integers. We are going to prove that

$$\sum_{\mathbf{n}\in\mathbf{N}^d} \frac{1}{b^2(\mathbf{n})} \int\limits_{|x|<b(\mathbf{n})} x^2\,dF(x) < \infty. \tag{9.21}$$

Denote the series on the left-hand side of (9.21) by ρ. Then

$$\begin{aligned}\rho &= \sum_{k=1}^{\infty} \frac{a_k}{k^2} \int\limits_{|x|<k} x^2\,dF(x) = \sum_{k=1}^{\infty} \frac{a_k}{k^2} \sum_{j=1}^{k} \int\limits_{j-1\le|x|<j} x^2\,dF(x)\\ &= \sum_{j=1}^{\infty} \int\limits_{j-1\le|x|<j} x^2\,dF(x) \sum_{k=j}^{\infty} \frac{a_k}{k^2}.\end{aligned}$$

Now we use the summation by parts formula: if $N > j$, then

$$\begin{aligned}\sum_{k=j}^{N} \frac{a_k}{k^2} &= \sum_{k=j}^{N} \frac{A_k - A_{k-1}}{k^2} = \sum_{k=j}^{N} \frac{A_k}{k^2} - \sum_{k=j-1}^{N-1} \frac{A_k}{(k+1)^2}\\ &\le \frac{A_N}{N^2} + 2\sum_{k=j}^{N-1} \frac{A_k}{k^3} = \frac{A_N}{N^2} + O\left(\frac{A_j}{j^2}\right)\end{aligned}$$

by condition (9.10). Since $A_N/N^2 \to 0$, we see that

$$\sum_{k=j}^{\infty} \frac{a_k}{k^2} \le \text{const}\, \frac{A_j}{j^2}$$

by passing to the limit as $N \to \infty$ (see relation (9.72) in the proof of Proposition 9.1), whence

$$\begin{aligned}\rho &\le \text{const} \sum_{j=1}^{\infty} A_j \mathsf{P}(j-1 \le |X| < j)\\ &= \text{const}\left[A_1 + \sum_{j=2}^{\infty} A_j \mathsf{P}(j-1 \le |X| < j)\right].\end{aligned}$$

Since $A_{j+1} = O(A_j)$ (see (9.74)), we get

$$\sum_{j=2}^{\infty} A_j \mathsf{P}(j-1 \le |X| < j) = O(1) \sum_{j=2}^{\infty} A_{j-1} \mathsf{P}(j-1 \le |X| < j)$$
$$= O(1) \sum_{j=1}^{\infty} a_j \mathsf{P}(|X| \ge j)$$
$$= O(1) \sum_{\mathbf{n} \in \mathbf{N}^d} \mathsf{P}(|X| \ge b(\mathbf{n})) < \infty.$$

Therefore condition (9.21) holds.

Consider the truncated random variables $Y(\mathbf{n}) = X(\mathbf{n}) \mathbb{I}(|X(\mathbf{n})| < b(\mathbf{n}))$, where $\mathbb{I}(A)$ is the indicator of a random event A. Since

$$\mathsf{E}\left[(Y(\mathbf{n}) - \mathsf{E}[Y(\mathbf{n})])^2\right] \le \int\limits_{|x|<b(\mathbf{n})} x^2 \, dF(x),$$

relation (9.21) together with the Kolmogorov strong law of large numbers for random fields (Corollary 8.1) implies that

$$\lim(\max) \frac{1}{b(\mathbf{n})} \sum_{\mathbf{k} \le \mathbf{n}} (Y(\mathbf{k}) - \mathsf{E}[Y(\mathbf{k})]) = 0 \qquad \text{a. s.} \tag{9.22}$$

Condition (9.15) is equivalent to $\sum \mathsf{P}(X(\mathbf{n}) \ne Y(\mathbf{n})) < \infty$, whence we deduce that

$$\mathsf{P}(X(\mathbf{n}) \ne Y(\mathbf{n}) \text{ i. o.}) = 0$$

by the Borel–Cantelli lemma (Proposition A.30). Now we obtain from equality (9.22) that

$$\lim(\max) \frac{1}{b(\mathbf{n})} \sum_{\mathbf{k} \le \mathbf{n}} (X(\mathbf{k}) - \mathsf{E}[Y(\mathbf{k})]) = 0 \qquad \text{a. s.} \tag{9.23}$$

Remark 9.2 We derived equality (9.23) from condition (9.15) by using (9.10) and the property $A_N/N^2 \to 0$ that follows from (9.10). We will use this fact in the proof of Theorem 9.6.

To complete the proof of Theorem 9.5 for integer-valued normalizing fields $\{b(\mathbf{n})\}$ we will show that

$$\lim(\max) \frac{1}{b(\mathbf{n})} \sum_{\mathbf{k} \le \mathbf{n}} \mathsf{E}[Y(\mathbf{k})] = 0. \tag{9.24}$$

According to condition (9.16) and since

$$\mathsf{E}[Y(\mathbf{k})] = \int\limits_{|x|<b(\mathbf{n})} x \, dF(x) - \int\limits_{b(\mathbf{k}) \le |x| < b(\mathbf{n})} x \, dF(x)$$

for $\mathbf{k} \le \mathbf{n}$, relation (9.24) follows from

$$\lim(\max)\frac{1}{b(\mathbf{n})}\sum_{\mathbf{k}\le\mathbf{n}}\int\limits_{b(\mathbf{k})\le|x|<b(\mathbf{n})} x\,dF(x) = 0. \tag{9.25}$$

Put $m = b(\mathbf{n})$. Then

$$\begin{aligned}\sum_{\mathbf{k}\le\mathbf{n}}\int\limits_{b(\mathbf{k})\le|x|<b(\mathbf{n})} |x|\,dF(x) &\le \sum_{i=1}^{m-1} a_i \int\limits_{i\le|x|<m} |x|\,dF(x)\\ &= \sum_{i=1}^{m-1} a_i \sum_{j=i}^{m-1}\int\limits_{j\le|x|<j+1} |x|\,dF(x)\\ &= \sum_{j=1}^{m-1} A_j \int\limits_{j\le|x|<j+1} |x|\,dF(x).\end{aligned} \tag{9.26}$$

Defining

$$\mu_i = \int\limits_{i\le|x|<i+1} |x|\,dF(x)$$

we are going to prove that the series $\sum_{i=1}^{\infty}\mu_i A_i/i$ converges. Indeed, condition (9.15) implies that

$$\begin{aligned}\sum_{i=1}^{\infty}\mu_i\frac{A_i}{i} &\le 2\sum_{i=1}^{\infty} A_i \mathsf{P}(i \le |X| < i+1) = 2\sum_{k=1}^{\infty} a_k \mathsf{P}(|X| \ge k)\\ &= 2\sum_{\mathbf{n}\in\mathbf{N}^d} \mathsf{P}(|X| \ge b(\mathbf{n})) < \infty.\end{aligned}$$

Since the series $\sum_{i=1}^{\infty}\mu_i A_i/i$ converges, we establish

$$\frac{1}{m}\sum_{i=1}^{m}\mu_i A_i \to 0, \quad m \to \infty,$$

by Kronecker's lemma. Thus we derive from bound (9.26) that

$$\lim(\max)\frac{1}{b(\mathbf{n})}\sum_{\mathbf{k}\le\mathbf{n}}\int\limits_{b(\mathbf{k})\le|x|<b(\mathbf{n})} |x|\,dF(x) = 0,$$

whence (9.25) follows, that is, conditions (9.15)–(9.16) are sufficient for the strong law of large numbers in the case of integer-valued fields $\{b(\mathbf{n})\}$.

Now let the values of a field $\{b(\mathbf{n})\}$ be general (not necessarily integer). Put $b'(\mathbf{n}) = [b(\mathbf{n})] + 1$, where $[z]$ is the integer part of a number z. It turns out that the assumptions of Theorem 9.5 hold for the field $\{b'(\mathbf{n}), \mathbf{n} \in \mathbf{N}^d\}$ as well. Using the inequality $b(\mathbf{n}) \le b'(\mathbf{n})$ we obtain

$$\sum_{\mathbf{n}\in\mathbf{N}^d} \mathsf{P}(|X| \ge b'(\mathbf{n})) \le \sum_{\mathbf{n}\in\mathbf{N}^d} \mathsf{P}(|X| \ge b(\mathbf{n})) < \infty,$$

that is, condition (9.15) does indeed hold for $\{b'(\mathbf{n}), \mathbf{n} \in \mathbf{N}^d\}$. Further,

$$\sum_{\mathbf{k}\le\mathbf{n}} \mathsf{P}(|X| \ge b(\mathbf{k})) - \sum_{k_1=1}^{[n_1/2]} \cdots \sum_{k_d=1}^{[n_d/2]} \mathsf{P}(|X| \ge b(\mathbf{k})) \ge \text{const}\,|\mathbf{n}|\mathsf{P}(|X| \ge b(\mathbf{n})) \tag{9.27}$$

and thus (9.15) implies that $\lim(\max)|\mathbf{n}|\mathsf{P}(|X| \ge b(\mathbf{n})) = 0$, since

$$\lim(\max)\left[\sum_{\mathbf{k}\le\mathbf{n}} \mathsf{P}(|X| \ge b(\mathbf{k})) - \sum_{k_1=1}^{[n_1/2]} \cdots \sum_{k_d=1}^{[n_d/2]} \mathsf{P}(|X| \ge b(\mathbf{k}))\right] = 0.$$

Note that

$$\begin{aligned}\frac{|\mathbf{n}|}{b'(\mathbf{n})} \int\limits_{|x|<b'(\mathbf{n})} x\,dF(x) &= \frac{b(\mathbf{n})}{b'(\mathbf{n})} \frac{|\mathbf{n}|}{b(\mathbf{n})} \int\limits_{|x|<b(\mathbf{n})} x\,dF(x)\\ &\quad + \frac{|\mathbf{n}|}{b'(\mathbf{n})} \int\limits_{b(\mathbf{n})\le|x|<b'(\mathbf{n})} x\,dF(x)\end{aligned}$$

and

$$\left|\frac{|\mathbf{n}|}{b'(\mathbf{n})} \int\limits_{b(\mathbf{n})\le|x|<b'(\mathbf{n})} x\,dF(x)\right| \le |\mathbf{n}|\mathsf{P}(|X| \ge b(\mathbf{n})).$$

Thus

$$\begin{aligned}&\lim(\max)\frac{|\mathbf{n}|}{b'(\mathbf{n})}\left|\int\limits_{|x|<b'(\mathbf{n})} x\,dF(x)\right|\\ &\quad\le \lim(\max)\frac{b(\mathbf{n})}{b'(\mathbf{n})} \lim(\max)\frac{|\mathbf{n}|}{b(\mathbf{n})}\left|\int\limits_{|x|<b(\mathbf{n})} x\,dF(x)\right| = 0.\end{aligned}$$

Thus all conditions (9.15)–(9.16) hold for the field $\{b'(\mathbf{n})\}$, too. It remains to check condition (9.10) to apply Theorem 9.5 with the field $\{b'(\mathbf{n})\}$. We construct the sequences $\{a'_k\}$ and $\{A'_k\}$ from the field $\{b'(\mathbf{n})\}$ in the same way as that used to construct the sequences $\{a_k\}$ and $\{A_k\}$ from the field $\{b(\mathbf{n})\}$. It is clear that $A_{k-1} \le A'_k \le A_k$ and thus the relation

$$\sum_{k=m}^{\infty} \frac{A'_k}{k^3} = \left(\frac{A'_m}{m^2} \right)$$

is equivalent to (9.10), since $A_m = O(A_{m-1})$. The part of the theorem proved for fields assuming only integer values gives us

$$\lim(\max) \frac{S(\mathbf{n})}{b'(\mathbf{n})} = 0 \qquad \text{a. s.}$$

The latter relation is equivalent to the strong law of large numbers (9.14). The proof of Theorem 9.5 is complete. □

Proof of Theorem 9.6 Condition (9.17) implies condition (9.10) (see bound (9.19)), thus we derive the implication (9.14) $\Longrightarrow$ (9.15) from Theorem 9.5. It remains to prove the inverse implication (9.15) $\Longrightarrow$ (9.14).

First we treat the case where all the numbers $b(\mathbf{n})$ are integers. Reasoning as in the proof of Theorem 9.5, we prove (9.23) (see Remark 9.2). Note that the relation $A_N/N^2 \to 0$ needed to reach such a conclusion follows from (9.17).

Now we prove equality (9.24). Below are necessary bounds similar to those obtained in the proof of Theorem 9.5:

$$\begin{aligned}
\sum_{i=1}^{\infty} \frac{a_i}{i} \int\limits_{|x|<i} |x|\, dF(x) &= \sum_{i=1}^{\infty} \frac{a_i}{i} \sum_{j=1}^{i} \int\limits_{j-1 \le |x| < j} |x|\, dF(x) \\
&= \sum_{j=1}^{\infty} \int\limits_{j-1 \le |x| < j} |x|\, dF(x) \sum_{i=j}^{\infty} \frac{a_i}{i} \\
&= \sum_{j=1}^{\infty} \int\limits_{j-1 \le |x| < j} |x|\, dF(x) \sum_{i=j}^{\infty} \frac{A_i - A_{i-1}}{i} \\
&= \sum_{j=1}^{\infty} \int\limits_{j-1 \le |x| < j} |x|\, dF(x) \lim_{N\to\infty} \left[\sum_{i=j}^{N} \frac{A_i}{i} - \sum_{i=j-1}^{N-1} \frac{A_i}{i+1} \right] \\
&\le \sum_{j=1}^{\infty} \int\limits_{j-1 \le |x| < j} |x|\, dF(x) \sum_{i=j}^{\infty} \frac{A_i}{i(i+1)}.
\end{aligned}$$

We used the relation $A_N/N \to 0$ in the latter inequality, which is true in view of

$$\sum_{k=n}^{\infty} \frac{A_k}{k^2} \geq A_n \sum_{k=n}^{\infty} \frac{A_k}{k^2} \geq \frac{A_n}{n}.$$

Hence

$$\begin{aligned}
\sum_{i=1}^{\infty} \frac{a_i}{i} \int\limits_{|x|<i} |x|\, dF(x) &\leq \text{const} \sum_{j=1}^{\infty} \frac{A_j}{j} \int\limits_{j-1\leq|x|<j} |x|\, dF(x) \\
&\leq \text{const} \sum_{j=1}^{\infty} A_j \int\limits_{j-1\leq|x|<j} dF(x) \\
&\leq \text{const} \sum_{j=1}^{\infty} A_{j-1} \int\limits_{j-1\leq|x|<j} dF(x).
\end{aligned}$$

Next we use the relation $A_{m+1} = O(A_m)$ that follows from the following reasoning: since

$$\sum_{i=m}^{\infty} \frac{a_i}{i} = o\left(\frac{A_m}{m}\right), \tag{9.28}$$

we have

$$\frac{a_{m+1}}{m+1} = o\left(\frac{A_m}{m}\right),$$

that is $a_{m+1} = O(A_m)$ or $A_{m+1} = O(A_m)$. It remains to prove (9.28):

$$\begin{aligned}
\sum_{i=m}^{\infty} \frac{a_i}{i} &= \lim_{N\to\infty} \sum_{i=m}^{N} \frac{A_i - A_{i-1}}{i} = \lim_{N\to\infty} \left[\sum_{i=m}^{N} \frac{A_i}{i} - \sum_{i=m-1}^{N-1} \frac{A_i}{i+1} \right] \\
&\leq \lim_{N\to\infty} \sum_{i=m}^{N-1} \frac{A_i}{i(i+1)} \leq \sum_{i=m}^{\infty} \frac{A_i}{i^2} = o\left(\frac{A_m}{m}\right)
\end{aligned}$$

(we again use the property $A_N/N \to 0$). Therefore,

$$\begin{aligned}
\sum_{i=1}^{\infty} \frac{a_i}{i} \int\limits_{|x|<i} |x|\, dF(x) &\leq \text{const} \sum_{i=1}^{\infty} a_i \mathsf{P}(|X| \geq i) \\
&= \text{const} \sum_{\mathbf{n}\in\mathbf{N}^d} \mathsf{P}(|X| \geq b(\mathbf{n})) < \infty.
\end{aligned}$$

By Kronecker's lemma

$$\frac{1}{m}\sum_{i=1}^{m} a_i \int\limits_{|x|<i} |x|\, dF(x) \to 0, \quad m \to \infty.$$

Let $b(\mathbf{n}) = m$. Then

$$\sum_{\mathbf{k}\le\mathbf{n}} \int\limits_{|x|<b(\mathbf{n})} |x|\, dF(x) \le \sum_{i=1}^{m} a_i \int\limits_{|x|<i} |x|\, dF(x),$$

whence

$$\lim(\max)\frac{1}{b(\mathbf{n})}\sum_{\mathbf{k}\le\mathbf{n}} \int\limits_{|x|<b(\mathbf{n})} |x|\, dF(x) = 0,$$

that is, relation (9.24) is proved. Relations (9.23), and (9.24) together with Kolmogorov's strong law of large numbers for multi-indexed sums (Corollary 8.1) imply Theorem 9.6 for integer-valued fields $\{b(\mathbf{n})\}$.

Now we treat the case of a general field $\{b(\mathbf{n})\}$ whose values are not necessarily integers. Put $b'(\mathbf{n}) = [b(\mathbf{n})] + 1$. It is clear that

$$\sum_{\mathbf{n}\in\mathbf{N}^d} \mathsf{P}(|X| \ge b'(\mathbf{n})) \le \sum_{\mathbf{n}\in\mathbf{N}^d} \mathsf{P}(|X| \ge b(\mathbf{n})) < \infty.$$

We construct the sequences $\{a_i'\}$ and $\{A_n'\}$ from the field $\{b'(\mathbf{n})\}$ in the same way as the sequences $\{a_i\}$ and $\{A_n\}$ are constructed from the field $\{b(\mathbf{n})\}$. Since $A_{k-1} \le A_k' \le A_k$, the limit relation

$$\sum_{k=m}^{\infty} \frac{A_k'}{k^2} = O\left(\frac{A_m'}{m}\right)$$

follows from (9.17) in view of $A_m = O(A_{m-1})$. Now we apply the part of Theorem 9.6 already proved to the field $\{b'(\mathbf{n})\}$ and get

$$\lim(\max)\frac{S(\mathbf{n})}{b'(\mathbf{n})} = 0 \qquad \text{a. s.},$$

whence we conclude

$$\lim(\max)\frac{S(\mathbf{n})}{b(\mathbf{n})} = \lim(\max)\frac{S(\mathbf{n})}{b'(\mathbf{n})}\,\lim(\max)\frac{b'(\mathbf{n})}{b(\mathbf{n})} = 0 \qquad \text{a. s.}$$

The proof of Theorem 9.6 is complete. □

9.3 The Strong Law of Large Numbers for the Marcinkiewicz–Zygmund Normalizing Field

Let L be a positive function slowly varying at infinity (see, for example, [368]). A particular class of normalizing fields $\{b(\mathbf{n}), \mathbf{n} \in \mathbf{N}^d\}$ for which condition (9.10) holds is presented by

$$b(\mathbf{n}) = n_1^{t_1} \dots n_d^{t_d} L\left(n_1^{t_1} \dots n_d^{t_d}\right) \tag{9.29}$$

with $\min(t_1, \dots, t_d) > \frac{1}{2}$. Moreover, if $\min(t_1, \dots, t_d) > 1$, then the field (9.29) satisfies condition (9.17). A partial case corresponds to the function $L(x) \equiv 1$. This case coincides with (9.6) if $t_1 = \dots = t_d = 1$, or with (9.7) if $t_1 = \dots = t_d > \frac{1}{2}$, or with (9.8) if $\min(t_1, \dots, t_d) > \frac{1}{2}$. Other examples of fields (9.29) correspond to $L(x) = (\log x)^\beta$, $\beta \in \mathbf{R}$; $L(x) = (\log\log x)^\beta$, $\beta \in \mathbf{R}$, or

$$L(x) = \exp\left\{(\log x)^\beta\right\}, \quad 0 < \beta < 1.$$

Note that the latter function $L(x)$ grows faster than any power of $\log x$.

To prove condition (9.10) for a field (9.29) in the case $\min(t_1, \dots, t_d) > \frac{1}{2}$ of for a field satisfying condition (9.17) in the case $\min(t_1, \dots, t_d) > 1$, we first estimate A_k for such a field $\{b(\mathbf{n})\}$ (see (9.9)). Put $R(x) = xL(x)$ and assume that R is a monotone continuous function. Then

$$\begin{aligned} A_k &= \operatorname{card}\left(\mathbf{n} \in \mathbf{N}^d : R\left(n_1^{t_1} \dots n_d^{t_d}\right) < k+1\right) \\ &= \operatorname{card}\left(\mathbf{n} \in \mathbf{N}^d : n_1^{t_1} \dots n_d^{t_d} < R^{-1}(k+1)\right), \end{aligned}$$

where $R^{-1}(\cdot)$ is the inverse function. According to Proposition A.20, we get

$$A_k \asymp \left(R^{-1}(k+1)\right)^{1/t} \left(\log R^{-1}(k+1)\right)^{c-1}, \tag{9.30}$$

where $t = \min(t_1, \dots, t_d)$ and $c = \operatorname{card}(i : t_i = t)$.

According to Proposition A.15, the inverse function admits the representation

$$R^{-1}(x) = xL^*(x), \tag{9.31}$$

where L^* is a slowly varying function, whence we conclude that

$$\left(R^{-1}(k+1)\right)^{1/t} \left(\log R^{-1}(k+1)\right)^{c-1} = k^{1/t} \widetilde{L}(k) (\log k)^{c-1}. \tag{9.32}$$

Here the symbol $\widetilde{L}$ stands for a slowly varying function that is expressed via L^*.

Note that

$$\sum_{k=n}^{\infty} \frac{\widetilde{L}(k)}{k^{1+\rho}} \le \sum_{k=n}^{\infty} r_k \int_{k}^{k+1} \frac{\widetilde{L}(t)}{t^{1+\rho}}\, dt,$$

where $r_k = 2^{1+\rho} \sup_{k \le t \le k+1} \widetilde{L}(k)/\widetilde{L}(t)$. The uniform convergence theorem for slowly varying functions (see Proposition A.16) yields

$$\lim_{x\to\infty} \frac{\widetilde{L}(\lambda x)}{\widetilde{L}(x)} = 1$$

uniformly with respect to $\lambda \in [a, b]$ whatever $0 < a < b < \infty$. Hence $\sup_{k\ge1} r_k < \infty$, that is

$$\sum_{k=n}^{\infty} \frac{\widetilde{L}(k)}{k^{1+\rho}} \le \text{const} \int_{n}^{\infty} \frac{\widetilde{L}(t)}{t^{1+\rho}}\, dt.$$

By Karamata's theorem for sums (see Proposition A.14),

$$\sum_{k=n}^{\infty} \frac{\widetilde{L}(k)}{k^{1+\rho}} \le \text{const}\, \frac{\widetilde{L}(n)}{n^{\rho}}.$$

Therefore, if $\min(t_1, \dots, t_d) > \frac{1}{2}$, then

$$\sum_{k=n}^{\infty} \frac{A_k}{k^3} \asymp \sum_{k=n}^{\infty} \frac{\widetilde{L}(k)\,(\log k)^{c-1}}{k^{3-1/t}} \le \text{const}\, \frac{\widetilde{L}(n)\,(\log n)^{c-1}}{n^{2-1/t}},$$

since $3 - \frac{1}{t} > 1$. These relations coincide with condition (9.10) in view of (9.30)–(9.32).

If $\min(t_1, \dots, t_d) > 1$, then

$$\sum_{k=n}^{\infty} \frac{A_k}{k^2} \asymp \sum_{k=n}^{\infty} \frac{\widetilde{L}(k)\,(\log k)^{c-1}}{k^{2-1/t}} \le \text{const}\, \frac{\widetilde{L}(n)\,(\log n)^{c-1}}{n^{1-1/t}}.$$

These relations coincide with condition (9.17) in view of (9.30)–(9.32).

The above reasoning allows one to provide several corollaries to the d-parameter Feller theorem.

Corollary 9.1 *Let $\{X(\mathbf{n}), \mathbf{n} \in \mathbf{N}^d\}$ be a field of independent identically distributed random variables. Assume that $\min(t_1, \dots, t_d) > \frac{1}{2}$. Let L be a positive slowly varying function such that $R(x) = xL(x)$ is an increasing continuous function. Denote by R^{-1} its inverse function. Put*

$$b(\mathbf{n}) = n_1^{t_1} \dots n_d^{t_d} L\left(n_1^{t_1} \dots n_d^{t_d}\right), \quad t = \min(t_1, \dots, t_d), \quad c = \operatorname{card}(i : t_i = t).$$

The strong law of large numbers (9.14) *is equivalent to the condition*

$$\mathsf{E}\left[\left(R^{-1}(|X|)\right)^{1/t}\left(\log^+|X|\right)^{c-1}\right] < \infty \tag{9.33}$$

if $t > 1$, where the random variable X has the same distribution function as all other random variables $X(\mathbf{n})$.

Further, if $\frac{1}{2} < t \le 1$, then the strong law of large numbers (9.14) *is equivalent to the set of conditions* (9.33) *and* (9.16).

Remark 9.3 Corollary 9.1 holds in the general case even if the function R is not monotone or is not continuous. Condition (9.33) in this case is written in terms of the inverse of an increasing and continuous function $R^{\uparrow}$, being asymptotically equivalent to R (see Proposition A.10).

Proof of Corollary 9.1 As proved above, condition (9.10) holds for the field $\{b(\mathbf{n}), \mathbf{n} \in \mathbf{N}^d\}$ and thus one can apply Theorem 9.5. To complete the proof of Corollary 9.1, we prove the equivalences (9.33) $\Longleftrightarrow$(9.15) and (9.15) $\Longrightarrow$ (9.16) for the case of $t > 1$.

First we prove the equivalence (9.33) $\Longleftrightarrow$ (9.15). Note that

$$\rho \stackrel{\text{def}}{=} \sum_{\mathbf{n}\in\mathbf{N}^d} \mathsf{P}(|X| \ge b(\mathbf{n})) = \sum_{\mathbf{n}\in\mathbf{N}^d} \mathsf{P}\left(R^{-1}(|X|) \ge n_1^{t_1} \dots n_d^{t_d}\right).$$

Putting $Y = R^{-1}(|X|)$, we obtain

$$\sum_{k=1}^{\infty} a_k \mathsf{P}(Y \ge k) \ge \rho \ge \sum_{k=1}^{\infty} a_k \mathsf{P}(Y \ge k+1),$$

where $a_k = \operatorname{card}\left\{\mathbf{n} : k \le n_1^{t_1} \dots n_d^{t_d} < k+1\right\}$ and

$$A_j = \operatorname{card}\left\{\mathbf{n} : n_1^{t_1} \dots n_d^{t_d} < j+1\right\}.$$

Thus

$$\sum_{j=1}^{\infty} A_j \mathsf{P}(j \le Y < j+1) \ge \rho \ge \sum_{j=1}^{\infty} A_{j-1} \mathsf{P}(j \le Y < j+1).$$

Proposition A.20 implies that

$$A_j \asymp A_{j-1} \asymp j^{1/t}\left(\log^+ j\right)^{c-1},$$

that is, condition (9.15) is equivalent to the convergence of the series

$$\sum_{j=1}^{\infty} j^{1/t} \left(\log^+ j\right)^{c-1} \mathsf{P}(j \le Y < j+1).$$

In turn, the convergence of the latter series is equivalent to the condition

$$\mathsf{E}\left[\left(R^{-1}(|X|)\right)^{1/t} \left(\log^+ R^{-1}(|X|)\right)^{c-1}\right] < \infty.$$

Representation (9.31) means in particular that the function $\log^+ R^{-1}(x)$ is asymptotically equivalent to the function $\log^+ x$ as $x \to \infty$. This observation proves the equivalence of conditions (9.33) and (9.15).

Now we prove the implication (9.15) $\Longrightarrow$ (9.16) for the case of $t > 1$. In view of the equivalence (9.15) $\Longleftrightarrow$ (9.33), we have $\mathsf{E}\left[\left(R^{-1}(|X|)\right)^{1/t}\right] < \infty$. Further, representation (9.31) implies that $\left(R^{-1}(x)\right)^{1/t} = x^{1/t}\widetilde{L}(x)$, where $\widetilde{L}(\cdot)$ is a slowly varying function. Moreover, R^{-1} is increasing by the inverse function theorem and hence

$$\begin{aligned}\sum_{k=1}^{\infty} \frac{\left(R^{-1}(k-1)\right)^{1/t}}{k} \int\limits_{k-1\le|x|<k} |x|\, dF &\le \sum_{k=1}^{\infty} \left(R^{-1}(k-1)\right)^{1/t} \int\limits_{k-1\le|x|<k} dF \\ &\le \sum_{k=1}^{\infty} \int\limits_{k-1\le|x|<k} \left(R^{-1}(|x|)\right)^{1/t} dF.\end{aligned}$$

The series on the right-hand side equals $\mathsf{E}\left[\left(R^{-1}(|X|)\right)^{1/t}\right]$ and thus converges. Now $\left(R^{-1}(x)\right)^{1/t}/x \to 0$ as $x \to \infty$, since $t > 1$. Putting

$$a_k = \int\limits_{k-1\le|x|<k} |x|\, dF, \qquad b_k = \frac{k}{\left(R^{-1}(k-1)\right)^{1/t}},$$

we derive from Lemma 9.1 below and Proposition A.10 that

$$\frac{\left(R^{-1}(n)\right)^{1/t}}{n} \int\limits_{|x|<n} |x|\, dF(x) \to 0, \quad x \to \infty. \tag{9.34}$$

Lemma 9.1 *Let $\{a_k\}$ and $\{b_k\}$ be two sequences of real numbers such that $a_k \ge 0$, $b_k > 0$, and $b_n \to \infty$ as $n \to \infty$. If*

$$b_n \sim b'_n, \quad n \to \infty, \tag{9.35}$$

for some non-decreasing sequence $\{b'_n\}$ *and*

$$\sum_{k=1} \frac{a_k}{b_k} < \infty,$$

then

$$\frac{a_1 + \cdots + a_n}{b_n} \to 0, \quad n \to \infty.$$

The proof of Lemma 9.1 is given below; now we use its result to complete the proof of Corollary 9.1. By the properties of regularly varying functions we establish a continuous parameter analog of relation (9.34), namely

$$\lim_{y\to\infty} \frac{\left(R^{-1}(y)\right)^{1/t}}{y} \int\limits_{|x|<y} |x|\, dF(x) = 0 \quad \text{or} \quad \lim_{z\to\infty} \frac{z^{1/t}}{R(z)} \int\limits_{|x|<R(z)} |x|\, dF(x) = 0.$$

Put $z = n_1^{t_1} \dots n_d^{t_d}$, then $z \geq |\mathbf{n}|^t$ and

$$\left| \frac{|\mathbf{n}|}{b(\mathbf{n})} \int\limits_{|x|<b(\mathbf{n})} x\, dF(x) \right| \leq \frac{z^{1/t}}{R(z)} \int\limits_{|x|<R(z)} |x|\, dF(x),$$

which means that condition (9.16) holds in the case of $t > 1$. □

Proof of Lemma 9.1 If the sequence $\{b_n\}$ itself is non-decreasing, then Lemma 9.1 coincides with Kronecker's lemma (see Proposition A.9). The general case follows from Proposition A.9 and condition (9.35). □

Corollary 9.2 *Let* $\{X(\mathbf{n}), \mathbf{n} \in \mathbf{N}^d\}$ *be a field of independent identically distributed random variables and let* $\min(t_1, \dots, t_d) > \frac{1}{2}$. *The strong law of large numbers*

$$\lim(\max) \frac{S(\mathbf{n})}{n_1^{t_1} \dots n_d^{t_d}} = 0 \quad a.\ s. \tag{9.36}$$

is equivalent to the following set of two conditions

$$\mathsf{E}\left[|X|^{1/t} \left(\log^+ |X|\right)^{c-1}\right] < \infty, \tag{9.37}$$

$$\mathsf{E}[X] = 0, \quad \textit{if} \quad t \leq 1, \tag{9.38}$$

where the random variable X *has the same distribution function as all other* $X(\mathbf{n})$.

Proof of Corollary 9.2 We apply Corollary 9.1 with $L(x) \equiv 1$. First of all, Corollaries 9.1 and 9.2 coincide if $t > 1$. It remains to treat the case $t \le 1$ for which we need to clarify the role of condition (9.38).

According to Corollary 9.1, the strong law of large numbers (9.36) implies conditions (9.33) and (9.16). The equivalence (9.33) $\Longleftrightarrow$ (9.37) is obvious. Since $t \le 1$, condition (9.33) yields the existence of the first moment.

Further, let $t_i = t$ for some i, $1 \le i \le d$. The Cauchy lim(max) convergence criterion (see Corollary A.1) together with (9.16) implies that

$$\lim_{n_i \to \infty} n_i^{1-t_i} \int\limits_{|x|<n_i^{t_i}} x\, dF(x) = 0$$

(we choose $\mathbf{n}_k = (\underbrace{1, \dots, 1}_{i-1}, k, \underbrace{1, \dots, 1}_{d-i})$ in Corollary A.1). The latter condition may hold only if $\mathsf{E}[X] = 0$, that is, condition (9.38) is satisfied.

On the other hand, let conditions (9.37)–(9.38) hold. Relation (9.37) implies that the moment $\mathsf{E}\left[|X|^{1/t}\right]$ is finite. In turn, this implies that

$$z^{\frac{1}{t}-1} \int\limits_{|x|\ge z} |x|\, dF(x) \to 0, \quad z \to \infty.$$

For $\mathbf{n} \in \mathbf{N}^d$, put $z = n_1^{t_1} \dots n_d^{t_d}$. It is clear that $|\mathbf{n}| \le z^{1/t}$ and thus

$$\frac{|\mathbf{n}|}{b(\mathbf{n})} \int\limits_{|x|\ge b(\mathbf{n})} |x|\, dF(x) \le z^{\frac{1}{t}-1} \int\limits_{|x|\ge z} |x|\, dF(x),$$

whence (9.16) follows. □

The simplest case of the preceding corollary corresponds to $t_1 = \dots = t_d \stackrel{\text{def}}{=} t$.

Corollary 9.3 *Let $\{X(\mathbf{n}), \mathbf{n} \in \mathbf{N}^d\}$ be a field of independent identically distributed random variables and let $\min(t_1, \dots, t_d) > \frac{1}{2}$. The strong law of large numbers*

$$\lim(\max) \frac{S(\mathbf{n})}{(n_1 \dots n_d)^t} = 0 \quad a.\,s.$$

is equivalent to the following set of two conditions

$$\mathsf{E}\left[|X|^{1/t} \left(\log^+ |X|\right)^{d-1}\right] < \infty,$$
$$\mathsf{E}[X] = 0, \quad \text{if } t \le 1.$$

The following particular case of Corollary 9.3 coincides with Smythe's Theorem 9.4.

Corollary 9.4 *Let* $\{X(\mathbf{n}), \mathbf{n} \in \mathbf{N}^d\}$ *be a field of independent identically distributed random variables. The strong law of large numbers*

$$\lim(\max)\frac{S(\mathbf{n})}{|\mathbf{n}|} = 0 \quad a.\ s.$$

is equivalent to the following set of two conditions

$$\mathsf{E}\left[|X| \left(\log^+ |X|\right)^{d-1}\right] < \infty, \quad \mathsf{E}[X] = 0.$$

Theorem 9.5 allows us to find the normalizing fields for which the strong law of large numbers holds under the classical moment assumptions.

Corollary 9.5 *Let* $\{X(\mathbf{n}), \mathbf{n} \in \mathbf{N}^d\}$ *be a field of independent identically distributed random variables. Further let* $t \overset{\text{def}}{=} \min(t_1, \dots, t_d) > \frac{1}{2}$ *and* $c = \operatorname{card}\{i : t_i = t\}$. *The strong law of large numbers*

$$\lim(\max)\frac{S(\mathbf{n})}{n_1^{t_1} \dots n_d^{t_d} / (\log |\mathbf{n}|)^{c-1}} = 0 \quad a.\ s.$$

is equivalent to the following set of two conditions

$$\mathsf{E}\left[|X|^{1/t}\right] < \infty,$$
$$\mathsf{E}[X] = 0, \quad \text{if} \ \ t \le 1,$$

where the random variable X *has the same distribution function as all other* $X(\mathbf{n})$.

Corollary 9.6 *Let* $\{X(\mathbf{n}), \mathbf{n} \in \mathbf{N}^d\}$ *be a field of independent identically distributed random variables and let* $t > \frac{1}{2}$. *The strong law of large numbers*

$$\lim(\max)\frac{S(\mathbf{n})}{|\mathbf{n}|^t / (\log |\mathbf{n}|)^{d-1}} = 0 \quad a.\ s.$$

is equivalent to the following set of two conditions

$$\mathsf{E}\left[|X|^{1/t}\right] < \infty,$$
$$\mathsf{E}[X] = 0, \quad \text{if} \ \ t \le 1,$$

where the random variable X *has the same distribution function as all other* $X(\mathbf{n})$.

Corollary 9.7 *Let $\{X(\mathbf{n}), \mathbf{n} \in \mathbf{N}^d\}$ be a field of independent identically distributed random variables. The strong law of large numbers*

$$\lim(\max)\frac{S(\mathbf{n})}{|\mathbf{n}|/(\log|\mathbf{n}|)^{d-1}} = 0 \qquad a.\ s.$$

is equivalent to the following set of two conditions

$$\mathsf{E}[|X|] < \infty, \qquad \mathsf{E}[X] = 0,$$

where the random variable X has the same distribution function as all other $X(\mathbf{n})$.

Another case that fits Theorem 9.5 is presented by the fields

$$b(\mathbf{n}) = \left(n_1^{t_1} + \cdots + n_d^{t_d}\right)^r L\left(\left(n_1^{t_1} + \cdots + n_d^{t_d}\right)^r\right), \tag{9.39}$$

where $L(\cdot)$ is a slowly varying function. If $d = 1$ and $L(x) \equiv 1$, then this is the Marcinkiewicz–Zygmund normalization; moreover if $d = 1$ and $t_1 = 1$, then (9.39) is the Kolmogorov normalization. In any case, (9.39) is a natural generalization of the classical normalizations for the case of $d > 1$.

Let, as above, $R(x) = xL(x)$ be an increasing continuous function and let $R^{-1}(\cdot)$ be its inverse function. By Proposition A.22, we have

$$\begin{aligned} A_k &= \operatorname{card}\left\{\mathbf{n} \in \mathbf{N}^d : R\left(\left(n_1^{t_1} + \cdots + n_d^{t_d}\right)^r\right) < k+1\right\} \\ &= \operatorname{card}\left\{\mathbf{n} \in \mathbf{N}^d : \left(n_1^{t_1} + \cdots + n_d^{t_d}\right)^r < R^{-1}(k+1)\right\} \\ &\asymp \left[R^{-1}(k+1)\right]^a, \quad \text{where} \quad a = \frac{1}{r}\left(\frac{1}{t_1} + \cdots + \frac{1}{t_d}\right). \end{aligned}$$

Representation (9.31) implies that

$$A_k \asymp k^a \left(L^*(k)\right)^a.$$

Therefore condition (9.10) holds with $a < 2$. Applying Theorem 9.5 and reasoning in the same way as in the proof of Corollaries 9.1–9.2, we arrive at the following results.

Corollary 9.8 *Let $\{X(\mathbf{n}), \mathbf{n} \in \mathbf{N}^d\}$ be a field of independent identically distributed random variables. Further, let $\min(t_1, \dots, t_d) > 0$, $r > 0$, and L be a positive slowly varying function. Let $R(x) = xL(x)$ be an increasing continuous function and R^{-1} be its inverse function. Define the field $\{b(\mathbf{n}), \mathbf{n} \in \mathbf{N}^d\}$ according to (9.39). If*

$$1 \le \frac{1}{r}\left(\frac{1}{t_1} + \cdots + \frac{1}{t_d}\right) < 2,$$

then the strong law of large numbers (9.14) *is equivalent to the set of conditions* (9.16) *and*

$$\mathsf{E}\left[\left(R^{-1}(|X|)\right)^a\right] < \infty, \quad a = \frac{1}{r}\left(\frac{1}{t_1} + \cdots + \frac{1}{t_d}\right), \tag{9.40}$$

where the random variable X *has the same distribution function as all other* $X(\mathbf{n})$.

Moreover, if

$$0 < \frac{1}{r}\left(\frac{1}{t_1} + \cdots + \frac{1}{t_d}\right) < 1,$$

then the strong law of large numbers (9.14) *is equivalent to condition* (9.40).

Corollary 9.9 *Let* $\{X(\mathbf{n}), \mathbf{n} \in \mathbf{N}^d\}$ *be a field of independent identically distributed random variables and let* $\min(t_1, \ldots, t_d) > 0$ *and* $r > 0$. *If*

$$1 \le \frac{1}{r}\left(\frac{1}{t_1} + \cdots + \frac{1}{t_d}\right) < 2,$$

then the strong law of large numbers

$$\lim(\max) \frac{S(\mathbf{n})}{\left(n_1^{t_1} + \cdots + n_d^{t_d}\right)^r} = 0 \quad \text{a. s.} \tag{9.41}$$

is equivalent to the set of conditions (9.16) *and*

$$\mathsf{E}\left[|X|^a\right] < \infty, \quad a = \frac{1}{r}\left(\frac{1}{t_1} + \cdots + \frac{1}{t_d}\right), \tag{9.42}$$

where the random variable X *has the same distribution function as all other* $X(\mathbf{n})$.

Moreover, if

$$0 < \frac{1}{r}\left(\frac{1}{t_1} + \cdots + \frac{1}{t_d}\right) < 1,$$

then the strong law of large numbers (9.41) *is equivalent to condition* (9.42).

Corollary 9.10 *Let* $\{X(\mathbf{n}), \mathbf{n} \in \mathbf{N}^d\}$ *be a field of independent identically distributed random variables and let* $t > 0$ *and* $r > 0$. *If*

$$\frac{d}{2} < rt \le d,$$

then the strong law of large numbers

$$\lim(\max) \frac{S(\mathbf{n})}{\left(n_1^t + \cdots + n_d^t\right)^r} = 0 \quad \text{a. s.} \tag{9.43}$$

is equivalent to the set of conditions (9.16) *and*

$$\mathsf{E}\left[|X|^{d/rt}\right] < \infty, \tag{9.44}$$

where the random variable X has the same distribution function as all other $X(\mathbf{n})$.

Moreover, if

$$0 < d < rt,$$

then the strong law of large numbers (9.43) *is equivalent to condition* (9.44).

Corollary 9.11 *Let $\{X(\mathbf{n}), \mathbf{n} \in \mathbf{N}^d\}$ be a field of independent identically distributed random variables and $r > 0$. If*

$$\frac{d}{2} < r \le d,$$

then the strong law of large numbers

$$\lim(\max)\frac{S(\mathbf{n})}{(n_1 + \cdots + n_d)^r} = 0 \qquad a.\ s. \tag{9.45}$$

is equivalent to the set of conditions (9.16) *and*

$$\mathsf{E}\left[|X|^{d/r}\right] < \infty, \tag{9.46}$$

where the random variable X has the same distribution function as all other $X(\mathbf{n})$.

Moreover, if

$$0 < d < r,$$

then the strong law of large numbers (9.45) *is equivalent to condition* (9.46).

Corollary 9.12 *Let $\{X(\mathbf{n}), \mathbf{n} \in \mathbf{N}^d\}$ be a field of independent identically distributed random variables. The strong law of large numbers*

$$\lim(\max)\frac{S(\mathbf{n})}{(n_1 + \cdots + n_d)^d} = 0 \qquad a.\ s.$$

is equivalent to the following moment condition

$$\mathsf{E}[|X|] < \infty, \qquad \mathsf{E}[X] = 0.$$

Remark 9.4 In the weak law of large numbers presented in Theorem 4.12, the rectangular sums of independent identically distributed random variables are also normalized by the field $b(\mathbf{n}) = n_1^{t_1} \dots n_d^{t_d}$. A particular case of Theorem 4.12 tells us that if $\min(t_1, \dots, t_d) = \frac{1}{2}$, then the weak law of large numbers holds for $\lim(\max)$-convergence only in the case of degenerate random variables $X(\mathbf{n})$.

Another particular case of Theorem 4.12 tells us that, for the normalizing field $\{b(\mathbf{n})\}$ with $\min(t_1, \dots, t_d) = \frac{1}{2}$ and $\max(t_1, \dots, t_d) > \frac{1}{2}$, the weak law of large numbers holds for lim(min)-convergence if $\mathsf{E}\left[X^2\right] < \infty$. This result seems impossible "in view of the central limit theorem" and this is indeed the case for almost sure convergence. We are now going to show that the strong law of large numbers for such a normalizing field cannot be true even for lim(min)-convergence if the common distribution function of the random variables $X(\mathbf{n})$ is nondegenerate.

We consider the case of $\max(t_1, \dots, t_d) > \frac{1}{2}$, since otherwise the central limit theorem contradicts even the weak law of large numbers. Without loss of generality we assume that $\frac{1}{2} = t_1 = \dots = t_c < t_{c+1} \le \dots t_d$ for some $1 \le c < d$. If the strong law of large numbers holds for lim(min) convergence, that is if

$$\lim(\min) \frac{S(\mathbf{n})}{n_1^{t_1} \dots n_d^{t_d}} = 0 \qquad \text{a. s.,}$$

then

$$\lim_{k\to\infty} \frac{S(\mathbf{n}_k)}{n_{1k}^{t_1} \dots n_{dk}^{t_d}} = 0 \qquad \text{a. s.} \tag{9.47}$$

for all sequences of multi-indices $\mathbf{n}_k = (n_{1k}, \dots, n_{dk})$, $k \ge 1$, of the space $\mathbf{N}^d$ such that $\min\{n_{1k}, \dots, n_{dk}\} \to \infty$ (see Corollary A.2). Put

$$a = \frac{1}{2(t_{c+1} + \dots + t_d) - d + c}, \qquad b = \frac{a(d-c)}{c}$$

and define the subsequence of multi-indices $\{\mathbf{n}_k, k \ge 1\}$ such that

$$n_{1k} = \dots = n_{ck} = \left[\frac{k^{1/c}}{\left(\log^+ \log^+ k\right)^b} \right], \qquad n_{c+1,k} = \dots = n_{dk} = \left[\log^+ \log^+ k\right]^a.$$

It is clear that $n_{1k} \dots n_{dk} \sim k$ and

$$n_{1k}^{1/2} \dots n_{ck}^{1/2} n_{c+1,k}^{t_{c+1}} \dots n_{dk}^{t_d} \sim \sqrt{k \log\log k} \qquad \text{as } k \to \infty.$$

Therefore, the strong law of large numbers (9.47) implies that

$$\lim_{k\to\infty} \frac{S(\mathbf{n}_k)}{\sqrt{|\mathbf{n}_k| \log\log |\mathbf{n}_k|}} = 0 \qquad \text{a. s.} \tag{9.48}$$

for the subsequence defined above. Since $\mathbf{n}_k \le \mathbf{n}_{k+1}$, one can treat $\{S(\mathbf{n}_k), k \ge 1\}$ as a subsequence of the sequence $\{\zeta_j, j \ge 1\}$ of usual cumulative sums of independent identically distributed random variables $\{\xi_i, i \ge 1\}$ where the distribution functions of the random variables $X(\mathbf{n})$ and ξ_i are the same (see Section A.17).

The subsequence $\{\zeta_{m_k}, k \geq 1\}$ corresponding to $\{S(\mathbf{n}_k), k \geq 1\}$ is such that $m_k = |\mathbf{n}_k|$ and $\liminf m_k/m_{k+1} > 0$. By the law of the iterated logarithm for subsequences,

$$\limsup_{k\to\infty} \frac{\zeta_{m_k} - m_k \mathsf{E}[X_1]}{\sqrt{m_k \log\log m_k}} = \sqrt{2\sigma^2} \quad \text{a. s.,}$$
$$\liminf_{k\to\infty} \frac{\zeta_{m_k} - m_k \mathsf{E}[X_1]}{\sqrt{m_k \log\log m_k}} = -\sqrt{2\sigma^2} \quad \text{a. s.} \tag{9.49}$$

(see [136]). Relations (9.48) and (9.49) may hold simultaneously only if $\sigma^2 = 0$, that is, if the distribution is degenerate.

9.4 The Strong Law of Large Numbers for Indices in a Restricted Domain

One of the most challenging problems for multi-indices is related to limit theorems for "subsequences". Gut [133] was the first to consider the strong law of large numbers for multi-indexed sums whose indices satisfy certain restrictions. Since not all sums are involved in the definition of the limit, it seems reasonable that the conditions for the limit theorems in this case are weaker than those used when all sums are considered.

Let $0 < \theta < 1$. Define the sector A_θ by

$$A_\theta = \{(n_1, \ldots, n_d) : \theta n_i \leq n_j \leq \theta^{-1} n_i \quad \text{for all } 1 \leq i < j \leq d\}.$$

Theorem 9.7 (A. Gut (1978)) *Let $\{X(\mathbf{n})\}$ be independent identically distributed random variables. The limit*

$$\lim_{\substack{|\mathbf{n}|\to\infty \\ \mathbf{n}\in A_\theta}} \frac{S(\mathbf{n})}{|\mathbf{n}|} \tag{9.50}$$

exists and is finite almost surely if and only if the first moment of the random variables $X(\mathbf{n})$ exists. If limit (9.50) *exists, then it almost surely equals the first moment.*

There are several complications in the proof of the existence of limit (9.50) as compared to the proof of the strong law of large numbers for the whole space $\mathbf{N}^d$. We show some of these complications in the course of the proof of necessity of condition (9.2) in Theorem 9.7.

A usual way to prove the necessity of condition (9.5) in Theorem 9.4 is to represent each term $X(\mathbf{n})$ as a finite linear combination of sums $S(\mathbf{n}-\boldsymbol{\varepsilon})$ with $\boldsymbol{\varepsilon} \in \mathscr{E}_d$ (see (9.20)) and then to apply the strong law of large numbers to each of the sums, whence we get $X(\mathbf{n})/|\mathbf{n}| \to$ almost surely. Then we use the Borel–Cantelli lemma for independent random events and obtain the moment condition needed.

Fig. 9.1 "Boundary" multi-indices

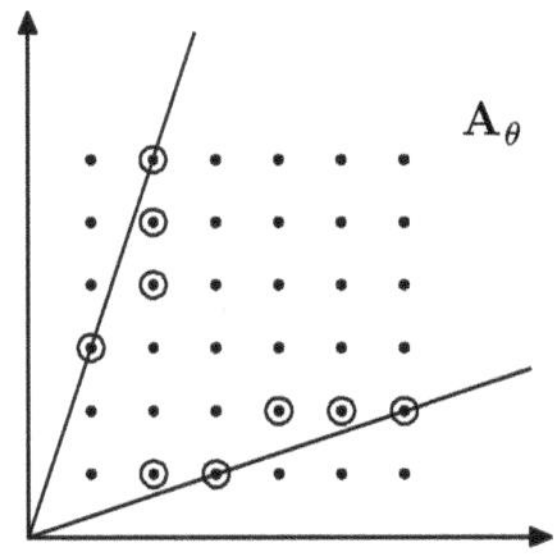

An attempt to use the same method to prove the corresponding part of Theorem 9.7 does not lead to success, since not all sums $S(\mathbf{n}-\boldsymbol{\varepsilon})$ in representation (9.20) belong to the set A_θ and thus the strong law of large numbers may fail for some of them. For example, some elements $(m, n) \in A_\theta$ are shown in Fig. 9.1 (circled points) for which representation (9.20) of the corresponding random variables $X(m, n)$ contains the sums $S(m-\varepsilon_1, n-\varepsilon_2)$ whose indices $(m-\varepsilon_1, n-\varepsilon_2)$ do not belong to the sector A_θ.

This problem is overcome by considering a tighter sector $A_{\theta'}$ with $\theta < \theta' < 1$. The convergence of the averaged sums on A_θ implies their convergence on $A_{\theta'}$. Now all variables $X(\mathbf{n})$, $\mathbf{n} \in A_{\theta'}$, except a finite number of them, can be represented as a linear combination of at most 2^d sums $S(\mathbf{n})$, $\mathbf{n} \in A_\theta$, and thus the strong law of large numbers can be applied to each of the sums in the representation. Unfortunately, this method does not work for other types of constraints (not necessarily linear constraints, as in the case of A_θ) imposed on the multi-indices.

9.4.1 Sectors with Curvilinear Boundaries

It seems reasonable to extend Theorem 9.7 to other subsets of multi-indices. In this section, we consider sectors with curvilinear boundaries. In what follows we need the following definition.

Definition 9.1 Let $A \subseteq \mathbf{N}^d$ be an infinite subset of multi-indices. For a field of real numbers $\{c(\mathbf{n})\}$, we write

$$\lim\nolimits_{\mathrm{A}} c(\mathbf{n}) = c,$$

if, for all $\varepsilon > 0$, the inequality $|c(\mathbf{n}) - c| < \varepsilon$ holds for all multi-indices $\mathbf{n} \in A \backslash A'$, where A' is a certain finite subset of $\mathbf{N}^d$.

In what follows we consider the case $d = 2$ only. We assume that f is a non-decreasing function such that

$$f(x) \geq x \quad \text{for all } x \geq 1. \tag{9.51}$$

The set A is defined by

$$A = \{(i, j) : i \le j \le f(i)\}. \tag{9.52}$$

Theorem 9.8 *Consider a field of independent identically distributed random variables $\{X(i, j)\}$ and the field of their double sums $\{S(m, n)\}$:*

$$S(m, n) = \sum_{i=1}^{m} \sum_{j=1}^{n} X(i, j).$$

Let X be a copy of $X(i, j)$. The distribution function of the random variable X is denoted by F. The strong law of large numbers

$$\lim_{\mathrm{A}} \frac{S(m, n)}{mn} = \mu \quad a.\ s., \tag{9.53}$$

where the set A is defined by (9.52)*, an increasing function f satisfies* (9.51)*, and where μ is a certain non-random constant, holds if and only if*

$$\mathsf{E}[|X|] < \infty, \qquad \mathsf{E}[X] = \mu, \tag{9.54}$$

$$\sum_{(i,j)\in A} \mathsf{P}(|X| \ge ij) < \infty. \tag{9.55}$$

Remark 9.5 Let $\tau_k, k \ge 1$, be the Dirichlet sequence for the set A, that is the number of solutions of the equation $ij = k$ such that $(i, j) \in A$. In other words,

$$\tau_k = \tau_k^A = \operatorname{card}\{(i, j) \in A : ij = k\}.$$

Then condition (9.55) can be rewritten as follows

$$\sum_{k=1}^{\infty} \tau_k \mathsf{P}(|X| \ge k) < \infty. \tag{9.56}$$

Put $T_1 = \tau_1$,

$$T_k = T_k^A = \tau_1 + \cdots + \tau_k, \quad k \ge 2. \tag{9.57}$$

Then condition (9.56) is equivalent to the following

$$\sum_{k=1}^{\infty} T_k \mathsf{P}(k \le |X| < k+1) < \infty.$$

Introducing the function $T(x) = T_{[x]}$, condition (9.55) follows from

$$\mathsf{E}[T(|X|)] < \infty.$$

If T does not vary quickly in the sense that $T(x+1) = O(T(x))$, then the latter condition is equivalent to (9.55).

We need two auxiliary results to prove Theorem 9.8. For $m \geq 1$ and $n \geq 1$, put

$$\Pi(m,n) = \{(i,j) : 1 \leq i \leq m, 1 \leq j \leq n\}.$$

We say that a family $\mathscr{A}$ of sets $\{A_\alpha\}$ is *linearly ordered* (with respect to the inclusion of sets) if, for each pair $A_\alpha, A_\beta \in \mathscr{A}$, at least one of the inclusions $A_\alpha \subseteq A_\beta$ or $A_\beta \subseteq A_\alpha$ holds.

Lemma 9.2 *Let an increasing function f satisfy* (9.51). *Put*

$$B = \{(i,j) : j > f(i)\}, \qquad C = \{(i,j) : j < i\}.$$

Then both families

$$\mathscr{B} = \{B \cap \Pi(m,n); (m,n) \in A\},$$
$$\mathscr{C} = \{C \cap \Pi(m,n); (m,n) \in A\}$$

are linearly ordered where A is defined according to (9.52).

We say that a family $\mathscr{V}$ of points $\{(m_\alpha, n_\alpha)\}$, $(m_\alpha, n_\alpha) \in \mathbf{N}^2$, is *linearly ordered* if, for each pair of points $(m_\alpha, n_\alpha), (m_\beta, n_\beta) \in \mathscr{V}$, at least one of the following two sets of relations $m_\alpha \leq m_\beta$ and $n_\alpha \leq n_\beta$ or $m_\beta \leq m_\alpha$ and $n_\beta \leq n_\alpha$ holds.

Let $D \subseteq \mathbf{N}^2$. Define the inner part $\operatorname{int}(D)$ and boundary $\partial(D)$ of a set D as follows

$$\operatorname{int}(D) = \{(i,j) \in D : \ (i-1,j) \in D, \ (i,j-1) \in D, \ (i-1,j-1) \in D\},$$
$$\partial(D) = D \setminus \operatorname{int}(D).$$

Lemma 9.3 *Let an increasing function f satisfy* (9.51). *Put*

$$B = \{(i,j) : j \geq f(i)\} \text{ and } C = \{(i,j) : i \geq j\}.$$

Then both families of points $\partial(B)$ and $\partial(C)$ are linearly ordered.

Proof of Lemma 9.2 Consider the family $\mathscr{B}$. Note that $B \cap \Pi(m,n) = B \cap \Pi(m^*,n)$ for all $(m,n) \in A$, where $m^* = \max\{m : f(m) \leq n\}$.

Now let (m_1,n_1) and (m_2,n_2) be two points in A and let m_1^* and m_2^* be defined as above for (m_1,n_1) and (m_2,n_2), respectively. Put $B_i = B \cap \Pi(m_i,n_i)$ for $i = 1,2$. We are going to prove that $B_2 \subseteq B_1$ if $n_2 \leq n_1$. If this is not the case, then there exists a point $(i,j) \in B_2$ such that $(i,j) \notin B_1$, that is

$$i \leq m_2^*, \quad j \leq n_2, \quad j > f(i), \quad \text{and} \quad i > m_1^* \quad \text{or} \quad j > n_1.$$

The inequality $j > n_1$ contradicts $j \le n_2$ and thus does not hold. In turn, the inequality $i > m_1^*$ does not hold, as well, since $i \ge m_1^* + 1$ in this case, whence we conclude that $f(i) \ge f(m_1^* + 1)$ in view of the monotonicity of the function f. Therefore, $j > f(i) \ge f(m_1^* + 1) > n_1 \ge n_2$ in accordance with the definition of the number m_1^*, whence $j > n_2$. This contradiction proves that $B_2 \subseteq B_1$.

The proof for the family $\mathscr{C}$ is analogous. □

Proof of Lemma 9.3 Consider the case of $\partial(B)$. Note that

$$(i-1, j) \in B \quad \text{for all } (i, j) \in B. \tag{9.58}$$

If (9.58) does not hold for some point $(i, j) \in B$, then $j < f(i-1) \le f(i)$, that is $(i, j) \notin B$.

Moreover,

$$(i, j-1) \notin B \quad \text{for all } (i, j) \in \partial(B). \tag{9.59}$$

If (9.59) does not hold for some point $(i, j) \in \partial(B)$, then $(i, j-1) \in B$. Applying property (9.58), we prove that $(i-1, j-1) \in B$, whence $(i, j) \in B$, $(i-1, j) \in B$, and $(i, j-1)$, $(i-1, j-1) \in B$, but this contradicts $(i, j) \in \partial(B)$.

Now let (m_1, n_1) and (m_2, n_2) be two points of the boundary $\partial(B)$. Without loss of generality, assume that $m_1 \le m_2$. We shall prove that $n_1 \le n_2$ in this case. Assume the converse, that is, $n_1 > n_2$. Then

$$n_2 \le n_1 - 1 < f(m_1) \le f(m_2),$$

which contradicts the inclusion $(m_2, n_2) \in B$. The second inequality in the latter relation follows from $(m_1, n_1 - 1) \notin B$, which in turn follows from (9.59).

The proof for $\partial(C)$ is analogous. □

Proof of Theorem 9.8 *Necessity*. Since $(k, k) \in A$ for all $k \ge 1$, we get $S(k, k)/k^2 \to \mu$ almost surely as $k \to \infty$. The sequence $\{S(k, k); k \ge 1\}$ can be treated as a subsequence of sums $\{S_{k^2}, k \ge 1\}$ of independent identically distributed random variables $\{X_k, k \ge 1\}$ with the distribution function F. Then Proposition A.28 implies that $\mathsf{E}[|X|] < \infty$. According to Theorem 9.2, this means that $S_n/n \to \mathsf{E}[X]$ almost surely as $n \to \infty$, that is, $\mu = \mathsf{E}[X]$. Thus (9.54) is proved.

It remains to prove (9.55). Clearly

$$X(i, j) = S(i, j) - S(i-1, j) - S(i, j-1) + S(i-1, j-1)$$

(cf. (A.6)). Thus the strong law of large numbers (9.53) implies that the inequality $|X(i, j)| \ge ij$ may hold only for a finite number of pairs $(i, j) \in \operatorname{int}(A)$. By the Borel–Cantelli lemma for the independent random events $\{\omega : |X(i, j)| \ge ij\}$ we obtain

$$\sum_{(i,j)\in\operatorname{int}(A)} \mathsf{P}(|X| \ge ij) < \infty. \tag{9.60}$$

Now let the sets B and C be defined in the same way as in Lemma 9.3. It is clear that $\partial(A) = \partial(B) \cup \partial(C)$. Moreover, $\partial(C) = \{(i, j) : i = j\}$. Hence

$$\sum_{(i,j)\in\partial(C)} \mathsf{P}(|X| \geq ij) < \infty \tag{9.61}$$

in view of condition (9.54). The set of points $\partial(B)$ is linearly ordered by Lemma 9.3. Now Theorem 9.2 together with condition (9.54) yields

$$\sum_{(i,j)\in\partial(B)} \mathsf{P}(|X| \geq ij) < \infty. \tag{9.62}$$

Combining (9.60)–(9.62) we prove (9.55). □

Proof of Theorem 9.8 Sufficiency Let conditions (9.54)–(9.55) hold. Without loss of generality, assume that $\mu = 0$. Define

$$S_D(m, n) = \sum_{(i,j)\in D\cap\Pi(m,n)} X(i, j)$$

for an arbitrary set $D \subset \mathbf{N}^2$ (we agree that $\sum_{(i,j)\in\varnothing} X(i, j) = 0$). Then

$$S(m, n) = S_A(m, n) + S_B(m, n) + S_C(m, n)$$

for all $m \geq 1$ and $n \geq 1$, where the sets B and C are defined in Lemma 9.2. The strong law of large numbers for sums S_B and S_C follows from Theorem 9.2 and (9.54) (recall that $\mu = 0$). Indeed, both families of sets

$$\{B \cap \Pi(m, n); (m, n) \in A\} \quad \text{and} \quad \{C \cap \Pi(m, n); (m, n) \in A\}$$

are linearly ordered. This, in particular, means that both families of random variables $\{S_B(m, n); (m, n) \in A\}$ and $\{S_C(m, n); (m, n) \in A\}$ can be viewed as two subsequences of the cumulative sums

$$\{S_{m_k}, k \geq 1\} \quad \text{and} \quad \{S_{n_k}, k \geq 1\}$$

of independent identically distributed random variables $\{X_i; i \geq 1\}$ with the distribution function F. Note that $n_k = k(k + 1)/2$ for sums S_C. Thus

$$\lim\nolimits_{\mathrm{A}} \frac{S_B(m, n)}{mn} = 0, \quad \lim\nolimits_{\mathrm{A}} \frac{S_C(m, n)}{mn} = 0 \quad \text{a. s.,} \tag{9.63}$$

since $\operatorname{card}(B \cap \Pi(m, n)) \leq mn$ and $\operatorname{card}(C \cap \Pi(m, n)) \leq mn$.

Now we prove the strong law of large numbers for sums S_A. Introduce the truncated random variables

$$Y(i, j) = \begin{cases} 0, & (i, j) \notin A, \\ 0, & (i, j) \in A \text{ and } |X(i, j)| \geq ij, \\ X(i, j), & (i, j) \in A \text{ and } |X(i, j)| < ij \end{cases}$$

and their partial sums

$$U(m, n) = \sum_{i=1}^{m} \sum_{j=1}^{n} Y(i, j).$$

The random variables $\{Y(i, j)\}$ are independent and possess finite second moments. Moreover,

$$\sum_{m=1}^{\infty} \sum_{n=1}^{\infty} \frac{\operatorname{var}[Y(m, n)]}{(mn)^2} < \infty \tag{9.64}$$

according to Lemma 9.4, to be proved below. Considering (9.64), we derive from the strong law of large numbers for multi-indexed sums of independent non-identically distributed random variables (see Corollary 8.1) that

$$\mathsf{P}(|U(m, n) - \mathsf{E}[U(m, n)]| \geq \varepsilon mn \text{ i. o.}) = 0 \quad \text{for all } \varepsilon > 0. \tag{9.65}$$

Note that

$$\mathsf{E}[Y(i, j)] = \begin{cases} 0, & (i, j) \notin A, \\ -\int\limits_{|X| \geq ij} X \, d\mathsf{P}, & (i, j) \in A, \end{cases}$$

since $\mu = 0$. Therefore,

$$|\mathsf{E}[U(m, n)]| \leq \sum_{(i,j) \in A \cap \Pi(m,n)} \int\limits_{|X| \geq ij} |X| \, d\mathsf{P},$$

whence we conclude that

$$|\mathsf{E}[U(m, n)]| \geq \varepsilon mn \quad \text{i. o. for all } \varepsilon > 0. \tag{9.66}$$

Indeed, fix $\varepsilon > 0$. Then the inequality $\int\limits_{|X| \geq ij} |X| \, d\mathsf{P} \geq \varepsilon$ holds only for a finite number of pairs (i, j). If the total number of such pairs is denoted by $l(\varepsilon)$, then

$$\frac{|\mathsf{E}[U(m, n)]|}{mn} \leq \frac{1}{mn} \sum_{i=1}^{m} \sum_{j=1}^{n} \int\limits_{|X| \geq ij} |X| \, d\mathsf{P} \leq \varepsilon + \frac{l(\varepsilon)}{mn}.$$

This proves (9.66).

To complete the proof, note that

$$\sum_{(i,j)\in A} \mathsf{P}(X(i,j) \neq Y(i,j)) = \sum_{(i,j)\in A} \mathsf{P}(|X| \geq ij) < \infty$$

in view of condition (9.55). Applying the Borel–Cantelli lemma for independent random events we make sure that, with probability 1, only a finite number of events $\{\omega : X(i,j) \neq Y(i,j)\}$ occur. This means that, with probability 1, the sums $S_A(m,n)$ and $U(m,n)$ differ by a finite number of terms. Combining this result with (9.65)–(9.66) we deduce that

$$\lim\nolimits_{\mathrm{A}} \frac{S_A(m,n)}{mn} = 0 \qquad \text{a. s.},$$

whence (9.53) follows by (9.63). □

The assumption that f is a non-decreasing function is necessary for the proof of Theorem 9.8, while it can be omitted in the following result.

Lemma 9.4 *Let X be a random variable with finite expectation. Let a function f satisfy condition* (9.51) *and let the set A be defined by equality* (9.52). *If*

$$\sum_{(i,j)\in A} \int\limits_{\{\omega:|X(\omega)|\geq ij\}} d\mathsf{P} < \infty, \tag{9.67}$$

then

$$\sum_{(i,j)\in A} \frac{1}{(ij)^2} \int\limits_{\{\omega:|X(\omega)|<ij\}} X^2\, d\mathsf{P} < \infty. \tag{9.68}$$

Proof of Lemma 9.4 Put $p_k = \int\limits_{k-1\leq|X|<k} d\mathsf{P}$ for $k \geq 1$ and

$$R_1 = \sum_{(i,j)\in A} \int\limits_{\{\omega:|X(\omega)|\geq ij\}} d\mathsf{P}.$$

It is clear that R_1 is the series on the right-hand side of (9.67). We need the following bounds for R_1:

$$\begin{aligned} R_1 &= \sum_{(i,j)\in A} \sum_{k:\, k\geq ij} p_{k+1} \geq \sum_{(i,j)\in A} \sum_{k:\, ij\leq k\leq if(i)} p_{k+1} \\ &= \sum_{i=1}^{\infty} \sum_{j:\, i\leq j\leq f(i)} \sum_{k:\, ij\leq k\leq if(i)} p_{k+1} = \sum_{i=1}^{\infty} \sum_{k:\, i^2\leq k\leq if(i)} \sum_{j:\, i\leq j\leq \frac{k}{i}} p_{k+1} \end{aligned}$$

$$= \sum_{i=1}^{\infty} \sum_{k:\, i^2 \le k \le i f(i)} p_{k+1} \left(\left[\frac{k}{i} \right] - i + 1 \right) \ge \sum_{i=1}^{\infty} \sum_{k:\, i^2 \le k \le i f(i)} p_{k+1} \left(\frac{k}{i} - i \right)$$

$$= \sum_{i=1}^{\infty} \sum_{k:\, i^2 \le k \le i f(i)} \frac{k}{i} p_{k+1} - \sum_{i=1}^{\infty} \sum_{k:\, i^2 \le k \le i f(i)} i p_{k+1}.$$

The second term on the right-hand side does not exceed $\mathsf{E}[|X|]$. Indeed,

$$\sum_{i=1}^{\infty} \sum_{k:\, i^2 \le k \le i f(i)} i p_{k+1} \le \sum_{i=1}^{\infty} \sum_{k:\, k \ge i^2} i p_{k+1}$$
$$= \sum_{k=1}^{\infty} p_{k+1} \sum_{i:\, i^2 \le k} i \le \sum_{k=1}^{\infty} k \int_{k \le |X| < k+1} d\mathsf{P} \le \mathsf{E}[|X|].$$

Now we define the sequence $\{\lambda_k, k \ge 1\}$ by

$$\lambda_k = \sum_{i:\, i^2 \le k \le i f(i)} \frac{1}{i}.$$

As proved above,

$$R_1 + \mathsf{E}[|X|] \ge \sum_{k=1}^{\infty} k \lambda_k p_{k+1}. \tag{9.69}$$

We also put

$$\sigma_k^2 = \int_{k-1 \le |X| < k} X^2 \, d\mathsf{P}, \quad R_2 = \sum_{(i,j) \in A} \frac{1}{(ij)^2} \int_{\{\omega: |X(\omega)| < ij\}} X^2 \, d\mathsf{P}.$$

It is clear that R_2 is the series on the left had side of (9.68). We need the following bounds for R_2:

$$R_2 = \sum_{(i,j) \in A} \frac{1}{(ij)^2} \sum_{k=1}^{ij} \sigma_k^2 \le \sum_{(i,j) \in A} \frac{1}{(ij)^2} \sum_{k=1}^{i^2} \sigma_k^2 + \sum_{(i,j) \in A} \frac{1}{(ij)^2} \sum_{k=i^2}^{ij} \sigma_k^2.$$

The terms on the right-hand side are denoted by R_{21} and R_{22}, respectively. The first of them does not exceed $4\mathsf{E}[|X|]$. Indeed,

$$R_{21} = \sum_{i=1}^{\infty} \sum_{j:i\le j\le f(i)} \sum_{k=1}^{i^2} \frac{\sigma_k^2}{(ij)^2} = \sum_{i=1}^{\infty} \sum_{k=1}^{i^2} \frac{\sigma_k^2}{i^2} \sum_{j:i\le j\le f(i)} \frac{1}{j^2} \le \sum_{i=1}^{\infty} \sum_{k=1}^{i^2} \frac{\sigma_k^2}{i^2} \sum_{j=i}^{\infty} \frac{1}{j^2}$$

$$\le 2 \sum_{i=1}^{\infty} \sum_{k=1}^{i^2} \frac{\sigma_k^2}{i^3} = 2 \sum_{k=1}^{\infty} \sigma_k^2 \sum_{i:i^2\ge k} \frac{1}{i^3} \le 4 \sum_{k=1}^{\infty} \frac{\sigma_k^2}{k} \le 4\mathsf{E}\left[|X|\right].$$

The term R_{22} is estimated as follows:

$$R_{22} \le \sum_{i=1}^{\infty} \sum_{j:i\le j\le f(i)} \frac{1}{(ij)^2} \sum_{k:\, i^2\le k\le ij} \sigma_k^2 \le \sum_{i=1}^{\infty} \sum_{k:\, i^2\le k\le if(i)} \frac{\sigma_k^2}{i^2} \sum_{j:\frac{k}{i}\le j\le f(i)} \frac{1}{j^2}$$

$$\le 2 \sum_{i=1}^{\infty} \sum_{k:\, i^2\le k\le if(i)} \frac{\sigma_k^2}{ik} \le 2 \sum_{k=1}^{\infty} k p_k \lambda_k.$$

Combining the above bounds we get

$$R_2 \le 4\mathsf{E}\left[|X|\right] + 2 \sum_{k=1}^{\infty} k p_k \lambda_k. \tag{9.70}$$

Comparing (9.69) and (9.70) we see that the proof of the lemma is complete after we compare the series

$$\sum_{k=1}^{\infty} k\lambda_k p_{k+1} \quad \text{and} \quad \sum_{k=1}^{\infty} k\lambda_k p_k.$$

Put

$$\lambda_k' = \sum_{\substack{i:\, i^2\le k+1, \\ if(i)\ge k}} \frac{1}{i}.$$

Note that

$$\lambda_k' - \lambda_k = \begin{cases} \frac{1}{m}, & \text{if } k+1 = m^2 \text{ for some } m \ge 1, \\ 0, & \text{otherwise.} \end{cases}$$

Hence

$$\sum_{k=1}^{\infty} (k+1) p_{k+1} \left(\lambda_k' - \lambda_k\right) = \sum_{m=1}^{\infty} m^2 p_{m^2} \frac{1}{m} \le \mathsf{E}\left[\sqrt{|X|+1}\right] \stackrel{\text{def}}{=} C.$$

On the other hand, $\lambda_k' \ge \lambda_{k+1}$ and thus

$$\sum_{k=1}^{\infty}(k+1)p_{k+1}\lambda_{k+1} \le \sum_{k=1}^{\infty}(k+1)p_{k+1}\lambda_k + \sum_{k=1}^{\infty}(k+1)p_{k+1}\left(\lambda_k' - \lambda_k\right),$$

whence

$$\sum_{k=1}^{\infty} kp_k\lambda_k \le 1 + \sum_{k=1}^{\infty} kp_{k+1}\lambda_k + \sum_{k=1}^{\infty} p_{k+1}\lambda_k + C.$$

Finally, $\lambda_k \le \sqrt{k}$ and we establish

$$\sum_{k=1}^{\infty} p_{k+1}\lambda_k \le \mathsf{E}\left[|X|^{1/2}\right] \le C,$$

whence

$$\sum_{k=1}^{\infty} kp_k\lambda_k \le 1 + \sum_{k=1}^{\infty} kp_{k+1}\lambda_k + 2C.$$

Combining the latter bound with (9.69)–(9.70) we get

$$\begin{aligned} R_2 &\le 4\mathsf{E}\left[|X|\right] + 2\sum_{k=1}^{\infty} kp_k\lambda_k \le 4\mathsf{E}\left[|X|\right] + 2 + 2\sum_{k=1}^{\infty} kp_{k+1}\lambda_k + 4C \\ &\le 6\mathsf{E}\left[|X|\right] + 2 + 2R_1 + 4C. \end{aligned}$$

This proves (9.68). The proof of Lemma 9.4 is complete. □

9.4.2 Corollaries to Theorem 9.8

Theorem 9.8 can easily be extended to the case where both boundaries of the set A are curvilinear, that is, to the case of

$$A = \{(i, j) : g_1(i) \le j \le g_2(i)\}.$$

Corollary 9.13 *Let g_1 and g_2 be two non-decreasing functions such that*

$$g_1(x) \le x \le g_2(x) \quad \text{for all} \quad x \ge 1$$

and let $A = \{(i, j) : g_1(i) \le j \le g_2(i)\}$. The strong law of large numbers (9.53) *on the set A is equivalent to conditions* (9.54)–(9.55).

Remark 9.6 The method used to prove Theorem 9.8 can be applied not only for double sums; with minor changes the method works for multi-indices of higher dimension (that is, for the case $d \ge 3$).

Below we provide some applications of Corollary 9.13 for the case $g_2(x) = x$. To outline the proofs we use the sequence T_n, $n \geq 1$, defined in Remark 9.5.

Corollary 9.14 *Let $\{X(i, j)\}$ be independent identically distributed random variables. Assume that $\mathsf{E}[X] = 0$. Consider the functions $g_1(x) = 1$ and $g_2(x) = x$. The strong law of large numbers on the set*

$$A = \{1 \leq j \leq i\}$$

is equivalent to the condition

$$\mathsf{E}\left[|X| \log^+ |X|\right] < \infty.$$

Corollary 9.14 follows from the "rough" asymptotics in the Dirichlet divisors problem:

$$T_n \asymp n \log n$$

(see bounds (A.41)). Thus condition 9.55 is equivalent to the existence of the first moment of the random variable $|X| \log^+ |X|$. By symmetry, we derive Corollary 9.4 from here if $d = 2$.

Corollary 9.15 *Let $\{X(i, j)\}$ be independent identically distributed random variables and let X be a copy of them. Let $0 < \theta < 1$ and $g_1(x) = \theta x$. Then the strong law of large numbers in the set $A = \{(i, j) : \theta i \leq j \leq j\}$ is equivalent to the condition*

$$\mathsf{E}[X] = 0.$$

Corollary 9.15 follows from the asymptotic relation

$$T_n \asymp n.$$

Theorem 9.7 for $d = 2$ follows from Corollary 9.15 by symmetry.

Corollary 9.15 shows that the strong law of large numbers on a sector with linear boundaries follows if the first moment exists. Other power boundaries lead to different moment conditions.

Corollary 9.16 *Let $\{X(i, j)\}$ be independent identically distributed random variables and let X be a copy of them. Assume that $\mathsf{E}[X] = 0$. Let $1 > \delta > 0$ and $g_1(x) = x^{1-\delta}$. Then the strong law of large numbers in the set*

$$A = \left\{(i, j) : i^{1-\delta} \leq j \leq i\right\}$$

is equivalent to the condition

$$\mathsf{E}[|X|] \log^+ |X| < \infty.$$

Corollary 9.16 follows from the asymptotic relation $T_n \asymp n \log n$.

The moment condition changes if the boundary becomes closer to the diagonal than any power function.

Corollary 9.17 *Let $\{X(i, j)\}$ be independent identically distributed random variables and let X be a copy of them. Assume that $\mathsf{E}[X] = 0$. Let $\delta > 0$, $g_1(x) = x/(\ln x)^\delta$, and $g_2(x) = x$. Then the strong law of large numbers on the set*

$$A = \left\{(i, j) : i/(\ln i)^\delta \le j \le i\right\}$$

is equivalent to the condition

$$\mathsf{E}\left[|X| \log^+ \log^+ |X|\right] < \infty.$$

Corollary 9.17 follows from the asymptotic relation $T_n \asymp n \log \log n$.

A variety of possible sufficient conditions for the strong law of large numbers is further exhibited in the following result.

Corollary 9.18 *Let $\{X(i, j)\}$ be independent identically distributed random variables and let X be a copy of them. Assume that $\mathsf{E}[X] = 0$. Let $1 > \delta \ge 0$, $g_1(x) = x \exp\{-(\log x)^\delta\}$, $g_2(x) = x$. Then the strong law of large numbers on the set*

$$A = \left\{(i, j) : i e^{-(\log i)^\delta} \le j \le i\right\}$$

is equivalent to the condition

$$\mathsf{E}\left[|X| \left(\log^+ |X|\right)^\delta\right] < \infty.$$

Corollary 9.18 follows from the asymptotic relation $T_n \asymp n (\log n)^\delta$.

9.4.3 Proof of Proposition 9.1

First we show that condition (9.10) is equivalent to the following: *there exists a constant $C > 0$ such that*

$$\sum_{i=m}^{\infty} \frac{a_i}{i^2} \le C \frac{A_m}{m^2} \tag{9.71}$$

for all $m \ge 1$.

The implication (9.71) $\Longrightarrow$ (9.10) is proved first. Note that if the series $\sum a_i/i^2$ converges, then

$$\frac{a_1 + \cdots + a_N}{N^2} \to 0, \qquad N \to \infty, \tag{9.72}$$

by Kronecker's lemma. With the help of the partial summation formula, we get for $N > m$, that

$$\sum_{k=m}^{N} \frac{a_k}{k^2} = \sum_{k=m}^{N} \frac{A_k - A_{k-1}}{k^2} = \frac{A_N}{N^2} - \frac{A_{m-1}}{m^2} + \sum_{k=m}^{N-1} A_k \left(\frac{1}{k^2} - \frac{1}{(k+1)^2} \right).$$

Passing to the limit as $N \to \infty$, we conclude that

$$\frac{A_{m-1}}{m^2} + \sum_{k=m}^{\infty} \frac{a_k}{k^2} = \sum_{k=m}^{\infty} \frac{A_k(2k+1)}{k^2(k+1)^2}, \tag{9.73}$$

whence condition (9.10) follows.

To prove the implication (9.10) $\Longrightarrow$ (9.71) note that

$$\sum_{k=N}^{\infty} \frac{A_k}{k^3} \geq \frac{A_N}{2N^2}$$

(see inequality (9.11)). The latter relation yields (9.72). Reasoning as above, we prove equality (9.73), whence (9.71) follows.

To complete the proof of Proposition 9.1, it remains to show that (9.71) $\Longleftrightarrow$ (9.12). First we consider the case where all members of the sequence $\{b_n\}$ are integers.

Assume that condition (9.12) holds. Define the set $\mathcal{M}_m = \{k : b_k = m\}$ for every m such that $a_m \neq 0$. Then $A_m = \max \mathcal{M}_m$.

Let m be such that $a_m \neq 0$. For an arbitrary $n \in \mathcal{M}_m$, we have

$$\sum_{k=n}^{\infty} \frac{1}{b_k^2} = \sum_{k=n}^{A_m} \frac{1}{b_k^2} + \sum_{k=A_m+1}^{\infty} \frac{1}{b_k^2} = \frac{A_m - n + 1}{m^2} + \sum_{i=m+1}^{\infty} \frac{a_i}{i^2}.$$

Therefore condition (9.12) implies that

$$\begin{aligned}\sum_{i=m}^{\infty} \frac{a_i}{i^2} &= \frac{a_m}{m^2} + \sum_{i=m+1}^{\infty} \frac{a_i}{i^2} \leq \frac{a_m}{m^2} + \sum_{k=n}^{\infty} \frac{1}{b_k^2} = O\left(\frac{A_m}{m^2}\right) + O\left(\frac{n}{b_n^2}\right) \\ &= O\left(\frac{A_m}{m^2}\right) + O\left(\frac{n}{m^2}\right) = O\left(\frac{A_m}{m^2}\right)\end{aligned}$$

and thus condition (9.71) follows from (9.12), indeed, for m such that $a_m \neq 0$. To prove the same implication for m such that $a_m = 0$, we define

$$m_1 = \min\{k > m : a_k \neq 0\}, \quad n = \min\{k : b_k = m_1\}.$$

It is clear that $A_m = n - 1$. Thus

$$\sum_{i=m}^{\infty} \frac{a_i}{i^2} = \sum_{i=m_1}^{\infty} \frac{a_i}{i^2} = \sum_{k=n}^{\infty} \frac{1}{b_k^2} = O\left(\frac{n}{b_n^2}\right) = O\left(\frac{A_m}{m_1^2}\right) = O\left(\frac{A_m}{m^2}\right).$$

Therefore (9.71) follows from (9.12) in this case, too.

Now we prove the inverse implication (9.71) $\Longrightarrow$ (9.12). First we note that bound (9.71) yields

$$A_{m+1} = O(A_m). \tag{9.74}$$

Indeed,

$$\sum_{i=m}^{\infty} \frac{a_i}{i^2} = O\left(\frac{A_m}{m^2}\right), \quad \text{whence} \quad \frac{a_{m+1}}{(m+1)^2} = O\left(\frac{A_m}{m^2}\right),$$

that is

$$a_{m+1} = O\left(A_m \frac{(m+1)^2}{m^2}\right) \quad \text{or} \quad A_{m+1} = O(A_m).$$

Now let n be arbitrary and $m = b_n$. It is clear that $A_{m-1} < n \le A_m$. Hence

$$\sum_{k=n}^{\infty} \frac{1}{b_k^2} \le \sum_{i=m}^{\infty} \frac{a_i}{i^2} = O\left(\frac{A_m}{m^2}\right) = O\left(\frac{n}{b_n^2}\right)$$

which proves (9.12).

Finally, we treat the case of a general sequence $\{b_n\}$ whose values are not necessarily integers. For this, we construct the sequences $\{b'_n, n \ge 1\}$, $\{a'_n, n \ge 1\}$, and $\{A'_n, n \ge 1\}$ as follows

$$b'_n = [b_n], \quad a'_n = \operatorname{card}\{k : b'_k = n\}, \quad A'_1 = a'_1, \quad A'_n = a'_1 + \cdots + a'_n, \quad n > 1.$$

It is clear that Feller's condition (9.12) is equivalent to

$$\sum_{k=n}^{\infty} \frac{1}{(b'_n)^2} = O\left(\frac{n}{(b'_n)^2}\right).$$

In turn, since $a_n = a'_n$, condition (9.71) is equivalent to the condition

$$\sum_{i=m}^{\infty} \frac{a'_i}{i^2} = O\left(\frac{A'_m}{m^2}\right).$$

Since all members of the sequence $\{b'_n\}$ are integers, the latter two conditions are equivalent. This proves Proposition 9.1 for general sequences $\{b_n, n \ge 1\}$.

9.5 Comments

The statement of Theorem 9.1 is given by Borel [28]. While the main idea of the proof in [28] is correct, Borel's reasoning contains a false assertion (see [19]). Kolmogorov's Theorem 9.2 is stated without proof in the German original of the monograph [257] in 1933. The last Russian edition of the monograph [257] contains the complete proof of this result. Theorem 9.3 is proved by Marcinkiewicz and Zygmund [301].

More general normalizing sequences are studied for the case $d = 1$ in the papers by Feller [101] and Petrov and Martikainen [310].

Necessary and sufficient conditions for the strong law of large numbers for arbitrary normalizing sequences are obtained by Buldygin [35], Petrov and Martikainen [309], and Volodin and Nagaev [404] in the case $d = 1$.

In a sense, the study of strong laws of large numbers for multi-indexed sums was initiated by Wiener [418], who established an ergodic theorem for non-commutative groups of transformations preserving a measure in a measurable space. In the language of the modern theory of probability, his result is the "strong law of large numbers" for "stationary random variables". Independent identically distributed random variables are stationary and thus the Wiener setting (more precisely, its probabilistic counterpart) is more general than that studied in this chapter, since it deals with stationary rather than independent random variables. It is also worth mentioning that specific probabilistic methods developed during recent decades allow us to establish the strong law of large numbers in an even more general setting than that considered by Wiener.

The strong law of large numbers for $d > 1$ (Theorem 9.4) is proved by Smythe [373] for the case (9.6) and by Gut [129] for the case (9.7). The proof of Theorem 9.4 in [373] is based on a technique of reversed martingales; another method is presented in [129].

There are also further differences between the Wiener and Smythe settings, namely

(i) Wiener studied a problem for continuous time and thus considered integrals rather than sums;
(ii) Wiener chose balls as the domains of integration, while Smythe considered sums over rectangles.

Yet another difference between the two settings is important: Wiener studied a one-parameter family of balls in $\mathbf{R}^d$ depending on a growing radius, while (9.4) is a "d-parameter" result. Surprisingly, the Wiener one-parameter setting is not reduced to the well-known Birkhoff and von Neumann ergodic theorems (see [418]).

Wiener's [418] result is generalized by Zygmund [440] who studied the continuous time setting but with d-parametric rectangles as the integration domains instead of balls, as in Wiener's setting. In this sense, Zygmund's setting is closer to what we consider in this chapter. The sufficient condition for the ergodic theorem for multiparameter groups of transformations is

$$\int_S |f(x)|\, d\mu < \infty \tag{9.75}$$

for Wiener's [418] setting and

$$\int_S |f(x)| \left(\log^+ |f(x)|\right)^{d-1} d\mu < \infty \tag{9.76}$$

for Zygmund's [440] setting. Zygmund's condition (9.76) looks similar to (9.5), while Wiener's condition (9.75) is identical to Kolmogorov's condition (9.2).

Dunford [79], independently of Zygmund [440], and at the same time, studied an ergodic theorem for discrete time. The result resembles Theorem 9.4. In the probabilistic language, Dunford [79] proved that condition (9.5) implies that the limit

$$\lim_{\substack{n_1\to\infty \\ \cdots \\ n_d\to\infty}} \frac{S(\mathbf{n})}{|\mathbf{n}|}$$

exists almost surely if $\{X(\mathbf{n})\}$ is a field of stationary random variables satisfying condition (9.5).

Section 9.1 Feller [101] was the first to use condition (9.12) for $d = 1$. Condition (9.18) for $d = 1$ is considered by Petrov and Martikainen [310]. Theorems 9.5 and 9.6 are proved in [221] (the final form of these results is obtained in [230]). Adler in [4] and [6] discusses the strong laws of large numbers for random fields occurring in the solution of the St. Petersburg paradox.

Section 9.2 The method for the proof of the strong law of large numbers is presented in [221] and [230]. Similar methods for $d = 1$ are used by Petrov and Martikainen [310] (see also [338]) who generalized Feller's theorem [101]. The latter result for $d = 1$ and its proof can also be found in the book [338].

Section 9.3 Corollary 9.2 is contained in [221]. Corollary 9.2 is stated by other authors, too: Thanh [395] states it for double sums, later [396] he states it for multi-indexed sums as well; Gut and Stadtmüller [143] also state it in the general case $d > 1$. Rosalsky and Thanh [358] consider an analog of this result for double sums in Banach spaces. Ko [249] proves the strong law of large numbers with the same normalization for negatively associated random variables, Wang and Jin [414] obtained the same result for "negatively quadrant dependent" random fields. Zang and Tien [431] consider some partial cases. Fazekas [93] studies conditions for the strong law of large numbers and [94] rates of convergence for the strong law of large numbers for Banach spaces.

Corollaries 9.3 and 9.4 are proved by Gut [129] and Smythe [373], respectively. Hong and Hwang [166] generalize Corollary 9.3 for the case of pairwise independent random variables for $d = 2$. Azarnoosh [14] proves an analogous result for $d = 2$ for "negatively dependent" random variables. Zhang and Wen [432] use a certain mixing condition in Corollary 9.3 instead of assuming the independence of random variables.

Hong and Volodin [165] prove Corollary 9.3 for double sums. Giang [124], Lagodowski [272], and Su and Taylor [387] obtain the strong laws of large numbers in Banach spaces for normalizations like those in Corollary 9.3.

Section 9.4 Theorem 9.7 is proved in [133]. Gabriel [116] studies the convergence of martingales whose indices are restricted to a sector. Smythe [374] considers sums of independent random variables whose indices belong to more general partially ordered sets.

Theorem 9.8 is proved in [182], where an earlier result of [243] is improved; an additional assumption that the function $x/f(x)$ is increasing is used in [243]. Similar results are later obtained by Lagodowski and Matula [274] and Matula and Seweryn [313]. Kendzhaev [194] studies the rate of convergence in the law of iterated logarithm for multi-indexed sums whose indices belong to a sector with linear boundaries.

Chapter 10
The Law of the Iterated Logarithm

10.1 The Classical Law of the Iterated Logarithm

The first law of the iterated logarithm is proved for symmetric Bernoulli random variables, that is, for independent random variables $\{X_n\}$ such that

$$\mathsf{P}(X_n = -1) = \mathsf{P}(X_n = 1) = \frac{1}{2}.$$

Theorem 10.1 (A.Ya. Khintchine) *Let $\{X_n, n \geq 1\}$ be a sequence of independent Bernoulli random variables with success probability $\frac{1}{2}$. Let S_n be the number of successes in the first n trials. Then*

$$\limsup_{n\to\infty} \frac{S_n}{\sqrt{2n \log\log n}} = 1 \qquad a.\ s.$$

Despite the totally probabilistic statement of Khintchine's theorem, its result solves a problem of analytic number theory on the frequency of digits 0 and 1 in the binary representation of real numbers that had been studied by many authors before Khintchine. For example, Hardy and Littlewood (see p. 185 in [154]) established that

$$\limsup_{n\to\infty} \frac{\left|\nu(n) - \frac{n}{2}\right|}{\sqrt{n \log n}} < \infty, \tag{10.1}$$

for almost all (with respect to the Lebesgue measure) numbers $x \in [0, 1]$, where $\nu(n)$ is the number of zeros among the first n digits in the binary decomposition of a number x. At the same time, Khintchine's Theorem 10.1 implies that

$$\limsup_{n\to\infty} \frac{\left|\nu(n) - \frac{n}{2}\right|}{\sqrt{n \log\log n}} = \sqrt{2}$$

O. Klesov, *Limit Theorems for Multi-Indexed Sums of Random Variables*,
Probability Theory and Stochastic Modelling 71, DOI 10.1007/978-3-662-44388-0_10

for almost all numbers $x \in [0, 1]$, which is a much better result than (10.1). Another application of Theorem 10.1 can be found in the algorithmic theory of information (see [405]). Many other applications are surveyed by Bingham [25].

Khintchine's theorem can essentially be generalized for the case of bounded independent random variables (which includes the case of Bernoulli trials).

Theorem 10.2 (A.N. Kolmogorov) *Let $\{X_n, n \geq 1\}$ be a sequence of independent random variables. Assume that*

$$|X_n| \leq M_n \qquad a.\ s.$$

for all $n \geq 1$, where $\{M_n\}$ is a sequence of non-random numbers. Let $\mathsf{E}[X_n] = 0$ and $B_n \stackrel{\text{def}}{=} \mathsf{var}[S_n] = \sum_{k=1}^{n} \mathsf{var}[X_k]$. If

$$M_n^2 = o\left(\frac{B_n}{\log^+ \log^+ B_n}\right), \tag{10.2}$$

then

$$\limsup_{n\to\infty} \frac{S_n}{\sqrt{2B_n \log^+ \log^+ B_n}} = 1 \qquad a.\ s. \tag{10.3}$$

Recall that $\log^+ z = \log(3 + |z|)$ for $z \in \mathbf{R}$.

The expression *law of the iterated logarithm* appeared in a paper of Kolmogorov written in German in 1929. Following Kolmogorov, we refer to relation (10.3) as the *law of the iterated logarithm.*

Kolmogorov's Theorem 10.2 can be used to find a sufficient condition for the law of the iterated logarithm in the case of identically distributed random variables.

Theorem 10.3 (P. Hartman and A. Wintner) *Let a random variable X be such that*

$$\mathsf{E}[X] = 0, \qquad \mathsf{E}\left[X^2\right] = 1. \tag{10.4}$$

Assume that X, $\{X_n, n \geq 1\}$ is a sequence of independent identically distributed random variables. Let $\{S_n, n \geq 1\}$ be the sequence of partial sums constructed from $\{X_n\}$: $S_1 = X_1$ and $S_n = X_1 + \cdots + X_n$ for $n \geq 2$. Then

$$\liminf_{n\to\infty} \frac{S_n}{\sqrt{2n \log\log n}} = -1, \quad \limsup_{n\to\infty} \frac{S_n}{\sqrt{2n \log\log n}} = 1 \qquad a.\ s. \tag{10.5}$$

It turns out that the Hartman–Wintner moment condition (10.4) is, in fact, necessary and sufficient for the law of the iterated logarithm (10.5).

Theorem 10.4 (V. Strassen) *Let X, $\{X_n, n \geq 1\}$ be a sequence of independent identically distributed random variables. If the law of the iterated logarithm* (10.5) *holds, then moment condition* (10.4) *holds, as well.*

More results concerning the law of the iterated logarithm for cumulative sums can be found in the book [338].

Two of Kolmogorov's celebrated exponential bounds are used in the proofs of this Chapter. Both inequalities are given in Chap. 7 (see Theorems 7.7 and 7.8 therein). The upper bound given in Theorem 7.7 fits the purposes of the current Chapter, while the lower exponential inequality needs minor changes as in Theorem 10.6. We repeat the statement of these results for the reader's convenience.

Theorem 10.5 (Kolmogorov's upper exponential bound) *Let $X_1, \dots, X_n$ be a sequence of independent bounded random variables, that is, there are non-random numbers $M_1, \dots, M_n$ such that*

$$|X_k| \le M_k \qquad a.\ s.$$

for all $1 \le k \le n$. Let $\mathsf{E}[X_k] = 0$ for all $1 \le k \le n$, $S_n = \sum_{k=1}^n X_k$, and $B_n = \mathsf{var}[S_n]$. If x is a real number such that $0 \le xM_n \le B_n$, then

$$\mathsf{P}(S_n \ge x) \le \exp\left\{-\frac{x^2}{2B_n}\left(1 - \frac{xM_n}{B_n}\right)\right\}. \tag{10.6}$$

Theorem 10.6 (Kolmogorov's lower exponential bound) *Let $X_1, \dots, X_n$ be a sequence of bounded independent random variables, that is, there are non-random numbers $M_1, \dots, M_n$ such that*

$$|X_k| \le M_k \qquad a.\ s.$$

for all $1 \le k \le n$. Assume that $\mathsf{E}[X_k] = 0$ for all $1 \le k \le n$, $S_n = \sum_{k=1}^n X_k$, and $B_n = \mathsf{var}[S_n]$. Let $\eta > 0$ and x be a positive real number. If x, M_n, and B_n are such that

$$\frac{xM_n}{B_n} \text{ is sufficiently small,} \qquad \frac{x^2}{B_n} \text{ is sufficiently large,} \tag{10.7}$$

then

$$\mathsf{P}(S_n \ge x) \ge \exp\left\{-\frac{x^2}{2B_n}(1+\eta)\right\}. \tag{10.8}$$

Kolmogorov's original statement of Theorem 10.6 is given in a slightly different form, where condition (10.7) is used in an asymptotic sense, namely Kolmogorov assumes that

$$\frac{xM_n}{B_n} \to 0, \qquad \frac{x^2}{B_n} \to \infty \tag{10.9}$$

instead of (10.7). Under condition (10.9), Kolmogorov proves that inequality (10.8) holds for sufficiently large numbers n.

All modern publications concerning the law of the iterated logarithm (see, for example, [336] or [57]) contain Kolmogorov's condition in its original form (10.9). However, this condition is not useful for the case of multi-indices, since, in this case, it would imply that an analog of Theorem 10.6 for multi-indexed sums may fail for infinitely many multi-indexed sums $\{S(\mathbf{n})\}$.

Analyzing the proof of Kolmogorov's lower exponential bound (see, for example, [336]) we see that condition (10.7) can indeed be substituted for (10.9) without any change in the proof of Theorem 10.6. Moreover, a restriction on both ratios in (10.7) can be made precise in terms of the constant η (in other words, one can explicitly indicate the lower bound n_0 for which (10.8) holds if $n \geq n_0$). We do not provide a proof for this, since the precise restriction for terms in (10.7) is not relevant to what we discuss below.

10.2 Kolmogorov's Law of the Iterated Logarithm for Multi-Indexed Sums

In this section, we show how Theorems 10.2, 10.3, and 10.4 can be generalized to the case of multi-indexed sums of random variables.

Relation (10.3) for lim sup in the case of $d > 1$ can be written in two similar ways depending on the mode of convergence considered in the law of the iterated logarithm, i.e. lim(min)- or lim(max)-convergence. For example, one can study the law of the iterated logarithm

$$\operatorname{limsup(min)} \frac{S(\mathbf{n})}{\sqrt{2B(\mathbf{n}) \log\log B(\mathbf{n})}} = \alpha_d \qquad \text{a. s.} \tag{10.10}$$

or

$$\operatorname{limsup(max)} \frac{S(\mathbf{n})}{\sqrt{2B(\mathbf{n}) \log\log B(\mathbf{n})}} = \beta_d \qquad \text{a. s.} \tag{10.11}$$

The left-hand side of (10.10) is well-defined for every elementary random event $\omega \in \Omega$ if we put

$$\operatorname{limsup(min)} s(\mathbf{n}) \stackrel{\text{def}}{=} \lim_{k\to\infty} \left(\sup_{\mathbf{n}\in A_k} s(\mathbf{n}) \right) \stackrel{\text{def}}{=} \overline{s}_{\min}$$

for every field of non-random numbers $\{s(\mathbf{n})\}$, where

$$A_k \stackrel{\text{def}}{=} \left\{ \mathbf{m} \in \mathbf{N}^d : m_1 \geq k, \dots, m_d \geq k \right\}.$$

Similarly we introduce the notation lim sup for the convergence lim(max), namely

$$\operatorname{limsup}(\max) s(\mathbf{n}) \stackrel{\text{def}}{=} \lim_{k\to\infty}\left(\sup_{\mathbf{n}\in B_k} s(\mathbf{n})\right) \stackrel{\text{def}}{=} \overline{s}_{\max}$$

for all fields of non-random numbers $\{s(\mathbf{n})\}$, where $B_k \stackrel{\text{def}}{=} \{\mathbf{m} \in \mathbf{N}^d : |\mathbf{m}| \geq k\}$. It is clear that

$$\overline{s}_{\min} \leq \overline{s}_{\max}.$$

In contrast to the case of $d = 1$, the inequality $\overline{s}_{\min} < \overline{s}_{\max}$ may hold if $d > 1$.

In view of the latter observation, one can expect that $\alpha_d < \beta_d$, where the numbers α_d and β_d are defined by relations (10.10) and (10.11), respectively. Surprisingly, this never happens in the law of the iterated logarithm for independent random variables under assumption (10.13).

Theorem 10.7 *Let $\{X(\mathbf{n}), \mathbf{n} \in \mathbf{N}^d\}$ be independent random variables such that $\mathsf{E}[X(\mathbf{n})] = 0$ for all $\mathbf{n} \in \mathbf{N}^d$. Assume that*

$$|X(\mathbf{n})| \leq M(\mathbf{n}) \qquad a.\ s.$$

for some field of non-random numbers $\{M(\mathbf{n}), \mathbf{n} \in \mathbf{N}^d\}$. Assume further that

$$\lim(\max) B(\mathbf{n}) = \infty, \qquad \lim(\max) \frac{B(\mathbf{n}+\mathbf{1})}{B(\mathbf{n})} = 1, \tag{10.12}$$

where $B(\mathbf{n}) = \mathsf{var}[S(\mathbf{n})]$. If

$$\lim(\max) M^2(\mathbf{n}) \cdot \frac{\log^+ \log^+ B(\mathbf{n})}{B(\mathbf{n})} = 0 \tag{10.13}$$

and

$$\lim(\max) \frac{B(\mathbf{n})}{b_{n_1}^{(1)} \dots b_{n_d}^{(d)}} = 1 \tag{10.14}$$

for some increasing sequences $\left\{b_n^{(1)}\right\}, \dots, \left\{b_n^{(d)}\right\}$ such that

$$\lim_{n\to\infty} b_n^{(1)} = \infty, \qquad \dots, \qquad \lim_{n\to\infty} b_n^{(d)} = \infty, \tag{10.15}$$

then relations (10.10) *and* (10.11) *hold and $\alpha_d = \beta_d = \sqrt{d}$. Moreover,*

$$\operatorname{liminf}(\min) \frac{S(\mathbf{n})}{\sqrt{2B(\mathbf{n}) \log\log B(\mathbf{n})}} = -\sqrt{d} \qquad a.\ s., \tag{10.16}$$

$$\liminf(\max) \frac{S(\mathbf{n})}{\sqrt{2B(\mathbf{n}) \log \log B(\mathbf{n})}} = -\sqrt{d} \qquad a.\,s. \tag{10.17}$$

Remark 10.1 The numbers $\liminf(\min) s(\mathbf{n})$ and $\liminf(\max) s(\mathbf{n})$ for non-random fields $\{s(\mathbf{n})\}$ are defined similarly to $\limsup(\min) s(\mathbf{n})$ and $\limsup(\max) s(\mathbf{n})$. The difference is that the "sup" in the definition is changed to "inf".

Remark 10.2 Equality (10.13) is called *Kolmogorov's condition for the law of the iterated logarithm for multi-indexed sums*. It holds, for example, if random variables are uniformly bounded, that is, if $M(\mathbf{n}) = M$ for all $\mathbf{n} \in \mathbf{N}^d$.

Remark 10.3 Theorem 10.7 differs from Theorem 10.2 by an extra condition (10.14) which is meaningful only for $d > 1$. Condition (10.14) holds, for example, if random variables $X(\mathbf{n})$ have equal second moments, that is, if $\operatorname{var}[X(\mathbf{n})] \stackrel{\text{def}}{=} \sigma^2$. Thus Theorem 10.7 implies the law of the iterated logarithm for multi-indexed sums if the random variables are independent, identically distributed, and bounded. We will show in Theorem 10.9 that the boundedness here can be changed by condition (10.46).

Proof of Theorem 10.7 Let $\chi(x) = \sqrt{2x \log^+ \log^+ x}$ for $x > 0$. Without loss of generality, we assume that $\operatorname{var}[X(\mathbf{1})] > 0$ and $b_1^{(i)} \le 1$ for all $1 \le i \le d$.

Step 1. First we show that

$$\limsup(\max) \frac{S(\mathbf{n})}{\chi(B(\mathbf{n}))} \le \sqrt{d} \qquad \text{a. s.} \tag{10.18}$$

Let $\varepsilon > 0$. Choose the parameters $\theta > 1$, $0 < \gamma < 1$, and $0 < \nu < 1$ such that

$$1 + \varepsilon > \theta^{d/2} \left(1 + \frac{\varepsilon}{2}\right), \tag{10.19}$$

$$\frac{1-\gamma}{\theta^d \log \theta} \quad \text{is not an integer,} \tag{10.20}$$

$$(1-\nu)\left(1 + \frac{\varepsilon}{2}\right)^2 > 1. \tag{10.21}$$

Without loss of generality, we assume that $b_1^{(i)} = 1$ for all $1 \le i \le d$. Conditions (10.13)–(10.14) imply that

$$\lim_{n\to\infty} \frac{b_{n+1}^{(i)}}{b_n^{(i)}} = 1, \qquad 1 \le i \le d.$$

Now we define the sequences $\left\{n_k^{(i)}, k \ge 1\right\}$, $1 \le i \le d$, by putting

$$n_k^{(i)} = \max\left\{n : b_n^{(i)} \le \theta^k\right\}. \tag{10.22}$$

If k is a sufficiently large number, then the interval $(\theta^{k-1}, \theta^k]$ contains at least one member of each of the sequences $\left(b_n^{(i)}, n \ge 1\right)$, $1 \le i \le d$. For definiteness, we assume that this property holds for $k \ge k_i^*$. Put $k_0 = \max\{k_1^*, \dots, k_d^*\}$ and $\mathbf{k}_0 = (k_0, \dots, k_0)$. For $\mathbf{k} = (k_1, \dots, k_d)$, we also put $\mathbf{n}(\mathbf{k}) = \left(n_{k_1}^{(1)}, \dots, n_{k_d}^{(d)}\right)$. Therefore the elements $\mathbf{n}(\mathbf{k}) \in \mathbf{N}^d$ are defined for $\mathbf{k} \ge \mathbf{k}_0$.

By condition (10.14), the number k_0 can be increased so that

$$1 - \gamma \le \frac{B(\mathbf{n}(\mathbf{k}))}{b_{n_{k_1}^{(1)}} \dots b_{n_{k_d}^{(d)}}} \le 1 + \gamma, \qquad \mathbf{k} \ge \mathbf{k}_0.$$

Hence

$$\frac{1-\gamma}{\theta^d} \theta^{k_1 + \dots + k_d} \le B(\mathbf{n}(\mathbf{k})) \le (1+\gamma)\, \theta^{k_1+\dots+k_d}, \qquad \mathbf{k} \ge \mathbf{k}_0. \tag{10.23}$$

Put $\chi(\mathbf{k}) = \chi(B(\mathbf{n}(\mathbf{k})))$. Increase k_0, if necessary, so that the following condition

$$(\sqrt{d} + \varepsilon)\chi(\mathbf{k}) - d\sqrt{2B(\mathbf{n}(\mathbf{k}+\mathbf{1}))} \ge \left(\sqrt{d} + \frac{\varepsilon}{2}\right) \chi(\mathbf{k}+\mathbf{1}), \qquad \mathbf{k} \ge \mathbf{k}_0, \tag{10.24}$$

holds along with all other properties mentioned above (this can be done in view of (10.19)).

Finally, we find a number k_0 such that all the above conditions hold and additionally

$$\nu\, B(\mathbf{n}(\mathbf{k}+\mathbf{1})) > \left(\sqrt{d} + \frac{\varepsilon}{2}\right) \chi(\mathbf{k}+\mathbf{1}) M(\mathbf{n}(\mathbf{k}+\mathbf{1})), \quad \mathbf{k} \ge \mathbf{k}_0, \tag{10.25}$$

(this can be done in view of (10.13)).

Then we use Kolmogorov's inequality (2.34) for multi-indexed sums:

$$\begin{aligned}
&\mathsf{P}\left(\max_{\mathbf{n} \le \mathbf{n}(\mathbf{k}+\mathbf{1})} S(\mathbf{n}) \ge (\sqrt{d} + \varepsilon)\chi(\mathbf{k}) \right) \\
&\quad \le 2^d \mathsf{P}\left(S(\mathbf{n}(\mathbf{k}+\mathbf{1})) \ge (\sqrt{d}+\varepsilon)\chi(\mathbf{k}) - d\sqrt{2B(\mathbf{n}(\mathbf{k}+\mathbf{1}))} \right) \\
&\quad \le 2^d \mathsf{P}\left(S(\mathbf{n}(\mathbf{k}+\mathbf{1})) \ge (\sqrt{d}+\varepsilon/2)\chi(\mathbf{k}+\mathbf{1}) \right), \quad \mathbf{k} \ge \mathbf{k}_0.
\end{aligned}$$

The latter inequality follows from (10.24). Next we apply Kolmogorov's upper exponential bound (Theorem 10.5) with $x = (\sqrt{d} + \frac{\varepsilon}{2})\chi(\mathbf{k}+\mathbf{1})$. Condition (10.25) justifies that this result can be applied, indeed:

$$\begin{aligned}
&\mathsf{P}\left(S(\mathbf{n}(\mathbf{k}+\mathbf{1})) \ge (\sqrt{d}+\varepsilon/2)\chi(\mathbf{k}+\mathbf{1}) \right) \\
&\quad \le \exp\left\{ -(1-\nu)\left(\sqrt{d} + \frac{\varepsilon}{2}\right)^2 \log^+ \log^+ B(\mathbf{n}(\mathbf{k}+\mathbf{1})) \right\}, \quad \mathbf{k} \ge \mathbf{k}_0,
\end{aligned}$$

where inequality (10.25) has been used to estimate the second term in the exponent on the right-hand side of (10.6). Then (10.23) implies that

$$\mathsf{P}\left(S(\mathbf{n}(\mathbf{k}+\mathbf{1})) \ge (\sqrt{d}+\varepsilon/2)\chi(\mathbf{k}+\mathbf{1})\right) \le \frac{1}{\left(\frac{1-\gamma}{\theta^d} + (k_1+\dots+k_d)\log\theta\right)^{(1-\nu)\left(\sqrt{d}+\frac{\varepsilon}{2}\right)^2}}, \quad \mathbf{k} \ge \mathbf{k}_0.$$

Collecting together the bounds obtained above and using (10.20)–(10.21) we prove that

$$\sum_{\mathbf{k}\ge\mathbf{k}_0} \mathsf{P}\left(\max_{\mathbf{n}\le\mathbf{n}(\mathbf{k}+\mathbf{1})} S(\mathbf{n}) \ge (\sqrt{d}+\varepsilon)\chi(\mathbf{k})\right) < \infty,$$

since the generalized harmonic series

$$\sum_{\mathbf{k}\in\mathbf{N}^d} \frac{1}{(k_1+\dots+k_d)^{(1+\delta)d}} \quad \begin{cases} \text{converges,} & \text{if } \delta > 0, \\ \text{diverges,} & \text{if } \delta \le 0. \end{cases} \tag{10.26}$$

The Borel–Cantelli lemma yields

$$\mathsf{P}\left(\max_{\mathbf{n}\le\mathbf{n}(\mathbf{k}+\mathbf{1})} S(\mathbf{n}) \ge (\sqrt{d}+\varepsilon)\chi(\mathbf{k}) \text{ i. o. in the domain } \mathbf{k} \ge \mathbf{k}_0\right) = 0. \tag{10.27}$$

Since the function χ is monotone, the latter result implies that

$$\mathsf{P}\left(\max_{\mathbf{n}\le\mathbf{n}(\mathbf{k}+\mathbf{1})} S(\mathbf{n}) \ge (\sqrt{d}+\varepsilon)\chi(\mathbf{k}+\mathbf{1}) \text{ i. o. in the domain } \mathbf{k} \ge \mathbf{k}_0\right) = 0. \tag{10.28}$$

Every multi-index $\mathbf{n} > \mathbf{n}(\mathbf{k}_0)$ belongs to a unique "block"

$$\mathfrak{B}^\circ_{\mathbf{k}} \stackrel{\text{def}}{=} \left\{\mathbf{m} : n^{(1)}_{k_1} < m_1 \le n^{(1)}_{k_1+1}, \dots, n^{(1)}_{k_d} < m_d \le n^{(1)}_{k_d+1}\right\}. \tag{10.29}$$

Moreover, if $\mathbf{n} \in \mathfrak{B}^\circ_{\mathbf{k}}$, then

$$\frac{S(\mathbf{n})}{\chi(B(\mathbf{n}))} \le \max\left\{\frac{1}{\chi(\mathbf{k})}\max_{\mathbf{n}\le\mathbf{n}(\mathbf{k}+\mathbf{1})} S(\mathbf{n}), \frac{1}{\chi(\mathbf{k}+\mathbf{1})}\max_{\mathbf{n}\le\mathbf{n}(\mathbf{k}+\mathbf{1})} S(\mathbf{n})\right\}.$$

Thus relations (10.27) and (10.28) imply that

$$\mathsf{P}\left(S(\mathbf{n}) \ge (\sqrt{d}+\varepsilon)\chi(B(\mathbf{n})) \text{ i. o. in the domain } \mathbf{n} > \mathbf{n}(\mathbf{k}_0)\right) = 0,$$

whence

$$\operatorname{limsup}(\min) \frac{S(\mathbf{n})}{\chi(B(\mathbf{n}))} \le \sqrt{d} + \varepsilon \qquad \text{a. s.}$$

Since $\varepsilon > 0$ is an arbitrary number,

$$\operatorname{limsup}(\min) \frac{S(\mathbf{n})}{\chi(B(\mathbf{n}))} \le \sqrt{d} \qquad \text{a. s.} \tag{10.30}$$

To derive (10.18) from (10.30) we consider a set $I \subseteq \{1, \dots, d\}$ such that $I \ne \varnothing$ and $I \ne \{1, \dots, d\}$. Given an arbitrary $\mathbf{n} \in \mathbf{N}^d$, denote by $\mathbf{n}_I$ the multi-index formed by the coordinates n_i with $i \in I$. It is clear that $\mathbf{n}_I \in \mathbf{N}^{\operatorname{card}(I)}$.

Let n_i^0, $i \notin I$, be fixed positive integers. Introduce the following random variables

$$X_I(\mathbf{n}_I) = \sum_{i \notin I} \sum_{1 \le n_i \le n_i^0} X(\mathbf{n}).$$

It is clear that the random variables $\{X_I(\mathbf{n}_I), \mathbf{n}_I \in \mathbf{N}^{\operatorname{card}(I)}\}$ are independent and satisfy all assumptions of Theorem 10.7. Put

$$S_I(\mathbf{n}_I) = \sum_{\mathbf{k}_I \le \mathbf{n}_I} X_I(\mathbf{k}_I).$$

Then relation (10.30) implies that

$$\limsup_{\mathbf{n}_I}{}_{\min} \frac{S_I(\mathbf{n}_I)}{\chi(B(\mathbf{n}_I))} \le \sqrt{\operatorname{card}(I)} \le \sqrt{d} \qquad \text{a. s.}$$

Since this result holds for an arbitrary set I and arbitrary n_i^0, $i \notin I$, it implies (10.18).

Changing $X(\mathbf{n})$ to $-X(\mathbf{n})$ in the above argument, we get

$$\operatorname{limsup}(\max) \left(-\frac{S(\mathbf{n})}{\chi(B(\mathbf{n}))} \right) \le \sqrt{d} \qquad \text{a. s.}$$

which together with (10.18) yields

$$\operatorname{limsup}(\max) \frac{|S(\mathbf{n})|}{\chi(B(\mathbf{n}))} \le \sqrt{d} \qquad \text{a. s.} \tag{10.31}$$

Step 2. Now we prove the converse inequality

$$\operatorname{limsup}(\min) \frac{S(\mathbf{n})}{\chi(B(\mathbf{n}))} \ge \sqrt{d} \qquad \text{a. s.} \tag{10.32}$$

For the proof of (10.32), we choose $\theta > 1$ and construct multi-indices $\mathbf{n}(\mathbf{k})$ according to the rule (10.22). As above, the multi-indices $\mathbf{n}(\mathbf{k})$ are defined for $\mathbf{k} \geq \mathbf{k}_0$ where $\mathbf{k}_0 \in \mathbf{N}^d$ is a certain multi-index. Given a block $\mathfrak{B}_{\mathbf{k}}^{\circ}$ defined by (10.29) put

$$S^{\circ}(\mathbf{k}) = \sum_{\mathbf{n}\in\mathfrak{B}_{\mathbf{k}}^{\circ}} X(\mathbf{n}), \quad B^{\circ}(\mathbf{k}) \stackrel{\text{def}}{=} \mathsf{var}\left[S^{\circ}(\mathbf{k})\right], \quad \chi^{\circ}(\mathbf{k}) \stackrel{\text{def}}{=} \chi(B^{\circ}(\mathbf{k})).$$

Lemma 10.1 *Suppose, for all $\theta > 1$ and $0 < \varepsilon < 1$, there exists a $\mathbf{k}_0$ such that*

$$\sum_{\mathbf{k}\geq\mathbf{k}_1} \mathsf{P}(S^{\circ}(\mathbf{k}) \geq (\sqrt{d} - \varepsilon)\chi^{\circ}(\mathbf{k})) = \infty \tag{10.33}$$

for all $\mathbf{k}_1 \geq \mathbf{k}_0$. Then inequality (10.32) *holds.*

We will prove Lemma 10.1 later, but now we show how to use it to complete the proof of Theorem 10.7. Our current aim is to prove that (10.33) follows from the assumptions of Theorem 10.7.

Let $0 < \varepsilon < 1$. Choose $\nu > 0$ such that $(1+\nu)(\sqrt{d} - \varepsilon)^2 < \sqrt{d}$. Then Kolmogorov's lower exponential bound (Theorem 10.6) implies that there exists a $\mathbf{k}_0$ such that

$$\mathsf{P}(S^{\circ}(\mathbf{k}) \geq (\sqrt{d} - \varepsilon)\chi^{\circ}(\mathbf{k})) \geq \left(\log B^{\circ}(\mathbf{k})\right)^{-(1+\nu)(\sqrt{d}-\varepsilon)^2}, \qquad \mathbf{k} \geq \mathbf{k}_0.$$

Conditions (10.12) and (10.14) together with definition (10.22) imply that

$$\lim_{\mathbf{k}}(\min) \frac{B(\mathbf{n}(\mathbf{k}))}{\theta^{k_1+\cdots+k_d}} = 1.$$

The variance of $B^{\circ}(\mathbf{k})$ can be represented as a linear combination of at most 2^d sums $B(\mathbf{n}(\mathbf{k}+\mathbf{1}-\boldsymbol{\delta}))$, where $\boldsymbol{\delta}$ is a multi-index whose coordinates are either 0 or 1. Thus

$$0 < \lim_{\mathbf{k}}(\min) \frac{B^{\circ}(\mathbf{k})}{B(\mathbf{n}(\mathbf{k}))} < \infty$$

for sufficiently large $\theta > 1$. Obviously condition (10.33) for all $\mathbf{k}_1 \geq \mathbf{k}_0$ follows from the latter result (see (10.26)). □

Proof of Lemma 10.1 In view of (10.33), the Borel–Cantelli lemma for independent random events implies that

$$\mathsf{P}(S^{\circ}(\mathbf{k}) \geq (\sqrt{d} - \varepsilon)\chi^{\circ}(\mathbf{k}) \text{ i. o. in the domain } \mathbf{k} \geq \mathbf{k}_1) = 1$$

for all $\mathbf{k}_1 \geq \mathbf{k}_0$. The sum $S(\mathbf{n}(\mathbf{k}+\mathbf{1}))$ can be represented as a linear combination of $S^\circ(\mathbf{k})$ and at most $2^d - 1$ sums $S(\mathbf{n}(\mathbf{k}+\mathbf{1}-\boldsymbol{\delta}))$, where $\boldsymbol{\delta} \neq \mathbf{0}$ is a multi-index whose coordinates are either 0 or 1. Put $\chi(\mathbf{k}) = \chi(B(\mathbf{n}(\mathbf{k})))$. Therefore the probability that the inequality

$$\frac{S(\mathbf{n}(\mathbf{k}+\mathbf{1}))}{\chi^\circ(\mathbf{k})} = \frac{S^\circ(\mathbf{k})}{\chi^\circ(\mathbf{k})} + \sum_{\boldsymbol{\delta}\neq\mathbf{0}} \pm \frac{S(\mathbf{n}(\mathbf{k}+\mathbf{1}-\boldsymbol{\delta}))}{\chi^\circ(\mathbf{k})}$$
$$\geq (\sqrt{d}-\varepsilon) - \sum_{\boldsymbol{\delta}\neq\mathbf{0}} \left| \frac{S(\mathbf{n}(\mathbf{k}+\mathbf{1}-\boldsymbol{\delta}))}{\chi(\mathbf{k}+\mathbf{1}-\boldsymbol{\delta})} \right| \cdot \frac{\chi(\mathbf{k}+\mathbf{1}-\boldsymbol{\delta})}{\chi^\circ(\mathbf{k})}$$

occurs infinitely often equals one.

Note that

$$\lim_{\mathbf{k}}(\min) \frac{\chi(\mathbf{k}+\mathbf{1}-\boldsymbol{\delta})}{\chi^\circ(\mathbf{k})} \leq \sqrt{\frac{d}{\theta}}$$

for all $\boldsymbol{\delta} \neq \mathbf{0}$. Without loss of generality, one can assume that $\mathbf{k}_0$ is such that

$$\frac{\chi(\mathbf{k}+\mathbf{1}-\boldsymbol{\delta})}{\chi^\circ(\mathbf{k})} \leq \frac{\sqrt{d}+\varepsilon}{\sqrt{\theta}}, \qquad \mathbf{k} \geq \mathbf{k}_0,$$

whence we establish that the probability that the inequality

$$\frac{S(\mathbf{n}(\mathbf{k}+\mathbf{1}))}{\chi^\circ(\mathbf{k})} \geq (\sqrt{d}-\varepsilon) - \frac{\sqrt{d}+\varepsilon}{\sqrt{\theta}} \sum_{\boldsymbol{\delta}\neq\mathbf{0}} \left| \frac{S(\mathbf{n}(\mathbf{k}+\mathbf{1}-\boldsymbol{\delta}))}{\chi(\mathbf{k}+\mathbf{1}-\boldsymbol{\delta})} \right|$$

occurs infinitely often equals one. Considering bound (10.31) proved above, we conclude that the probability that the inequality

$$\frac{S(\mathbf{n}(\mathbf{k}+\mathbf{1}))}{\chi(\mathbf{k}+\mathbf{1})} \geq \frac{\chi^\circ(\mathbf{k})}{\chi(\mathbf{k}+\mathbf{1})} \cdot \left[(\sqrt{d}-\varepsilon) - 2^d \frac{(\sqrt{d}+\varepsilon)^2}{\sqrt{\theta}} \right]$$

occurs infinitely often with respect to multi-indices $\mathbf{k}$ equals one. Finally,

$$\lim_{\theta\to\infty} \lim_{\mathbf{k}}(\min) \frac{\chi^\circ(\mathbf{k})}{\chi(\mathbf{k}+\mathbf{1})} = 1,$$

which, together with the preceding result, completes the proof. □

10.3 The Law of the Iterated Logarithm for Multi-Indexed Sums of Weighted Independent Identically Distributed Random Variables

Consider a field $\{X(\mathbf{n}), \mathbf{n} \in \mathbf{N}^d\}$ of independent identically distributed random variables and the field $\{S(\mathbf{n}), \mathbf{n} \in \mathbf{N}^d\}$ of their partial sums

$$S(\mathbf{n}) = \sum_{\mathbf{k} \le \mathbf{n}} w(\mathbf{k}) X(\mathbf{k}), \tag{10.34}$$

where $\{w(\mathbf{n})\}$ are some real numbers called the *weights*. In this section, we study the law of the iterated logarithm for the field $\{S(\mathbf{n})\}$.

The assumptions for the main result of this section are given in terms of two auxiliary non-random fields $\{W^2(\mathbf{n}), \mathbf{n} \in \mathbf{N}^d\}$ and $\{\varphi(\mathbf{n}), \mathbf{n} \in \mathbf{N}^d\}$. The first of them is defined via the squared weights

$$W^2(\mathbf{n}) = \sum_{\mathbf{k} \le \mathbf{n}} w^2(\mathbf{k}).$$

In what follows we assume that this field is unbounded, that is

$$\lim(\max)\, W^2(\mathbf{n}) = \infty. \tag{10.35}$$

Condition (10.35) means that, for an arbitrary number L, the inequality $W^2(\mathbf{n}) < L$ holds only finitely often (see Sect. A.2.2).

Moreover, we assume that there are d increasing sequences of real numbers $\{W_n^{(1)}, n \ge 1\}, \dots, \{W_n^{(d)}, n \ge 1\}$ (each of them depends on a one-dimensional index) such that

$$\lim(\max) \frac{W^2(\mathbf{n})}{W_{n_1}^{(1)} \dots W_{n_d}^{(d)}} = 1. \tag{10.36}$$

Condition (10.36) means that, for all $\varepsilon > 0$, the inequality

$$\left| \frac{W^2(\mathbf{n})}{W_{n_1}^{(1)} \dots W_{n_d}^{(d)}} - 1 \right| > \varepsilon$$

occurs only finitely often (see Definition A.2).

We further assume that each of these sequences grows not very fast in the sense that

$$\lim_{n\to\infty} \frac{W_{n+1}^{(j)}}{W_n^{(j)}} = 1 \tag{10.37}$$

for each $1 \le j \le d$.

Note also that relations (10.35) and (10.36) imply that $\lim_{n\to\infty} W_n^{(j)} = \infty$ for all $1 \le j \le d$.

As far as the field $\{\varphi(\mathbf{n}), \mathbf{n} \in \mathbf{N}^d\}$ is concerned we assume that it is monotone and unbounded, that is

$$\varphi(\mathbf{k}) \le \varphi(\mathbf{n}) \quad \text{for} \quad \mathbf{k} \le \mathbf{n},$$
$$\lim(\max)\, \varphi(\mathbf{n}) = \infty. \tag{10.38}$$

We also assume that

$$\varphi(\mathbf{n}+\mathbf{1}) = O\left(\varphi(\mathbf{n})\right), \quad \mathbf{n} \in \mathbf{N}^d. \tag{10.39}$$

As above, put $\chi(x) = \sqrt{2x \log^+\left(\log^+(x)\right)}$. For the sake of brevity, we write $\log^+ x$ and $\log^+\log^+ x$ instead of $\log^+(x)$ and $\log^+\left(\log^+(x)\right)$, respectively.

Theorem 10.8 *Let X be a random variable with the distribution function F and such that $\mathsf{E}[X] = 0$ and $\mathsf{E}\left[X^2\right] = 1$. Let $\{X(\mathbf{n}), \mathbf{n} \in \mathbf{N}^d\}$ be a field of independent identically distributed random variables with the same distribution function F. Further, let the field $\{W^2(\mathbf{n}), \mathbf{n} \in \mathbf{N}^d\}$ satisfy conditions* (10.35)–(10.37) *and let a field $\{\varphi(\mathbf{n}), \mathbf{n} \in \mathbf{N}^d\}$ satisfy conditions* (10.38)–(10.39). *If*

$$\sum_{\mathbf{n}\in\mathbf{N}^d} \mathsf{P}(|X| \ge \varphi(\mathbf{n})) < \infty \tag{10.40}$$

and

$$\sum_{\mathbf{n}\in\mathbf{N}^d} \left(\frac{w^2(\mathbf{n})}{\chi^2\left(W^2(\mathbf{n})\right)}\right)^{1+\frac{\delta}{2}} \int\limits_{|x|<\varphi(\mathbf{n})} |x|^{2+\delta}\, dF(x) < \infty \tag{10.41}$$

for some $0 < \delta \le 1$, then the law of the iterated logarithm holds, that is

$$\liminf(\max) \frac{S(\mathbf{n})}{\sqrt{2W^2(\mathbf{n}) \log^+\log^+ W^2(\mathbf{n})}} = -\sqrt{d} \quad \textit{a. s.}, \tag{10.42}$$

$$\limsup(\max) \frac{S(\mathbf{n})}{\sqrt{2W^2(\mathbf{n}) \log^+\log^+ W^2(\mathbf{n})}} = \sqrt{d} \quad \textit{a. s.}, \tag{10.43}$$

where the multi-indexed sums $S(\mathbf{n})$ are defined by (10.34).

Remark 10.4 It is also easy to show that, under the assumptions of Theorem 10.8, the law of the iterated logarithm holds not only for $\lim(\max)$ convergence but also for $\lim(\min)$ convergence, that is

$$\liminf(\min) \frac{S(\mathbf{n})}{\sqrt{2W^2(\mathbf{n}) \log^+\log^+ W^2(\mathbf{n})}} = -\sqrt{d} \quad \text{a. s.},$$

$$\limsup(\min) \frac{S(\mathbf{n})}{\sqrt{2W^2(\mathbf{n}) \log^+ \log^+ W^2(\mathbf{n})}} = \sqrt{d} \qquad \text{a. s.}$$

Remark 10.5 If the first two moments of the random variables $\{X(\mathbf{n})\}$ in Theorem 10.8 equal μ and σ^2, respectively, then the law of the iterated logarithm is written as follows

$$\liminf(\max) \frac{S(\mathbf{n}) - \mu|\mathbf{n}|}{\sqrt{2W^2(\mathbf{n}) \log^+ \log^+ W^2(\mathbf{n})}} = -\sigma\sqrt{d} \qquad \text{a. s.,}$$

$$\limsup(\max) \frac{S(\mathbf{n}) - \mu|\mathbf{n}|}{\sqrt{2W^2(\mathbf{n}) \log^+ \log^+ W^2(\mathbf{n})}} = \sigma\sqrt{d} \qquad \text{a. s.}$$

instead of (10.42)–(10.43).

The simplest, still important, case of Theorem 10.8 corresponds to the weights $w(\mathbf{n}) = 1$ for all $\mathbf{n} \in \mathbf{N}^d$. In this case, one can obtain necessary and sufficient conditions for the law of the iterated logarithm for multi-indexed sums of independent identically distributed random variables. This result is an analog of Theorem 10.3 for $d > 1$.

Theorem 10.9 *Let X and $\{X(\mathbf{n}), \mathbf{n} \in \mathbf{N}^d\}$ be a field of independent identically distributed random variables. Let, as usual, $S(\mathbf{n}) = \sum_{\mathbf{k} \le \mathbf{n}} X(\mathbf{k})$. Then the law of the iterated logarithm*

$$\liminf(\max) \frac{S(\mathbf{n})}{\sqrt{2|\mathbf{n}| \log^+ \log^+ |\mathbf{n}|}} = -\sqrt{d}, \qquad a.\ s. \tag{10.44}$$

$$\limsup(\max) \frac{S(\mathbf{n})}{\sqrt{2|\mathbf{n}| \log^+ \log^+ |\mathbf{n}|}} = \sqrt{d} \qquad a.\ s. \tag{10.45}$$

is equivalent to the set of conditions (10.4) *and*

$$\mathsf{E}\left[X^2 \frac{\left(\log^+ |X|\right)^{d-1}}{\log^+ \log^+ |X|}\right] < \infty. \tag{10.46}$$

Condition (10.46) is called *Wichura's condition* after the mathematician who was the first to prove Theorem 10.9 for $d > 1$. Note that (10.46) implies that the second moment exists if $d > 1$. In contrast, if $d = 1$, then (10.46) does not in general imply that the second moment is finite.

Proof of Theorem 10.8 To avoid technicalities we assume that $w^2(\mathbf{1}) \ge e^e$ (this assumption does not restrict the generality of the result).

Consider the "truncated" random variables

$$Y(\mathbf{n}) = \begin{cases} X(\mathbf{n}), & \text{if } |X(\mathbf{n})| < \varphi(\mathbf{n}), \\ 0, & \text{otherwise.} \end{cases}$$

Condition (10.40) together with the Borel–Cantelli lemma for independent random events implies that

$$\mathsf{P}(X(\mathbf{n}) \neq Y(\mathbf{n}) \text{ i. o.}) = 0,$$

where the abbreviation " i. o." stands for "infinitely often". Thus the law of the iterated logarithm (10.42) –(10.43) follows from

$$\lim(\max) \frac{1}{\chi\left(W^2(\mathbf{n})\right)} \sum_{\mathbf{k}\le\mathbf{n}} w(\mathbf{k})\mu(\mathbf{k}) = 0, \tag{10.47}$$

$$\limsup(\max) \frac{1}{\chi\left(W^2(\mathbf{n})\right)} \sum_{\mathbf{k}\le\mathbf{n}} w(\mathbf{k})\left(Y(\mathbf{k}) - \mu(\mathbf{k})\right) = \sqrt{d} \quad \text{a. s.,} \tag{10.48}$$

$$\liminf(\max) \frac{1}{\chi\left(W^2(\mathbf{n})\right)} \sum_{\mathbf{k}\le\mathbf{n}} w(\mathbf{k})\left(Y(\mathbf{k}) - \mu(\mathbf{k})\right) = -\sqrt{d} \quad \text{a. s.,} \tag{10.49}$$

where

$$\mu(\mathbf{k}) = \mathsf{E}\left[Y(\mathbf{k})\right] = \int\limits_{|x|<\varphi(\mathbf{k})} x\, dF(x).$$

To prove equality (10.47) we apply the Cauchy–Schwarz inequality:

$$\left|\frac{1}{\chi\left(W^2(\mathbf{n})\right)} \sum_{\mathbf{k}\le\mathbf{n}} w(\mathbf{k})\mu(\mathbf{k})\right| \le \left(\frac{1}{2\log^+\log^+ W^2(\mathbf{n})} \sum_{\mathbf{k}\le\mathbf{n}} \mu^2(\mathbf{k})\right)^{1/2}.$$

Thus condition (10.47) follows from the convergence of the multi-indexed series $\sum \mu^2(\mathbf{n})$, since condition (10.35) holds. So we need to prove that the series $\sum \mu^2(\mathbf{n})$ does indeed converge.

Consider the set of numbers $\Psi = \{\varphi(\mathbf{n}), \mathbf{n} \in \mathbf{N}^d\}$. Let $\{\psi_k, k \ge 1\}$ be the members of the set Ψ written in ascending order (equal numbers are also counted). If, for some z, there are exactly l numbers in the set Ψ equal to z, then the sequence $\{\psi_k, k \ge 1\}$ contains exactly l members that are equal to z. It is clear that condition (10.40) is equivalent to the convergence of the series $\sum_{k\ge1} \mathsf{P}(|X| \ge \psi_k)$. Put $\mu_k = \int_{|x|<\psi_k} x\, dF(x)$. We are going to prove that $\sum_{k\ge1} \mu_k^2 < \infty$, which obviously means that the series $\sum_{\mathbf{n}\in\mathbf{N}^d} \mu^2(\mathbf{n})$ converges. The equality $\mathsf{E}[X] = 0$ implies that

$$|\mu_n| \le \int\limits_{|x|\ge\psi_n} |x|\, dF(x) = \sum_{i=n}^{\infty} \alpha_i, \quad \text{where} \quad \alpha_i = \int\limits_{\psi_i\le|x|<\psi_{i+1}} |x|\, dF(x).$$

It is also clear that

$$\sum_{n=1}^{\infty} \mu_n^2 \le \sum_{n=1}^{\infty} |\mu_n| \sum_{i=n}^{\infty} \alpha_i = \sum_{i=1}^{\infty} \alpha_i \sum_{n=1}^{i} |\mu_n| \le \sum_{i=1}^{\infty} \alpha_i \sum_{n=1}^{i} \sum_{j=n}^{\infty} \alpha_j$$
$$\le \sum_{i=1}^{\infty} i\alpha_i \sum_{j=i+1}^{\infty} \alpha_j + \sum_{i=1}^{\infty} \alpha_i \sum_{j=1}^{i} j\alpha_j \le 2 \sum_{m=1}^{\infty} \alpha_m \sum_{n=1}^{m} n\alpha_n.$$

The first part of the latter inequality is estimated as follows

$$\sum_{m=1}^{\infty} \alpha_m \sum_{n=1}^{m} n\alpha_n \le \sum_{m=1}^{\infty} \int\limits_{\psi_m \le |x| < \psi_{m+1}} \frac{x^2}{\psi_m} \, dF(x) \sum_{n=1}^{m} n\alpha_n$$
$$\le \left(\sup_{m\ge 1} \frac{1}{\psi_m} \sum_{n=1}^{m} n\alpha_n \right) \int\limits_{|x|\ge \psi_1} x^2 \, dF(x).$$

Now equality (10.39) implies that $\psi_{n+1} = O(\psi_n)$. Thus

$$\sup_{m\ge 1} \frac{1}{\psi_m} \sum_{n=1}^{m} n\alpha_n \le \sum_{n=1}^{\infty} \frac{n\alpha_n}{\psi_n} \le \sum_{n=1}^{\infty} \frac{\psi_{n+1}}{\psi_n} n \mathsf{P}(\psi_n \le |X| < \psi_{n+1})$$
$$\le \text{const} \sum_{n=1}^{\infty} \mathsf{P}(|X| \ge \psi_n) = \text{const} \sum_{\mathbf{n}\in\mathbf{N}^d} \mathsf{P}(|X| \ge \varphi(\mathbf{n})).$$

By condition (10.40), we finally obtain

$$\sum_{n=1}^{\infty} \mu_n^2 \le \text{const}\, \mathsf{E}\left[X^2\right] \sum_{\mathbf{n}\in\mathbf{N}^d} \mathsf{P}(|X| \ge \varphi(\mathbf{n})) < \infty.$$

This proves that the series $\sum \mu_k^2$ converges. Therefore $\sum \mu^2(\mathbf{n}) < \infty$ whence we conclude that relation (10.47) holds.

Now we prove relation (10.48). Put

$$\widehat{W}^2(\mathbf{n}) = \sum_{\mathbf{k}\le\mathbf{n}} w^2(\mathbf{k}) \mathsf{var}\,[Y(\mathbf{k})]\,. \tag{10.50}$$

First of all we show that

$$\lim(\max) \frac{\widehat{W}^2(\mathbf{n})}{W^2(\mathbf{n})} = 1. \tag{10.51}$$

Indeed,

$$|\mathsf{var}\,[Y(\mathbf{n})] - \mathsf{var}\,[X]\,| \le \int\limits_{|x|\ge\varphi(\mathbf{n})} x^2\,dF + \mu^2(\mathbf{n})$$

and thus the inequality $|\mathsf{var}\,[Y(\mathbf{n})] - \mathsf{var}\,[X]\,| > \varepsilon$ occurs only finitely often for an arbitrary $\varepsilon > 0$, since

$$\lim(\max)\,\varphi(\mathbf{n}) = \infty \quad \text{and} \quad \lim(\max)\,\mu^2(\mathbf{n}) = 0$$

in view of assumption (10.4).

We denote by $W_0^2 = \sum w^2(\mathbf{n})$ the sum with respect to those multi-indices $\mathbf{n}$ for which $|\mathsf{var}\,[Y(\mathbf{n})] - \mathsf{var}\,[X]\,| > \varepsilon$. Then

$$\left|\widehat{W}^2(\mathbf{n}) - W^2(\mathbf{n})\right| \le W_0^2 + \varepsilon W^2(\mathbf{n})$$

(recall that $\mathsf{var}\,[X] = 1$). This implies that (10.51) holds in view of (10.35), since the number $\varepsilon > 0$ is arbitrary.

Relations (10.51) together with assumption (10.36) implies that

$$\lim(\max)\,\frac{\widehat{W}^2(\mathbf{n})}{W_{n_1}^{(1)}\ldots W_{n_d}^{(d)}} = 1. \tag{10.52}$$

Fix a real number $\theta > 1$. For a given θ, introduce d sequences of integers $\left\{n_k^{(i)}, k \ge 1\right\}$, $i = 1, \ldots, d$, such that

$$W_{n_k^{(i)}}^{(i)} \le \theta^k < W_{n_k^{(i)}+1}^{(i)}, \quad i = 1, \ldots, d. \tag{10.53}$$

Remark 10.6 The numbers $n_k^{(i)}$ are well-defined by the property (10.53) only for sufficiently large k, say for $k > k_0^{(i)}$, where the ratio $W_{n+1}^{(i)}/W_n^{(i)}$, $n \ge n_{k_0^{(i)}}$, becomes (and continues to be) less than θ (condition (10.37) establishes the existence of such a number $k_0^{(i)}$). The precise values of $n_k^{(i)}$ for $k \le k_0^{(i)}$ do not play an essential role in the remaining argument and thus can be chosen arbitrarily. For example, $n_k^{(i)} = 1$, $k \le k_0^{(i)}$. For convenience and without loss of generality we will assume that $W_{n+1}^{(i)}/W_n^{(i)} \le \theta$ for all $n \ge 1$ and thus condition (10.53) well-defines all members of the sequences $\left\{n_k^{(i)}, k \ge 1\right\}$, $1 \le i \le d$.

Properties (10.53) and (10.37) imply that

$$\lim_{k\to\infty} \frac{W_{n_k^{(i)}}^{(i)}}{\theta^k} = 1 \quad \text{for all} \quad 1 \le i \le d. \tag{10.54}$$

This result together with (10.53) yields the existence of a constant $c_i > 0$ for each $1 \le i \le d$ such that

$$c_i \theta^k \le W^{(i)}_{n^{(i)}_k} \le \theta^k \quad \text{for all} \quad k \ge 1. \tag{10.55}$$

As in the proof of Theorem 10.7, we introduce the sequence of multi-indices $\mathbf{n}(\mathbf{k}) = (n^{(1)}_{k_1}, \dots, n^{(d)}_{k_d})$. According to property (10.36), there exists a constant $C < \infty$ and an index $\mathbf{k}_0 \in \mathbf{N}^d$ such that

$$W^2(\mathbf{n}(\mathbf{k})) \le C\theta^{k_1+\cdots+k_d} \quad \text{for all} \quad \mathbf{k} \ge \mathbf{k}_0. \tag{10.56}$$

In what follows we use the following non-uniform bound for the rate of convergence in the central limit theorem called the *Bikyalis inequality* [24].

Theorem 10.10 (Bikyalis inequality) *Let $X_1, \dots, X_n$ be independent random variables with zero mathematical expectations and finite second moments. Put $V_n = \sum_{k=1}^n \mathsf{var}[X_k]$ and*

$$F_n(x) = \mathsf{P}\left(\frac{1}{\sqrt{V_n}} \sum_{k=1}^n X_k < x\right).$$

Further, let $0 < \delta \le 1$ and $\mathsf{E}\left[|X_k|^{2+\delta}\right] < \infty$, $1 \le k \le n$. Then there exists a universal constant $A = A_\delta$, that does not depend on n or on the distribution functions of the random variables $X_1, \dots, X_n$, such that

$$|F_n(x) - \Phi(x)| \le \frac{A}{V_n^{1+\frac{\delta}{2}}(1+|x|)^{2+\delta}} \sum_{k=1}^n \mathsf{E}\left[|X_k|^{2+\delta}\right] \tag{10.57}$$

for all $x \in \mathbf{R}$, where Φ is a standard normal distribution function.

The Bikyalis inequality (10.57) is applied to the random variables $Y(\mathbf{k}) - \mu(\mathbf{k})$ in the proof of Theorem 10.8. Put

$$\zeta(\mathbf{n}) \stackrel{\text{def}}{=} \sum_{\mathbf{k} \le \mathbf{n}} w(\mathbf{k})\big[Y(\mathbf{k}) - \mu(\mathbf{k})\big], \quad C(\mathbf{n}) \stackrel{\text{def}}{=} \chi\left(\widehat{W}^2(\mathbf{n})\right). \tag{10.58}$$

It is clear that $\widehat{W}^2(\mathbf{n}) = \mathsf{var}[\zeta(\mathbf{n})]$. We assume, without loss of generality, that $\widehat{W}^2(\mathbf{1}) \ge e^e$.

Let $\gamma > 0$. Putting $x = \sqrt{2(d+\gamma)\log^+\log^+\widehat{W}^2(\mathbf{n})}$ in the Bikyalis inequality (10.57), we obtain the following bound

$$\begin{aligned} &\left|\mathsf{P}\left(\zeta(\mathbf{n}) \ge x\widehat{W}(\mathbf{n})\right) - \Phi(-x)\right| \\ &\le \frac{\text{const}}{\left(C(\mathbf{n})\right)^{2+\delta}} \sum_{\mathbf{k} \le \mathbf{n}} |w(\mathbf{k})|^{2+\delta} \int\limits_{|x|<\varphi(\mathbf{k})} |x - \mu(\mathbf{k})|^{2+\delta}\, dF(x). \end{aligned}$$

Since $|u+v|^{2+\delta} \le 2^{1+\delta}\left(|u|^{2+\delta}+|v|^{2+\delta}\right)$ for all $u, v \in \mathbf{R}$, we get

$$|x-\mu(\mathbf{k})|^{2+\delta} \le 4\left(|x|^{2+\delta}+|\mu(\mathbf{k})|^{2+\delta}\right).$$

By Hölder's inequality, we get

$$|\mu(\mathbf{k})| = \left|\int\limits_{|x|<\varphi(\mathbf{k})} x\,dF\right| \le \left(\int\limits_{|x|<\varphi(\mathbf{k})} |x|^{2+\delta}\,dF\right)^{\frac{1}{2+\delta}} \cdot \left(\int\limits_{|x|<\varphi(\mathbf{k})} dF\right)^{\frac{1+\delta}{2+\delta}}. \tag{10.59}$$

Therefore

$$\begin{aligned}&\left|\mathsf{P}\left(\zeta(\mathbf{n}) \ge x\widehat{W}(\mathbf{n})\right) - \Phi(-x)\right| \\ &\quad\le \frac{\text{const}}{\left(C(\mathbf{n})\right)^{2+\delta}} \sum_{\mathbf{k}\le\mathbf{n}} |w(\mathbf{k})|^{2+\delta} \int\limits_{|x|<\varphi(\mathbf{k})} |x|^{2+\delta}\,dF(x).\end{aligned}$$

We write the latter inequality for $\mathbf{n} = \mathbf{n}(\mathbf{k})$:

$$\begin{aligned}&\left|\mathsf{P}\left(\zeta(\mathbf{n}(\mathbf{k})) \ge x\widehat{W}(\mathbf{n}(\mathbf{k}))\right) - \Phi(-x)\right| \\ &\quad\le \frac{\text{const}}{\left(C(\mathbf{n}(\mathbf{k}))\right)^{2+\delta}} \sum_{\mathbf{m}\le\mathbf{n}(\mathbf{k})} |w(\mathbf{m})|^{2+\delta} \int\limits_{|x|<\varphi(\mathbf{m})} |x|^{2+\delta}\,dF(x).\end{aligned} \tag{10.60}$$

Next we prove that

$$\sum_{\mathbf{k}\in\mathbf{N}^d} \frac{1}{\left(C(\mathbf{n}(\mathbf{k}))\right)^{2+\delta}} \sum_{\mathbf{m}\le\mathbf{n}(\mathbf{k})} |w(\mathbf{m})|^{2+\delta} \int\limits_{|x|<\varphi(\mathbf{m})} |x|^{2+\delta}\,dF(x) < \infty. \tag{10.61}$$

It follows from (10.52) and (10.55) that there exists a constant $c>0$ such that

$$\widehat{W}^2(\mathbf{n}(\mathbf{k})) \ge c\theta^{k_1+\cdots+k_d} \quad \text{for all} \quad \mathbf{k}\in\mathbf{N}^d. \tag{10.62}$$

Since the function χ is monotone,

$$C(\mathbf{n}(\mathbf{k})) = \chi\left(\widehat{W}^2(\mathbf{n}(\mathbf{k}))\right) \ge \text{const}\,\chi\left(\theta^{k_1+\cdots+k_d}\right) \quad \text{for all} \quad \mathbf{k}\in\mathbf{N}^d. \tag{10.63}$$

Put

$$a(\mathbf{m}) = \sum_{\mathbf{n}(\mathbf{m}-\mathbf{1})<\mathbf{j}\le\mathbf{n}(\mathbf{m})} |w(\mathbf{j})|^{2+\delta} \int\limits_{|x|<\varphi(\mathbf{j})} |x|^{2+\delta}\,dF(x),$$

where $\mathbf{1} = (1, \dots, 1)$. We agree that $\mathbf{n}(\mathbf{m} - \mathbf{1}) = \mathbf{0}$ if $\min\{m_1, \dots, m_d\} = 1$ in the latter equality. Moreover, we agree that $w(\mathbf{j}) = 0$ if $\min\{j_1, \dots, j_d\} = 0$. These agreements mean, for example, that

$$a(\mathbf{1}) = \sum_{\mathbf{1} \le \mathbf{j} \le \mathbf{n}(\mathbf{1})} |w(\mathbf{j})|^{2+\delta} \int_{|x|<\varphi(\mathbf{j})} |x|^{2+\delta} \, dF(x).$$

By inequality (10.63),

$$\begin{aligned}
&\sum_{\mathbf{k}\in\mathbf{N}^d} \frac{1}{\left(C(\mathbf{n}(\mathbf{k}))\right)^{2+\delta}} \sum_{\mathbf{m}\le\mathbf{n}(\mathbf{k})} |w(\mathbf{m})|^{2+\delta} \int_{|x|<\varphi(\mathbf{m})} |x|^{2+\delta} \, dF(x) \\
&\le \text{const} \sum_{\mathbf{k}\in\mathbf{N}^d} \frac{1}{\chi^{2+\delta}(\theta^{k_1+\dots+k_d})} \sum_{\mathbf{m}\le\mathbf{k}} a(\mathbf{m}) \\
&= \text{const} \sum_{\mathbf{m}\in\mathbf{N}^d} a(\mathbf{m}) \sum_{\mathbf{k}\ge\mathbf{m}} \frac{1}{\chi^{2+\delta}(\theta^{k_1+\dots+k_d})} \\
&\le \text{const} \sum_{\mathbf{m}\in\mathbf{N}^d} a(\mathbf{m}) \frac{1}{\chi^{2+\delta}(\theta^{m_1+\dots+m_d})} \\
&\le \text{const} + \text{const} \sum_{\mathbf{n}\in\mathbf{N}^d} \left(\frac{w^2(\mathbf{n})}{\chi^2(W(\mathbf{n}))}\right)^{1+\frac{\delta}{2}} \int_{|x|<\varphi(\mathbf{n})} |x|^{2+\delta} \, dF(x).
\end{aligned}$$

The latter inequality follows from relation (10.56), since the function χ is monotone. Now (10.61) follows from condition (10.41).

Applying a standard bound for the tail of the normal distribution, we derive from (10.62) that

$$\sum_{\mathbf{k}\in\mathbf{N}^d} \Phi\left(-\sqrt{2(d+\gamma)\log\log \widehat{W}^2(\mathbf{n}(\mathbf{k}))}\right) < \infty$$

for all $\gamma > 0$. Thus (10.60) and (10.61) imply that

$$\sum_{\mathbf{k}\in\mathbf{N}^d} \mathsf{P}\left(\zeta(\mathbf{n}(\mathbf{k})) \ge \sqrt{d+\gamma}\, C(\mathbf{n}(\mathbf{k}))\right) < \infty \tag{10.64}$$

for all $\gamma > 0$. By Kolmogorov's inequality for multi-indexed sums (Corollary 2.7),

$$\mathsf{P}\left(\max_{n\le\mathbf{n}(\mathbf{k})} \zeta(\mathbf{n}) \ge x\right) \le 2^d \mathsf{P}\left(\zeta(\mathbf{n}(\mathbf{k})) \ge x - d\sqrt{2\mathsf{var}\left[\zeta(\mathbf{n}(\mathbf{k}))\right]}\right)$$

for all $x > 0$. Then, in view of (10.64),

$$\sum_{\mathbf{k}\in\mathbf{N}^d} \mathsf{P}\left(\max_{n\le\mathbf{n}(\mathbf{k})} \zeta(\mathbf{n}) \ge \sqrt{d+\gamma}C(\mathbf{n}(\mathbf{k}))\right) < \infty$$

for all $\gamma > 0$. Applying the Borel–Cantelli lemma, we conclude that

$$\text{limsup(max)}\,\frac{\zeta(\mathbf{n})}{\chi\left(\widehat{W}^2(\mathbf{n})\right)} \le \sqrt{d+\gamma} \qquad \text{a. s.} \tag{10.65}$$

for all $\gamma > 0$. Since $\gamma > 0$ is arbitrary, the latter result holds for $\gamma = 0$, as well.

Property (10.36) and (10.51) imply that

$$\text{limsup(max)}\,\frac{\zeta(\mathbf{n})}{\chi(W^2(\mathbf{n}))} \le \sqrt{d} \qquad \text{a. s.} \tag{10.66}$$

Note that the assumptions of Theorem 10.8 are valid for the random variables $\{-X(\mathbf{n})\}$. Applying the part of Theorem 10.8 just proved to the random variables $\{-X(\mathbf{n})\}$ we derive from (10.66) that

$$\text{limsup(max)}\,\frac{-\zeta(\mathbf{n})}{\chi(W^2(\mathbf{n}))} \le \sqrt{d} \qquad \text{a. s.,}$$

whence

$$\text{limsup(max)}\,\frac{|\zeta(\mathbf{n})|}{\chi(W^2(\mathbf{n}))} \le \sqrt{d} \qquad \text{a. s.} \tag{10.67}$$

We will prove that, in fact, the inequality in (10.66) is an equality. For the sequences $\{n_k^{(i)}, k \ge 1\}$, $1 \le i \le d$, defined by inequalities (10.55), put

$$\begin{aligned}
\mathfrak{B}_{\mathbf{k}}^{\circ} &= \left\{\mathbf{m} : n_{k_1-1}^{(1)} < m_1 \le n_{k_1}^{(1)}, \dots, n_{k_d-1}^{(d)} < m_d \le n_{k_d}^{(d)}\right\},\\
\mathscr{S}(\mathbf{k}) &= \sum_{\mathbf{m}\in\mathfrak{B}_{\mathbf{k}}^{\circ}} w(\mathbf{m})(Y(\mathbf{m}) - \mu(\mathbf{m})),\\
\mathscr{W}^2(\mathbf{k}) &= \sum_{\mathbf{m}\in\mathfrak{B}_{\mathbf{k}}^{\circ}} w^2(\mathbf{m})\mathsf{var}\,[Y(\mathbf{m})]\,, \qquad \mathscr{C}(\mathbf{k}) \stackrel{\text{def}}{=} \chi\left(\mathscr{W}^2(\mathbf{k})\right).
\end{aligned}$$

The sums $\mathscr{S}(\mathbf{k})$ are well-defined for all $\mathbf{k} \in \mathbf{N}^d$ if we agree that if a certain coordinate k_j, $1 \le j \le d$, of a multi-index $\mathbf{k}$ equals 1, then the corresponding coordinate of the multi-index $\mathbf{n}(\mathbf{k} - \mathbf{1})$ equals 0. Moreover, we agree that $w(\mathbf{m}) = 0$ if $\min\{m_1, \dots, m_d\} = 0$. Finally, we let the sum equal zero if the set of multi-indices for the summation is empty.

Below we make use of the following properties of the blocks $\mathfrak{B}^\circ_{\mathbf{k}}$:

(i) the sets $\mathfrak{B}^\circ_{\mathbf{k}}$, $\mathbf{k} \in \mathbf{N}^d$, are mutually disjoint;
(ii) there exists a multi-index $\mathbf{k}_0$ such that all the sets $\mathfrak{B}^\circ_{\mathbf{k}}$, $\mathbf{k} \ge \mathbf{k}_0$, are non-empty;
(iii) the union of all sets $\mathfrak{B}^\circ_{\mathbf{k}}$, $\mathbf{k} \in \mathbf{N}^d$, equals $\mathbf{N}^d$.

It is also clear that $\mathscr{W}^2(\mathbf{k}) = \operatorname{var}[\mathscr{S}(\mathbf{k})]$.

For those multi-indices $\mathbf{k}$ for which the sets $\mathfrak{B}^\circ_{\mathbf{k}}$ are non-empty we apply the Bikyalis inequality (10.57) once again, now with $0 < \gamma < d$:

$$\left|\mathsf{P}\left(\mathscr{S}(\mathbf{k}) \ge \sqrt{d-\gamma}\,\mathscr{C}(\mathbf{k})\right) - \Phi\left(-\sqrt{2(d-\gamma)\log\log \mathscr{W}^2(\mathbf{k})}\right)\right| \\ \le \frac{\text{const}}{\left(\mathscr{C}(\mathbf{k})\right)^{2+\delta}} \sum_{\mathbf{m}\in\mathfrak{B}^\circ_{\mathbf{k}}} \mathsf{E}\left[|w(\mathbf{m})(Y(\mathbf{m})-\mu(\mathbf{m}))|^{2+\delta}\right]. \tag{10.68}$$

There exists a $\mathbf{k}_0 \in \mathbf{N}^d$ such that

$$\sum_{\mathbf{k}\ge\mathbf{k}_0} \frac{1}{\left(\mathscr{C}(\mathbf{k})\right)^{2+\delta}} \sum_{\mathbf{m}\in\mathfrak{B}^\circ_{\mathbf{k}}} |w(\mathbf{m})|^{2+\delta}\mathsf{E}\left[|(Y(\mathbf{m})-\mu(\mathbf{m}))|^{2+\delta}\right] < \infty.$$

This property follows from

$$\sum_{\mathbf{k}\ge\mathbf{k}_0} \frac{1}{\left(\mathscr{C}(\mathbf{k})\right)^{2+\delta}} \sum_{\mathbf{m}\in\mathfrak{B}^\circ_{\mathbf{k}}} |w(\mathbf{m})|^{2+\delta}\mathsf{E}\left[|Y(\mathbf{m})|^{2+\delta}\right] < \infty, \tag{10.69}$$

since $|\mu(\mathbf{m})|^{2+\delta} \le \mathsf{E}\left[|(Y(\mathbf{m})|^{2+\delta}\right]$ (see (10.59)). In turn, property (10.69) follows from

$$\lim_{\mathbf{k}}(\min) \frac{\mathscr{W}^2(\mathbf{k})}{\theta^{k_1+\cdots+k_d}} = \left(\frac{\theta-1}{\theta}\right)^d. \tag{10.70}$$

Indeed, the asymptotics (10.70) yields the existence of a number $c > 0$ and a multi-index $\mathbf{k}_0$ such that property (ii) of blocks $\mathfrak{B}^\circ_{\mathbf{k}}$ holds and

$$\mathscr{C}(\mathbf{k}) \ge c\chi\left(\theta^{k_1+\cdots+k_d}\right), \quad \mathbf{k} \ge \mathbf{k}_0.$$

Substituting this bound in the left-hand side of (10.69) and applying (10.56) we derive (10.69) from condition (10.41).

To prove (10.70) note that $\mathscr{W}^2(\mathbf{k})$ are the increments of the field $\widehat{W}^2(\mathbf{n}(\mathbf{m}))$ defined by equality (10.50) (see Section A.5), whence

$$\mathscr{W}^2(\mathbf{m}) = \sum_{(\delta_1,\ldots,\delta_d)\in\mathscr{E}_d} (-1)^{\delta_1+\cdots+\delta_d}\,\widehat{W}^2(n^{(1)}_{m_1}-\delta_1,\ldots,n^{(d)}_{m_d}-\delta_d)$$

(see Proposition A.5). Recall that $\mathscr{E}_d$ denotes the set of d-dimensional multi-indices whose coordinates equal either 0 or 1. Since

$$\left(\frac{\theta-1}{\theta}\right)^d = \sum_{(\delta_1,\dots,\delta_d)\in\mathscr{E}_d} (-1)^{\delta_1+\dots+\delta_d}\theta^{-(\delta_1+\dots+\delta_d)},$$

we obtain

$$\begin{aligned}&\left|\frac{\mathscr{W}^2(\mathbf{m})}{\theta^{m_1+\dots+m_d}} - \left(\frac{\theta-1}{\theta}\right)^d\right| \\ &\qquad\le \sum_{(\delta_1,\dots,\delta_d)\in\mathscr{E}_d}\left|\frac{\widehat{W}^2\left(n_{m_1}^{(1)}-\delta_1,\dots,n_{m_d}^{(d)}-\delta_d\right)}{\theta^{m_1+\dots+m_d}} - \theta^{-(\delta_1+\dots+\delta_d)}\right|.\end{aligned}$$

Property (10.52) and asymptotics (10.54) imply that the limit of each term in the latter sum as $m_1\to\infty,\dots,m_d\to\infty$ equals 0. Since the latter sum contains 2^d terms, the latter result proves (10.70).

On the other hand, $\widehat{W}^2(\mathbf{n}) \le W^2(\mathbf{n})$ and thus (10.56) implies that

$$\sum_{\mathbf{k}\ge\mathbf{k}_1} \Phi\left(-\sqrt{2(d-\gamma)\log\log\widehat{W}^2(\mathbf{n}(\mathbf{k}))}\right) = \infty$$

for all $0<\gamma<d$ whatever the multi-index $\mathbf{k}_1 \ge \mathbf{k}_0$. Now we derive from relations (10.64) and (10.69) that

$$\sum_{\mathbf{k}\ge\mathbf{k}_1} \mathsf{P}\left(\mathscr{S}(\mathbf{k}) \ge \sqrt{d-\gamma}\,\mathscr{C}(\mathbf{k})\right) = \infty$$

for all $0<\gamma<d$ and all $\mathbf{k}_1\ge\mathbf{k}_0$. Applying the Borel–Cantelli lemma for independent random events we establish

$$\mathsf{P}\left(\mathscr{S}(\mathbf{k}) \ge \sqrt{d-\gamma}\,\mathscr{C}(\mathbf{k}) \text{ i. o. in the domain } \mathbf{k}\ge\mathbf{k}_1\right) = 1 \tag{10.71}$$

for all $0<\gamma<d$ and all $\mathbf{k}_1\ge\mathbf{k}_0$.

Then we introduce the following two random events

$$E_1 \stackrel{\text{def}}{=} \{\omega\in\Omega : S(\mathbf{k}) \ge \sqrt{d+\gamma}\,C(\mathbf{k}) \text{ i. o.}\} \quad\text{and}\quad E_2 = \bigcap_{\mathbf{k}_1\ge\mathbf{k}_0} E_{\mathbf{k}_1},$$

where

$$E_{\mathbf{k}_1} \stackrel{\text{def}}{=} \{\omega\in\Omega : \mathscr{S}(\mathbf{k}) \ge \sqrt{d-\gamma}\,\mathscr{C}(\mathbf{k}) \text{ i. o. in the domain } \mathbf{k}\ge\mathbf{k}_1\}.$$

We have proved above that $\mathsf{P}(E_1) = 0$ and $\mathsf{P}(E_2) = 1$. Fix an arbitrary elementary random event $\omega \in \overline{E}_1 \cap E_2$ and construct the sequence $\{\mathbf{k}_j\}$ as follows: as $\mathbf{k}_1$, we choose an arbitrary multi-index such that $\mathscr{S}(\mathbf{k}_1; \omega) \geq \sqrt{d-\gamma}\,\mathscr{C}(\mathbf{k}_1)$. To ease the description of the procedure we denote by $(\mathbf{n})_i$ the ith coordinate of a multi-index $\mathbf{n}$. Put $\varkappa \stackrel{\text{def}}{=} \max\{(\mathbf{k}_1)_1, \ldots, (\mathbf{k}_1)_d\}$, and, as $\mathbf{k}_2$, we choose an arbitrary multi-index such that $\mathbf{k}_2 > (\varkappa, \ldots, \varkappa)$ and $\mathscr{S}(\mathbf{k}_2; \omega) \geq \sqrt{d-\gamma}\,\mathscr{C}(\mathbf{k}_2)$. All other members of the sequence are chosen according to the following rule: if a multi-index $\mathbf{k}_j$ is already defined, then we put $\varkappa \stackrel{\text{def}}{=} \max\{(\mathbf{k}_j)_1, \ldots, (\mathbf{k}_j)_d\}$, and, as $\mathbf{k}_{j+1}$, we choose an arbitrary multi-index such that

$$\mathbf{k}_{j+1} > (\varkappa, \ldots, \varkappa) \quad \text{and} \quad \mathscr{S}(\mathbf{k}_{j+1}; \omega) \geq \sqrt{d-\gamma}\,\mathscr{C}(\mathbf{k}_{j+1}).$$

For every subset $\Lambda \subset \{1, \ldots, d\}$, $\Lambda \neq \{1, \ldots, d\}$, and every integer $j \geq 1$ we define the multi-index $\mathbf{k}_j^\Lambda$ whose coordinates are

$$\left(\mathbf{k}_j^\Lambda\right)_i = \begin{cases} (\mathbf{k}_j)_i, & \text{if } i \in \Lambda, \\ (\mathbf{k}_{j-1})_i & \text{if } i \notin \Lambda. \end{cases}$$

For each $j \geq 1$, the sum $\zeta(\mathbf{k}_j)$ defined by equality (10.58) is a linear combination of the random variable $\mathscr{S}(\mathbf{k}_j)$ and at most $2^d - 1$ sums $\zeta(\mathbf{k}_j^\Lambda)$, $\Lambda \neq \{1, \ldots, d\}$. Therefore, $\zeta(\mathbf{k}_j) = \mathscr{S}(\mathbf{k}_j) + \sum_\Lambda \zeta(\mathbf{k}_j^\Lambda)$ (the number of terms in the sum does not exceed $2^d - 1$). It is clear that

$$\lim_{j\to\infty} \frac{C(\mathbf{k}_j)}{\mathscr{C}(\mathbf{k}_j)} = \left(\frac{\theta}{\theta-1}\right)^{d/2}, \qquad \lim_{j\to\infty} \frac{C(\mathbf{k}_j^\Lambda)}{\mathscr{C}(\mathbf{k}_j)} \leq \frac{1}{\sqrt{\theta^{d-\operatorname{card}(\Lambda)}}} \left(\frac{\theta-1}{\theta}\right)^{d/2}.$$

Thus relations (10.66) and (10.71) imply that

$$\begin{aligned} \liminf_{j\to\infty} \frac{\zeta(\mathbf{k}_j)}{C(\mathbf{k}_j)} &\geq \left[\liminf_{j\to\infty} \frac{\mathscr{S}(\mathbf{k}_j)}{C(\mathbf{k}_j)} - \frac{1}{C(\mathbf{k}_j)} \sum_\Lambda \frac{|\zeta(\mathbf{k}_j^\Lambda)|}{\mathscr{C}(\mathbf{k}_j^\Lambda)} \cdot \mathscr{C}(\mathbf{k}_j^\Lambda)\right] \\ &\geq \left(\frac{\theta-1}{\theta}\right)^{d/2} \left[\sqrt{d-\gamma} - (2^d-1)\sqrt{\frac{d}{\theta}}\right]. \end{aligned}$$

Now we conclude that

$$\operatorname{limsup(min)} \frac{\zeta(\mathbf{n}; \omega)}{C(\mathbf{n})} \geq \left(\frac{\theta-1}{\theta}\right)^{d/2} \left[\sqrt{d-\gamma} - (2^d-1)\sqrt{\frac{d}{\theta}}\right] \tag{10.72}$$

for every elementary random event $\omega \in \overline{E}_1 \cap E_2$. Completing the proof, denote by $\mathbf{Q}$ the set of positive rational numbers and introduce the following random event

$$E \stackrel{\text{def}}{=} \bigcap_{\gamma\in\mathbf{Q}\cap(0,d)} \bigcap_{\theta\in\mathbf{Q}\cap(1,\infty)} \left(\overline{E}_1 \cap E_2\right).$$

The probability of this event equals 1, that is $\mathsf{P}(E) = 1$. Moreover, for each elementary random event $\omega \in E$ we deduce from (10.72) that

$$\operatorname{limsup}(\min) \frac{\zeta(\mathbf{n};\omega)}{C(\mathbf{n})} \geq \sqrt{d}.$$

This inequality together with (10.66) implies equality (10.48) in view of (10.51). Relation (10.49) is proved similarly to (10.48).

Therefore the proof of Theorem 10.8 is complete. □

Prior the proof of Theorem 10.9 we provide two auxiliary results which will often be used throughout this section.

Lemma 10.2 *Let $\delta > 0$ and let X be a random variable with a distribution function F. Further, let φ be a continuous increasing function and denote by φ^{-1} its inverse. Put $\varphi(\mathbf{n}) = \varphi(|\mathbf{n}|)$. Then condition* (10.40) *is equivalent to*

$$\mathsf{E}\left[\varphi^{-1}(|X|)\left(\log^+ \varphi^{-1}(|X|)\right)^{d-1}\right] < \infty. \tag{10.73}$$

Lemma 10.3 *Let g be a non-decreasing regularly varying function of index $\rho > 1$. Then*

$$\sum_{k=j}^{\infty} \frac{\tau_k}{g(k)} \leq C \cdot \frac{j\left(\log^+ j\right)^{d-1}}{g(j)}$$

for some constant C that does not depend on j.

Proof of Lemma 10.2 Since $\varphi(\mathbf{n})$ is, in fact, a function of a single argument (φ depends on $|\mathbf{n}|$), condition (10.40) can be rewritten in terms of the Dirichlet sequence $\{\tau_k\}$, namely

$$\sum_{k=1}^{\infty} \tau_k \mathsf{P}(|X| \geq \varphi(k)) < \infty, \tag{10.74}$$

where τ_k is the number of solutions of the equation $|\mathbf{n}| = k$. It is also clear that condition (10.74) is equivalent to

$$\sum_{k=1}^{\infty} T_k \mathsf{P}(\varphi(k) \leq |X| < \varphi(k+1)) < \infty,$$

where $T_1 = \tau_1$ and $T_k = \tau_1 + \cdots + \tau_k$, $k \ge 2$. Using the rough asymptotics (A.41) for the Dirichlet function,

$$T_k \asymp k \left(\log^+ k\right)^{d-1}, \tag{10.75}$$

we prove that condition (10.74) is equivalent to

$$\sum_{k=1}^{\infty} k (\log k)^{d-1} \mathsf{P}(\varphi(k) \le |X| < \varphi(k+1)) < \infty$$

which, in turn, is equivalent to (10.73). □

Proof of Lemma 10.3 We use the sequences $\{\tau_k\}$ and $\{T_n\}$ defined in the proof of Lemma 10.2. Let $T_0 = 0$. Since $\tau_k = T_k - T_{k-1}$, $k \ge 1$,

$$\sum_{k=j}^{\infty} \frac{\tau_k}{g(k)} = -\frac{T_{j-1}}{g(j)} + \sum_{k=j}^{\infty} T_k \left(\frac{1}{g(k)} - \frac{1}{g(k+1)} \right)$$

by the partial summation formula. Using the rough asymptotics (10.75) we get

$$\begin{aligned} \sum_{k=j}^{\infty} \frac{\tau_k}{g(k)} &\le \text{const} \sum_{k=j}^{\infty} k (\log k)^{d-1} \left(\frac{1}{g(k)} - \frac{1}{g(k+1)} \right) \\ &\le \text{const} \, \frac{j (\log j)^{d-1}}{g(j)} + \text{const} \sum_{k=j}^{\infty} \frac{\delta_k}{g(k)}, \end{aligned}$$

where δ_k are the increments of the sequence $\{k \left(\log^+ k\right)^{d-1}\}$, that is

$$\delta_k \stackrel{\text{def}}{=} k (\log k)^{d-1} - (k-1) (\log(k-1))^{d-1}.$$

Since $\delta_k \asymp (\log k)^{d-1}$, one needs to show that

$$\sum_{k=j}^{\infty} \frac{1}{h(k)} \le \text{const} \, \frac{j}{h(j)}, \qquad \text{where} \quad h(x) \stackrel{\text{def}}{=} \frac{g(x)}{\left(\log^+ x\right)^{d-1}}$$

to complete the proof. It is clear that h is a regularly varying function of index ρ. Rewrite h in the form $h(x) = x^{\rho} \ell(x)$, where ℓ is a slowly varying function. Now we choose a number $0 < r < \rho - 1$ and consider a monotone version $h_0(x) = x^r \ell_0(x)$ of the function h, where $\ell_0(x) \sim \ell(x)$, $x \to \infty$. Such a function exists in view of Proposition A.10. Then

$$\sum_{k=j}^{\infty} \frac{1}{h(k)} \leq \text{const} \frac{1}{h_0(j)} \sum_{k=j}^{\infty} \frac{1}{k^{\rho-r}} \leq \text{const} \frac{1}{h_0(j) j^{\rho-r-1}} \leq \text{const} \frac{j}{h(j)}$$

and Lemma 10.3 is proved. □

Proof of Theorem 10.9 We use Lemma 10.2 with $\varphi = \chi$. Since

$$\chi^{-1}(x) \asymp \frac{x^2}{\log^+ \log^+ x}, \tag{10.76}$$

condition (10.73) and, as a consequence, condition (10.40) reduces to Wichura's condition (10.46).

If $w(\mathbf{n}) = 1$, $\mathbf{n} \in \mathbf{N}^d$, then $W^2(\mathbf{n}) = |\mathbf{n}|$, $\mathbf{n} \in \mathbf{N}^d$, and thus condition (10.41) is equivalent to

$$\sum_{k=1}^{\infty} \frac{\tau_k}{\chi^{2+\delta}(k)} \int\limits_{|x|<\chi(k)} |x|^{2+\delta} \, dF(x) < \infty. \tag{10.77}$$

Then

$$\sum_{k=1}^{\infty} \frac{\tau_k}{\chi^{2+\delta}(k)} \int\limits_{|x|<\chi(k)} |x|^{2+\delta} \, dF(x)$$
$$= \sum_{j=1}^{\infty} \int\limits_{\chi(j-1)\leq|x|<\chi(j)} |x|^{2+\delta} \, dF(x) \sum_{k=j}^{\infty} \frac{\tau_k}{\chi^{2+\delta}(k)}.$$

The tail of the series on the right-hand side is estimated with the help of Lemma 10.3 with $g = \chi^{2+\delta}$:

$$\sum_{k=1}^{\infty} \frac{\tau_k}{\chi^{2+\delta}(k)} \int\limits_{|x|<\chi(k)} |x|^{2+\delta} \, dF(x)$$
$$\leq \text{const} \sum_{j=1}^{\infty} \frac{j \left(\log^+ j\right)^{d-1}}{\chi^{2+\delta}(j)} \int\limits_{\chi(j-1)\leq|x|<\chi(j)} |x|^{2+\delta} \, dF(x)$$
$$\leq \text{const} \sum_{j=1}^{\infty} j \left(\log^+ j\right)^{d-1} \mathsf{P}\left(\chi(j-1) \leq |X| < \chi(j)\right)$$
$$\leq \text{const} \, \mathsf{E}\left[X^2 \frac{\left(\log^+ |X|\right)^{d-1}}{\log^+ \log^+ |X|} \right].$$

The latter inequality follows from (10.76). Thus condition (10.41) is a consequence of Wichura's condition (10.46). We restrict ourselves in the proof below only to the case $d > 1$, since the case $d = 1$ is already considered in Theorem 10.4.

We derive from Theorem 10.8 that the law of the iterated logarithm (10.42)–(10.43) holds if conditions (10.4) and (10.46) are satisfied.

Conversely, conditions (10.4) and (10.46) follow from the law of the iterated logarithm (10.45). Indeed, every random variable $X(\mathbf{n})$ is a linear combination of at most 2^d sums $S(\mathbf{m})$ whose multi-indices $\mathbf{m}$ are such that $\mathbf{n} - \mathbf{1} \le \mathbf{m} \le \mathbf{n}$ (see Proposition A.5). Taking this result into consideration, we conclude from (10.46) that the inequality

$$|X(\mathbf{n})| > 2^d \sqrt{d+\gamma}\, \chi(|\mathbf{n}|)$$

occurs finitely often for all $\gamma > 0$ and for almost all elementary random events $\omega \in \Omega$. Since the random variables $X(\mathbf{n})$ are jointly independent, the Borel–Cantelli lemma implies that

$$\sum_{\mathbf{n}\in\mathbf{N}^d} \mathsf{P}(|X(\mathbf{n})| \ge b\chi(|\mathbf{n}|)) < \infty,$$

where $b = 2^d\sqrt{d+\gamma}$. Since the random variables $X(\mathbf{n})$ are identically distributed, the preceding result means that

$$\sum_{k\ge 1} \tau_k \mathsf{P}(|X| \ge b\chi(k)) < \infty.$$

Now Lemma 10.2 with $\varphi(x) = b\chi(x)$ implies condition (10.73). In turn, condition (10.73) is equivalent to Wichura's condition (10.46) in view of (10.76). In particular, the latter result means that the two first moments are finite, since $d > 1$. It remains to prove that they are equal to 0 and 1, respectively. Put $\mu \stackrel{\text{def}}{=} \mathsf{E}[X]$, $\sigma^2 \stackrel{\text{def}}{=} \mathsf{var}[X]$, and

$$Y(\mathbf{n}) \stackrel{\text{def}}{=} \frac{1}{\sigma}(X(\mathbf{n}) - \mu), \qquad S_Y(\mathbf{n}) = \sum_{\mathbf{k}\le\mathbf{n}} Y(\mathbf{k}).$$

It is clear that the first two moments of the random variables $Y(\mathbf{k})$ are equal to 0 and 1, respectively. Moreover, Wichura's condition (10.46) holds for $\{Y(\mathbf{n})\}$. Now the part of Theorem 10.9 already proved implies that

$$\sqrt{d} = \limsup(\max) \frac{S_Y(\mathbf{n})}{\chi(|\mathbf{n}|)} = \limsup(\max) \frac{S(\mathbf{n}) - \mu|\mathbf{n}|}{\sigma\chi(|\mathbf{n}|)} \qquad \text{a. s.}$$

This relation holds simultaneously with (10.44) only if $\mu = 0$ and $\sigma^2 = 1$. The proof is complete. $\square$

10.4 Corollaries of the Law of the Iterated Logarithm

If the function φ does not change for arguments $\mathbf{n}$ with identical $|\mathbf{n}|$, Theorem 10.8 can be simplified for many particular cases. For example, condition (10.40) with $\varphi(x) = \sqrt{x}$ is equivalent to

$$\mathsf{E}\left[X^2 \left(\log^+ |X|\right)^{d-1}\right] < \infty \tag{10.78}$$

in view of Lemma 10.2. If $\varphi(x) = \sqrt{x \log^+ \log^+ x}$, then condition (10.40) is equivalent to

$$\mathsf{E}\left[X^2 \frac{\left(\log^+ |X|\right)^{d-1}}{\log^+ \log^+ |X|}\right] < \infty. \tag{10.79}$$

We provide two corollaries of Theorem 10.8 corresponding to the cases (10.78) and (10.79).

Corollary 10.1 *Let $X, \{X(\mathbf{n}), \mathbf{n} \in \mathbf{N}^d\}$ be independent identically distributed random variables with the distribution function F. Assume that both conditions* (10.4) *and* (10.78) *hold.*

Further, let there exist $d \geq 1$ sequences of real numbers $\{W_j^{(1)}, j \geq 1\}, \dots, \{W_j^{(d)}, j \geq 1\}$ for which conditions (10.36) *and* (10.37) *are satisfied. If*

$$\sum_{\mathbf{n} \in \mathbf{N}^d} \left(\frac{w^2(\mathbf{n})}{\chi^2(W(\mathbf{n}))}\right)^{1+\frac{\delta}{2}} \int\limits_{|x| < \sqrt{|\mathbf{n}|}} |x|^{2+\delta} \, dF(x) < \infty$$

for some $\delta \in (0, 1]$, then the law of the iterated logarithm (10.42)–(10.43) *prevails.*

Remark 10.7 If $d = 1$, then condition (10.78) means the existence of the second moment. However, condition (10.78) is stronger than Wichura's condition (10.46) if $d > 1$.

Corollary 10.2 *Let $X, \{X(\mathbf{n}), \mathbf{n} \in \mathbf{N}^d\}$ be independent identically distributed random variables with the distribution function F. Assume that both conditions* (10.4) *and* (10.79) *hold.*

Further, let there exist $d \geq 1$ sequences of real numbers $\{W_j^{(1)}, j \geq 1\}, \dots, \{W_j^{(d)}, j \geq 1\}$ for which conditions (10.36) *and* (10.37) *are satisfied.*

If

$$\sum_{\mathbf{n} \in \mathbf{N}^d} \left(\frac{w^2(\mathbf{n})}{\chi^2(W(\mathbf{n}))}\right)^{1+\frac{\delta}{2}} \int\limits_{|x| < \sqrt{|\mathbf{n}| \log^+ \log^+ |\mathbf{n}|}} |x|^{2+\delta} \, dF(x) < \infty$$

for some $\delta \in (0, 1]$, *then the law of the iterated logarithm* (10.42)–(10.43) *prevails.*

Remark 10.8 Condition (10.79) coincides with Wichura's condition (10.46) if $d > 1$. Condition (10.79) is even weaker than the existence of the second moment if $d = 1$.

The assumptions of Corollaries 10.1 and 10.2 can be simplified if the moment of order $2 + \delta$ exists.

Corollary 10.3 *Let* $\{X(\mathbf{n}), \mathbf{n} \in \mathbf{N}^d\}$ *be independent identically distributed random variables with the distribution function* $F(x)$. *Assume that condition* (10.4) *holds and*

$$\mathsf{E}\left[|X|^{2+\delta}\right] < \infty \tag{10.80}$$

for some $\delta \in (0, 1]$.

Further, let there exist $d \geq 1$ *sequences of real numbers* $\{W_j^{(1)}, j \geq 1\}, \dots, \{W_j^{(d)}, j \geq 1\}$ *for which conditions* (10.36) *and* (10.37) *are satisfied. Then the law of the iterated logarithm* (10.42) –(10.43) *prevails if*

$$\sum_{\mathbf{n}\in\mathbf{N}^d} \left(\frac{w^2(\mathbf{n})}{\chi^2(W(\mathbf{n}))}\right)^{1+\frac{\delta}{2}} < \infty. \tag{10.81}$$

Remark 10.9 Condition (10.80) is more restrictive than the existence of the second moment as well as Wichura's condition (10.46). At the same time, the restrictions on weights $\{w(\mathbf{n})\}$ become simpler under assumption (10.80) (see condition (10.81)).

Below are some examples of weights $\{w(\mathbf{n}), \mathbf{n} \in \mathbf{N}^d\}$ for which the checking of the above conditions is comparatively easy.

10.4.1 The Chow–Teicher Condition

One of the conditions on the weights in the law of the iterated logarithm used by several authors in the case of cumulative sums (that is, in the case $d = 1$) of independent identically distributed random variables with weights is written below

$$w^2(\mathbf{n}) = O\left(\frac{W(\mathbf{n})}{|\mathbf{n}|}\right). \tag{10.82}$$

We refer to (10.82) as the *Chow–Teicher condition.*

Theorem 10.8 implies the law of the iterated logarithm under an even more general condition:

$$w^2(\mathbf{n}) = O\left(\frac{W(\mathbf{n}) \log\log W(\mathbf{n})}{|\mathbf{n}|}\right). \tag{10.83}$$

We refer to (10.83) as the *extended Chow–Teicher condition*.

The corresponding result is stated below.

Corollary 10.4 *Let X and $\{X(\mathbf{n}), \mathbf{n} \in \mathbf{N}^d\}$ be independent identically distributed random variables with the distribution function $F(x)$. Assume that conditions* (10.4) *and* (10.46) *hold and that a field $\{w(\mathbf{n}), \mathbf{n} \in \mathbf{N}^d\}$ of real numbers satisfies the extended Chow–Teicher condition* (10.83).

Further assume that there are $d \geq 1$ sequences of real numbers $\{W_j^{(1)}, j \geq 1\}, \dots, \{W_j^{(d)}, j \geq 1\}$ for which conditions (10.36) *and* (10.37) *are satisfied.*

Then the law of the iterated logarithm (10.42) –(10.43) *prevails.*

Proof of Corollary 10.4 Put $\varphi(x) = x^{1/2} \log^+ \log^+ x$. Considering condition (10.83) we have

$$\sum_{\mathbf{n}\in\mathbf{N}^d} \left(\frac{w^2(\mathbf{n})}{\chi^2(W(\mathbf{n}))}\right)^{3/2} \int\limits_{|x|<\varphi(|\mathbf{n}|)} |x|^3 \, dF(x)$$

$$\leq \text{const} \sum_{\mathbf{n}\in\mathbf{N}^d} \frac{1}{|\mathbf{n}|^{3/2}} \int\limits_{|x|<\varphi(|\mathbf{n}|)} |x|^3 \, dF(x)$$

$$= \text{const} \sum_{k=1}^{\infty} \frac{\tau_k}{k^{3/2}} \int\limits_{|x|<\varphi(k)} |x|^3 \, dF(x),$$

where τ_k is the number of solutions of the equation $|\mathbf{n}| = k$ (see Section A.10). The preceding bound can be continued as follows

$$\sum_{\mathbf{n}\in\mathbf{N}^d} \left(\frac{w^2(\mathbf{n})}{\chi^2(W(\mathbf{n}))}\right)^{3/2} \int\limits_{|x|<\varphi(|\mathbf{n}|)} |x|^3 \, dF(x)$$

$$\leq \text{const} \sum_{k=1}^{\infty} \frac{\tau_k}{k^{3/2}} \sum_{j=1}^{k} \int\limits_{\varphi(j-1)\leq|x|<\varphi(j)} |x|^3 \, dF(x)$$

$$\leq \text{const} \sum_{j=1}^{\infty} \int\limits_{\varphi(j-1)\leq|x|<\varphi(j)} |x|^3 \, dF(x) \sum_{k=j}^{\infty} \frac{\tau_k}{k^{3/2}}.$$

Then Lemma 10.3 with $g(x) = x^{3/2}$ implies the following bound for the tail series

$$\sum_{k=j}^{\infty} \frac{\tau_k}{k^{3/2}} \leq C \frac{\left(\log^+ j\right)^{d-1}}{j^{1/2}},$$

whence

$$\sum_{\mathbf{n}\in\mathbf{N}^d}\left(\frac{w^2(\mathbf{n})}{\chi^2(W(\mathbf{n}))}\right)^{3/2}\int\limits_{|x|<\varphi(|\mathbf{n}|)}|x|^3\,dF(x)$$

$$\le \text{const}\sum_{j=1}^{\infty}\frac{\left(\log^+ j\right)^{d-1}}{j^{1/2}}\int\limits_{\varphi(j-1)\le|x|<\varphi(j)}|x|^3\,dF(x)$$

$$\le \text{const}\,\mathsf{E}\left[X^2\frac{\left(\log^+|X|\right)^{d-1}}{\log^+\log^+|X|}\right]<\infty,$$

since

$$\varphi^{-1}(x)\asymp\left(\frac{x}{\log^+\log^+ x}\right)^2.$$

Note that condition (10.40) for such a function φ is equivalent to condition (10.46). Thus Corollary 10.4 follows from Theorem 10.8 with $\delta=1$. □

10.4.2 The Martikainen Condition

Another condition imposed on weights $\{w(\mathbf{n})\}$ is given by

$$w^2(\mathbf{n})=O\left(\frac{W(\mathbf{n})\log^+|\log^+W(\mathbf{n})|}{|\mathbf{n}|\log^+\log^+|\mathbf{n}|}\right). \tag{10.84}$$

We refer to (10.84) as the *Martikainen condition*, since Martikainen used it in the case $d=1$. It is clear that condition (10.84) is less restrictive than the Chow–Teicher condition (10.82). At the same time, condition (10.84) is more restrictive than the extended Chow–Teicher condition (10.83). Thus Corollary 10.4 remains true even if condition (10.84) substitutes condition (10.83).

Condition (10.84) holds, in particular, for weights $w(\mathbf{n})=n_1^{r_1}\dots n_d^{r_d}$, where $r_1,\dots,r_d$ are real numbers greater than $-\frac{1}{2}$.

Corollary 10.5 *Let X and $\{X(\mathbf{n}),\mathbf{n}\in\mathbf{N}^d\}$ be independent identically distributed random variables with the distribution function $F(x)$. Assume that conditions* (10.4) *and* (10.46) *hold. Let*

$$\min\{r_1,\dots,r_d\}>-\frac{1}{2}.$$

Put

$$\zeta(\mathbf{n}) = \sum_{\mathbf{k}\le\mathbf{n}} k_1^{r_1}\dots k_d^{r_d} X(\mathbf{k}).$$

Then

$$\limsup_{\max} \frac{\zeta(\mathbf{n})}{\sqrt{2n_1^{2r_1+1}\dots n_d^{r_d+1}\log^+\log^+|\mathbf{n}|}} = \sqrt{\frac{d}{(2r_1+1)\dots(2r_d+1)}} \quad a.\ s. \tag{10.85}$$

Proof of Corollary 10.5 All the assumptions of Corollary 10.4 are satisfied for the field $w(\mathbf{n}) = n_1^{r_1}\dots n_d^{r_d}$ and thus the law of the iterated logarithm (10.42) –(10.43) holds. Since

$$\lim(\min) \frac{1}{n_1^{2r_1+1}\dots n_d^{2r_d+1}} \sum_{k_1\le n_1}\cdots\sum_{k_d\le n_d} k_1^{2r_1}\dots k_d^{2r_d} = \frac{1}{(2r_1+1)\dots(2r_d+1)},$$

relation (10.85) follows from (10.42) –(10.43). □

If $d > 1$ and $r_1 = \dots = r_d$, then Corollary 10.5 can be viewed as an assertion on the necessary and sufficient conditions for the law of the iterated logarithm. The same result for $d = 1$ can also be obtained but the proof is different.

Corollary 10.6 *Let X and $\{X(\mathbf{n}), \mathbf{n} \in \mathbf{N}^d\}$ be independent identically distributed random variables with the distribution function $F(x)$. Assume that $r > -1/2$ and put*

$$\zeta(\mathbf{n}) = \sum_{\mathbf{k}\le\mathbf{n}} |\mathbf{k}|^r X(\mathbf{k}).$$

Then the law of the iterated logarithm

$$\limsup_{\max} \frac{\zeta(\mathbf{n})}{\sqrt{2|\mathbf{n}|^{2r+1}\log^+\log^+|\mathbf{n}|}} = \sqrt{\frac{d}{(2r+1)^d}} \quad a.\ s. \tag{10.86}$$

is equivalent to the set of two conditions (10.4) *and* (10.46).

Proof of Corollary 10.6 The implication (10.4), (10.46) $\Longrightarrow$ (10.86) is true in view of Corollary 10.5. It remains to prove that (10.86) $\Longrightarrow$ (10.4), (10.46). We restrict our attention to the case $d > 1$. The law of the iterated logarithm (10.86) implies that

$$\limsup_{\max} \frac{|\Delta\zeta(\mathbf{n})|}{\sqrt{|\mathbf{n}|^{2r+1}\log^+\log^+|\mathbf{n}|}} < \infty \quad \text{a. s.}$$

Since $\Delta\zeta(\mathbf{n}) = |\mathbf{n}|^r X(\mathbf{n})$, the preceding inequality is equivalent to

$$\limsup_{\max} \frac{|X(\mathbf{n})|}{\sqrt{|\mathbf{n}| \log^+ \log^+ |\mathbf{n}|}} < \infty \qquad \text{a. s.}$$

Recalling the 0–1 law, the latter relation means that there exists a non-random constant C such that

$$\mathsf{P}\left(|X(\mathbf{n})| > C\sqrt{|\mathbf{n}| \log^+ \log^+ |\mathbf{n}|} \text{ i. o.}\right) = 0.$$

Now we conclude that

$$\sum_{\mathbf{n}\in\mathbf{N}^d} \mathsf{P}\left(|X(\mathbf{n})| > C\sqrt{|\mathbf{n}| \log^+ \log^+ |\mathbf{n}|}\right) < \infty$$

by the Borel–Cantelli lemma for independent random events (see Proposition A.30). The convergence of the latter series is equivalent to condition (10.46) in view of Lemma 10.2. In particular, $\mathsf{E}\left[X^2\right] < \infty$ and, of course, $\mathsf{E}\left[|X|\right] < \infty$. Assuming that $\mathsf{E}[X] \neq 0$, we use Corollary 10.5 for the random variables $Y(\mathbf{n}) = X(\mathbf{n}) - \mathsf{E}[X]$ and prove that

$$\limsup_{\max} \mathsf{E}[X] \sqrt{\frac{|\mathbf{n}|}{\log^+ \log^+ |\mathbf{n}|}} < \infty$$

which is possible only if $\mathsf{E}[X] = 0$. An analogous reasoning leads to a conclusion that $\mathsf{E}\left[X^2\right] = 1$. The proof of Corollary 10.6 is complete. □

One of the simplest cases of Corollary 10.6 corresponds to $r = 1$.

Corollary 10.7 *Let X and $\{X(\mathbf{n}), \mathbf{n} \in \mathbf{N}^d\}$ be independent identically distributed random variables with the distribution function $F(x)$. Put*

$$\zeta(\mathbf{n}) = \sum_{\mathbf{k}\le\mathbf{n}} |\mathbf{k}| X(\mathbf{k}).$$

Then the law of the iterated logarithm

$$\limsup_{\max} \frac{\zeta(\mathbf{n})}{\sqrt{2|\mathbf{n}| \log^+ \log^+ |\mathbf{n}|}} = \sqrt{\frac{d}{3^d}} \qquad \textit{a. s.} \tag{10.87}$$

is equivalent to the set of two conditions (10.4) *and* (10.46).

Remark 10.10 There are many different sufficient conditions (imposed on the field $\{w(\mathbf{n})\}$) for the law of the iterated logarithm for sums of weighted terms in the case of $d = 1$. The Chow–Teicher condition (10.82) and Martikainen condition (10.84) are among them. Theorem 10.8 allows us to prove the law of the iterated logarithm under less restrictive conditions even in the case $d = 1$. A condition of this kind is used in Corollary 10.4. Another possibility is to use the following restriction

$$b_n^2 = O\left(\frac{B_n \log^+ \log^+ B_n}{n^\gamma}\right). \tag{10.88}$$

Corollary 10.8 *Let $\{X_n, n \geq 1\}$ be independent identically distributed random variables, $\mathsf{E}[X_1] = 0$, and $\mathsf{E}\left[X_1^2\right] = 1$. Let $\{b_n, n \geq 1\}$ be a sequence of real numbers and let $B_n = \sum_{k=1}^n b_k^2$. Assume that $\frac{2}{3} < \gamma \leq 1$ and $\mathsf{E}\left[|X_1|^{2/\gamma}\right] < \infty$. If condition* (10.88) *is satisfied, then the law of the iterated logarithm* (10.89) *prevails.*

Proof of Corollary 10.8 We apply Theorem 10.8 with $\varphi(n) = n^{\gamma/2}$. In this case, condition (10.40) is equivalent to the existence of the moment $\mathsf{E}\left[|X_1|^{2/\gamma}\right]$, while condition (10.41) follows from $\mathsf{E}\left[|X_1|^{2/\gamma}\right] < \infty$. Thus Corollary 10.8 follows directly from Theorem 10.8. □

A similar reasoning concerning conditions for the law of the iterated logarithm leads to an analog of (10.88) for a wider class of functions ψ_1 and ψ_2, namely

$$b_n^2 = O\left(\frac{\psi_1(B_n)}{\psi_2(n)}\right).$$

10.4.3 A Relationship Between Kolmogorov's and Egorov's Conditions

We restrict our attention to the case $d = 1$ in this section.

Let $\{X_n, n \geq 1\}$ be a sequence of independent identically distributed random variables and let $\{b_n, n \geq 1\}$ be a sequence of real numbers. Put $S_n = \sum_{k=1}^n b_k X_k$, $B_n = b_1^2 + \cdots + b_n^2$, and $\chi_n = \chi(B_n)$. If $\operatorname{ess\,sup}|X_1| < \infty$ almost surely, then Theorem 10.2 implies the law of the iterated logarithm

$$\liminf_{n\to\infty} \frac{S_n}{\chi_n} = -1, \qquad \limsup_{n\to\infty} \frac{S_n}{\chi_n} = 1 \qquad \text{a. s.} \tag{10.89}$$

if

$$b_n^2 = o\left(\frac{B_n}{\log^+ \log^+ B_n}\right). \tag{10.90}$$

Moreover, Egorov's theorem [83] implies that if Kolmogorov's condition (10.90) is changed to

$$b_n^2 = O\left(\frac{B_n}{\log^+ \log^+ B_n}\right), \tag{10.91}$$

then one can prove that

$$\limsup_{n\to\infty} \frac{|S_n|}{\chi_n} < \infty \qquad \text{a. s.} \tag{10.92}$$

rather than (10.89). Kolmogorov's and Egorov's theorems hold for non-identically distributed random variables with weights.

On the other hand, if the terms are of a specific structure, then similar results can be proved for more general conditions. One possible result in this direction is Corollary 10.3 which reads as follows in the case $d = 1$.

Corollary 10.9 *Let $\{X_n, n \geq 1\}$ be independent identically distributed random variables, $\mathsf{E}[X_1] = 0$, and $\mathsf{E}\left[X_1^2\right] = 1$. Let $0 < \delta \leq 1$ and $\mathsf{E}\left[|X_1|^{2+\delta}\right] < \infty$. Let $\{b_n, n \geq 1\}$ be a sequence of real numbers, $B_n = \sum_{k=1}^n b_k^2$, and moreover $B_n \to \infty$ and $B_{n+1}/B_n \to 1$ as $n \to \infty$. If*

$$\sum_{n=1}^{\infty} \left(\frac{b_n^2}{B_n \log^+ \log^+ B_n} \right)^{1+\frac{\delta}{2}} < \infty, \tag{10.93}$$

then the law of the iterated logarithm (10.89) *prevails.*

The condition that the terms $w_n X_n$ are bounded is changed to $\mathsf{E}\left[|X_1|^{2+\delta}\right] < \infty$ and (10.93) in Corollary 10.9. The existence of the moment $2 + \delta$ is a much weaker restriction than the boundedness. Moreover, condition (10.93) allows us to consider unbounded weights, for example $b_k^2 = k^r$, $r > 0$.

Remark 10.11 Condition (10.93) holds, for example, if the sequence $\{b_n^2\}$ is bounded and $B_n \to \infty$. In this case, the series in (10.93) is bounded from above by the series

$$\text{const} \sum_{n=1}^{\infty} \frac{b_n^2}{B_n^{1+\frac{\delta}{2}}}$$

that converges in view of the Dini lemma (see, [107], Chapter XI, Sect. 375, Proposition 4 or [247], p. 290, Proposition 173). An analogous proof allows us to change the condition on the boundedness of the sequence $\{b_n^2\}$ to

$$b_n^2 = O\left(\log^+ \log^+ B_n\right)^{1+\gamma} \qquad \text{for some} \quad \gamma > 0.$$

An even more general restriction under which condition (10.93) holds is

$$b_n^2 = O\left(\frac{B_n}{(\log B_n)^{2/\delta} \log^+ \log^+ B_n} \right). \tag{10.94}$$

Egorov's condition (10.91) holds for weights $b_k^2 = k^r$. Let us show that there are sequences $\{b_n, n \ge 1\}$ satisfying condition (10.93) but for which Egorov's condition (10.91) does not hold (of course, Kolmogorov's condition (10.90) also does not hold in this case).

Example 10.1 Let $\delta > 0$ and $\gamma_k = \sqrt{k}$. Put $b_i^2 = i^{-1}$ for $1 \le i \le n_1$, where $n_1 \stackrel{\text{def}}{=} \min\{n : \log^+ \log^+ B_n \ge 1\}$. Then we put $b_{n_1+1}^2 \stackrel{\text{def}}{=} B_{n_1} / \log^+ \log^+ B_{n_1}$. Define the numbers n_k, $k > 1$, and b_i^2, $i > n_1 + 1$, recursively. If the numbers n_{k-1} and b_i^2, $i \le n_{k-1} + 1$, are defined for some $k > 1$, then we put

$$b_i^2 = \frac{1}{i}, \qquad n_{k-1} + 1 < i \le n_k,$$

where $n_k \stackrel{\text{def}}{=} \min\{n > n_{k-1} : \log^+ \log^+ B_n \ge k\}$. Moreover, we set

$$b_{n_k+1}^2 \stackrel{\text{def}}{=} \frac{\gamma_k B_{n_k}}{\log^+ \log^+ B_{n_k}}.$$

Let $\mathcal{N} \stackrel{\text{def}}{=} \{n_k, k \ge 1\}$. It is clear that $B_n \to \infty$, $n \to \infty$. Moreover, $B_{n+1}/B_n \to 1$, $n \to \infty$. The latter property follows from

$$\frac{b_{n+1}^2}{B_n} = \begin{cases} \frac{1}{(n+1)B_n}, & \text{if } n \notin \mathcal{N}, \\ \frac{\gamma_k}{\log^+ \log^+ B_{n_k}} \le \frac{1}{\sqrt{k}}, & \text{if } n \in \mathcal{N}, n = n_k, \end{cases}$$

since $\log^+ \log^+ B_{n_k} \ge k$. Similarly,

$$\frac{b_n^2}{B_n \log^+ \log^+ B_n} \le \begin{cases} \frac{1}{n}, & \text{if } n-1 \notin \mathcal{N}, \\ \frac{\gamma_k}{\left(\log^+ \log^+ B_{n_k}\right)^2} \le \frac{1}{k^{3/2}}, & \text{if } n-1 \in \mathcal{N}, n = n_k + 1, \end{cases}$$

whence (10.93) follows. At the same time, Egorov's condition (10.91) does not hold, since $B_{n+1}/B_n \to 1$ and hence

$$b_{n_k+1}^2 \cdot \frac{\log^+ \log^+ B_{n_k+1}}{B_{n_k+1}} = \gamma_k \cdot \frac{B_{n_k} \log^+ \log^+ B_{n_k+1}}{B_{n_k+1} \log^+ \log^+ B_{n_k}} \to \infty, \qquad k \to \infty.$$

10.4.4 A Relationship Between Petrov's Condition and the Law of the Iterated Logarithm

A relationship between the law of the iterated logarithm and the rate of convergence in the central limit theorem for cumulative sums of independent random variables is studied by Petrov in [336]. Petrov's condition written for the case of weighted identically distributed terms reads as follows

$$\sum_{n=1}^{\infty} \frac{b_n^2}{B_n} \int\limits_{|xb_n| \geq \varepsilon (B_n/\log\log B_n)^{1/2}} x^2 dF(x) < \infty \tag{10.95}$$

for all $\varepsilon > 0$ (we use the notation introduced in Sect. 10.4.3). Here the symbol F stands for the distribution function of the random variable X_1.

An advantage of Petrov's result as compared to Theorem 10.2 is that it holds not only for bounded random variables. If $b_n^2 = 1$ for all $n \geq 1$, then Petrov's result concerns the sums of independent identically distributed random variables, but condition (10.95) is more restrictive in this case than the Hartman–Wintner condition (Theorem 10.3). Nevertheless, condition (10.95) is close to the conditions used in the corollaries above.

Example 10.2 We show that condition (10.95) follows from (10.93) if (10.94) holds and if $\mathsf{E}\left[|X|^{2+\delta}\right] < \infty$ for some $\delta > 0$. Note that condition (10.94) is "almost" necessary for (10.95).

Indeed, the series on the left-hand side of (10.95) can be estimated from above as follows

$$\sum_{n=1}^{\infty} \int\limits_{\log B_n \leq \text{const}\, |x|^\delta < \log B_{n+1}} x^2\, dF(x) \sum_{k=1}^{n} \frac{b_k^2}{B_k}.$$

If a sequence $\{n_k, k \geq 1\}$ is chosen such that $n_0 = 0$, $n_k = \max\{n : B_n \leq 2^k\}$, then

$$\sum_{k=1}^{n} \frac{b_k^2}{B_k} \leq \sum_{k=1}^{n_i} \frac{b_k^2}{B_k} = \sum_{j=0}^{i-1} \sum_{n_j < k \leq n_{j+1}} \frac{b_j^2}{B_j} \leq \sum_{j=0}^{i-1} 2^{-j} 2^{j+1} \leq \text{const} \log B_n$$

for $n_{i-1} < n \leq n_i$. Thus the left-hand side of (10.95) does not exceed

$$\text{const} \sum_{n=1}^{\infty} \log B_n \int\limits_{\log B_n \leq \text{const}\, |x|^\delta < \log B_{n+1}} x^2\, dF(x) \leq \text{const}\, \mathsf{E}\left[|X_1|^{2+\delta}\right].$$

Therefore condition (10.94) is sufficient for the law of the iterated logarithm by Petrov's theorem if $\mathsf{E}\left[|X_1|^{2+\delta}\right] < \infty$.

10.5 Chover's Law of the Iterated Logarithm for Multi-Indexed Sums

The rate of increase of multi-indexed sums of independent random variables can be found even if the second moment is infinite. One of the cases that can be treated is related to stable terms, however the result itself is stated in a different form as compared to (10.45).

The result of Theorem 10.11 stated below is called *Chover's law of the iterated logarithm.*

Theorem 10.11 *Let independent random variables $\{X(\mathbf{n})\}$ have an identical stable symmetric distribution function with parameter $\alpha \in (0, 2)$. Then*

$$\operatorname{limsup}(\max) \left| \frac{S(\mathbf{n})}{|\mathbf{n}|^{1/\alpha}} \right|^{1/\log\log|\mathbf{n}|} = e^{d/\alpha} \quad a.\ s. \tag{10.96}$$

Remark 10.12 The Gaussian distribution function also is stable and corresponds to the parameter $\alpha = 2$. Note, however, that Gaussian random variables have finite second moment and thus Chover's law of the iterated logarithm (10.96) does not hold in the Gaussian case in view of Theorem 10.9.

Proof of Theorem 10.11 To prove relation (10.96) one needs to show that

$$\mathsf{P}(|S(\mathbf{n})| \geq |\mathbf{n}|^{1/\alpha} (\log|\mathbf{n}|)^{(1+\varepsilon)d/\alpha} \text{ i. o.}) = 0, \tag{10.97}$$

$$\mathsf{P}(|S(\mathbf{n})| > |\mathbf{n}|^{1/\alpha} (\log|\mathbf{n}|)^{(1-\varepsilon)d/\alpha} \text{ i. o.}) = 1 \tag{10.98}$$

for all $0 < \varepsilon < 1$.

Let $\theta > 1$ and $\mathbf{n}(\mathbf{k}) = \left(\left[\theta^{k_1}\right], \dots, \left[\theta^{k_d}\right]\right)$. For $|\mathbf{n}| = n_1 \dots n_d$, let

$$U_\varepsilon(\mathbf{k}) = |\mathbf{n}(\mathbf{k})|^{1/\alpha} (\log|\mathbf{n}(\mathbf{k})|)^{(1+\varepsilon)d/\alpha} \tag{10.99}$$

and

$$A_1(\mathbf{k}) = \{\omega : |S(\mathbf{n}(\mathbf{k}))| \geq U_\varepsilon(\mathbf{k})\},$$
$$A_2(\mathbf{k}) = \{\omega : |S(\mathbf{n}(\mathbf{k}+\mathbf{1}))| \geq U_\varepsilon(\mathbf{k})\},$$
$$B(\mathbf{k}) = \left\{\omega : \max_{\mathbf{n}(\mathbf{k}) \leq \mathbf{n} \leq \mathbf{n}(\mathbf{k}+\mathbf{1})} |S(\mathbf{n})| \geq U_\varepsilon(\mathbf{k})\right\}.$$

Denote the distribution function of a symmetric stable law with parameter α by F_α and let ξ be a stable random variable with the distribution function F_α. Then

$$x^\alpha (F_\alpha(-x) + 1 - F_\alpha(x)) \to \varkappa_\alpha, \qquad x \to \infty, \tag{10.100}$$

for some constant $0 < \varkappa_\alpha < \infty$. Note that the latter relation holds even in the Gaussian case but with $\varkappa_2 = 0$. Relation (10.100) is first mentioned in Lévy's book [283] (see also [126], §36, or [104], formula (5.13)).

Using the Lévy inequality for multi-indexed sums of symmetric random variables (inequality (2.27) in Corollary 2.4) we get $\mathsf{P}(B(\mathbf{k})) \leq 2^d \mathsf{P}(A_2(\mathbf{k}))$. The random variable $S(\mathbf{n}(\mathbf{k}+\mathbf{1}))/|\mathbf{n}(\mathbf{k}+\mathbf{1})|^{1/\alpha}$ is stable with parameter α, hence

$$\mathsf{P}(B(\mathbf{k})) \leq 2^d \mathsf{P}(|\xi| \geq x(\mathbf{k})), \qquad x(\mathbf{k}) \stackrel{\text{def}}{=} \frac{U_\varepsilon(\mathbf{k})}{|\mathbf{n}(\mathbf{k}+\mathbf{1})|^{1/\alpha}}. \tag{10.101}$$

Since

$$\frac{|\mathbf{n}(\mathbf{k})|}{|\mathbf{n}(\mathbf{k}+\mathbf{1})|} \geq \left(\frac{\theta-1}{\theta^2}\right)^d \stackrel{\text{def}}{=} a, \qquad \frac{|\mathbf{n}(\mathbf{k})|}{\theta^{k_1+\cdots+k_d}} \geq \left(\frac{\theta-1}{\theta}\right)^d \stackrel{\text{def}}{=} e^b$$

for all $\mathbf{k} \in \mathbf{N}^d$, we get

$$x(\mathbf{k}) \geq a^{1/\alpha}(b + c(k_1 + \cdots + k_d))^{(1+\varepsilon)d/\alpha} \tag{10.102}$$

with $c \stackrel{\text{def}}{=} \log\theta$ and $\varepsilon > 0$. The asymptotic result (10.100) implies that

$$\sup_{\mathbf{k}\in\mathbf{N}^d} \mathsf{P}(|\xi| \geq x(\mathbf{k}))(b + c(k_1 + \cdots + k_d))^{(1+\varepsilon)d} < \infty.$$

Thus relations (10.26), (10.101), and (10.102) imply the convergence of the series $\sum \mathsf{P}(B(\mathbf{k}))$. By the Borel–Cantelli lemma this means that $\mathsf{P}(B(\mathbf{k}) \text{ i. o.}) = 0$, which proves (10.97), since $\mathsf{P}(A_1(\mathbf{k}) \text{ i. o.}) \leq \mathsf{P}(B(\mathbf{k}) \text{ i. o.})$.

Equality (10.98) will be proved by contradiction. We assume that, for some $0 < \varepsilon < 1$ and some random event Ω_1 such that $\mathsf{P}(\Omega_1) > 0$, we have

$$|S(\mathbf{n};\omega)| \leq |\mathbf{n}|^{1/\alpha} (\log|\mathbf{n}|)^{(1-\varepsilon)d/\alpha} \tag{10.103}$$

for all elementary random events $\omega \in \Omega_1$ and for all multi-indices $\mathbf{n} \in \mathbf{N}^d$ except a finite number of them.

Fix $\theta > 1$ and let $\mathbf{n}(\mathbf{k}) = \left(\left[\theta^{k_1}\right], \ldots, \left[\theta^{k_d}\right]\right)$. We use the numbers $U_\varepsilon(\mathbf{k})$ defined by (10.99) and let

$$\mathfrak{B}^\circ(\mathbf{k}) = \left\{\mathbf{m} : \left[\theta^{k_1-1}\right] < m_1 \leq \left[\theta^{k_1}\right], \ldots, \left[\theta^{k_d-1}\right] < m_d \leq \left[\theta^{k_d}\right]\right\}, \qquad \mathbf{k} \geq \mathbf{0},$$

$$S^\circ(\mathbf{k}) = \sum_{\mathbf{m}\in\mathfrak{B}^\circ(\mathbf{k})} X(\mathbf{m}).$$

Finally, let ξ be a random variable with the symmetric stable distribution function with parameter α. The random variable $S^\circ(\mathbf{k})/\left(\operatorname{card}(\mathfrak{B}^\circ(\mathbf{k}))\right)^{1/\alpha}$ has the same distribution function if $\operatorname{card}(\mathfrak{B}^\circ(\mathbf{k})) > 0$, whence we conclude

$$\mathsf{P}(|S^\circ(\mathbf{k})| \geq U_{-\varepsilon/2}(\mathbf{k})) = \mathsf{P}(|\xi| \geq y(\mathbf{k})), \qquad y(\mathbf{k}) \stackrel{\text{def}}{=} \frac{U_{-\varepsilon/2}(\mathbf{k})}{(\operatorname{card}(\mathfrak{B}^\circ(\mathbf{k})))^{1/\alpha}}.$$

In what follows we choose the number θ such that $\theta^2 > 1 + \theta$. Since

$$\inf_{\mathbf{k}} y(\mathbf{k}) \geq a\,(1 + k_1 + \cdots + k_d)^{(1-\varepsilon/2)d/\alpha}$$

for some positive constant a, we establish

$$\mathsf{P}(|S^\circ(\mathbf{k})| \ge U_{-\varepsilon/2}(\mathbf{k})) \le \mathsf{P}\left(|\xi| \ge a\,(1+k_1+\cdots+k_d)^{(1-\varepsilon/2)d/\alpha}\right).$$

Taking into account (10.100) and (10.26), we prove that

$$\sum \mathsf{P}(|S^\circ(\mathbf{k})| \ge U_{-\varepsilon/2}(\mathbf{k})) = \infty.$$

Since the random variables $S^\circ(\mathbf{k})$ are jointly independent, the latter result together with the Borel–Cantelli lemma implies that

$$\mathsf{P}(\Omega_2) = 1, \qquad \Omega_2 \stackrel{\text{def}}{=} \left\{|S^\circ(\mathbf{k})| \ge U_{-\varepsilon/2}(\mathbf{k}) \text{ i. o.}\right\}.$$

Every sum $S(\mathbf{n}(\mathbf{k}))$ can be represented in the form of a linear combination of $S^\circ(\mathbf{k})$ and at most $2^d - 1$ sums $S(\mathbf{n}(\mathbf{m}))$, $\mathbf{m} < \mathbf{k}$. Thus assumption (10.103) implies that

$$\begin{aligned}\left|\frac{S(\mathbf{n}(\mathbf{k});\,\omega)}{|\mathbf{n}(\mathbf{k})|^{1/\alpha}}\right| &\ge (\log|\mathbf{n}(\mathbf{k})|)^{(1-\varepsilon/2)d/\alpha} - (2^d-1)\,(\log|\mathbf{n}(\mathbf{k})|)^{(1-\varepsilon)d/\alpha}\\ &> (\log|\mathbf{n}(\mathbf{k})|)^{(1-\varepsilon)d/\alpha}\end{aligned}$$

for $\omega \in \Omega_1 \cap \Omega_2$ and for multi-indices $\mathbf{n}(\mathbf{k})$ such that $|S^\circ(\mathbf{k};\,\omega)| \ge U_{-\varepsilon/2}(\mathbf{k})$ if $|\mathbf{n}(\mathbf{k})|$ is sufficiently large. The latter result contradicts assumption (10.103). Theorem 10.11 is proved. □

10.6 Comments

Section 10.1 Theorem 10.1 is proved by Khintchine [195]. Theorem 10.3 is proved by Hartman and Wintner [156]. Different proofs of Theorem 10.3 are found by Heyde [160], Strassen [384], de Acosta [2]. Theorem 10.3 is proved by Strassen [385].

An earlier history of investigations related to the law of the iterated logarithm is presented by Feller [103]. The first half-century of existence of the law of the iterated logarithm is surveyed by Stout [383]. A more recent survey is given by Bingham [25] (see also the book by Petrov [338]).

The inverse statement of the law of the iterated logarithm (Theorem 10.4) is proved by Strassen [385]. Other proofs of this result are found by Heyde [159] and Steiger and Zaremba [380]. Rosalsky [355] and Martikainen [306] prove that condition (10.4) follows from any single condition in (10.5) (from the one-sided law of the iterated logarithm).

Exponential bounds (Theorems 10.5 and 10.6) are proved by Kolmogorov [253].

Theorem 10.7 is proved in [215]. A conjecture stated in [215] that one can omit condition (10.14) is disproved by Martikainen [307]. In [307], it is also described how the normalization in the law of the iterated logarithm for multi-indexed sums is changed if condition (10.14) does not hold. Theorem 10.8 for $d = 1$ is proved in [222]. Theorem 10.9 was first proved by Wichura [420] by a method different from that shown above. The law of the iterated logarithm for multi-indexed sums with other normalizations are considered by Chen [47].

Some results related to the law of the iterated logarithm for multi-indexed sums are obtained by Adler [3], Kholmuradov [203], Li and Wu [291], Lin [294]. The so-called bounded law of the iterated logarithm for multi-indexed sums is considered in the paper by Li, Rao, and Wang [288]. Another (the so-called *compact*) law of the iterated logarithm for multi-indexed sums is investigated by Li and Tomkins [290], Zinchenko [434].

An application of the law of the iterated logarithm for multi-indices for uniformly distributed sequences is given by Aisleitner [9]. Many other applications in various areas of mathematics are listed by Bingham [25].

Section 10.2 Theorem 10.2 is proved by Kolmogorov [253]. Kolmogorov's condition (10.2) in Theorem 10.2 has been studied by many authors for $d = 1$. If the symbol o in (10.2) is changed to O, then the examples due to Marcinkiewicz and Zygmund [302] and Weiss [417] show that the inequality in (10.3) may become an inequality (both cases "<" and ">" may happen). Egorov [83] shows that if o in (10.2) is changed to O, then the upper limit in (10.3) is still almost surely finite but not necessarily equal to 1. A relationship between the limit constants in (10.2) (with "O" instead of "o") and (10.3) is investigated by Tomkins [400]. His results are improved by Rozovskiĭ [361, 362].

Section 10.3 Theorem 10.8 is proved in [222] for $d = 1$ and in [235] for $d > 1$. Necessary and sufficient conditions for the law of the iterated logarithm in Theorem 10.9 are due to Hartman and Wintner [156] and Strassen [385] in the case $d = 1$ and Wichura [420] for $d > 1$.

Section 10.4 The Chow–Teicher condition (10.82) is introduced in [56], while the Martikainen condition (10.84) is studied in [305] (both papers deal with the case $d = 1$).

Section 10.5 The case $d = 1$ in Theorem 10.11 is proved by Chover [55]. It is worth mentioning that Chover himself noted in his paper [55], with reference to a communication by Strassen, that this law of the iterated logarithm can be obtained from some results of Khintchine [202] concerning the upper and lower functions for stochastic processes with independent increments. The case $d > 1$ is considered in [212]. The same result was later rediscovered by Chen and Shan [51]. Closely related results for random fields are obtained by Yamamuro [423], Watanabe and Yamamuro [415], Zinchenko [435, 436]. The rate of convergence in the Chover law of the iterated logarithm for multi-indexed sums is considered by Zinchenko [438].

The so-called law of a single logarithm for random fields is studied by Steinebach [381], Pfuhl and Steinebach [339], Zinchenko [437], Frolov [110, 111], Shcherbakova [365], Gut and Stadtmüller [141, 142, 144], Kabluchko and Munk [187], Lin, Choi, and Hwang [295].

Chapter 11
Renewal Theorems for Random Walks with Multi-Dimensional Time

In this chapter, we consider the problems on the asymptotic behavior of the renewal function and process constructed from multi-indexed sums of independent identically distributed random variables. The asymptotic behavior depends on the dimension d of the space of multi-indices and differs from the classical case. It is worth mentioning that more complicated questions on the rate of convergence of renewal functions and processes constructed from multi-indexed sums of random variables still have no definite answers, since they depend on the Riemann hypothesis.

11.1 Some Renewal Theory for $d = 1$

One of the most useful probabilistic objects in the case $d = 1$ is the renewal process and its important characteristic called the renewal function. A number of applications of renewal processes is discussed in [104], Chap. VI, Sect. 6–8. Many new applications of renewal processes have appeared just recently. For example, the renewal processes is the main ingredient in many models of actuarial mathematics.

11.1.1 An Insurance Model

One of the simplest models of insurance mathematics can be described as follows. Let an insurance company possess an initial capital t used to reimburse losses of insured persons in the case of insurance events. Denote by $X_1, X_2, \dots$ the losses of insured persons. Then $S_n = X_1 + \dots + X_n$ is the total amount paid by the company after n insurance events. The number

$$N(t) = \max\{n \colon S_n < t\} \tag{11.1}$$

O. Klesov, *Limit Theorems for Multi-Indexed Sums of Random Variables*,
Probability Theory and Stochastic Modelling 71, DOI 10.1007/978-3-662-44388-0_11

is the moment when the last insurance event occurs prior the total expenses of the company used to reimburse the losses of insured persons exceed the level t. Therefore $N(t)$ is the moment when the last insurance event occurs prior to the ruin of the company. More realistic models assume that the capital of the company increases when insured persons pay premiums. Even in this case, the process $N(t)$ is of principal importance.

The variables $\{X_n, n \geq 1\}$ are random in probability theory and $N(t)$, as a function of the argument t, is called the *renewal process*. Its expectation $U(t) = \mathsf{E}[N(t)]$ is called the *renewal function*. The main problem in renewal theory deals with the asymptotic behavior of $N(t)$ and $U(t)$ as $t \to \infty$.

11.1.2 Asymptotic Behavior of Renewal Functions and Processes

Below are some results from the classical theory that will be used throughout this chapter.

Theorem 11.1 (elementary renewal theorem) *Let $\{X_n, n \geq 1\}$ be independent identically distributed random variables. If $X_n \geq 0$ almost surely and $0 < \mu \stackrel{\text{def}}{=} \mathsf{E}[X_1] < \infty$, then*

$$\lim_{t\to\infty} \frac{U(t)}{t} = \frac{1}{\mu}. \tag{11.2}$$

Theorem 11.2 (SLLN for renewal processes) *Let $\{X_n, n \geq 1\}$ be independent identically distributed random variables. If $X_n \geq 0$ almost surely and $0 < \mu \stackrel{\text{def}}{=} \mathsf{E}[X_1] < \infty$, then*

$$\lim_{t\to\infty} \frac{N(t)}{t} = \frac{1}{\mu} \quad \text{a. s.} \tag{11.3}$$

Theorem 11.3 (rate of convergence in SLLN for renewal processes) *Let $1 \leq r < 2$ and let $\{X_n, n \geq 1\}$ be independent identically distributed random variables. If $X_n \geq 0$ almost surely, $0 < \mu \stackrel{\text{def}}{=} \mathsf{E}[X_1] < \infty$, and $\mathsf{E}\left[X_1^r\right] < \infty$, then*

$$\lim_{t\to\infty} \frac{N(t) - \frac{t}{\mu}}{t^{1/r}} = 0 \quad \text{a. s.} \tag{11.4}$$

11.2 An Example for $d > 1$

A straightforward attempt to define the renewal process for $d > 1$ via equality (11.1) is hopeless, since the function "max" is not well defined for all its arguments in $\mathbf{N}^d$. This is explained by the fact that the order $\leq$ in the space $\mathbf{N}^d$ is not complete if $d > 1$. An alternative approach is the following.

For simplicity, we discuss the case of the space $\mathbf{N}^2$ (although a similar construction is valid for other spaces $\mathbf{N}^d$). Consider a uniform lattice $\mathcal{L}$ in the space $\mathbf{N}^2$ that consists of the points $\{P(m,n), \in \mathbf{N}^2\}$, where $P(m,n) = (hm, hn)$ and h is a certain positive number. Assume that a "local" load $X(m,n)$ acts at every point $P(m,n)$. Assume that the "cumulative" load $S(P)$ at every point $P = P(m,n)$ is additive, that is, it equals the sum of "local" loads $X(i,j)$ acting at inner points of the rectangle $\Pi(m,n) = \{P(i,j) : i \le m, j \le n\}$, that is

$$S(P(m,n)) \stackrel{\text{def}}{=} S(m,n) = \sum_{i=1}^{m}\sum_{j=1}^{n} X(i,j).$$

If t is the level of "safe" loads, then

$$A = \{P \in \mathcal{L}\colon S(P) < t\}$$

can be viewed as the set of points where the loads are "safe". Similarly, $B = \{P \in \mathcal{L}\colon S(P) \ge t\}$ is the set of points where the loads are "dangerous". The structure and characteristics of the sets with safe and dangerous loads as well as the variation of their configuration with growing t are of obvious interest in practice.

One of the most important characteristics of sets in the plane is their measure. The Lebesgue measure of sets whose structure is similar to A or B can be evaluated in terms of the total number of their inner points that belong to the lattice $\mathcal{L}$. The corresponding result is called the *Pick formula*:

$$\operatorname{meas}(A) = \operatorname{card}(P \in \mathcal{L}\colon P \in \operatorname{int}(A)) + \frac{1}{2}\operatorname{card}(P \in \mathcal{L}\colon P \in \partial(A)) - 1, \quad (11.5)$$

where $\operatorname{int}(A)$ is the set of inner points of the set A that belong to the lattice $\mathcal{L}$ and $\partial(A)$ is the set of boundary points of the set A that belong to the lattice $\mathcal{L}$. Since our consideration is very schematic, we follow our intuition and agree that the number of boundary points is "much" less than the total number of inner points. It is also clear that $P \in A \iff \mathbb{I}_{\{S(P)<t\}} = 1$, where the indicator $\mathbb{I}_{\{\cdot\}}$ equals 1 if the condition written in the braces occurs and equals 0 otherwise. This leads to the conclusion

$$\operatorname{meas}(A) \approx \operatorname{card}(P \in \mathcal{L}\colon P \in A) = \sum_{i=1}^{\infty}\sum_{j=1}^{\infty} \mathbb{I}_{\{S(i,j)<t\}}.$$

If $X(m,n)$ are random variables, then we get by passing to the limit that

$$\mathsf{E}\,[\operatorname{meas}(A)] \approx \sum_{i=1}^{\infty}\sum_{j=1}^{\infty} \mathsf{P}(S(i,j) < t). \quad (11.6)$$

Therefore the renewal process and renewal function for $d = 2$ are approximations to the square and expectation of the square of the set with safe loads.

Similar conclusions are true for other dimensions d, as well. Such a motivation leads to new settings even in the classical case $d = 1$.

The double series on the right-hand side of (11.6) is called the *renewal function constructed from a random walk with two dimensional time*. The reason for this name becomes clear from the case of $d = 1$ if another representation is used in this case for a renewal process $N(t)$ instead of (11.1). Indeed,

$$N(t) = \sum_{n=1}^{\infty} \mathbb{I}_{\{S_n < t\}} \tag{11.7}$$

and thus the renewal function $U(t) \stackrel{\text{def}}{=} \mathsf{E}[N(t)]$ constructed from a random walk $\{S_n, n \geq 1\}$ can be rewritten as follows

$$U(t) = \mathsf{E}[N(t)] = \sum_{n=1}^{\infty} \mathsf{P}(S_n < t). \tag{11.8}$$

Since the field $\{S(m,n), m \geq 1, n \geq 1\}$ is often called the *random walk with two dimensional time*, the right-hand sides of (11.6) and (11.8) are similar and thus the right-hand side of (11.6) can indeed be called the renewal function by analogy with the case $d = 1$.

11.3 Definition and Properties of Renewal Processes and Functions

Let $\{X(\mathbf{n}), \mathbf{n} \in \mathbf{N}^d\}$ be independent identically distributed random variables. Assume that $X(\mathbf{n}) \geq 0$ almost surely and that the distribution function of the random variables $X(\mathbf{n})$ is not concentrated at 0, that is, $\mathsf{P}(X(\mathbf{n}) > 0) > 0$.

Definition 11.1 The stochastic process

$$N(t) = \sum_{\mathbf{n} \in \mathbf{N}^d} \mathbb{I}_{\{S(\mathbf{n}) < t\}} \tag{11.9}$$

is called the *renewal process constructed from a random walk with d-dimensional time*.

Note that the series on the right-hand side of (11.9) converges almost surely, that is, the process $N(t)$ is well defined almost surely for all $t > 0$.

Lemma 11.1 *Let* $\{X(\mathbf{n}), \mathbf{n} \in \mathbf{N}^d\}$ *be independent identically distributed random variables. Assume that* $X(\mathbf{n}) \geq 0$ *almost surely and* $\mathsf{P}(X(\mathbf{n}) > 0) > 0$. *Then* $N(t) < \infty$ *almost surely for all* $t > 0$.

Proof We show that, given a fixed $t > 0$, only a finite number of the inequalities $S(\mathbf{n}) < t$ occurs almost surely. This means that only a finite number of terms of the series on the right-hand side of (11.9) is non-zero almost surely and thus the series converges almost surely. First, note that

$$S(n, 1, \ldots, 1) \to \infty, \quad S(1, n, 1, \ldots, 1) \to \infty, \quad \ldots, \quad S(1, \ldots, 1, n) \to \infty \tag{11.10}$$

almost surely as $n \to \infty$. Indeed, each of the random variables in (11.10) is a sum of n independent identically distributed random variables that are non-negative almost surely and not totally concentrated at 0. If the preceding relation does not hold for some subsequence of partial sums, then the corresponding series converges almost surely by the 0–1 law. However, this is not possible by the Kolmogorov three series theorem.

Fix $t > 0$. Let $\mathbf{n} = (n_1, \ldots, n_d)$. For every random elementary event ω for which relation (11.10) holds, we find positive integers $m_1, \ldots, m_d$ such that

$$\begin{aligned} &S(n_1, 1, \ldots, 1) \geq t, &&\text{for all } n_1 \geq m_1, \\ &S(1, n_2, 1, \ldots, 1) \geq t, &&\text{for all } n_2 \geq m_2, \\ &\ldots\ldots\ldots\ldots\ldots\ldots\ldots\ldots\ldots\ldots\ldots\ldots \\ &S(1, \ldots, 1, n_d) \geq t, &&\text{for all } n_d \geq m_d. \end{aligned}$$

If $|\mathbf{n}| \geq m_1 \ldots m_d$, then $n_i \geq m_i$ for at least one index $1 \leq i \leq d$ and thus

$$S(\mathbf{n}) \geq \max\{S(n_1, 1, \ldots, 1), S(1, n_2, 1, \ldots, 1), \ldots, S(1, \ldots, 1, n_d)\} \geq t.$$

Therefore the series on the right-hand side of (11.9) converges almost surely, since it contains only a finite number of non-zero terms for all $t > 0$. □

Definition 11.2 The expectation of the renewal process, $U(t) = \mathsf{E}[N(t)]$, is called the *renewal function constructed from a random walk with d-dimensional time*.

Note that the renewal function is finite for all positive t and

$$U(t) = \sum_{\mathbf{n} \in \mathbf{N}^d} \mathsf{P}(S(\mathbf{n}) < t). \tag{11.11}$$

Lemma 11.2 *Let* $\{X(\mathbf{n}), \mathbf{n} \in \mathbf{N}^d\}$ *be independent identically distributed random variables. Assume that* $X(\mathbf{n}) \geq 0$ *almost surely and* $\mathsf{P}(X(\mathbf{n}) > 0) > 0$. *Then* $U(t) < \infty$ *for all* $t > 0$ *and equality* (11.11) *holds.*

Proof Indeed, passing to the expectations on both sides of (11.9) and interchanging the summation and integration with respect to the probability measure we prove (11.11). To prove that the renewal function is finite, we rewrite (11.11) as follows

$$U(t) = \sum_{k=1}^{\infty} \tau_k \, \mathsf{P}(S_k < t), \tag{11.12}$$

where $\{\tau_n\}$ is the Dirichlet sequence (see Sect. A.10) and where $\{S_n\}$ is the sequence of sums associated with the field of random variables $\{X(\mathbf{n}), \mathbf{n} \in \mathbf{N}^d\}$. Recall that the associated sequence means that $\{S_n\}$ are the sums of independent identically distributed random variables with the same distribution function as that of any $X(\mathbf{n})$, see Sect. A.17.1.

Since $S_k \to \infty$ almost surely as $k \to \infty$, for each fixed $t > 0$ there exists a number $r \geq 1$ such that $p_r(t) \overset{\text{def}}{=} \mathsf{P}(S_r < t) < 1$. Since $p_{ir}(t) \leq (p_r(t))^i$, the monotonicity of the sequence $\{S_j\}$ implies that

$$U(t) \leq T_{r-1} + \sum_{i=1}^{\infty} \sum_{ri \leq j < r(i+1)} \tau_j \, \mathsf{P}(S_j < t) \leq T_{r-1} + \sum_{i=1}^{\infty} p_r^i(t) T_{r(i+1)},$$

where $T_m = \tau_1 + \cdots + \tau_m$. We apply the rough asymptotic bound in the Dirichlet divisors problem, $T_m \leq \text{const}\ m\,(\log m)^{d-1}$ (see (A.41)). Putting $i_r = i(r+1)$, we finally obtain

$$U(t) \leq \text{const} \left[1 + \sum_{i=1}^{\infty} i_r \, (\log i_r)^{d-1} \, p_r^i(t) \right] < \infty.$$

□

Lemma 11.3 *Let $\{X(\mathbf{n}), \mathbf{n} \in \mathbf{N}^d\}$ be independent identically distributed random variables. Assume that $X(\mathbf{n}) \geq 0$ almost surely and $\mathsf{P}(X(\mathbf{n}) > 0) > 0$. Then the process $N(t)$ increases almost surely, while the function $U(t)$ increases. Moreover, $N(t) \to \infty$ almost surely and $U(t) \to \infty$ as $t \to \infty$.*

Proof The stochastic process $N(t)$ and function $U(t)$ are unbounded if $d = 1$, thus they are unbounded if $d > 1$, too. Indeed,

$$N(t) \geq \max\{n\colon S(n, 1, \dots, 1) < t\}, \qquad U(t) \geq \sum_{k=1}^{\infty} \mathsf{P}(S(k, 1, \dots, 1) < t),$$

whence we conclude that

$$\lim_{t\to\infty} N(t) = +\infty \quad \text{a. s.}, \qquad \lim_{t\to\infty} U(t) = +\infty.$$

□

All the properties of the renewal processes and functions discussed above are collected in the following result.

Theorem 11.4 *Let* $\{X(\mathbf{n}), \mathbf{n} \in \mathbf{N}^d\}$ *be independent identically distributed random variables. Assume that* $X(\mathbf{n}) \geq 0$ *almost surely and* $\mathsf{P}(X(\mathbf{n}) \neq 0) > 0$. *Then*

(1) $N(t) < \infty$ *a. s. for all* $t > 0$;
(2) the function $N(t)$ *increases and moreover* $\lim_{t\to\infty} N(t) = \infty$ *a. s.*;
(3) $U(t) < \infty$ *for all* $t > 0$;
(4) the function $U(t)$ *is increasing and* $\lim_{t\to\infty} U(t) = \infty$;
(5) renewal processes and functions admit the representations (11.11) *and* (11.12).

11.4 Asymptotic Behavior of Renewal Functions Constructed from Multi-Indexed Sums

Below is the counterpart of the elementary renewal theorem for $d > 1$.

Theorem 11.5 *Let* $\{X(\mathbf{n}), \mathbf{n} \in \mathbf{N}^d\}$ *be independent identically distributed random variables. Assume that* $X(\mathbf{n}) \geq 0$ *almost surely.*

If $0 < \mu \overset{\text{def}}{=} \mathsf{E}[X] < \infty$, *then*

$$\lim_{t\to\infty} \frac{U(t)}{t(\log t)^{d-1}} = \frac{1}{\mu} \cdot \frac{1}{(d-1)!}. \tag{11.13}$$

Remark 11.1 It follows from (11.13) that $\frac{U(t)}{t}$ approaches ∞ in the case of $d > 1$ in contrast to the case of $d = 1$ where $\frac{U(t)}{t}$ approaches $\frac{1}{\mu}$.

We recall the main result in the Dirichlet divisors problem (see equality (A.32)) to be used below in the proof of Theorem 11.5, namely there exists a polynomial

$$\mathscr{P}(z) = \varkappa_0 + \varkappa_1 z + \cdots + \varkappa_{d-1} z^{d-1}, \qquad \varkappa_{d-1} = \frac{1}{(d-1)!} \tag{11.14}$$

such that

$$T_k = k\mathscr{P}(\log k) + \rho_k, \qquad |\rho_k| = o\left(k^\lambda\right), \tag{11.15}$$

for some $0 < \lambda < 1$, where $T_k = \tau_1 + \cdots + \tau_k$ and where τ_k is the total number of positive integer solutions of the equation $n_1 \dots n_d = k$.

If higher moments exist, then the behavior of $U(t)$ can be described more precisely. It turns out (see Theorem 11.6) that $\frac{U(t)}{t}$ has an additive logarithmic trend $(\log t)^{d-1}$ if $d > 1$. The precise expression for the trend can be written via the coefficients of the polynomial $\mathscr{P}$ in (11.14). We recall that the trend does not exist if $d = 1$.

Theorem 11.6 *Let $\{X(\mathbf{n}), \mathbf{n} \in \mathbf{N}^d\}$ be independent identically distributed random variables. Assume that $X(\mathbf{n}) \geq 0$ almost surely. Assume further that $0 < \mu \stackrel{\text{def}}{=} \mathsf{E}[X] < \infty$ and*

$$t(\log t)^{2(d-1)}\, \mathsf{P}(X \geq t) \to 0, \qquad t \to \infty. \tag{11.16}$$

Then

$$\lim_{t\to\infty} \left[\frac{U(t)}{t} - \frac{1}{\mu}\mathscr{P}\left(\log\frac{t}{\mu}\right)\right] = 0. \tag{11.17}$$

Remark 11.2 Theorem 11.5 shows that the asymptotic behavior of the renewal function constructed from a random walk with multi-dimensional time differs from that of the classical renewal function. Moreover, the behavior varies with $d > 1$ and differs from the classical behavior by a multiplicative factor $(\log t)^{d-1}$. Theorem 11.6 improves this result under the additional assumption (11.16). This condition guarantees that the ratio $\frac{U(t)}{t}$ has an additive trend of order $(\log t)^{d-1}$ if $d > 1$. Note that the constant on the right-hand side of (11.13) also changes with the dimension of the multi-indices.

Remark 11.3 Condition (11.16) for $d = 1$ follows from the existence of the expectation μ. Thus Theorem 11.6 coincides with the elementary renewal theorem if $d = 1$.

Proof of Theorem 11.5 First we consider the case of $\mu = 1$. Represent the renewal function as a sum of two terms

$$U(t) = \sum_{|\mathbf{n}|\leq t} \mathsf{P}(S(\mathbf{n}) < t) + \sum_{|\mathbf{n}|>t} \mathsf{P}(S(\mathbf{n}) < t). \tag{11.18}$$

Representation (11.15) implies that equality (11.13) is equivalent to

$$\lim_{t\to\infty} \frac{U(t) - T_{[t]}}{t\,(\log t)^{d-1}} = 0$$

which, in view of (11.18), follows from the following two relations

$$\lim_{t\to\infty} \frac{1}{t\,(\log t)^{d-1}} \sum_{|\mathbf{n}|\leq t} \mathsf{P}(S(\mathbf{n}) \geq t) = 0, \tag{11.19}$$

$$\lim_{t\to\infty} \frac{1}{t\,(\log t)^{d-1}} \sum_{|\mathbf{n}|>t} \mathsf{P}(S(\mathbf{n}) < t) = 0. \tag{11.20}$$

Choose and fix two constants c and C such that $0 < c < 1 < C < \infty$. It is clear that (11.19) follows from relations (11.21)–(11.22), while (11.20) follows from relations (11.23)–(11.24):

$$\lim_{t\to\infty} \frac{1}{t\,(\log t)^{d-1}} \sum_{|\mathbf{n}|\le ct} \mathsf{P}(S(\mathbf{n}) \ge t) = 0, \tag{11.21}$$

$$\lim_{c\uparrow 1} \limsup_{t\to\infty} \frac{1}{t\,(\log t)^{d-1}} \sum_{ct<|\mathbf{n}|\le t} \mathsf{P}(S(\mathbf{n}) \ge t) = 0, \tag{11.22}$$

$$\lim_{t\to\infty} \frac{1}{t\,(\log t)^{d-1}} \sum_{t<|\mathbf{n}|\le Ct} \mathsf{P}(S(\mathbf{n}) < t) = 0, \tag{11.23}$$

$$\lim_{C\downarrow 1} \limsup_{t\to\infty} \frac{1}{t\,(\log t)^{d-1}} \sum_{|\mathbf{n}|>Ct} \mathsf{P}(S(\mathbf{n}) < t) = 0. \tag{11.24}$$

To prove (11.21) note that $\frac{t}{|\mathbf{n}|} \ge \frac{1}{c} > 1$ if $|\mathbf{n}| \le ct$ and thus the probabilities in the sum (11.21) can be estimated as follows

$$\mathsf{P}(S(\mathbf{n}) \ge t) \le \mathsf{P}\left(\left|\frac{S(\mathbf{n})}{|\mathbf{n}|} - \mu\right| \ge \delta\right)$$

with $\delta \stackrel{\text{def}}{=} \frac{1}{c} - 1$. Recall that we are dealing with the case of $\mu = 1$. By the law of large numbers (Theorem 4.11),

$$\lim(\max)\, \mathsf{P}\left(\left|\frac{S(\mathbf{n})}{|\mathbf{n}|} - \mu\right| \ge \delta\right) = 0$$

for all $\delta > 0$ and thus relation (11.21) follows. Note also that relation (11.21) holds for all $c < 1$.

To prove (11.22) we observe that

$$\limsup_{t\to\infty} \frac{1}{t\,(\log t)^{d-1}} \sum_{ct<|\mathbf{n}|\le t} \mathsf{P}(S(\mathbf{n}) \ge t) \le \lim_{t\to\infty} \frac{T_{[t]} - T_{[ct]}}{t\,(\log t)^{d-1}} = \frac{1-c}{(d-1)!}.$$

Since $c > 1$ is arbitrary, we derive (11.22) from the latter result.

Our reasoning is similar when proving relation (11.23): since

$$\lim_{t\to\infty} \frac{1}{t\,(\log t)^{d-1}} \sum_{t<|\mathbf{n}|\le Ct} \mathsf{P}(S(\mathbf{n}) < t) \le \lim_{t\to\infty} \frac{T_{[Ct]} - T_{[t]}}{t\,(\log t)^{d-1}} = \frac{C-1}{(d-1)!}$$

for all $C > 1$, relation (11.23) follows.

It remains to prove (11.24). Choose a number a such that

$$\mu_a \stackrel{\text{def}}{=} \int_{x<a} x\,dF(x) > \frac{1}{C},$$

where $F(\cdot)$ denotes the distribution function of the random variables $X(\mathbf{n})$. Put

$$X_a(\mathbf{n}) = X(\mathbf{n})\mathbb{I}_{\{X(\mathbf{n})<a\}}, \qquad S_a(\mathbf{n}) = \sum_{\mathbf{k}\le\mathbf{n}} X_a(\mathbf{k}).$$

It is clear that $\mu_a = \mathsf{E}[X_a(\mathbf{n})]$, $C\mu_a > 1$, and $\mathsf{P}(S(\mathbf{n}) < t) \le \mathsf{P}(S_a(\mathbf{n}) < t)$ for all $\mathbf{n} \in \mathbf{N}^d$ and $t > 0$. Moreover, the Chebyshev–Markov inequality implies that

$$\mathsf{P}(S_a(\mathbf{n}) < t) \le \mathsf{P}\left(\left|\frac{S_a(\mathbf{n})}{|\mathbf{n}|} - \mu_a\right| \ge \delta\right) \le \frac{\mathsf{E}\left[S_a(\mathbf{n}) - |\mathbf{n}|\mu_a^4\right]}{\delta^4|\mathbf{n}|^4} \le \frac{\text{const}}{|\mathbf{n}|^2}$$

for all $\delta \overset{\text{def}}{=} 1 - \frac{1}{C}$. Hence

$$\lim_{t\to\infty} \frac{1}{t\,(\log t)^{d-1}} \sum_{|\mathbf{n}|>Ct} \mathsf{P}(S(\mathbf{n}) < t) \le \text{const} \lim_{t\to\infty} \sum_{|\mathbf{n}|>Ct} \frac{1}{|\mathbf{n}|^2} = 0.$$

Therefore Theorem 11.5 is proved for $\mu = 1$.

The case of the general μ in Theorem 11.5 is considered as follows. Put

$$Y(\mathbf{n}) = \frac{X(\mathbf{n})}{\mu}, \qquad S_Y(\mathbf{n}) = \sum_{\mathbf{k}\le\mathbf{n}} Y(\mathbf{k}), \qquad U_Y(t) = \sum_{\mathbf{n}\in\mathbf{N}^d} \mathsf{P}(S_Y < t).$$

Then $\mathsf{E}[Y(\mathbf{n})] = 1$ and the part of Theorem 11.5 already proved can be applied to the function $U_Y(t)$:

$$\lim_{t\to\infty} \frac{U_Y(t)}{t(\log t)^{d-1}} = \frac{1}{(d-1)!}.$$

Since $U_Y(t) = U_X(\mu t) \overset{\text{def}}{=} U(\mu t)$, Theorem 11.5 is proved in the general case. □

The proof of Theorem 11.6 requires some auxiliary results.

Lemma 11.4 *Let $\{X_n\}$ be independent identically distributed random variables and $S_n = \sum_{k=1}^n X_k$. Assume that $X_n \ge 0$ almost surely and that $0 < \mathsf{E}[X_1] < \infty$. If a sequence of real numbers $\{a_n, n \ge 1\}$ decreases to 0, then*

$$\lim_{t\to\infty} \frac{1}{t} \sum_{k=1}^{\infty} a_k \,\mathsf{P}(S_k < t) = 0.$$

Proof It is clear that $t^{-1}\sum_{k\le[t]} a_k \to 0$, $t \to \infty$. Thus

$$\frac{1}{t}\sum_{k=1}^{\infty} a_k \,\mathsf{P}(S_k < t) \le \frac{1}{t}\sum_{k\le[t]} a_k + \frac{a_{[t]}}{t}\sum_{k=1}^{\infty} \mathsf{P}(S_k < t) \to 0, \qquad t \to \infty,$$

by the elementary renewal Theorem 11.1 for one-dimensional time. □

Lemma 11.5 *Assume that all the assumptions of Theorem 11.5 hold. Denote by* $R_k \stackrel{\text{def}}{=} \Delta\big[k\mathscr{P}(\log k)\big]$ *the increments of the sequence* $\{k\mathscr{P}(\log k), k \ge 1\}$, *where the polynomial* $\mathscr{P}$ *is defined by* (11.14), *that is*

$$R_0 = 0, \qquad R_1 = 1,$$
$$R_k = k\mathscr{P}(\log k) - (k-1)\mathscr{P}(\log(k-1)), \qquad k > 1.$$

Then

$$U(t) = U_1(t) + o(t),$$

where

$$U_1(t) = \sum_{k=1}^{\infty} R_k \, \mathsf{P}(S_k < t). \tag{11.25}$$

Proof The series on the right-hand side of (11.25) converges. This result can be proved in the same way as Lemma 11.2.

We further use the partial summation formula: let $\{a_k, 1 \le k \le n\}$ and $\{b_k, 1 \le k \le n\}$ be two sequences of real numbers such that $a_0 = 0$; then, for all $N \ge 1$,

$$\sum_{k=1}^{N}(a_k - a_{k-1})b_k = a_N b_N + \sum_{k=1}^{N-1} a_k(b_k - b_{k+1}) = a_N b_N + \sum_{k=1}^{N-1} a_k \nabla\left[b_k\right],$$

where $\nabla\left[b_k\right] = b_k - b_{k+1}$. We also put $P_k(t) \stackrel{\text{def}}{=} \mathsf{P}(S_k < t)$. Observe that the sequence of numbers $P_k(t)$ is decreasing with respect to k for all fixed $t > 0$. Putting $T_0 = 0$ and using the partial summation formula we prove for all $N \ge 1$ that

$$\begin{aligned}
\sum_{k=1}^{N} \tau_k P_k(t) &= \sum_{k=1}^{N}(T_k - T_{k-1})P_k(t) = T_N P_N(t) + \sum_{k=1}^{N-1} T_k(P_k(t) - P_{k+1}(t)) \\
&= \big[N\mathscr{P}(\log N) + \rho_N\big]P_N(t) + \sum_{k=1}^{N-1}\big[k\mathscr{P}(\log k) + \rho_k\big](P_k(t) - P_{k+1}(t)) \\
&= \sum_{k=1}^{N} \Delta\big[k\mathscr{P}(\log k)\big]\, P_k(t) + \rho_N P_N(t) + \sum_{k=1}^{N-1} \rho_k \nabla\left[P_k(t)\right].
\end{aligned}$$

The latter two terms are estimated with the help of representation (11.15):

$$|\rho_N| P_N(t) + \sum_{k=1}^{N-1} |\rho_k| \nabla [P_k(t)] \le \text{const} \left[N^\lambda P_N(t) + \sum_{k=1}^{N-1} k^\lambda \nabla [P_k(t)] \right]$$
$$= \text{const} \sum_{k=1}^{N} \Delta \left[k^\lambda\right] P_k(t).$$

Passing to the limit as $N \to \infty$ we conclude that $U(t) = U_1(t) + U_2(t)$, where

$$U_2(t) = \lim_{N\to\infty} \left(\rho_N P_N(t) + \sum_{k=1}^{N-1} \rho_k \nabla [P_k(t)] \right).$$

Moreover,

$$|U_2(t)| \le \text{const} \sum_{k=1}^{\infty} \Delta \left[k^\lambda\right] P_k(t).$$

Now Lemma 11.4 implies that $U_2(t) = o(t)$, since $0 < \lambda < 1$, and thus the sequence $\{\Delta \left[k^\lambda\right]\}$ is decreasing. Lemma 11.5 is proved. □

Lemma 11.6 *Let $\{\xi_n, n \ge 1\}$ be independent identically distributed random variables and $\zeta_n = \sum_{k=1}^{n} \xi_k$. If $\alpha \ge 0$ and $\beta \ge 0$, then the relation*

$$(\log n)^\alpha \mathsf{P}\left(|\zeta_n| \ge \varepsilon \frac{n}{(\log n)^\beta} \right) \to 0, \qquad n \to \infty, \qquad \forall \varepsilon > 0,$$

follows from the set of the following conditions:

$$n (\log n)^{\alpha+\beta} \mathsf{P}(|\xi| \ge n) \to 0, \qquad n \to \infty, \tag{11.26}$$

$$(\log n)^\beta \int_{-n}^{n} x \, dF(x) \to 0, \qquad n \to \infty, \tag{11.27}$$

where the random variable ξ has the same distribution function $F(\cdot)$ as all other random variables ξ_n.

Remark 11.4 If $\alpha = \beta = 0$, then Lemma 11.6 is the law of large numbers for the sequence $\{\xi_n\}$. If $\beta = 0$ but $\alpha > 0$, then Lemma 11.6 is a result on the rate of convergence in the law of large numbers. For other α and β, this result also describes the rate of convergence in the law of large numbers for the normalization $n/(\log n)^\beta$.

Proof Note that (11.26)–(11.27) imply that

$$\lim_{t\to\infty} t (\log t)^{\alpha+\beta} \mathsf{P}(|X| \ge t) = 0, \qquad \lim_{t\to\infty} (\log t)^\beta \int_{-t}^{t} x \, dF(X) = 0,$$

where we pass to the limit for the real parameter t. These relations are proved with the help of the following bounds: if $t \geq 2$, then

$$t\,(\log t)^{\alpha+\beta}\,\mathsf{P}(|X| \geq t) \leq O(1)[t]\,(\log[t])^{\alpha+\beta}\,\mathsf{P}(|X| \geq [t]),$$

$$(\log t)^{\beta}\left|\int_{-t}^{t} x\,dF(X)\right| \leq O(1)\,(\log[t])^{\beta}\left(\left|\int_{|x|<[t]} x\,dF(X)\right| + [t]\,\mathsf{P}(|X| \geq [t])\right),$$

where $[t]$ denotes the integer part of a number t.

For $k \leq n$, put

$$\xi_{kn} = \xi_k\,\mathbb{I}_{\{|\xi_k| \leq n(\log n)^{-\beta}\}}, \qquad \zeta_{nn} = \sum_{k\leq n} \xi_{kn}.$$

It is clear that, for all $\varepsilon > 0$ and $n \geq 1$,

$$\left\{\omega : |\zeta_n| \geq \varepsilon\frac{n}{(\log n)^{\beta}}\right\} \subseteq \left\{\omega : \max_{k\leq n} |\xi_{kn}| \geq \frac{n}{(\log n)^{\beta}}\right\} \cup \left\{\omega : |\zeta_{nn}| \geq \varepsilon\frac{n}{(\log n)^{\beta}}\right\}.$$

Fix $\varepsilon > 0$. Condition (11.27) implies that, for all sufficiently large n,

$$\left|\int_{|x|\leq n(\log n)^{-\beta}} x\,dF(x)\right| \leq \frac{\varepsilon}{2\,(\log n)^{\beta}}.$$

Thus

$$\mathsf{P}\left(|\zeta_n| \geq \varepsilon\frac{n}{(\log n)^{\beta}}\right) \leq n\,\mathsf{P}\left(|\xi| \geq \frac{n}{(\log n)^{\beta}}\right) + \mathsf{P}\left(|\zeta_{nn} - n\,\mathsf{E}\,[\xi_{1n}]\,| \geq \frac{\varepsilon}{2}\frac{n}{(\log n)^{\beta}}\right).$$

The Chebyshev–Markov inequality yields

$$(\log n)^{\alpha}\,\mathsf{P}\left(|\zeta_n| \geq \varepsilon\frac{n}{(\log n)^{\beta}}\right) \leq n\,(\log n)^{\alpha}\,\mathsf{P}\left(|\xi| \geq \frac{n}{(\log n)^{\beta}}\right) + \frac{4}{\varepsilon^2 n}\,(\log n)^{\alpha+2\beta}\,\mathsf{var}\,[\xi_{1n}].$$

The first term on the right-hand side approaches 0 in view of condition (11.26). The second term also approaches 0, since, with $x_n = n/\,(\log n)^{\beta}$,

$$\frac{(\log n)^{\alpha+2\beta}}{n}\,\mathsf{var}\,[\xi_{1n}] \le \frac{(\log n)^{\alpha+2\beta}}{n}\,\mathsf{E}\left[\xi_{1n}^2\right] = O(1)\frac{(\log x_n)^{\alpha+\beta}}{x_n}\int\limits_{|x|\le x_n} x^2\,dF(x)$$

$$= O(1)\frac{(\log x_n)^{\alpha+\beta}}{x_n}\int\limits_0^{x_n} x\,\mathsf{P}(|\xi|\ge x)\,dx \to 0, \qquad n\to\infty.$$

We have used above the integration by parts formula (see Sect. A.19). Lemma 11.6 is proved. □

Proof of Theorem 11.6 First we consider the case of $\mu = 1$. In doing so we use the notation introduced in Lemma 11.5. By Lemma 11.5,

$$U(t) - \sum_{k\le t} R_k = -\sum_{k\le t} R_k\,\mathsf{P}(S_k \ge t) + \sum_{k>t} R_k\,\mathsf{P}(S_k < t) + o(t). \qquad (11.28)$$

Then we prove that

$$\lim_{t\to\infty}\frac{1}{t}\sum_{k\le t} R_k\,\mathsf{P}(S_k\ge t) = 0, \qquad (11.29)$$

$$\lim_{t\to\infty}\frac{1}{t}\sum_{k>t} R_k\,\mathsf{P}(S_k< t) = 0. \qquad (11.30)$$

Using representation (11.15) it is easy to show that there exist real numbers r_j, $0\le j\le d-1$, such that

$$R_k = \sum_{j=0}^{d-1} r_j\,(\log k)^j + o(1). \qquad (11.31)$$

Let $\varepsilon > 0$ be an arbitrary number and let $\phi(t) \stackrel{\text{def}}{=} t/\left(\log^+ t\right)^{d-1}$, $\phi_k \stackrel{\text{def}}{=} \phi(k)$. Fix $0\le j\le d-1$. Then, for sufficiently large t,

$$\begin{aligned}
&\sum_{k\le t}(\log k)^j\,\mathsf{P}(S_k\ge t)\\
&= \sum_{k\le t-\varepsilon\phi(t)}(\log k)^j\,\mathsf{P}(S_k\ge t) + \sum_{t-\varepsilon\phi(t)<k\le t}(\log k)^j\,\mathsf{P}(S_k\ge t)\\
&\le \sum_{k\le t-\varepsilon\phi(t)}(\log k)^j\,\mathsf{P}(S_k\ge t) + (\log t)^j\,\varepsilon\phi(t) \qquad (11.32)\\
&\le \sum_{k\le t-\varepsilon\phi(t)}(\log k)^j\,\mathsf{P}(|S_k-k|\ge \varepsilon\phi_k) + \varepsilon O(t)
\end{aligned}$$

$$\leq \sum_{k\leq t} (\log k)^{d-1} \, \mathsf{P}(|S_k - k| \geq \varepsilon\phi_k) + \varepsilon O(t).$$

Now we show that the first term on the right-hand side of (11.32) is $o(t)$ for all $\varepsilon > 0$. We will deduce this result from the asymptotic relation

$$(\log k)^{d-1} \, \mathsf{P}(|S_k - k| \geq \varepsilon\phi_k) = o(1), \qquad k \to \infty,$$

for all $\varepsilon > 0$. Recall that we are dealing with the case of $\mu = 1$. The latter relation follows from the law of large numbers if $d = 1$. Let $d > 1$. For this case, we check the assumptions of Lemma 11.6 with $\alpha = \beta = d - 1$ and random variables $\xi_k = X_k - 1$. It is clear that condition (11.26) (for $\xi \stackrel{\text{def}}{=} \xi_1$) follows from condition (11.16). Since $\mathsf{E}[\xi_k] = 0$, we rewrite condition (11.27) as follows

$$(\log n)^{d-1} \int\limits_{|x|\geq n} x \, dF(x) \to 0, \qquad n \to \infty, \tag{11.33}$$

where F is the distribution function of the random variable ξ. Hence

$$\left| \int\limits_{|x|\geq n} x \, dF(x) \right| \leq \sum_{k=n}^{\infty} (k+1) \, \mathsf{P}(k \leq |\xi| < k+1) = O(1) \sum_{k=n}^{\infty} \mathsf{P}(|\xi| \geq k).$$

Using (11.16) we obtain

$$\sum_{k=n}^{\infty} \mathsf{P}(|\xi| \geq k) \leq O(1) \sum_{k=n}^{\infty} \frac{1}{k \, (\log k)^{2(d-1)}} \asymp \frac{1}{(\log n)^{2d-1}}.$$

This result proves (11.33) for the case $d > 1$. Now the required assertion follows from Lemma 11.6.

Turning back to (11.32), we get

$$\sum_{k\leq t} (\log k)^j \, \mathsf{P}(S_k \geq t) = o(t) + \varepsilon t, \qquad t \to \infty. \tag{11.34}$$

We substitute decomposition (11.31) for R_k in (11.29) and rewrite the left-hand side of (11.29) as the sum of the corresponding $d+1$ sums. Then we apply asymptotic relation (11.34) to each of the first d sums and use Lemma 11.4 for the last sum that corresponds to $o(1)$ in (11.31). Finally, we get

$$\limsup_{t\to\infty} \frac{1}{t} \sum_{k\leq t} R_k \, \mathsf{P}(S_k \geq t) \leq O(\varepsilon).$$

Since ε is an arbitrary positive number, relation (11.29) is proved.

In order to prove (11.30), fix $C > 1$ and $\varepsilon > 0$. Then

$$\sum_{k>t} R_k \,\mathsf{P}(S_k < t) = \sum_{t<k\le t+\varepsilon\phi(t)} R_k \,\mathsf{P}(S_k < t) + \sum_{t+\varepsilon\phi(t)<k\le Ct} R_k \,\mathsf{P}(S_k < t) + \sum_{k>Ct} R_k \,\mathsf{P}(S_k < t). \tag{11.35}$$

We use equality (11.31) for the first term on the right-hand side of (11.35). Since probabilities do not exceed one, we see that this term is

$$O(1)\left(t_\varepsilon (\log t_\varepsilon)^{d-1} - t(\log t)^{d-1}\right), \qquad t_\varepsilon \overset{\text{def}}{=} t + \varepsilon\phi(t).$$

The Lagrange mean value theorem applied to the function $t\,(\log t)^{d-1}$ implies that

$$\frac{1}{t}\sum_{t<k\le t+\varepsilon\phi(t)} R_k \,\mathsf{P}(S_k < t) = O(\varepsilon).$$

Therefore

$$\limsup_{t\to\infty}\frac{1}{t}\sum_{t<k\le t+\varepsilon\phi(t)} R_k \,\mathsf{P}(S_k < t) \le O(\varepsilon).$$

The probabilities in the second term on the right-hand side of (11.35) are estimated as follows

$$\mathsf{P}(S_k < t) \le \mathsf{P}(|S_k - k\mu| \ge \varepsilon\phi(t)) \le \mathsf{P}\left(|S_k - k\mu| \ge \varepsilon\phi\left(\frac{k}{C}\right)\right).$$

We recall that we are dealing with the case of $\mu = 1$. By Lemma 11.6,

$$(\log k)^{d-1}\,\mathsf{P}\left(|S_k - k\mu| \ge \varepsilon\phi\left(\frac{k}{C}\right)\right) \to 0, \qquad k\to\infty,$$

for all $\varepsilon > 0$ and $C > 1$. Thus

$$\begin{aligned}\sum_{t+\varepsilon\phi(t)\le k<Ct} R_k \,\mathsf{P}(S_k < t) &\le O(1)\sum_{t+\varepsilon\phi(t)\le k<Ct}\frac{R_k}{(\log k)^{d-1}}\\ &\le \frac{O(1)}{(\log t)^{d-1}}\left[Ct\mathscr{P}(\log Ct) - t\mathscr{P}(\log t)\right]\\ &\le O(1)\cdot t\cdot(C-1).\end{aligned}$$

Therefore

$$\limsup_{t\to\infty} \frac{1}{t} \sum_{t+\phi(t)\le k<Ct} R_k \, \mathsf{P}(S_k < t) = \text{const} \cdot (C-1).$$

The third term on the right-hand side of (11.35) is estimated by using the method similar to that used in the proof of Theorem 11.5. Let $a > 0$ be a fixed number; put

$$X_k^a = X_k \, \mathbb{I}_{\{X_k < a\}}, \qquad S_n^a = \sum_{k\le n} X_k^a, \qquad \mu^a \stackrel{\text{def}}{=} \mathsf{E}\left[X_k^a\right].$$

Choose a number a in such a way that $\mu^a > C^{-1}$. If $k > Ct$, then

$$\mathsf{P}(S_k < t) \le \mathsf{P}(S_k^a < t) \le \mathsf{P}(|S_k - k\mu^a| \ge k\mu^a - t) \le \mathsf{P}(|S_k - k\mu^a| \ge k\delta),$$

where $\delta \stackrel{\text{def}}{=} \mu^a - C^{-1} > 0$. It follows from (11.31) that $R_k = O(k)$. Thus

$$\sum_{k>Ct} R_k \, \mathsf{P}(S_k < t) \le \text{const} \sum_{k>Ct} k \, \mathsf{P}(|S_k^a - k\mu^a| \ge \delta k).$$

The right-hand side is a tail of a convergent series (Theorem 13.14), since the random variables X_k^a are bounded and thus all the moments are finite; in particular, $\mathsf{E}\left[|X_k^a|^3\right] < \infty$. Finally,

$$\limsup_{t\to\infty} \frac{1}{t} \sum_{k>t} R_k \, \mathsf{P}(S_k < t) \le \text{const} \cdot (\varepsilon + C - 1).$$

Since $\varepsilon > 0$ and $C > 1$ are arbitrary numbers, we conclude that (11.30) follows in the case of $\mu = 1$.

The general case of $\mu \ne 1$ is proved as follows. Consider the random variables $Y(\mathbf{n}) \stackrel{\text{def}}{=} X(\mathbf{n})/\mu$. The renewal function constructed from $\{Y(\mathbf{n})\}$ is denoted by $U_Y(t)$. With $U_X(t) \stackrel{\text{def}}{=} U(t)$ we have $U_Y(t) = U_X(\mu t)$. The part of the theorem already proved implies that

$$\begin{aligned} 0 &= \lim_{t\to\infty} \left[\frac{U_Y(t)}{t} - \mathscr{P}(\log t)\right] = \lim_{t\to\infty} \left[\frac{U_X(\mu t)}{t} - \mathscr{P}(\log t)\right] \\ &= \lim_{s\to\infty} \left[\mu \frac{U_X(s)}{s} - \mathscr{P}\left(\log \frac{s}{\mu}\right)\right], \end{aligned}$$

whence (11.17) follows in the general case. □

11.5 The Asymptotic Behavior of Renewal Processes Constructed from Multi-Indexed Sums

We are going to study the asymptotic behavior of the renewal process for $d > 1$. In this case, there are several new problems as compared to the case of $d = 1$. One of them is that, in contrast to $d = 1$, we have for $d > 1$ that

$$\frac{N(t)}{t} \to \infty, \qquad t \to \infty,$$

almost surely (compare with the law of large numbers for renewal processes).

Since the limit behavior of the expectation of the renewal process changes with $d > 1$ (see Theorems 11.5 and 11.6), we expect analogous changes in the limit behavior of the renewal process itself. An investigation of the asymptotic behavior for $d > 1$ is not easy at all, since the known methods do not apply for multi-indices.

11.5.1 Duality in the Case d = 1

One of the powerful tools of renewal theory for $d = 1$ is the duality

$$S_{N(t)} < t \le S_{N(t)+1} \qquad \text{for all t} \ge 0, \tag{11.36}$$

which is a straightforward consequence of Definition 11.1. For example, if one wants to prove the strong law of large numbers for renewal processes,

$$\lim_{t\to\infty} \frac{N(t)}{t} = \frac{1}{\mu} \qquad \text{a. s.,} \tag{11.37}$$

one uses the above duality as follows

$$\frac{N(t)}{N(t)+1} \cdot \frac{N(t)+1}{S_{N(t)+1}} \le \frac{N(t)}{t} < \frac{N(t)}{S_{N(t)}}.$$

Having obtained these results, we establish the strong law of large numbers for renewal processes and hence, since $N(t) \to \infty$ almost surely, the strong law of large numbers holds for the sequence $\{S_n\}$.

An essential ingredient in the proof above is the relation $S_{N(t)} \approx t$ that, generally speaking, does not have any analog for $d > 1$, since the sums $S(\mathbf{n}) = S(n_1, \dots, n_d)$ depend on d parameters, while the renewal process $N(t)$ depends only on a single coordinate and can change only one parameter in the multi-index. Therefore we need to develop a new method for the strong law of large numbers for renewal processes that do not rely on the duality. The method proposed below may have value even in the case $d = 1$.

11.5.2 The Bound r_0 for the Rate of Convergence

The rate of convergence in the law of large numbers for renewal processes depends on a moment of a certain order r of random variables $\{X_n\}$. Moreover the rate of convergence increases with $1 \le r < 2$, since

$$\frac{N(t)}{t} - \frac{1}{\mu} = o\left(t^{\frac{1}{r}-1}\right) \qquad \text{a. s.}$$

by Theorem 11.3. This result fails if $r = 2$. The central limit theorem for renewal processes (see [138]) is the reason for this phenomenon. If r varies in the interval $(2, \infty)$, then the rate of convergence described by Theorem 11.3 does not change at all.

A similar situation occurs for $d > 1$, too, with the difference that the role of the upper bound for r, after which the rate of convergence stops changing, is played by a special number r_0 (rather than 2 for $d = 1$). The number r_0 is defined by

$$r_0 = \min\left\{2, \frac{1}{\lambda}\right\}, \tag{11.38}$$

where λ is the order of approximation of the Dirichlet function by the polynomial $\mathscr{P}$ (see (11.15)). The precise value of this number is unknown at the present moment: it depends on a solution of the Dirichlet divisors problem which is to find the precise order λ for the error of approximation of the Dirichlet function by polynomials. To highlight the fact that the numbers λ and r_0 depend on the dimension of multi-indices, we write λ_d and $r_{0;d}$ rather than λ and r_0.

Results from number theory allow us to conclude that $r_{0;d} = 2$ in some particular cases. For example, if $d = 1$, then there is no approximation error at all (that is $\lambda_1 = 0$ and thus $r_{0;1} = 2$). The Dirichlet asymptotics (A.28) for $d = 2$ implies that $r_{0;2} = 2$. The minimal λ in the Dirichlet divisors problem is not known even in the case of $d = 2$. It is worth mentioning, however, that further improvements of the Dirichlet result for $d = 2$ will not improve the rate of convergence for renewal functions and processes as compared to that described by (A.31).

The same situation occurs in the next two dimensions, $d = 3$ and $d = 4$ (see (A.33)). In contrast, existing bounds for λ_d to achieve the conjectured value $r_{0;d} = 2$ if $d > 4$ are not as nice as those for $d \le 4$. What is known is that $r_{0;5} = \frac{20}{11} \approx 1.89$ (see (A.33)). In the general case, it is known that $r_{0;d} > 1$ for all $d \ge 2$ (see (A.34)). This, by the way, implies that the interval $(1, r_0)$ involved in the assertions below is not empty.

For many specific values of d, there are better bounds for λ_d as compared to (A.34).

If the Lindelöf conjecture (A.38) or an equivalent Hardy conjecture (A.36) were true, then we would get $r_{0;d} = 2$ for all $d \ge 1$. From the point of view of limit theorems of probability theory, it is a natural conjecture that $r_{0;d} = 2$ for all $d \ge 1$.

11.5.3 The Asymptotic Behavior of Renewal Processes for $d > 1$

Theorem 11.7 is an analog of the strong law of large numbers for renewal processes (Theorem 11.2) for $d > 1$, while Theorem 11.8 is an analog of the result on the rate of convergence Theorem 11.3. We outline the main idea of the proof in Corollaries 11.2 and 11.1 corresponding to the case of $d = 2$. The case of a general dimension $d > 1$ differs from that of $d = 2$ only in some technical details.

Theorem 11.7 *Let $\{X(\mathbf{n}), \mathbf{n} \in \mathbf{N}^d\}$ be independent identically distributed random variables such that $X(\mathbf{n}) \ge 0$ almost surely and $0 < \mu \stackrel{\text{def}}{=} \mathsf{E}[X(\mathbf{n})] < \infty$. If*

$$\mathsf{E}\left[X(\log^+ X)^{d-1}\right] < \infty,$$

where X is a copy of the random variables $X(\mathbf{n})$, then

$$\lim_{t\to\infty} \frac{N(t)}{t(\log t)^{d-1}} = \frac{1}{\mu(d-1)!} \qquad a.\ s.$$

Theorem 11.8 *Let $\{X(\mathbf{n}), \mathbf{n} \in \mathbf{N}^d\}$ be independent identically distributed random variables such that $X(\mathbf{n}) \ge 0$ almost surely and $0 < \mu \stackrel{\text{def}}{=} \mathsf{E}[X(\mathbf{n})] < \infty$. Let the number r_0 be defined by equality* (11.38). *If*

$$\mathsf{E}\left[X^r(\log^+ X)^{d-1}\right] < \infty \tag{11.39}$$

for some $1 < r < r_0$, where X is a copy of random variables $X(\mathbf{n})$, then

$$\lim_{t\to\infty} \frac{N(t) - \frac{t}{\mu}\mathscr{P}\left(\log\left(\frac{t}{\mu}\right)\right)}{t^{1/r}(\log t)^{d-1}} = 0 \qquad a.\ s. \tag{11.40}$$

The polynomial $\mathscr{P}$ above is defined by equality (11.14).

Remark 11.5 The precise value of the number r_0 is important in obtaining the rate of convergence in the form of (11.40). On the other hand, if (11.40) is rewritten with the Dirichlet function $T(x) = \operatorname{card}\{\mathbf{n} \in \mathbf{N}^d : |\mathbf{n}| \le x\}$ instead of the polynomial $\mathscr{P}$, that is

$$\lim_{t\to\infty} \frac{N(t) - \frac{t}{\mu}T\left(\frac{t}{\mu}\right)}{t^{1/r}(\log t)^{d-1}} = 0 \qquad \text{a. s.,}$$

then one can take $r_0 = 2$ for all $d \ge 1$. On the other hand, $\mathscr{P}$ is a much nicer function than T, since all properties of $\mathscr{P}$ are clear.

Corollary 11.1 *Let $d = 2$. Let $\left\{X(m,n), (m,n) \in \mathbf{N}^2\right\}$ be independent identically distributed random variables such that $X(m,n) \ge 0$ almost surely and $0 < \mu \stackrel{\text{def}}{=}$*

$\mathsf{E}[X] < \infty$, *where* X *is a copy of the random variables* $X(m, n)$. *If*

$$\mathsf{E}\left[X \log^+ X\right] < \infty,$$

then

$$\lim_{t\to\infty} \frac{N(t)}{t \log t} = \frac{1}{\mu} \quad a.\ s.$$

Corollary 11.2 *Let* $d = 2$. *Let* $\left\{X(m, n), (m, n) \in \mathbf{N}^2\right\}$ *be independent identically distributed random variables such that* $X(m, n) \geq 0$ *almost surely and* $0 < \mu \stackrel{\text{def}}{=} \mathsf{E}[X] < \infty$, *where* X *is a copy of random variables* $X(m, n)$. *If*

$$\mathsf{E}\left[X^r \log^+ X\right] < \infty$$

for some $1 < r < 2$, *then*

$$\lim_{t\to\infty} \frac{N(t) - \frac{t\log t}{\mu} - \left(2\gamma - 1 + \frac{\log \mu}{\mu}\right) t}{t^{1/r} \log t} = 0 \quad a.\ s. \tag{11.41}$$

Here γ *is the so-called Euler–Mascheroni constant* (*see* (A.29)).

Remark 11.6 Corollary 11.2 is an assertion on the rate of convergence in the strong law of large numbers for renewal processes. In contrast to the discussion above, the statement of Corollary 11.2 does not mention the number r_0. This is explained by a sufficiently high rate of approximation by polynomials in the Dirichlet divisors problem for $d = 2$ which implies that the upper bound for r in (11.41) is $r_0 = 2$.

Remark 11.7 The random variable

$$\Lambda(\varepsilon) = \sum_{\mathbf{n}\in\mathbf{N}^d} \mathbb{I}_{\left\{\left|\frac{S(\mathbf{n})-\mu|\mathbf{n}|}{|\mathbf{n}|^{1/r}}\right| \geq \varepsilon\right\}} \tag{11.42}$$

occurs in the bounds obtained in the course of the proof below (here $\mathsf{E}[X(\mathbf{n})] = \mu$). It is clear that $\Lambda(\varepsilon) < \infty$ almost surely for all $\varepsilon > 0$ if and only if the Marcinkiewicz–Zygmund strong law of large numbers holds for $\{S(\mathbf{n})\}$. Recall that condition (11.39) is a criterion for this result (see Corollary 9.3).

Proof of Corollaries 11.1 *and* 11.2 Let T_k be the Dirichlet function, that is, the number of solutions of the inequality $mn \leq k$. We extend the domain of this function to the interval $(0, \infty)$ by putting $T(t) = T_n$ if $n \leq t < n + 1$.

First we consider the case of $\mu = 1$. The Dirichlet asymptotics (A.28) implies that we have to show that

$$\lim_{t\to\infty} \frac{N(t) - T(t)}{t^{1/r} \log t} = 0 \quad \text{a. s.} \tag{11.43}$$

Denote the double sums of the random variables $\{X(k,l)\}$ by $S(m,n)$. The renewal process (11.9) for $d=2$ can be written as a sum of two processes $N(t)=N_1(t)+N_2(t)$, where

$$N_1(t)=\sum_{mn\le[t]} \mathbb{I}_{\{S(m,n)<t\}} \qquad N_2(t)=\sum_{mn>[t]} \mathbb{I}_{\{S(m,n)<t\}}$$

(here $[t]$ is the integer part of t).

Consider the process $N_1(t)$. It is clear that

$$N_1(t)-T(t)=-\sum_{mn\le[t]} \mathbb{I}_{\{S(m,n)\ge t\}}.$$

Fix $\varepsilon>0$. Introduce the sets A_t and B_t as follows

$$A_t=\left\{m,n\colon t-mn\ge\varepsilon(mn)^{1/r},\, mn\le[t]\right\},$$
$$B_t=\left\{m,n\colon t-mn<\varepsilon(mn)^{1/r},\, mn\le[t]\right\}.$$

For the set A_t,

$$\frac{t-mn}{(mn)^{1/r}}\ge\varepsilon,$$

whence

$$\sum_{(m,n)\in A_t} \mathbb{I}_{\{S(m,n)\ge t\}}=\sum_{(m,n)\in A_t} \mathbb{I}_{\left\{\frac{S(m,n)-mn}{(mn)^{1/r}}\ge\frac{t-mn}{(mn)^{1/r}}\right\}}\le\sum_{(m,n)\in A_t} \mathbb{I}_{\left\{\left|\frac{S(m,n)-mn}{(mn)^{1/r}}\right|\ge\varepsilon\right\}}$$

(recall that we are dealing with the case of $\mu=1$). Moreover,

$$\sum_{(m,n)\in A_t} \mathbb{I}_{\{S(m,n)\ge t\}}\le\Lambda(\varepsilon),$$

where the random variable $\Lambda(\varepsilon)$ is defined by (11.42) for $d=2$, that is

$$\Lambda(\varepsilon)=\sum_{(m,n)\in\mathbf{N}^2} \mathbb{I}_{\left\{\frac{|S(m,n)-mn|}{(mn)^{1/r}}\ge\varepsilon\right\}}. \tag{11.44}$$

An obvious bound

$$\sum_{(m,n)\in B_t} \mathbb{I}_{\{S(m,n)\ge t\}}\le \operatorname{card}(B_t)$$

allows us to estimate $N_1(t)$ by

$$|N_1(t) - T(t)| \le \Lambda(\varepsilon) + \operatorname{card}(B_t). \tag{11.45}$$

Next we consider the process $N_2(t)$. The sets C_t and D_t are defined similarly to the sets A_t and B_t, namely

$$C_t = \left\{m, n\colon t - mn \le -\varepsilon(mn)^{1/r}, mn > [t]\right\},$$
$$D_t = \left\{m, n\colon t - mn > -\varepsilon(mn)^{1/r}, mn > [t]\right\}.$$

For the set C_t,

$$\frac{t - mn}{(mn)^{1/r}} \le -\varepsilon,$$

whence

$$\sum_{(m,n)\in C_t} \mathbb{I}_{\{S(m,n)<t\}} = \sum_{(m,n)\in C_t} \mathbb{I}_{\left\{\frac{S(m,n)-mn}{(mn)^{1/r}} < \frac{t-mn}{(mn)^{1/r}}\right\}}$$
$$\le \sum_{(m,n)\in C_t} \mathbb{I}_{\left\{\left|\frac{S(m,n)-mn}{(mn)^{1/r}}\right| \ge \varepsilon\right\}}.$$

Moreover,

$$\sum_{(m,n)\in C_t} \mathbb{I}_{\{S(m,n)<t\}} \le \Lambda(\varepsilon),$$

where the random variable $\Lambda(\varepsilon)$ is defined by equality (11.44). An obvious bound

$$\sum_{(m,n)\in D_t} \mathbb{I}_{\{S(m,n)<t\}} \le \operatorname{card}(D_t)$$

allows one to estimate $N_2(t)$ as follows

$$N_2(t) \le \Lambda(\varepsilon) + \operatorname{card}(D_t). \tag{11.46}$$

Inequalities (11.45) and (11.46) imply that

$$|N(t) - T(t)| \le 2\Lambda(\varepsilon) + \operatorname{card}(B_t) + \operatorname{card}(D_t), \tag{11.47}$$

that is

$$\limsup_{t\to\infty} \frac{|N(t) - T(t)|}{t^{1/r}\log t} \le \limsup_{t\to\infty} \frac{\operatorname{card}(B_t)}{t^{1/r}\log t} + \limsup_{t\to\infty} \frac{\operatorname{card}(D_t)}{t^{1/r}\log t} \tag{11.48}$$

almost surely. We estimate both terms on the right-hand side of (11.49). The set B_t belongs to the set

$$B_t' = \{m, n\colon\ f(t) < mn \le t\}, \tag{11.49}$$

where $f(\cdot)$ is the inverse function to $x + \varepsilon x^{1/r}$.

Below are some simple properties of the function f.

(f_1) f is increasing,
(f_2) $f(x) \le x$,
(f_3) if $r = 1$, then $f(x) = \frac{x}{1+\varepsilon}$,
(f_4) if $r > 1$, then $f(x) = x - \varepsilon x^{1/r} + o\left(x^{1/r}\right)$.

We will prove these properties after the proof of Corollaries 11.1 and 11.2 is complete.

Since $f(t) \le t$,

$$\begin{aligned}\operatorname{card}(B_t) &\le T(t) - T(f(t))\\ &= t\log t + (2\gamma - 1)t + o\left(\sqrt{t}\right)\\ &\quad - f(t)\log f(t) - (2\gamma - 1)f(t) - o\left(\sqrt{f(t)}\right)\\ &= (t - f(t))(1 + \log \xi_t) + (2\gamma - 1)(t - f(t)) + o\left(\sqrt{t}\right)\end{aligned}$$

according to (A.28). The number ξ_t is such that $f(t) \le \xi_t \le t$ and its existence is given by the Lagrange mean value theorem applied to the function $x\log x$ in the interval $[f(t), t]$.

We derive from properties (f_3) and (f_4) that

$$\operatorname{card}(B_t) \le O(\varepsilon)t^{1/r}\log t + o(t^{1/r})\log t + O\left(t^{1/r}\right) + o\left(\sqrt{t}\right),$$

whence

$$\limsup_{t\to\infty} \frac{\operatorname{card}(B_t)}{t^{1/r}\log t} \le O(\varepsilon).$$

Since ε is an arbitrary positive number, the latter result means

$$\lim_{t\to\infty} \frac{\operatorname{card}(B_t)}{t^{1/r}\log t} = 0.$$

The same reasoning leads to

$$\lim_{t\to\infty} \frac{\operatorname{card}(D_t)}{t^{1/r}\log t} = 0.$$

The preceding two equalities together with (11.49) prove Corollaries 11.1 and 11.2 for $\mu = 1$.

Now we consider the general case of $\mu \neq 1$. Define the random variables $Y(m,n) \stackrel{\text{def}}{=} X(m,n)/\mu$ and the corresponding renewal process $N_Y(T)$. The part of the theorem already proved implies that

$$\lim_{t\to\infty} \frac{N_Y(t) - t\log t - (2\gamma - 1)t}{t^{1/r}\log t} = 0 \qquad \text{a. s.,}$$

since $\mathsf{E}[Y(m,n)] = 1$. Substituting $N_Y(t) = N_X(\mu t)$, we complete the proof of Corollaries 11.1 and 11.2 in the general case. □

Proof of properties (f_1)–(f_4) By definition, $f^{-1}(x) = x + \varepsilon x^{1/r}$, $\varepsilon > 0$. Since f^{-1} is increasing, its inverse function f is increasing, too, and thus (f_1) is proved.

Further, $x = f\left(x + \varepsilon x^{1/r}\right)$, whence (f_2) follows in view of (f_1). Property (f_3) is clear.

To prove (f_4), we change x to $f(x)$ in the definition of f^{-1} and obtain $x = f(x) + \varepsilon\,(f(x))^{1/r}$ or $f(x) = x - \varepsilon\,(f(x))^{1/r}$. This equality applied to the second term yields $f(x) = x - \varepsilon\left(x - \varepsilon\,(f(x))^{1/r}\right)^{1/r}$. Thus,

$$f(x) = x - \varepsilon\,(x + o(x))^{1/r} = x - \varepsilon x^{1/r} + o\left(x^{1/r}\right)$$

by (f_2). □

11.6 Comments

The elementary renewal theorem (Theorem 11.1) follows from Theorem 2 in a paper by Feller [102]. The strong law of large numbers for renewal processes (Theorem 11.2) is proved by Doob [76]. Theorem 11.3 is an analog for renewal processes of the generalization of the Kolmogorov strong law of large numbers obtained by Marcinkiewicz and Zygmund (see, for example, Theorem 4.4 in the book by Gut [138]).

Section 11.2 Pick's formula (11.5) is proved in [340]. The asymptotic behavior of the double series on the right-hand side of (11.35) is studied as an analytic problem and without any probabilistic motivation by Ney and Wainger [330], Maejima and Mori [300], Galambos and Kátai [119–120], Galambos et al. [118]. These authors do not study the renewal process. Their primary interest was to find analytically an asymptotics of the function (11.8) or (11.11). A similar approach is followed by Hagwood [149], Abay [1], and Indlekofer, Kátai and Klesov [179].

Section 11.3 If the function $U(t)$ defined by (11.11) is rewritten in the form of (11.8), then it can be viewed as the so-called *weighted renewal function*. More general results than Theorem 11.5 are discussed by Smith [372] for a wider class of weighted renewal functions. The proof presented in Sect. 11.3 is simpler than that in [372].

Section 11.4 Theorems 11.4–11.6 are obtained in [225], Theorem 11.8 is proved in [245]. The equivalences between limit theorems for renewal processes and underlying sequences of cumulative sums are discussed in [139] for $d = 1$.

Section 11.5 Theorem 11.7 is proved in [225], while Theorem 11.8 is obtained in [245]. The moment condition (11.39) in Theorem 11.8 seems to be close to a necessary condition. However this is nothing more than a conjecture for $d > 1$ at the moment.

The asymptotic behavior of renewal processes and functions is studied in the papers [181] and [246] (also see [36]) for the case where the summation in formulas (11.9) and (11.11) is a certain unbounded set A instead of the whole space $\mathbf{N}^d$. The asymptotic behavior in this case is defined by the Dirichlet function in the set A, that is, by the function $D_A(x) \stackrel{\text{def}}{=} \operatorname{card}\{\mathbf{n} \in A : |\mathbf{n}| \le x\}$.

A different definition of the renewal process is introduced by Ivanoff and Merzbach [184] for $d > 1$. However, the asymptotic behavior of the processes defined in this way has not been studied yet.

The duality inequalities (11.36) are a starting point for investigations of comparative relationships between limit behaviors of functions f and g such that $f(g(x)) \sim x$. A number of similar problems are discussed in the monograph [37]. Speaking the language of [37], g is a right asymptotic inverse function for f, while f is a left asymptotic inverse function for g. The property $f(g(x)) \sim x$ alone is a source of a rich theory developed in [37]. In particular, the renewal process is a left asymptotic inverse function for the sequence of partial sums $\{S_n\}$ if $d = 1$.

Chapter 12
Existence of Moments of Suprema of Multi-Indexed Sums and the Strong Law of Large Numbers

One of the most well-known results in ergodic theory is the so-called *dominated ergodic theorem*. In fact, this name is attributed to an area of research in ergodic theory. The first result in this line, going back to 1939, was due to Wiener [418]. Note, however, that the same result was obtained two years earlier (expressed in probabilistic language for a less general case) by Marcinkiewicz and Zygmund [301].

Theorem 12.1 (J. Marcinkiewicz and A. Zygmund) *Let X and $\{X_n\}$ be independent identically distributed random variables*; $S_n = X_1 + \cdots + X_n$, $n > 1$, $S_1 = X_1$. *Let $q \geq 1$. If $q = 1$ and*

$$\mathsf{E}\left[|X| \log^+ |X|\right] < \infty \tag{12.1}$$

or if $q > 1$ and

$$\mathsf{E}\left[|X|^q\right] < \infty, \tag{12.2}$$

then

$$\mathsf{E}\left[\sup_{n\geq 1} \left|\frac{S_n}{n}\right|^q\right] < \infty. \tag{12.3}$$

It is clear that (12.2) follows from (12.3) for every $q \geq 1$ as $X_1 = S_1$. For a long time it was unknown whether or not (12.1) follows from (12.3) for $q = 1$, it was not until 1962 that Burkholder [41] proved this implication. The following result bears his name; however, Burkholder's real contribution [41] is his proof of the implication (12.3) $\Longrightarrow$ (12.1) in the case of $q = 1$.

Theorem 12.2 (D. Burkholder) *Let X and $\{X_n\}$ be independent identically distributed random variables*; $S_n = X_1 + \cdots + X_n$, $n > 1$, $S_1 = X_1$. *Assume that $q \geq 1$. Then* (12.3) $\Longleftrightarrow$ (12.1) *in the case of $q = 1$, and* (12.3) $\Longleftrightarrow$ (12.2) *in the case of $q > 1$.*

It is clear that Theorem 12.1 complements Kolmogorov's strong law of large numbers (Theorem 9.2). Note, however, that (12.3) implies a weaker result

O. Klesov, *Limit Theorems for Multi-Indexed Sums of Random Variables*, Probability Theory and Stochastic Modelling 71, DOI 10.1007/978-3-662-44388-0_12

$$\limsup_{n\to\infty}\left|\frac{S_n}{n}\right| < \infty \qquad \text{a. s.} \tag{12.4}$$

Relations similar to (12.3) are also studied in the literature for other normalizations. A general problem here is to find moment conditions for

$$\mathsf{E}\left[g\left(\sup_{n\geq 1}\left|\frac{S_n}{b_n}\right|\right)\right] < \infty \tag{12.5}$$

for a given function g and a given sequence of real numbers $\{b_n\}$. Under some constraints imposed on g and $\{b_n\}$, the latter relation implies either the strong law of large numbers

$$\lim_{n\to\infty}\frac{S_n}{b_n} = 0 \qquad \text{a. s.,}$$

or the law of the iterated logarithm

$$\limsup_{n\to\infty}\frac{S_n}{b_n} = 1 \qquad \text{a. s.}$$

An even more general relation than the strong law of large numbers (9.14) can be studied for $d > 1$, namely

$$\lim_{n_1\to\infty}\sup_{\substack{n_2\geq 1\\ \cdots\\ n_d\geq 1}}\frac{|S(\mathbf{n})|}{b(\mathbf{n})} = 0 \qquad \text{a. s.} \tag{12.6}$$

It is also possible to consider other limit relations, where some of indices tend to infinity and the remaining ones are used to evaluate the supremum. It is clear that the strong law of large numbers (9.14) follows from (12.6), while the converse is not obvious at all. Nevertheless, this implication does hold for a wide class of normalizing fields $\{b(\mathbf{n}), \mathbf{n}\in\mathbf{N}^d\}$.

In Sects. 12.1 and 12.2, we consider the Marcinkiewicz–Zygmund normalization and prove that (12.6) and the strong law of large numbers are equivalent. Another problem we are going to discuss below, closely related to the previous one, is to find moment conditions imposed on the individual random variables under which the mathematical expectation

$$\mathsf{E}\left[\sup_{\mathbf{n}\in\mathbf{N}^d}\left(\frac{|S(\mathbf{n})|}{b(\mathbf{n})}\right)^q\right] \tag{12.7}$$

exists for a given numerical field $\{b(\mathbf{n})\}$.

In Sects. 12.3 and 12.4, we find conditions ensuring that expectation (12.7) exists for the field $b(\mathbf{n}) = \sqrt{|\mathbf{n}| \log^+ \log^+ |\mathbf{n}|}$. We also obtain a stronger form of the law of the iterated logarithm in the Gaussian case.

12.1 The Existence of Moments of Multi-Indexed Sums for the Marcinkiewicz–Zygmund Normalization

For given real numbers $t_1 > 0, \dots, t_d > 0$, consider the field $b(\mathbf{n}) = n_1^{t_1} \cdots n_d^{t_d}$. As in the case of $d = 1$, we say that this is the *Marcinkiewicz–Zygmund* field.

Let $d > 1$ and let t and c be as defined in Corollary 9.1, that is

$$t = \min\{t_1, \dots, t_d\}, \qquad c = \operatorname{card}\{i : t_i = t\}. \tag{12.8}$$

Theorem 12.3 *Let X and $\{X(\mathbf{n}), \mathbf{n} \in \mathbf{N}^d\}$ be independent identically distributed random variables. Assume that $t_1, \dots, t_d$ are positive numbers and $\{b(\mathbf{n})\}$ is the Marcinkiewicz–Zygmund field. Finally, let the numbers t and c be as defined in* (12.8). *Suppose that $t > 1/2$ and $q > 0$. Then*

(12.9) $\Longleftrightarrow$ (12.10) *and* (12.11) $\Longleftrightarrow$ (12.10), (12.12),

where

$$\mathsf{E}\left[\sup_{\mathbf{n}\in\mathbf{N}^d}\left(\frac{|X(\mathbf{n})|}{n_1^{t_1}\cdots n_d^{t_d}}\right)^q\right] < \infty; \tag{12.9}$$

$$\begin{cases} \mathsf{E}\left[|X|^q\right] < \infty, & \text{if } tq > 1, \\ \mathsf{E}\left[|X|^q \left(\log^+ |X|\right)\right]^c < \infty, & \text{if } tq = 1, \\ \mathsf{E}\left[|X|^{1/t} \left(\log^+ |X|\right)^{c-1}\right] < \infty, & \text{if } tq < 1; \end{cases} \tag{12.10}$$

$$\mathsf{E}\left[\sup_{\mathbf{n}\in\mathbf{N}^d}\left(\frac{|S(\mathbf{n})|}{n_1^{t_1}\cdots n_d^{t_d}}\right)^q\right] < \infty; \tag{12.11}$$

$$\mathsf{E}[X] = 0, \qquad \text{if } t < 1. \tag{12.12}$$

Corollary 12.1 *Let X and $\{X(\mathbf{n}), \mathbf{n} \in \mathbf{N}^d\}$ be independent identically distributed random variables. Suppose that $t > 1/2$ and $q > 0$. Then*

(12.13) $\Longleftrightarrow$ (12.14) *and* (12.15) $\Longleftrightarrow$ (12.14), (12.12),

where

$$\mathsf{E}\left[\sup_{\mathbf{n}\in\mathbf{N}^d}\left(\frac{|X(\mathbf{n})|}{|\mathbf{n}|^t}\right)^q\right] < \infty; \tag{12.13}$$

$$\begin{cases} \mathsf{E}\left[|X|^q\right] < \infty, & \textit{if} \ \ tq > 1, \\ \mathsf{E}\left[|X|^q \left(\log^+ |X|\right)\right]^d < \infty, & \textit{if} \ \ tq = 1, \\ \mathsf{E}\left[|X|^{1/t} \left(\log^+ |X|\right)^{d-1}\right] < \infty, & \textit{if} \ \ tq < 1; \end{cases} \tag{12.14}$$

$$\mathsf{E}\left[\sup_{\mathbf{n}\in\mathbf{N}^d} \left(\frac{|S(\mathbf{n})|}{|\mathbf{n}|^t}\right)^q\right] < \infty. \tag{12.15}$$

Corollary 12.2 *Let X and $\{X(\mathbf{n}), \mathbf{n} \in \mathbf{N}^d\}$ be independent identically distributed random variables. Then the following three conditions are equivalent:*

$$\mathsf{E}\left[\sup_{\mathbf{n}\in\mathbf{N}^d} \frac{|X(\mathbf{n})|}{|\mathbf{n}|}\right] < \infty;$$

$$\mathsf{E}\left[\sup_{\mathbf{n}\in\mathbf{N}^d} \frac{|S(\mathbf{n})|}{|\mathbf{n}|}\right] < \infty;$$

$$\mathsf{E}\left[|X| \left(\log^+ |X|\right)^d\right] < \infty.$$

Proof of Theorem 12.3 Denote by F the common distribution function of the random variables $X(\mathbf{n})$.

Proof of the implication (12.9) $\Longrightarrow$ (12.10). *Since*

$$\sup_{\mathbf{n}\in\mathbf{N}^d} \left|\frac{X(\mathbf{n})}{b(\mathbf{n})}\right| \geq |X(\mathbf{1})|,$$

we have $\mathsf{E}\left[|X|^q\right] < \infty$ for all t and q. This completes the proof for the case when $tq > 1$.

Assume now that $tq < 1$. Formula (12.9) implies that

$$\sup_{\mathbf{n}\in\mathbf{N}^d} \frac{|X(\mathbf{n})|}{b(\mathbf{n})} < \infty \qquad \text{a. s.} \tag{12.16}$$

Now we place the random variables $X(\mathbf{n})/b(\mathbf{n})$ according to the linear ordering $\preccurlyeq$ in the space $\mathbf{N}^d$ (see Sect. A.1.2). Denote by $\{Y_k, k \geq 1\}$ the sequence of independent random variables obtained according to this ordering. Then (12.16) implies that $\limsup |Y_k| < \infty$ almost surely. By the zero–one law, there exists a non-random constant $M > 0$ such that $\limsup |Y_k| = M$ almost surely, that is

$$\limsup_{\mathbf{n}} {}_{\max} \frac{|X(\mathbf{n})|}{b(\mathbf{n})} = M \qquad \text{a. s.}$$

Therefore, with probability one, only a finite number of random events $\{|X(\mathbf{n})| \geq 2Mb(\mathbf{n})\}$ occur. Since these events are jointly independent and since the random variables $X(\mathbf{n})$ are identically distributed, the Borel–Cantelli lemma implies that

$$\sum_{\mathbf{n}\in\mathbf{N}^d} \mathsf{P}(|X| \geq 2Mb(\mathbf{n})) < \infty.$$

Put $Y = |X/2M|$,

$$a_k = \operatorname{card}\{\mathbf{n} : [b(\mathbf{n})] = k\}, \qquad A_k = \operatorname{card}\{\mathbf{n} : [b(\mathbf{n})] \leq k\}. \tag{12.17}$$

Proposition A.20 describing the asymptotic behavior of the sequence $\{A_k, k \geq 1\}$ implies that

$$\begin{aligned}\sum_{\mathbf{n}\in\mathbf{N}^d} \mathsf{P}(|X| \geq 2Mb(\mathbf{n})) &\geq \sum_{k=1}^{\infty} a_k \mathsf{P}(Y \geq k+1) = \sum_{k=2}^{\infty} A_{k-1} \mathsf{P}(k \leq Y < k+1)\\ &\geq \text{const} \sum_{k=2}^{\infty} k^{1/t} \left(\log^+ k\right)^{c-1} \mathsf{P}(k \leq Y < k+1).\end{aligned}$$

Then $\mathsf{E}\left[Y^{1/t}\left(\log^+ Y\right)^{c-1}\right] < \infty$ and thus $\mathsf{E}\left[|X|^{1/t}\left(\log^+ |X|\right)^{c-1}\right] < \infty$. The proof for $tq < 1$ is also complete.

Finally, let $tq = 1$. Without any loss of generality we assume that the first c numbers in the sequence $t_1, \dots, t_d$ are equal to t, that is, $t = t_1 = \dots = t_c < \min\{t_{c+1}, \dots, t_d\}$. Put $Y(n_1, \dots, n_c) = |X(n_1, \dots, n_c, 1, \dots, 1)|^q$. Condition (12.9) implies that

$$\mathsf{E}\left[\sup_{(n_1,\dots,n_c)\in\mathbf{N}^c} \frac{Y(n_1,\dots,n_c)}{n_1\dots n_c}\right] < \infty. \tag{12.18}$$

Lemma 12.1 *Let $\{X(\mathbf{n}), \mathbf{n} \in \mathbf{N}^d\}$ be independent identically distributed random variables. If*

$$\mathsf{E}\left[\sup_{\mathbf{n}\in\mathbf{N}^d}\left|\frac{X(\mathbf{n})}{|\mathbf{n}|}\right|\right] < \infty,$$

then

$$\mathsf{E}\left[|X|\left(\log^+ |X|\right)^d\right] < \infty.$$

In fact, Lemma 12.1 is a part of Corollary 12.2. Nevertheless, we provide below a separate proof of Lemma 12.1 to avoid any circularity in our argument.

First we finish the proof of Theorem 12.3. By Lemma 12.1, formula (12.18) adapted to our case means that $\mathsf{E}\left[|X|^q \left(\log^+ |X|\right)^c\right] < \infty$. This proves the implication (12.9) $\Longrightarrow$ (12.10).

Proof of the implication (12.10) $\Longrightarrow$ (12.9). Observe that $\mathsf{E}\left[|X|^q\right] < \infty$ for all t and q. Put $X'(\mathbf{n}) = X(\mathbf{n})\mathbb{I}(|X(\mathbf{n})| < b(\mathbf{n}))$ and $X''(\mathbf{n}) = X(\mathbf{n})\mathbb{I}(|X(\mathbf{n})| \ge b(\mathbf{n}))$. It is clear that

$$\mathsf{E}\left[\sup_{\mathbf{n}\in\mathbf{N}^d}\left(\frac{|X'(\mathbf{n})|}{b(\mathbf{n})}\right)^q\right] \le 1.$$

Since $|x+y|^q \le B_q(|x|^q + |y|^q)$ for $B_q = \max\{1, 2^{q-1}\}$ and any $x, y \in \mathbf{R}$, it remains to prove that

$$\mathsf{E}\left[\sup_{\mathbf{n}\in\mathbf{N}^d}\left(\frac{|X''(\mathbf{n})|}{b(\mathbf{n})}\right)^q\right] < \infty.$$

We have

$$\begin{aligned}\mathsf{E}\left[\sup_{\mathbf{n}\in\mathbf{N}^d}\left(\frac{|X''(\mathbf{n})|}{b(\mathbf{n})}\right)^q\right] &\le \sum_{\mathbf{n}\in\mathbf{N}^d}\frac{\mathsf{E}\left[|X''(\mathbf{n})|^q\right]}{b^q(\mathbf{n})}\\ &\le \sum_{k\ge 1}\frac{a_k}{k^q}\int_{|x|\ge k}|x|^q\,dF(x),\end{aligned} \tag{12.19}$$

where we have again used notation (12.17). It is clear that

$$\sum_{k\ge 1}\frac{a_k}{k^q}\int_{|x|\ge k}|x|^q\,dF(x) = \sum_{j\ge 1}\int_{j\le|x|<j+1}|x|^q\,dF(x)\sum_{k=1}^{j}\frac{a_k}{k^q}. \tag{12.20}$$

The Abel transform and the Lagrange mean value theorem imply

$$\sum_{k=1}^{j}\frac{a_k}{k^q} = \frac{A_j}{j^q} + \sum_{k=1}^{j-1}A_k\left[\frac{1}{k^q} - \frac{1}{(k+1)^q}\right] \le q\left[\frac{A_j}{j^q} + \sum_{k=1}^{j-1}\frac{A_k}{k^{q+1}}\right].$$

By Proposition A.20, we obtain

$$\begin{aligned}\sum_{k=1}^{j}\frac{a_k}{k^q} &\le \text{const}\left[j^{\frac{1}{t}-q}\left(\log^+ j\right)^{c-1} + \sum_{k=1}^{j} j^{\frac{1}{t}-q-1}\left(\log^+ j\right)^{c-1}\right]\\ &\le O(1)\begin{cases}1, & \text{if } tq > 1,\\ \left(\log^+ j\right)^c, & \text{if } tq = 1,\\ j^{\frac{1}{t}-q}\left(\log^+ j\right)^{c-1}, & \text{if } tq < 1.\end{cases}\end{aligned}$$

The latter estimates can explicitly be proved. Another way to prove them is to use Karamata's theorem (see Section A.9.2).

By substituting these estimates into (12.20) and by (12.19), we complete the proof of the implication (12.10) $\Longrightarrow$ (12.9).

Proof of the implication (12.11) $\Longrightarrow$ (12.9). Since the field $\{b(\mathbf{n}), \mathbf{n} \in \mathbf{N}^d\}$ is monotone, that is, $b(\mathbf{m}) \le b(\mathbf{n})$ for $\mathbf{m} \le \mathbf{n}$, bound (A.7) proves that

$$\sup_{\mathbf{n}\in\mathbf{N}^d} \frac{|X(\mathbf{n})|}{b(\mathbf{n})} \le 2^d \sup_{\mathbf{n}\in\mathbf{N}^d} \frac{|S(\mathbf{n})|}{b(\mathbf{n})},$$

giving the required implication.

Proof of the implication (12.11) $\Longrightarrow$ (12.12). Let $t < 1$. Without any loss of generality we assume that $t_1 = t$. We have already proved that (12.11) $\Longrightarrow$ (12.9) $\Longrightarrow$ (12.10). Therefore $\mathsf{E}\left[|X|^{1/t}\right] < \infty$ for any q and t. In particular $\mathsf{E}\left[|X|\right] < \infty$. Put $\mu = \mathsf{E}[X]$. The Marcinkiewicz–Zygmund strong law of large numbers (see Corollary 9.1 with $L(x) \equiv 1$) implies that

$$\lim(\max) \frac{S(\mathbf{n}) - |\mathbf{n}|\mu}{b(\mathbf{n})} = 0 \qquad \text{a. s.}$$

or

$$\sup_{\mathbf{n}\in\mathbf{N}^d} \left| \frac{S(\mathbf{n}) - |\mathbf{n}|\mu}{b(\mathbf{n})} \right| < \infty \qquad \text{a. s.}$$

By combining these results with (12.11), we obtain

$$\infty > \sup_{\mathbf{n}\in\mathbf{N}^d} |\mu| \frac{|\mathbf{n}|}{b(\mathbf{n})} \ge \sup_{n_1\ge 1} |\mu| n_1^{1-t_1}$$

and therefore $\mu = 0$.

Proof of the implication (12.9), (12.12) $\Longrightarrow$ (12.11). First, we consider the case where the terms $X(\mathbf{n})$ are symmetric. Put

$$X(\mathbf{k}, \mathbf{n}) = \begin{cases} \dfrac{X(\mathbf{k})}{b(\mathbf{n})}, & \text{if } \mathbf{k} \ge \mathbf{n}, \\ 0, & \text{otherwise.} \end{cases}$$

Given a multi-index $\mathbf{k}$, we arrange the random variables $X(\mathbf{k}, \mathbf{n})$ by the linear ordering $\preccurlyeq$ of the multi-indices $\mathbf{n}$ in the space $\mathbf{N}^d$ (see Sect. A.1.2). The infinite-dimensional vector obtained in this way is denoted by $\mathfrak{X}(\mathbf{k})$. We write for convenience

$$\mathfrak{X}(\mathbf{k}) = (X(\mathbf{k}, \mathbf{n}); \mathbf{n} \in \mathbf{N}^d).$$

A similar notation is used for other vectors, for example

$$\mathfrak{S}(\mathbf{k}) \stackrel{\text{def}}{=} \sum_{\mathbf{m}\le\mathbf{k}} \mathfrak{X}(\mathbf{m}) = \left(\frac{1}{b(\mathbf{n})} \sum_{\mathbf{m}\le\min\{\mathbf{k},\mathbf{n}\}} X(\mathbf{m}); \mathbf{n} \in \mathbf{N}^d\right),$$

where $\min\{\mathbf{k}, \mathbf{n}\} = (\min\{k_1, n_1\}, \dots, \min\{k_d, n_d\}) \in \mathbf{N}^d$. It is clear that $\mathfrak{X}(\mathbf{k})$, $\mathbf{k} \in \mathbf{N}^d$, are independent symmetric random elements taking values in the Banach space $\mathbf{R}^\infty$ of infinite sequences equipped with the norm $\|\mathfrak{x}\| = \sup_{\mathbf{n}\in\mathbf{N}^d} |x(\mathbf{n})|$, where $\mathfrak{x} = (x(\mathbf{n}); \mathbf{n} \in \mathbf{N}^d)$. It is clear that

$$\|\mathfrak{X}(\mathbf{k})\| = \max_{\mathbf{m}\le\mathbf{k}} \frac{|X(\mathbf{m})|}{b(\mathbf{k})}, \qquad \|\mathfrak{S}(\mathbf{k})\| = \max_{\mathbf{m}\le\mathbf{k}} \frac{|S(\mathbf{m})|}{b(\mathbf{m})}$$

and hence

$$\sup_{\mathbf{k}\in\mathbf{N}^d} \|\mathfrak{X}(\mathbf{k})\| = \sup_{\mathbf{n}\in\mathbf{N}^d} \frac{|X(\mathbf{n})|}{b(\mathbf{n})}, \qquad \sup_{\mathbf{k}\in\mathbf{N}^d} \|\mathfrak{S}(\mathbf{k})\| = \sup_{\mathbf{n}\in\mathbf{N}^d} \frac{|S(\mathbf{n})|}{b(\mathbf{n})}. \tag{12.21}$$

Note that (12.9) $\Longrightarrow$ (12.10) as we have already proved above. Condition (12.10) means, in particular, that $\mathsf{E}\left[|X|^{1/t}\left(\log^+|X|\right)^{c-1}\right] < \infty$ for any t and q. Therefore Corollary 9.2 implies that

$$\lim(\max) \frac{S(\mathbf{n})}{b(\mathbf{n})} = 0 \qquad \text{a. s.}$$

Then

$$\sup_{\mathbf{n}\in\mathbf{N}^d} \frac{|S(\mathbf{n})|}{b(\mathbf{n})} < \infty \qquad \text{a. s.}$$

In other words, the field $\|\mathfrak{S}(\mathbf{n})\|$ is bounded almost surely and therefore it is bounded in probability.

Since the function $\varphi(x) = |x|^q, q > 0$, is such that $\varphi(2x) \le C\varphi(x)$, the inequality

$$\mathsf{E}\left[\varphi\left(\sup_{\mathbf{n}\in\mathbf{N}^d} \|\mathfrak{X}(\mathbf{n})\|\right)\right] < \infty$$

implies that

$$\mathsf{E}\left[\varphi\left(\sup_{\mathbf{n}\in\mathbf{N}^d} \|\mathfrak{S}(\mathbf{n})\|\right)\right] < \infty.$$

The proof of this result is given in Chap. 6 for real-valued random variables. The proof for the case of Banach spaces follows the lines of that for $\mathbf{R}$ (see Remark 6.5). By recalling equalities (12.21), we complete the proof of the implication (12.9), (12.12) $\Longrightarrow$ (12.11) for symmetric random variables.

Now we come back to the general case of the theorem where the distribution of the random variables $X(\mathbf{n})$ is arbitrary (we drop the assumption that the distribution of random variables is symmetric). Assume that a field of independent identically distributed random variables $\{X'(\mathbf{n}), \mathbf{n} \in \mathbf{N}^d\}$ does not depend on $\{X(\mathbf{n}), \mathbf{n} \in \mathbf{N}^d\}$ and has the same distribution as $\{X(\mathbf{n}), \mathbf{n} \in \mathbf{N}^d\}$. Symmetrize this field by letting $X^{(s)}(\mathbf{n}) = X(\mathbf{n}) - X'(\mathbf{n})$ and consider the sums $S^{(s)}(\mathbf{n}) = \sum_{\mathbf{k} \le \mathbf{n}} X^{(s)}(\mathbf{k})$. Formula (12.9) implies that

$$\mathsf{E}\left[\left(\sup_{\mathbf{n}\in\mathbf{N}^d} \frac{|X^{(s)}(\mathbf{n})|}{b(\mathbf{n})}\right)^q\right] \le 2^{q+1}\,\mathsf{E}\left[\left(\sup_{\mathbf{n}\in\mathbf{N}^d} \frac{|X(\mathbf{n})|}{b(\mathbf{n})}\right)^q\right] < \infty.$$

By what we have already proved for symmetric distributions, we obtain

$$\mathsf{E}\left[\sup_{\mathbf{n}\in\mathbf{N}^d} \left(\frac{|S^{(s)}(\mathbf{n})|}{b(\mathbf{n})}\right)^q\right] < \infty.$$

Further, we use the symmetrization inequalities (A.78):

$$\mathsf{P}\left(\sup_{\mathbf{n}\in\mathbf{N}^d} |Y(\mathbf{n}) - \mu(Y(\mathbf{n}))| \ge x\right) \le 2\,\mathsf{P}\left(\sup_{\mathbf{n}\in\mathbf{N}^d} |Y^{(s)}(\mathbf{n})| \ge x\right),$$

where $x > 0$, $\{Y(\mathbf{n}), \mathbf{n} \in \mathbf{N}^d\}$ is an arbitrary field of random variables, $\mu(Y(\mathbf{n}))$ is a median of the random variable $Y(\mathbf{n})$, and where $\{Y^{(s)}(\mathbf{n}), \mathbf{n} \in \mathbf{N}^d\}$ is a symmetrization of the field $\{Y(\mathbf{n}), \mathbf{n} \in \mathbf{N}^d\}$. We multiply the latter inequality by x^{q-1} and integrate the result from 0 to ∞ with respect to x. We obtain

$$\mathsf{E}\left[\left(\sup_{\mathbf{n}\in\mathbf{N}^d} |Y(\mathbf{n}) - \mu(Y(\mathbf{n}))|\right)^q\right] \le 2\,\mathsf{E}\left[\left(\sup_{\mathbf{n}\in\mathbf{N}^d} |Y^{(s)}(\mathbf{n})|\right)^q\right]$$

(see Section A.19). In our case this bound means that

$$\mathsf{E}\left[\sup_{\mathbf{n}\in\mathbf{N}^d} \left|\left(\frac{S(\mathbf{n})}{b(\mathbf{n})} - \mu\left(\frac{S(\mathbf{n})}{b(\mathbf{n})}\right)\right)\right|^q\right] \le 2\,\mathsf{E}\left[\sup_{\mathbf{n}\in\mathbf{N}^d} \left(\frac{|S^{(s)}(\mathbf{n})|}{b(\mathbf{n})}\right)^q\right].$$

Therefore condition (12.11) follows from

$$\sup_{\mathbf{n}\in\mathbf{N}^d} \left|\mu\left(\frac{S(\mathbf{n})}{b(\mathbf{n})}\right)\right| < \infty. \tag{12.22}$$

Now our aim is to prove formula (12.22).

As we have already proved, (12.9) $\Longrightarrow$ (12.10) and condition (12.10) implies

$$\mathsf{E}\left[|X|^{1/t}\left(\log^+|X|\right)^{c-1}\right] < \infty$$

irrespective of a relation between the numbers $t > 0$ and $q > 0$. Similarly to the case of symmetric random variables, this result together with (12.12) enables us to conclude that

$$\lim(\max) \frac{S(\mathbf{n})}{b(\mathbf{n})} = 0 \quad \text{a. s.}$$

according to Corollary 9.1. This implies that

$$\lim(\max) \mu\left(\frac{S(\mathbf{n})}{b(\mathbf{n})}\right) = 0,$$

whence condition (12.22) follows.

This completes the proof of the theorem. □

Proof of Lemma 12.1 By Proposition A.23 (see implication (A.63)), the assumption of the lemma implies that

$$\sum_{k=1}^{\infty} \mathsf{P}\left(\sup_{\mathbf{n}\in\mathbf{N}^d} \left|\frac{|X(\mathbf{n})|}{|\mathbf{n}|}\right| \geq k\right) < \infty.$$

As usual, F denotes the common distribution function of the random variables $X(\mathbf{n})$. Then the above series converges if and only if

$$\sum_{k=1}^{\infty} \left[1 - \prod_{\mathbf{n}\in\mathbf{N}^d} F(k|\mathbf{n}|)\right] < \infty.$$

Without any loss of generality we assume that $F(1) > 0$. Note that if $0 < \theta_k \leq 1$ for all $k \geq 1$ and if the series $\sum(1 - \theta_k)$ converges, then the series $\sum \log \theta_k$ also converges, since $-\log z \sim 1 - z$ as $z \to 1$. Then the infinite product $\prod \theta_k$ is convergent. Therefore the double product

$$\prod_{k=1}^{\infty} \prod_{\mathbf{n}\in\mathbf{N}^d} F(k|\mathbf{n}|) \tag{12.23}$$

is also convergent. Denote by τ_i the number of solutions of the equation $k|\mathbf{n}| = i$. It is clear that $\{\tau_i, i \geq 1\}$ is a Dirichlet sequence for the space $\mathbf{N}^{d+1}$. With this notation, the convergence of product (12.23) is equivalent to that of the product

$$\prod_{i=1}^{\infty} F^{\tau_i}(i).$$

Since

$$\prod_{i=1}^{\infty} F^{\tau_i}(i) = \exp\left\{\sum_{i=1}^{\infty} \tau_i \log F(i)\right\},$$

the series $\sum_{i=1}^{\infty} \tau_i \left| \log F(i) \right|$ converges. Let $F^*(x) = 1 - F(x)$. Now $F^*(i) \to 0$, $i \to \infty$, and $\log(1-z) \sim -z$, $z \to 0$, whence we conclude that the series

$$\sum_{i=1}^{\infty} \tau_i F^*(i) \qquad \text{converges.}$$

An application of Lemma A.3 (for dimension $d+1$) completes the proof. □

12.2 A Generalized Strong Law of Large Numbers

Now we give an exact statement of the generalized strong law of large numbers written in the form of (12.6) (cf. Corollary 9.2).

Theorem 12.4 *Let X and $\{X(\mathbf{n}), \mathbf{n} \in \mathbf{N}^d\}$ be independent identically distributed random variables. Assume that $t_1, \dots, t_d$ are positive numbers and let t and c be as defined in* (12.8). *If $d > 1$, then the strong law of large numbers*

$$\lim_{n_1\to\infty} \frac{1}{n_1^{t_1}} \left[\sup_{\substack{n_2\ge 1\\ \dots\\ n_d \ge 1}} \frac{|S(\mathbf{n})|}{n_2^{t_2}\cdots n_d^{t_d}} \right] = 0 \quad \text{a. s.} \tag{12.24}$$

is equivalent to the condition

$$\mathsf{E}\left[|X|^{1/t} \left(\log^+ |X|\right)^{c-1} \right] < \infty \tag{12.25}$$

if $t > 1$, or to the set of conditions (12.25) *and*

$$\mathsf{E}[X] = 0 \tag{12.26}$$

if $1/2 < t \le 1$. Formula (12.24) *remains valid if the index n_1 is substituted by another index n_j, $1 < j \le d$, and the set of indices $n_2, \dots, n_d$ used to evaluate* sup *is changed accordingly.*

Theorem 12.5 *Let X and $\{X(\mathbf{n}), \mathbf{n} \in \mathbf{N}^d\}$ be independent identically distributed random variables. Assume that $t_1, \dots, t_d$ are positive numbers and let t and c be as defined in* (12.8). *If $d > 1$ and $1 \le d_0 < d$, then the strong law of large numbers*

$$\lim_{\substack{n_1\to\infty\\ \dots\\ n_{d_0}\to\infty}} \frac{1}{n_1^{t_1}\dots n_{d_0}^{t_{d_0}}}\left[\sup_{\substack{n_{d_0+1}\ge 1\\ \dots\\ n_d\ge 1}} \frac{|S(\mathbf{n})|}{n_{d_0+1}^{t_{d_0+1}}\cdots n_d^{t_d}}\right] = 0 \quad \text{a. s.} \tag{12.27}$$

is equivalent to condition (12.25) *if* $t > 1$ *or to the set of conditions* (12.25)–(12.26) *if* $1/2 < t \le 1$. *Formula* (12.27) *remains true if the indices* $n_1, \dots, n_{d_0}$ *are changed by any other collection of indices* $n_{i_1}, \dots, n_{i_{d_0}}$ *and the indices used to evaluate* sup *are changed accordingly.*

We drop the proof of Theorem 12.5, since it mimics that of Theorem 12.4.

Proof of Theorem 12.4 For a fixed k, we write the multi-indices $(k, k_2, \dots, k_d)$ according to the linear ordering $\preccurlyeq$ of the subindices $(k_2, \dots, k_d)$ in the space $\mathbf{N}^{d-1}$ (see Sect. A.1.2). The same order is used to write the coordinates of the vector

$$\xi_k = \left(n_2^{-t_2}\dots n_d^{-t_d}\sum_{k_2=1}^{n_2}\cdots\sum_{k_d=1}^{n_d} X(k, k_2, \dots, k_d);\, n_2 \ge 1, \dots, n_d \ge 1\right).$$

The random elements ξ_k, $k \ge 1$, are independent and identically distributed. Moreover, with probability one ξ_k belongs to the Banach space c_0 of sequences tending to 0 endowed with the uniform norm. This fact follows from Corollary 9.2.

Put

$$\zeta_n = \sum_{k=1}^{n} \xi_k.$$

It is clear that

$$\zeta_n = \left(n_2^{-t_2}\dots n_d^{-t_d} S(n, n_2, \dots, n_d);\, n_2 \ge 1, \dots, n_d \ge 1\right),$$

where the order of the coordinates is the same as that used for the elements ξ_k.

The Marcinkiewicz–Zygmund strong law of large numbers applied to the space c_0 (see Theorem 12.6 below) enables us to claim that if

$$\begin{cases} \mathsf{E}\left[\|\xi_1\|^{1/t_1}\right] < \infty, & \text{in the case of } t_1 > 1,\\ \mathsf{E}\left[\|\xi_1\|^{1/t_1}\right] < \infty,\ \mathsf{E}[\xi_1] = 0, & \text{in the case of } \frac{1}{2} < t_1 \le 1, \end{cases} \tag{12.28}$$

then

$$\lim_{n\to\infty} \frac{\|\zeta_n\|}{n^{t_1}} = 0 \quad \text{a. s.} \tag{12.29}$$

It is clear that the strong law of large numbers (12.29) is equivalent to (12.24), while the condition $\mathsf{E}\left[\|\xi_1\|^{1/t_1}\right] < \infty$ is equivalent to

$$\mathsf{E}\left[\sup_{\substack{n_2\ge 1\\ \dots\\ n_d\ge 1}}\left(\frac{|S(1,n_2,\dots,n_d)|}{n_2^{t_2}\dots n_d^{t_d}}\right)^{1/t_1}\right]<\infty. \tag{12.30}$$

Put $t' = \min\{t_2,\dots,t_d\}$ and $c' = \operatorname{card}\{k : t_k = t', 2 \le k \le d\}$. If $t' \ge 1$, then Theorem 12.3 applied to the space $\mathbf{N}^{d-1}$ with $q = 1/t_1$ implies that condition (12.30) is equivalent to

$$\begin{cases}\mathsf{E}\left[|X|^{1/t_1}\right]<\infty, & t_1<t',\\ \mathsf{E}\left[|X|^{1/t_1}\left(\log^+|X|\right)^{c'}\right]<\infty, & t_1=t',\\ \mathsf{E}\left[|X|^{1/t'}\left(\log^+|X|\right)^{c'-1}\right]<\infty, & t_1>t'.\end{cases} \tag{12.31}$$

Then condition (12.31) is equivalent to (12.25).

This proves the implication (12.25)–(12.26) $\Longrightarrow$ (12.24) for $\frac{1}{2} < t \le 1$ and the implication (12.25) $\Longrightarrow$ (12.24) for $t > 1$.

The inverse implications are immediate, since (12.24) yields the strong law of large numbers

$$\lim(\max)\frac{S(\mathbf{n})}{b(\mathbf{n})}=0 \quad \text{a. s.}$$

and, by Corollary 9.2, the latter is equivalent to condition (12.25) if $t > 1$ or to (12.25)–(12.26) if $\frac{1}{2} < t \le 1$. □

12.2.1 The Strong Law of Large Numbers in the Space c_0

Suppose that c_0 is the Banach space of infinite sequences of real numbers tending to 0. The norm in the space c_0 is given by

$$\|v\|=\sup_{i\ge 1}|v_i| \quad \text{for} \quad v=(v_1,\dots,v_n,\dots)\in c_0.$$

The space c_0 is separable, with separability set $C^1 \cup C^2 \cup \dots$, where C^n is the family of elements $v = (v_1,\dots,v_n,0,\dots)$ whose first n coordinates $v_1,\dots,v_n$ are rational. Moreover, the sequence of elements $\{\beta_i\} \in c_0$,

$$\beta_1=(1,0,\dots), \qquad \beta_i=(\underbrace{0,\dots,0}_{i-1\text{ zeros}},1,0,\dots), \quad i>1,$$

forms a Schauder basis in c_0. This means that, for each element $v \in c_0$, there exists a unique sequence of scalars $\{t_i\}$ such that

$$\lim_{m\to\infty}\left\|v-\sum_{i=1}^{m}t_i\beta_i\right\|=0.$$

Consider the sequence of coordinate functionals $\{f_i\}$:

$$f_i(v)=t_i,\quad v\in c_0,$$

and the sequence of tail functionals $\{Q_m\}$:

$$Q_m(v)=v-\sum_{i=1}^{m}f_i(v)\beta_i,\quad v\in c_0.$$

Theorem 12.6 *Suppose that* $\{\xi_k\}$ *are independent identically distributed* c_0*-valued random elements. Put* $\zeta_1=\xi_1$ *and* $\zeta_n=\xi_1+\cdots+\xi_n$ *for* $n\ge 2$. *If* $0<p<2$ *and if*

$$\mathsf{E}\left[\|\xi_1\|^p\right]<\infty$$

in the case of $p<1$ *or if*

$$\mathsf{E}\left[\|\xi_1\|^p\right]<\infty,\qquad \mathsf{E}[\xi_1]=0$$

in the case of $1\le p<2$, *then*

$$\lim_{n\to\infty}\left\|\frac{\zeta_n}{n^{1/p}}\right\|=0\quad\text{a. s.}\tag{12.32}$$

Theorem 12.6 follows from the results on the strong law of large numbers in general Banach spaces obtained by Azlarov and Volodin [15] for $0<p<1$, and by Adler et al. [7] for $1<p<2$. The case of $p=1$ is studied by Mourier [324].

Theorem 12.7 (T. A. Azlarov, N. A. Volodin) *Let* $\{V_n\}$ *be a sequence of independent identically distributed random elements taking values in a real-valued Banach space. If* $0<p<1$, *then the following two conditions are equivalent:*

$$\mathsf{E}\left[\|V_1\|^p\right]<\infty,$$

$$\frac{1}{n^{1/p}}\sum_{i=1}^{n}V_i\to 0\quad\text{a. s.}$$

Theorem 12.8 (A. Adler, A. Rosalsky, R. Taylor) *Let* $\{V_n\}$ *be a sequence of independent random elements taking values in a separable Banach space* E *having a Schauder basis* $\{\beta_i\}$. *Suppose that* $\{a_n\}$ *and* $\{b_n\}$ *are two sequences of real numbers such that*

$$
\begin{aligned}
0 < b_n &\uparrow \infty, \\
\sum_{j=1}^{n} |a_j| &= O(b_n), \\
\frac{|a_n|}{b_n} &= O\left(n^{-1/p}\right) \quad \text{for some } 1 \le p < 2.
\end{aligned}
$$

Assume that there exist two sequences of random variables $\{X_i\}$ *and* $\{Y_m\}$ *and a real constant* $0 < D < \infty$ *such that*

$$
\sup_{i \ge 1} \mathsf{E}\left[|X_i|^p\right] < \infty, \qquad \sup_{m \ge 1} \mathsf{E}\left[|Y_m|^p\right] < \infty
$$

and that, for all $t > 0$, $n \ge 1$, $m \ge 1$, *and* $i \ge 1$

$$
\begin{aligned}
\mathsf{P}(|f_i(V_n)| > t) &\le D\,\mathsf{P}(|DX_i| > t), \\
\mathsf{P}\left(\big|\|Q_m(V_n)\| - \mathsf{E}\left[\|Q_m(V_n)\|\right]\big| > t\right) &\le D\,\mathsf{P}(|DY_m| > t).
\end{aligned}
$$

If

$$
\lim_{m \to \infty} \sup_{n \ge 1} \mathsf{E}\left[\|Q_m(V_n)\|\right] = 0
$$

and if $\mathsf{E}[V_n] = 0$ *for all* $n \ge 1$, *then*

$$
\frac{1}{b_n} \sum_{i=1}^{n} a_i V_i \to 0 \qquad \text{a. s.}
$$

Proof of Theorem 12.6 The case of $0 < p < 1$ follows from Theorem 12.7. For $1 \le p < 2$, apply Theorem 12.8 with $a_i = 1$ and $b_n = n^{1/p}$. Since the random elements $\{V_n\}$ are identically distributed in the space c_0, all the assumptions of Theorem 12.8 hold. Therefore the strong law of large numbers (12.32) also holds for $1 \le p < 2$. □

12.3 Moments of the Supremum and the Law of the Iterated Logarithm

Put $b_k = \sqrt{k \log^+ \log^+ k}$ and $b(\mathbf{n}) = b_{|\mathbf{n}|}$. The normalization $\{b(\mathbf{n}), \mathbf{n} \in \mathbf{N}^d\}$ is used in the law of the iterated logarithm (see Theorem 10.9). The proof of the following result follows the lines of that of Theorem 12.3.

Theorem 12.9 *Let* $\{X(\mathbf{n})\}$ *be independent identically distributed symmetric random variables. If an increasing function* $\varphi(x)$ *is such that*

$$
\varphi(2x) \le C\varphi(x), \qquad x > 0, \tag{12.33}
$$

for some constant $C > 0$, *then the following two conditions are equivalent:*

$$\mathsf{E}\left[\varphi\left(\sup_{\mathbf{n}\in\mathbf{N}^d}\frac{|S(\mathbf{n})|}{\sqrt{|\mathbf{n}|\log^+\log^+|\mathbf{n}|}}\right)\right]<\infty, \tag{12.34}$$

$$\mathsf{E}\left[\varphi\left(\sup_{\mathbf{n}\in\mathbf{N}^d}\frac{|X(\mathbf{n})|}{\sqrt{|\mathbf{n}|\log^+\log^+|\mathbf{n}|}}\right)\right]<\infty. \tag{12.35}$$

An example of a function satisfying condition (12.33) is easy to construct, say $\varphi(x) = x^2$. For this function, one can write a moment condition on $X(\mathbf{n})$ equivalent to either of (12.34) and (12.35).

Corollary 12.3 *Let X and $\{X(\mathbf{n})\}$ be independent identically distributed symmetric random variables. The following three conditions are equivalent:*

$$\mathsf{E}\left[\sup_{\mathbf{n}\in\mathbf{N}^d}\left(\frac{|S(\mathbf{n})|)}{\sqrt{|\mathbf{n}|\log^+\log^+|\mathbf{n}|}}\right)^2\right]<\infty,$$

$$\mathsf{E}\left[\sup_{\mathbf{n}\in\mathbf{N}^d}\left(\frac{|X(\mathbf{n})|)}{\sqrt{|\mathbf{n}|\log^+\log^+|\mathbf{n}|}}\right)^2\right]<\infty,$$

$$\mathsf{E}\left[X^2\frac{\left(\log^+|X|\right)^d}{\log^+\log^+|X|}\right]<\infty.$$

Corollary 12.3 is quite easy to adapt to the general case of non-symmetric random variables $X(\mathbf{n})$. We omit the corresponding statements and leave them to an interested reader.

12.4 A Generalized Law of the Iterated Logarithm

We consider a special form of the law of the iterated logarithm in this section.

Let $d = 2$ and let $X(m, n)$ be independent Gaussian random variables with parameters 0 and 1. Put $\mathrm{L}_2(z) = \log^+\log^+ z$ and $\chi(z) = \sqrt{2z\mathrm{L}_2(z)}$ for $z > 0$. In what follows we show that Corollary 12.3 implies

$$\limsup_{m\to\infty}\frac{1}{\chi(m)}\left[\sup_{n\geq 1}\frac{|S(m,n)|}{\chi(n)}\right]=\sqrt{2}\quad \text{a. s.} \tag{12.36}$$

One can compare this result with the law of the iterated logarithm (Theorem 10.9). Observe that Theorem 10.9 implies that

$$\lim_{\substack{m\to\infty\\ n\to\infty}}\frac{|S(m,n)|}{\chi(m)\chi(n)}=0\quad \text{a. s.} \tag{12.37}$$

Indeed, we obtain from Theorem 10.9 that

$$0 < \limsup_{\substack{m\to\infty \\ n\to\infty}} \frac{|S(m,n)|}{\chi(mn)} < \infty \qquad \text{a. s.}$$

Since

$$\lim(\min) \frac{\mathrm{L}_2(mn)}{\mathrm{L}_2(m)\mathrm{L}_2(n)} = 0, \tag{12.38}$$

equality (12.37) is immediate. Comparing (12.36) and (12.37) we see that the random variables $\sup_{n\ge 1} \frac{|S(m,n)|}{\chi(n)}$ are "much larger" than $\frac{|S(m,n)|}{\chi(n)}$.

Proof of formula (12.36). Put

$$\xi_i = \left(\frac{1}{\sqrt{n\mathrm{L}_2(n)}} \sum_{j=1}^{n} X(i,j),\ n \ge 1 \right), \qquad \zeta_i = \left(\frac{1}{\sqrt{n\mathrm{L}_2(n)}} S(i,n),\ n \ge 1 \right).$$

It is clear that ξ_i, $i \ge 1$, are independent identically distributed Gaussian elements in the space ℓ_∞ of bounded sequences. Then, by Corollary 12.3, we have

$$\mathsf{E}\left[\sup_{n\ge 1} \left(\frac{|S(1,n)|}{\sqrt{n\mathrm{L}_2(n)}} \right)^2 \right] < \infty.$$

Fernique's inequality (see below) implies that

$$\mathsf{E}\left[e^{\alpha \|\xi_1\|^2} \right] < \infty$$

for all

$$\alpha < \frac{1}{2\sigma^2} = \frac{1}{2}, \quad \text{where} \quad \sigma^2 = \sup_{n\ge 1} \mathsf{E}\left[\left(\frac{|S(1,n)|}{\sqrt{n\mathrm{L}_2(n)}} \right)^2 \right].$$

Theorem 12.10 (Fernique's inequality) *Let $(E, \mathscr{B})$ be a measurable vector space, X a Gaussian vector in E, N a pseudo semi-norm in E, and $\{y_n\}$ a sequence of linear measurable forms in $(E, \mathscr{B})$ satisfying*

$$\mathsf{P}\left(N(X) = \sup_{n\in\mathbf{N}} |\langle X, y_n\rangle| < \infty \right) = 1.$$

Denote by σ the upper limit of the mean square deviations of the Gaussian random variables $\langle X, y_n\rangle$. Then

$$\mathsf{E}\left[e^{\alpha N^2(X)} \right] < \infty \quad \textit{for } \alpha < \frac{1}{2\sigma^2}. \tag{12.39}$$

Since the element $\zeta_m/\sqrt{m}$, $m \geq 1$, has the same distribution as ξ_1, we have

$$\mathsf{P}(\|\zeta_m\| \geq x\sqrt{m}) = \mathsf{P}(\|\xi_1\| \geq x) \leq e^{-\alpha x^2}\,\mathsf{E}\left[e^{\alpha\|\xi_1\|^2}\right]$$

for all $x > 0$ and $\alpha < 1/2$.

Now we apply a generalization of Lévy's inequality for symmetric random elements taking values in general vector spaces.

Theorem 12.11 (Lévy's inequality for Banach spaces) *Let $\{V_n\}$ be independent symmetric random elements taking values in a separable Banach space. Put*

$$S = \sum_{n=1}^{\infty} V_n, \qquad M = \max_{n\geq 1} \|V_1 + \cdots + V_n\|.$$

If the series S converges almost surely, then

$$\mathsf{P}(M \geq r) \leq 2\,\mathsf{P}(\|S\| \geq r) \tag{12.40}$$

for all $r > 0$.

Let $\theta > 1$, $n_k = [\theta^k]$, and $\varepsilon > 0$. We apply Theorem 12.11 for all $0 < \alpha < \frac{1}{2}$:

$$\begin{aligned}\mathsf{P}\left(\max_{i\leq n_k}\|\zeta_i\| \geq \sqrt{2(1+\varepsilon)n_k \mathrm{L}_2(n_k)}\right) &\leq 2\,\mathsf{P}\left(\|\zeta_{n_k}\| \geq \sqrt{2(1+\varepsilon)n_k\mathrm{L}_2(n_k)}\right)\\ &\leq \frac{\text{const}}{k^{2\alpha(1+\varepsilon)}}.\end{aligned} \tag{12.41}$$

If $\alpha < 1/2$ is such that $2\alpha(1+\varepsilon) > 1$, then

$$\sum_{k=1}^{\infty}\mathsf{P}\left(\max_{i\leq n_k}\|\zeta_i\| \geq \sqrt{2(1+\varepsilon)n_k\mathrm{L}_2(n_k)}\right) < \infty.$$

A standard argument used in the proof of the law of the iterated logarithm yields

$$\limsup_{m\to\infty}\frac{\|\zeta_m\|}{\sqrt{m\mathrm{L}_2(m)}} \leq \sqrt{2} \qquad \text{a. s.}$$

The inverse inequality is clear:

$$\limsup_{m\to\infty}\left[\sup_{n\geq 1}\frac{|S(m,n)|}{\sqrt{mn\mathrm{L}_2(m)\mathrm{L}_2(n)}}\right] \geq \limsup_{m\to\infty}\frac{|S(m,1)|}{\sqrt{m\mathrm{L}_2(m)}} = \sqrt{2} \qquad \text{a. s.}$$

by the law of the iterated logarithm. □

12.5 Comments

Different results on dominated convergence can be found in the monograph by Krengel [261]. The name given to these kinds of results is due to Fukamiya [113]. The fact that the expectations are finite in (12.5) for $d = 1$ is used, for example, in optimal stopping problems (see, for example, Robbins et al. [58]) and in sequential optimization problems (see Cairoli and Dalang [44]). Other proofs of Theorem 12.2 are due to McCabe and Shepp [303], Gundy [127], and Davis [66].

Gut [131] established conditions for (12.4) with n replaced by k_n, where $\{k_n\}$ is a subsequence of positive integers (also see Choi and Sung [53]); necessary and sufficient conditions in the case of a general subsequence $\{n_k\}$ are obtained in [227].

More general normalizations $\{c_n\}$ in (12.4) are studied by Choi and Sung [54], Li [286], and by Li and Huang [287].

Results like (12.5) have also been studied in different situations for dependent random variables. Chen and Gan [48] studied random variables with φ-mixing or with ρ-mixing [49] (the case of ρ^--mixing was considered by Liu and Liu [297]); Esseen and Janson [88] considered martingale differences; Wang et al. [407] focused on negatively associated random variables.

Section 12.1 Theorem 12.3 for $tq \geq 1$ is proved by Giang [123] under an additional assumption that $\mathsf{E}[X] = 0$ if the expectation exists. Corollary 12.1 is due to Gut [130], Corollary 12.2 is proved by Gabriel [115]. A different proof of Corollary 12.2 is given by Khoshnevisan [205]. Etemadi [91] extends Corollary 12.2 to pairwise independent random variables. Sufficient conditions for (12.5) in the case of random variables with multi-dimensional indices are studied by Kholmuradov [204] for some functions g. A link between (12.11) and (12.13) in the case of $d = 1, q = 1, 1 < t < 2$ is established by Klass [207].

Section 12.2 Theorem 12.4 is proved in [238]. De Acosta [2] has proved that the strong law of large numbers (12.32) follows from the law of large numbers.

Section 12.3 Theorem 12.9 is proved in [217]. Corollary 12.3 holds in general, that is, not only for symmetric random variables. Corollary 12.3 is true for $d \geq 1$ in general, that is, not only for symmetric random variables; for $d = 1$ and $\varphi(x) = |x|^q$, this result is considered by Siegmund [369] if $q \geq 2$ is an integer and by Teicher [394] if $q > 2$ is general; the case of $0 < q < 2$ is studied by LePage [284]. The existence of moments of normalized sums in Banach lattices is studied by Matsak [311] and Matsak and Plichko [312].

Section 12.4 The law of the iterated logarithm (12.36) is proved in [217]. Inequality (12.39) was proved at the same time and independently by Fernique [105] and Marcus and Shepp [304] (also see [106]). Lévy's inequality for Banach spaces is given in the book by Kahane [188]. A special case of Theorem 12.11 is obtained by Hoffman-Jørgensen [163]. More general results similar to the Lévy inequality are obtained by Buldygin [33].

Chapter 13
Complete Convergence

A sequence of random variables $\{U_n\}$ is said to *converge completely* to zero if

$$\sum_{n=1}^{\infty} \mathsf{P}(|U_n| \geq \varepsilon) < \infty \tag{13.1}$$

for all $\varepsilon > 0$. One can extend the definition to the case of non-zero limits, namely a sequence of random variables $\{U_n\}$ converges completely to a constant $c \in \mathbf{R}$ if $\{U_n - c\}$ converges completely to zero. We add to this definition that a sequence of random variables $\{U_n, n \geq 1\}$ is said to be *completely bounded* by a constant ε_0 if relation (13.1) holds for all $\varepsilon \geq \varepsilon_0$.

The following result is well known in the theory of limit theorems for cumulative sums of independent random variables.

Theorem 13.1 (P.L. Hsu, H. Robbins, P. Erdös) *Let $\{X_n, n \geq 1\}$ be independent identically distributed random variables and $S_1 = X_1$, $S_n = X_1 + \cdots + X_n, n \geq 2$. The sequence $\{S_n/n,\ n \geq 1\}$ converges completely to the common expectation of the random variables X_n if and only if*

$$\mathsf{E}\left[X_1^2\right] < \infty. \tag{13.2}$$

In other words, Theorem 13.1 asserts that the convergence for all $\varepsilon > 0$ of the series

$$\sum_{n=1}^{\infty} \mathsf{P}(|S_n| \geq \varepsilon n) \tag{13.3}$$

is equivalent to the set of two conditions

$$\mathsf{E}[X_1] = 0, \qquad \mathsf{E}\left[X_1^2\right] < \infty.$$

O. Klesov, *Limit Theorems for Multi-Indexed Sums of Random Variables*,
Probability Theory and Stochastic Modelling 71, DOI 10.1007/978-3-662-44388-0_13

Looking at series (13.3) one may arrive at the idea of studying series of the following type

$$\sum_{n=1}^{\infty} w_n \mathsf{P}(|S_n| \ge \varepsilon b_n) \tag{13.4}$$

for various weight and normalizing sequences $\{w_n, n \ge 1\}$ and $\{b_n, n \ge 1\}$. Most of the results of this chapter deal with the case $b_n = n^\alpha$, $\alpha > 0$, that is, we study the convergence of the series

$$\sum_{n=1}^{\infty} w_n \mathsf{P}(|S_n| \ge \varepsilon n^\alpha) \tag{13.5}$$

for various sequences of weight coefficients $\{w_n, n \ge 1\}$.

The following result generalizes Theorem 13.1.

Theorem 13.2 (F. Spitzer, L. Baum, M. Katz) *If* $\alpha > \frac{1}{2}$ *and* $\alpha r \ge 1$, *then the following two statements are equivalent*:

$$\sum_{n=1}^{\infty} n^{\alpha r-2} \mathsf{P}(|S_n| \ge n^\alpha \varepsilon) < \infty \textit{ for all } \quad \varepsilon > 0 \tag{13.6}$$

and

$$\mathsf{E}\left[|X_1|^r\right] < \infty,$$

$$\mathsf{E}[X_1] = 0 \;\; \textit{if} \;\; \alpha \le 1. \tag{13.7}$$

Series like (13.5) appear not only in the framework of complete convergence. For example, the convergence of such series indicates a rate of convergence in the law of large numbers for $\{S_n/n^\alpha\}$ if $w_n \to \infty$. Moreover, a criterion for the strong law of large numbers can also be expressed in terms of the convergence of series (13.5) with a specific sequence $\{w_n\}$.

Yet another application of series (13.5) is related to the complete convergence of random variables with multi-indices. We say that a field of random variables $\{U(\mathbf{n}), \mathbf{n} \in \mathbf{N}^d\}$ converges completely to a constant c if

$$\sum_{\mathbf{n} \in \mathbf{N}^d} \mathsf{P}(|U(\mathbf{n}) - c| \ge \varepsilon) < \infty$$

for all $\varepsilon > 0$. If $U(\mathbf{n}) = S(\mathbf{n})/|\mathbf{n}|$ where $S(\mathbf{n})$ are rectangular sums of independent identically distributed random variables, then the strong law of large numbers for complete convergence is equivalent to the convergence for all $\varepsilon > 0$ of the series

$$\sum_{\mathbf{n} \in \mathbf{N}^d} \mathsf{P}(|S(\mathbf{n})| \ge \varepsilon |\mathbf{n}|). \tag{13.8}$$

Putting $w_k = \operatorname{card}\{\mathbf{n} \in \mathbf{N}^d : |\mathbf{n}| = k\}$, we reduce series (13.8) to series (13.5) with weight coefficients w_n being equal to the values of the Dirichlet function. Thus studying series (13.5) we automatically get the corresponding results for the complete convergence of multi-indexed sums.

We start with some general results for series (13.5), whence we later derive some applications for multi-indexed sums.

13.1 Necessary Conditions for Complete Convergence

The first problem we solve is to find the class of weight coefficients $\{w_n, n \geq 1\}$ such that the series (13.5) converges for all $\varepsilon > 0$ for at least one sequence of random variables $\{X_n\}$ with a nondegenerate distribution function. In other words, we want to exclude those sequences $\{w_n\}$ for which there is no sequence of nondegenerate random variables $\{X_n\}$ such that series (13.5) converges for all $\varepsilon > 0$.

Note that series (13.5) converges for all $\varepsilon > 0$ and for all random variables $\{S_n\}$ if $\sum w_n < \infty$. This is why we restrict our attention to the case of divergent series $\sum w_n$.

13.1.1 Necessary Conditions for Weight Coefficients

The cases $\alpha \leq 1$ and $\alpha > 1$ are essentially different. For example, if $\alpha > 1$, then there *always* exists a sequence of nondegenerate random variables for which series (13.5) converges for all $\varepsilon > 0$, namely any sequence of bounded random variables $\{X_n\}$ can serve as an example in this case. Thus the problem mentioned above is non-trivial only if $\alpha \leq 1$.

Theorem 13.3 *Assume that $\{X_n, n \geq 1\}$ is a sequence of independent identically distributed random variables with a distribution function F. Let $w_n \geq 0$, $\alpha > 0$, and let series $\sum w_n$ diverge.*

(i) *If $\alpha \leq \frac{1}{2}$, then series* (13.5) *diverges for all $\varepsilon > 0$ and for any nondegenerate distribution function F, that is for all $F \neq \mathbb{I}_{(x_0,\infty)}$, $x_0 \in \mathbf{R}$.*

(ii) *If $\frac{1}{2} < \alpha \leq 1$, then the convergence for all $\varepsilon > 0$ of series* (13.5) *implies that*

$$\lim_{n\to\infty} \frac{\log(1 + w_n)}{n^{2\alpha-1}} = 0. \tag{13.9}$$

(iii) *If $\frac{1}{2} < \alpha \leq 1$ and F is a Gaussian distribution function, then the convergence for all $\varepsilon > 0$ of series* (13.5) *is equivalent to condition* (13.9).

(iv) *If $\alpha > 1$, then series* (13.5) *converges for all $\varepsilon > 0$ and for any distribution function F with a bounded support.*

As a consequence of statement (ii), there is no nondegenerate distribution function such that series (13.5) converges for all $\varepsilon > 0$ if $w_n = e^n$ and $\alpha = 1$.

Sufficient conditions for the convergence of series (13.5) for general distribution functions are studied in the rest of this chapter.

Remark 13.1 Let $\alpha > 0$. We say that a sequence of positive numbers $\{w_n\}$ belongs to the class $\mathscr{W}_\alpha$ if $\sum w_n = \infty$ and there exists a nondegenerate distribution function $F \neq \mathbb{I}_{(x_0,\infty)}$ such that series (13.5) converges for all $\varepsilon > 0$. Then Theorem 13.3 can be stated as follows.

Theorem 13.4 *Let $\{w_n\}$ be a sequence of non-negative numbers and $\alpha > 0$. If $\sum w_n = \infty$, then*

(1) $\mathscr{W}_\alpha = \varnothing$ *for* $\alpha \le \frac{1}{2}$;
(2) $\mathscr{W}_\alpha \neq \varnothing$ *for* $\frac{1}{2} < \alpha \le 1$ *if and only if condition* (13.9) *holds*;
(3) $\mathscr{W}_\alpha \neq \varnothing$ *for* $\alpha > 1$.

To estimate the probabilities in series (13.5) we need a generalized Kolmogorov exponential inequality for bounded random variables.

Lemma 13.1 (generalized Kolmogorov exponential lower bound) *Let $\{Y_n, n \ge 1\}$ be independent identically distributed random variables with a common nondegenerate symmetric distribution function. Put $T_n = \sum_{k=1}^n Y_k$ and assume that $|Y_k| \le a$ almost surely for some constant $a > 0$. If $\frac{1}{2} < \alpha \le 1$, then there are two universal constants $C_1 > 0$ and $C_2 > 0$ such that*

$$\mathsf{P}(|T_n| \ge \varepsilon n^\alpha) \ge C_1 \exp\left\{-C_2 \delta n^{2\alpha-1}\right\} \tag{13.10}$$

for all $\varepsilon > 0$ and $n \ge 1$, where $\delta = \varepsilon^2 / \operatorname{var}[X_1]$.

Proof of Lemma 13.1 The result follows from the lower Kolmogorov exponential inequality if $\alpha < 1$ (see Theorem 10.6). In contrast, the case $\alpha = 1$ is different, since the condition

$$\frac{x_n M_n}{B_n} \to 0, \qquad n \to \infty, \tag{13.11}$$

does not hold under assumptions of Lemma 13.1, where $M_n = a$ and $B_n = \operatorname{var}[T_n]$. Instead,

$$\frac{x_n M_n}{B_n} \to d, \qquad n \to \infty, \tag{13.12}$$

for some constant $d > 0$. Repeating the proof of the lower Kolmogorov exponential inequality (see, for example, [336]) but for the case of (13.12) rather than (13.11) we prove bound (13.10). We do not give the details of the proof, since it is quite similar to that of Theorem 10.6 (see, for example, [336], Chap. 10, Sect. 1, Lemma 2). □

We would like to stress that condition (13.12) implies (13.10) and does not imply the original Kolmogorov inequality (10.8) which is a slightly stronger result than (13.10). Nevertheless, bound (13.10) is sufficient for the purposes of the current section.

Proof of Theorem 13.3 Assume that series (13.5) converges for all $\varepsilon > 0$ and some distribution function $F \neq \mathbb{I}_{(x_0,\infty)}$. Denote by $\{X_n, n \geq 1\}$ a sequence of independent random variables for which F is the distribution function. Finally, let $\{X_n^{(s)}\}$ be the symmetrizations of the random variables $\{X_n\}$ and put $S_n^{(s)} = \sum_{i=1}^n X_i^{(s)}$.

Since $\mathsf{P}\left(|S_n^{(s)}| \geq x\right) \leq 2\mathsf{P}(|S_n| \geq x/2)$, the series $\sum w_n \mathsf{P}(|S_n^{(s)}| \geq \varepsilon n^\alpha)$ converges for all $\varepsilon > 0$. Fix $c > 0$ such that

$$\sigma^2(c) \stackrel{\text{def}}{=} \int_{-c}^{c} x^2 \, dF^{(s)}(x) > 0.$$

Such a constant c does not exist only if the distribution function F is degenerate, that is if $F = \mathbb{I}_{(x_0,\infty)}$ (in which case $F^{(s)}$ is concentrated at 0).

Now put $Y_n = X_n^{(s)} \mathbb{I}_{\{|X_n^{(s)}|<c\}}$ and $T_n = \sum_{i=1}^n Y_i$. Lemma 6.9 implies that

$$\mathsf{P}(T_n \geq x) \leq 2\mathsf{P}\left(S_n^{(s)} \geq x\right).$$

Thus we have proved that the series $\sum w_n \mathsf{P}(|T_n| \geq \varepsilon n^\alpha)$ converges for all $\varepsilon > 0$. The rest of the proof is done separately for the following three cases.

The case $0 < \alpha \leq \frac{1}{2}$. Since T_n are the cumulative sums of independent identically distributed nondegenerate uniformly bounded random variables, the central limit theorem implies that

$$\frac{T_n}{\sqrt{n\sigma^2(c)}} \stackrel{\mathrm{w}}{\longrightarrow} Z, \qquad n \to \infty,$$

where Z is a Gaussian $\mathscr{N}(0, 1)$ random variable. Thus

$$\mathsf{P}\left(|T_n| \geq \varepsilon n^\alpha\right) \geq \mathsf{P}\left(|T_n| \geq \varepsilon\sqrt{n}\right) \to \mathsf{P}(|Z| \geq \varepsilon/\sigma(c)) > 0,$$

whence we derive $\sum w_n < \infty$. This contradicts the assumption of Theorem 13.3, that is, $\mathscr{W}_\alpha = \varnothing$ in this case.

The case $\frac{1}{2} < \alpha \leq 1$. Put $x_n = \varepsilon n^\alpha$. Then inequality (13.10) implies that the series

$$\sum_{n=1}^{\infty} w_n \exp\left\{-C_2\varepsilon^2 n^{2\alpha-1}\right\} \tag{13.13}$$

converges for all $\varepsilon > 0$.

If condition (13.9) were not satisfied then there would exists a number $\delta_0 > 0$ and a subsequence $\{n_k, k \geq 1\}$ such that $\log(1 + w_{n_k}) \geq \delta_0 n_k^{2\alpha-1}$. This would mean that series (13.13) diverges for sufficiently small ε, namely for $\varepsilon < (\delta_0/C_2)^{1/2}$, and this proves (ii).

Statement (iii) for a Gaussian distribution function follows from the known bound for the tail of the Gaussian distribution function:

$$\mathsf{P}(|Z| \geq x) \asymp \frac{1}{x} e^{-x^2/2}.$$

Using this asymptotics we prove that the convergence of series (13.5) for a given $\varepsilon > 0$ is equivalent to

$$\sum_{n=1}^{\infty} \frac{w_n}{n^{\alpha-\frac{1}{2}}} e^{-\varepsilon^2 n^{2\alpha-1}/2} < \infty \quad \Longleftrightarrow \quad \sum_{n=1}^{\infty} e^{\log(1+w_n)-\left(\alpha-\frac{1}{2}\right)\log n - \varepsilon^2 n^{2\alpha-1}/2} < \infty.$$

Since $\varepsilon > 0$ is an arbitrary number, the latter condition is equivalent to (13.9).

The case $\alpha > 1$. If the random variables X_n are bounded almost surely, then $\mathsf{P}(|S_n| \geq \varepsilon n^\alpha) = 0$ for sufficiently large n. This proves (iv). □

13.1.2 Necessary Conditions for Distributions of Random Variables

Another necessary condition for the convergence of series (13.5) for all $\varepsilon > 0$ is written below

$$\sum_{n=1}^{\infty} n w_n \mathsf{P}(|X| \geq \varepsilon n^\alpha) < \infty \qquad \text{for all} \quad \varepsilon > 0. \tag{13.14}$$

Condition (13.14) is also sufficient in many particular cases.

Theorem 13.5 *Assume that* $\{X_n, n \geq 1\}$ *is a sequence of independent identically distributed random variables with a distribution function* F. *Let* $w_n \geq 0, \alpha > 0$, *and assume that the series* $\sum w_n$ *diverges. If series* (13.5) *converges for all* $\varepsilon > 0$ *and*

$$\sup_{n\geq 1} \left[n\mathsf{P}(|X| \geq \varepsilon n^\alpha) \right] < \infty, \tag{13.15}$$

then series (13.14) *converges for all* $\varepsilon > 0$.

Remark 13.2 Condition (13.15) holds, for example, if $\mathsf{E}\left[|X|^{1/\alpha}\right] < \infty$. Indeed, in this case

$$n\mathsf{P}(|X| \geq \varepsilon n^\alpha) \leq \frac{1}{\varepsilon^{1/\alpha}} \int\limits_{|X|\geq \varepsilon n^\alpha} |X|^{1/\alpha} \, d\mathsf{P} \to 0, \qquad n \to \infty.$$

The assumptions of Theorem 13.5 can be weakened for regularly varying sequences $\{w_n, n \geq 1\}$.

Corollary 13.1 *Let $w_n = n^\rho L(n)$, where $\rho > -1$ and $L(\cdot)$ is some positive regularly varying function. If series* (13.5) *converges for all $\varepsilon > 0$, then series* (13.14) *converges for all $\varepsilon > 0$.*

Remark 13.3 It is not difficult to show that if $L(n) \equiv 1$ and $\rho = \alpha r - 2$ in Corollary 13.1, then the convergence of series (13.14) for only one number $\varepsilon > 0$ is equivalent to the existence of the moment $\mathsf{E}\left[|X|^r\right]$, where $r = 1 + (1+\rho)/\alpha$. Thus the necessity in Theorem 13.2 for $\alpha r > 1$ follows from Corollary 13.1.

The simplest case of Theorem 13.5 is $\alpha r = 2$ and it corresponds to the necessity parts of the results of Hsu, Robbins, and Erdös (Theorem 13.1) and Baum, Katz, and Spitzer (Theorem 13.2).

Corollary 13.2 *Let $w_n = 1$ for all $n \geq 1$. If series* (13.5) *converges for all $\varepsilon > 0$, then*

$$\mathsf{E}\left[|X|^{2/\alpha}\right] < \infty \text{ and } \mathsf{E}[X] = 0, \quad \text{if} \quad \alpha \leq 1. \tag{13.16}$$

We need two auxiliary results to prove Theorem 13.5.

Lemma 13.2 *Let $\{Z_i, i \leq N\}$ be independent non-negative random variables. Then, for all $t > 0$,*

$$\mathsf{P}\left(\max_{i\leq N} Z_i \geq t\right) \geq \frac{\pi_N}{1+\pi_N}, \qquad \pi_N \overset{\text{def}}{=} \sum_{i=1}^{N} \mathsf{P}(Z_i \geq t).$$

Proof of Lemma 13.2 Recall that $1 - x \leq \exp\{-x\}$ and $1 - \exp\{-x\} \geq x/(1+x)$ for $x \geq 0$. Since the random variables are independent,

$$\mathsf{P}\left(\max_{i\leq N} Z_i \geq t\right) = 1-\prod_{i=1}^{N}[1-\mathsf{P}(Z_i \geq t)] \geq 1-\exp\left\{-\sum_{i=1}^{N}\mathsf{P}(Z_i \geq t)\right\} \geq \frac{\pi_N}{1+\pi_N}.$$

□

Lemma 13.3 *Let X, $\{X_n\}$ be independent identically distributed random variables and let $\{t_n\}$ be a sequence of real numbers such that*

$$a \overset{\text{def}}{=} \sup_{n\geq 1} n\mathsf{P}(|X^{(s)}| \geq 4t_n) < \infty,$$

where $X^{(s)}$ is the symmetrization of a random variable X. Then

$$n\mathsf{P}(|X^{(s)}| \geq 4t_n) \leq 8(1+a)\mathsf{P}(|S_n| \geq t_n).$$

Proof of Lemma 13.3 Lemma 13.2 with $t = 4t_n$ implies that

$$\frac{n\mathsf{P}(|X^{(s)}| \geq 4t_n)}{1+a} \leq \frac{n\mathsf{P}(|X^{(s)}| \geq 4t_n)}{1+n\mathsf{P}(|X^{(s)}| \geq 4t_n)} \leq \mathsf{P}\left(\max_{i \leq n} |X_i^{(s)}| \geq t\right).$$

Lévy's inequality for the maximum of cumulative sums of symmetric random variables (see Corollary 2.4 for $d = 1$) implies that

$$\begin{aligned}\mathsf{P}\left(\max_{i \leq n} |X_i^{(s)}| \geq t\right) &\leq 2\mathsf{P}\left(\max_{i \leq n} |S_i^{(s)}| \geq t/2\right) \leq 4\mathsf{P}\left(|S_n^{(s)}| \geq t/2\right) \\ &\leq 8\mathsf{P}\left(|S_n| \geq t/4\right).\end{aligned}$$

Combining the latter two results we complete the proof of Lemma 13.3. □

Proof of Theorem 13.5 Lemma 13.3 implies that

$$\sum_{n=1}^{\infty} nw_n \mathsf{P}(|X^{(s)}| \geq \varepsilon n^{\alpha}) < \infty \qquad \text{for all} \quad \varepsilon > 0.$$

Applying the symmetrization inequalities (see Sect. A.18 for $d = 1$) we prove that

$$\sum_{n=1}^{\infty} nw_n \mathsf{P}(|X - \operatorname{med}(X)| \geq \varepsilon n^{\alpha}) < \infty \qquad \text{for all} \quad \varepsilon > 0,$$

where $\operatorname{med}(X)$ is a median of the random variable X. Thus condition (13.14) follows. □

Proof of Corollary 13.1 Choose ρ_1 such that $-1 < \rho_1 < \rho$. Any regularly varying function of $\mathcal{RV}_\rho$ is such that $n^{\rho_1} = O(w_n)$, whence

$$\sum_{n=1}^{\infty} n^{\rho_1} \mathsf{P}(|S_n| \geq \varepsilon n^{\alpha}) < \infty \quad \text{for all} \quad \varepsilon > 0.$$

Theorem 13.2 implies that $\mathsf{E}\left[|X|^{(\rho_1+2)/\alpha}\right] < \infty$, that is $\mathsf{E}\left[|X|^{1/\alpha}\right] < \infty$. As in Remark 13.2, $n\mathsf{P}(|X| \geq \varepsilon n^{\alpha}) \to 0$, $n \to \infty$, for all $\varepsilon > 0$. Therefore condition (13.15) holds and thus Corollary 13.1 follows from Theorem 13.5. □

Proof of Corollary 13.2 In view of Corollary 13.1, we only need to prove that $\mathsf{E}[X] = 0$ in the case of $\alpha \leq 1$. This property is a corollary of the Marcinkiewicz–Zygmund strong law of large numbers (see Corollary 9.2 for $d = 1$). □

13.2 Sufficient Conditions for Complete Convergence

In this section, we study some sufficient conditions for the convergence of series (13.5) and try to write these conditions in terms of the existence of certain moments of random variables $\{X_n\}$. We start with a brief discussion of some "candidates" for such a condition.

Theorem 13.5 implies that if $w_n = n^{\alpha r-2}$, $\alpha r \geq 1$, then the convergence of series (13.5) for all $\varepsilon > 0$ implies the convergence of series (13.14) for all $\varepsilon > 0$ and the latter is equivalent to $\mathsf{E}\left[|X|^r\right] < \infty$.

Denote by $\mathcal{RV}_{\geq\rho_0}$ the class of sequences $\{n^\rho\}$ with $\rho \geq \rho_0$. As noted above, the minimal moment condition for the convergence of series (13.14) for all $\varepsilon > 0$ in the class $\mathcal{RV}_{\geq -1}$ of sequences $\{w_n\}$ is given by

$$\mathsf{E}\left[|X|^{1/\alpha}\right] < \infty. \tag{13.17}$$

If $\alpha \leq 1$, then (13.17) needs to be completed with the centering condition:

$$\mathsf{E}\left[X\right] = 0. \tag{13.18}$$

The latter condition is "almost" necessary, since otherwise the probabilities forming the series (13.5) are separated from 0 by the law of large numbers. More precisely, $\mathsf{P}(|S_n| \geq \varepsilon n^\alpha) > \delta > 0$ for all $\varepsilon > 0$ if $\alpha < 1$ or for sufficiently small $\varepsilon > 0$ if $\alpha = 1$. Therefore series (13.5) diverges if $\sum w_n = \infty$.

It is also clear that (13.17) implies

$$\lim_{n\to\infty} n\mathsf{P}(|X| \geq \varepsilon n^\alpha) = 0 \quad \text{for all} \quad \varepsilon > 0.$$

Hence condition (13.15) follows from (13.17).

For $t > 1/\alpha$, we use the following condition

$$\mathsf{E}\left[|X|^t\right] < \infty, \tag{13.19}$$

which is a minimal moment condition in the class $\mathcal{RV}_{\geq\alpha t-2}$.

Unfortunately there is a gap between the necessary conditions discussed in Sect. 13.1 and sufficient conditions given below. These conditions coincide if the weight sequence $\{w_n, n \geq 1\}$ is regularly varying but this is not the case for other classes of sequences $\{w_n\}$.

Theorem 13.6 *Let $\alpha > 1/2$ and $\theta > 0$. Assume that a sequence of positive numbers $\{w_n, n \geq 1\}$ satisfies the following condition*

$$\lim_{n\to\infty} \frac{w_n}{n^\theta} = 0. \tag{13.20}$$

If

(i) *series* (13.14) *converges for all* $\varepsilon > 0$,
(ii) *condition* (13.19) *holds for some* $t > 1/\alpha$,
(iii) *condition* (13.18) *holds if* $\alpha \le 1$,

then series (13.5) *converges for all* $\varepsilon > 0$.

Remark 13.4 It is clear that (13.20) is a more restrictive condition than (13.9). The following equivalent form of condition (13.20) is sometimes more convenient than the original one:

$$\sum_{n=1}^{\infty} \frac{w_n}{n^\theta} < \infty \text{ for some } \quad \theta > 0.$$

Conditions (13.14) and (13.18) are necessary. On the other hand, condition (13.19) is weaker than the necessary moment conditions in many cases. However, (13.19) is too restrictive for the class $\mathcal{RV}_{\ge -1}$.

The assumptions of Theorem 13.6 can be rewritten in a more convenient form if the sequence $\{w_n, n \ge 1\}$ is regularly varying.

Corollary 13.3 *Let* $\alpha > 1/2$, $w_n = n^\rho L(n)$, *where* $\rho > -1$, *and let* $L(\cdot)$ *be some positive regularly varying function. If*

$$\sum_{n=1}^{\infty} n w_n \mathsf{P}(|X| \ge n^\alpha) < \infty \tag{13.21}$$

and additionally condition (13.18) *holds if* $\alpha \le 1$, *then series* (13.5) *converges for all* $\varepsilon > 0$.

If $L(n) \equiv 1$, then the sufficient part of Theorem 13.2 for $\alpha r > 1$ implies Corollary 13.3. The necessity part of Theorem 13.2 implies Corollary 13.1.

For some cases, condition (13.19) is rather far of being optimal. For example, if $w_n = (\log n)^\nu / n$ for some $\nu \ge 0$, then the moment condition

$$\mathsf{E}\left[|X|^{1/\alpha} \left(\log^+ |X|\right)^\nu\right] < \infty$$

is a criterion of the convergence of series (13.14) for all $\varepsilon > 0$. Instead, it is assumed in Theorem 13.6 that condition (13.19) holds for $t > 1/\alpha$.

This disadvantage of Theorem 13.6 is partially removed in the following result, where more restrictions are imposed on the sequence $\{w_n, n \ge 1\}$ and then the "optimality" of the moment condition is achieved.

Theorem 13.7 *Let* $\alpha > 1/2$ *and let* $\{w_n, n \ge 1\}$ *be a sequence of non-negative numbers. Put*

$$W_n = \sum_{k=1}^{n} k w_k.$$

Assume that $W_n = O(W_{n-1})$ *and*

$$\sum_{n=m}^{\infty} \frac{w_n}{n^\theta} = O\left(\frac{W_m}{m^{\theta+1}}\right) \quad \text{for some} \quad \theta > 0. \tag{13.22}$$

If

(i) series (13.14) *converges for all* $\varepsilon > 0$;
(ii) condition (13.17) *holds*;
(iii) condition (13.18) *holds if* $\alpha \le 1$,

then series (13.5) *converges for all* $\varepsilon > 0$.

Condition (13.22) is more restrictive than (13.20). Condition (13.22) is, in fact, Feller's condition if $\theta = 2$ is used in the statements on the strong law of large numbers (see Chap. 9). The proof below uses the following simple result: *if* (13.22) *holds for* $\theta = \theta_0$, *then* (13.22) *holds for* $\theta > \theta_0$, *as well.* This result follows from the estimate

$$\sum_{n=m}^{\infty} \frac{w_n}{n^\theta} \le \frac{1}{m^{\theta-\theta_0}} \sum_{n=m}^{\infty} \frac{w_n}{n^{\theta_0}}.$$

Note also that the assumptions of Theorem 13.7 imply

$$\sum_{n\ge 1} W_n \mathsf{P}(n-1 \le |X|^{1/\alpha} < n) < \infty. \tag{13.23}$$

Indeed, we derive from condition (13.14) with $\varepsilon = 1$ that

$$\sum W_n \mathsf{P}(n \le |X|^{1/\alpha} < n+1) < \infty,$$

whence (13.23) follows in view of $W_n = O(W_{n-1})$.

All the above results become simpler for monotone sequences $\{w_n, n \ge 1\}$.

Corollary 13.4 *Let* $\alpha > 1/2$ *and let* $\{w_n, n \ge 1\}$ *be a decreasing sequence of positive numbers. Assume that condition* (13.17) *holds. In the case of* $\alpha \le 1$ *we additionally assume that condition* (13.18) *is satisfied. If series* (13.14) *converges for some* $\varepsilon > 0$, *then series* (13.5) *converges for all* $\varepsilon > 0$.

Prior to proving the main results stated above we are going to establish some auxiliary assertions.

Lemma 13.4 *Let the common distribution function* F *of random variables* $\{X_n\}$ *be symmetric. Assume that series* (13.14) *converges for all* $\varepsilon > 0$. *If*

$$\sum_{n=1}^{\infty} w_n \left[n\mathsf{P}(|X| \geq \varepsilon n^{\alpha})\right]^{2N} < \infty \qquad \textit{for all } \varepsilon > 0, \tag{13.24}$$

$$\sum_{n=1}^{\infty} w_n \left[n^{1-\alpha r} \int\limits_{|x|<n^{\alpha}} |x|^r \, dF(x) \right]^N < \infty \tag{13.25}$$

for some $0 < r \leq 2$ *and* $N = 2^j$, *then series* (13.5) *converges for all* $\varepsilon > 0$.

Proof of Lemma 13.4 First we apply the following particular case of the Buldygin–Hoffman-Jørgensen inequality (see Theorem 6.10): *for any* $j \geq 1$, *there are constants* $C_j > 0$ and $D_j > 0$ *such that*

$$\mathsf{P}(|S_n| \geq 3^j z) \leq C_j n \mathsf{P}(|X| \geq z) + D_j \left(\mathsf{P}(|S_n| \geq z)\right)^{2^j} \tag{13.26}$$

for all $n \geq 1$ *and* $z > 0$.

Put $X_{nk} = X_k \mathbb{I}(|X_k| < n^{\alpha})$ for $k \leq n$ and $S_{nn} = \sum_{k \leq n} X_{nk}$ for all $n \geq 1$. Now fix $\varepsilon > 0$ and introduce the random events

$$\begin{aligned}
A_n(\varepsilon) &= A_n \overset{\text{def}}{=} \{\omega : |S_n| \geq \varepsilon n^{\alpha}\}, \\
A_n^1(\varepsilon) &= A_n^1 \overset{\text{def}}{=} \left\{\omega : \max_{k \leq n} |X_k| \geq \varepsilon n^{\alpha}/2\right\}, \\
A_n^2(\varepsilon) &= A_n^2 \overset{\text{def}}{=} \{\omega : |S_{nn}| \geq \varepsilon n^{\alpha}/2\}, \\
A_n^3 &\overset{\text{def}}{=} \{\omega : \exists k_1 \leq n, k_2 \leq n : |X_{k_1}| \geq n^{\alpha}, |X_{k_2}| \geq n^{\alpha}\}.
\end{aligned}$$

We are going to show that $A_n \subseteq A_n^1 \cup A_n^2 \cup A_n^3$. Indeed, if $\omega \in A_n$ and $\omega \notin A_n^1 \cup A_n^3$, then

(i) either $|X_k| < n^{\alpha}$ for all $k \leq n$,
(ii) or $|X_k| \geq n^{\alpha}$ only for one index $k \leq n$.

In the first case, $S_n = S_{nn}$ and thus $\omega \in A_n^2$. In the second case, $S_n = S_{nn} + X_k$ and thus

$$|S_{nn}| \geq |S_n| - |X_k| \geq \varepsilon n^{\alpha} - \frac{\varepsilon n^{\alpha}}{2} = \frac{\varepsilon n^{\alpha}}{2},$$

whence $\omega \in A_n^2$. This proves the required result.

Now inequality (13.26) implies that

$$\begin{aligned}
\mathsf{P}(|S_n| \geq \varepsilon n^{\alpha}) = O(1)\Big[& n\mathsf{P}(|X| \geq \varepsilon_1 n^{\alpha}) \\
&+ \left(\mathsf{P}(A_n^1(\varepsilon_1)\right)^N + \left(\mathsf{P}(A_n^2(\varepsilon_1)\right)^N + \left(\mathsf{P}(A_n^3\right)^N \Big],
\end{aligned} \tag{13.27}$$

where $\varepsilon_1 = \varepsilon 3^{-j}$. Moreover,

$$\mathsf{P}(A_n^1(\varepsilon_1)) \le n\mathsf{P}(|X| \ge \varepsilon_1 n^\alpha/2), \qquad \mathsf{P}(A_n^3) \le \left[n\mathsf{P}(|X| \ge n^\alpha)\right]^2.$$

In order to estimate the probability $\mathsf{P}(A_n^2)$, we use the Chebyshev–Markov inequality and then the Bahr–Esseen inequality (2.6):

$$\mathsf{P}(A_n^2(\varepsilon_1)) \le O(1)\frac{\mathsf{E}\left[|S_{nn}|^r\right]}{n^{\alpha r}} \le O(1)n^{1-\alpha r}\int\limits_{|x|<n^\alpha} |x|^r\,dF(x)$$

(the constants in the latter inequality depend on ε but they do not depend on n). Substituting these bounds in (13.27) we complete the proof of Lemma 13.4. It only remains to explain why the series $\sum w_n\left[n\mathsf{P}(|X| \ge \varepsilon n^\alpha)\right]^N$ converges for all $\varepsilon > 0$. Put

$$\mathbf{N}_1 = \{n \in \mathbf{N} : n\mathsf{P}(|X| \ge \varepsilon n^\alpha) < 1\}.$$

It is clear that

$$\left[n\mathsf{P}(|X| \ge \varepsilon n^\alpha)\right]^N \le \begin{cases} n\mathsf{P}(|X| \ge \varepsilon n^\alpha), & n \in \mathbf{N}_1, \\ \left[n\mathsf{P}(|X| \ge \varepsilon n^\alpha)\right]^{2N}, & n \notin \mathbf{N}_1. \end{cases}$$

Thus the series $\sum w_n\left[n\mathsf{P}(|X| \ge \varepsilon n^\alpha)\right]^N$ converges in view of (13.14) and (13.24). □

Lemma 13.5 *Let $\{X_n\}$ be a sequence of independent identically distributed random variables and let $\{X_n^{(s)}\}$ be the sequence of their symmetrizations. Also let $\{w_n\}$ be a sequence of positive numbers. Assume that*

(i) condition (13.17) *holds,*
(ii) additionally, for the case of $\alpha \le 1$, condition (13.18) *is satisfied,*
(iii) moreover,

$$\sum_{n=1}^{\infty} w_n\mathsf{P}(|S_n^{(s)}| \ge \varepsilon n^\alpha) < \infty$$

for all $\varepsilon > 0$, where $S_n^{(s)} = \sum_{k\le n} X_k^{(s)}$. Then series (13.5) *converges for all $\varepsilon > 0$.*

Proof of Lemma 13.5 Since $\mathsf{P}(|S_n^{(s)}| \ge z) \ge \mathsf{P}(|S_n| \ge 2z)\mathsf{P}(|S_n| < z)$ for all $z > 0$, we get

$$\mathsf{P}(|S_n^{(s)}| \ge \varepsilon n^\alpha) \ge \frac{1}{2}\mathsf{P}(|S_n| \ge 2\varepsilon n^\alpha)$$

for $z = \varepsilon n^\alpha$ and sufficiently large n. The latter bound follows by assumptions (i)–(ii) and the Marcinkiewicz–Zygmund law of large numbers, that is, $\mathsf{P}(|S_n| < \varepsilon n^\alpha) \to 1$ as $n \to \infty$ (see Theorem 4.4). Lemma 13.5 is proved. □

Proof of Theorem 13.6 Without loss of generality we assume that $t \le 2$. First we consider the case of symmetric random variables $\{X_n, n \ge 1\}$. Since $\mathsf{E}\left[|X|^{1/\alpha}\right] \le \left(\mathsf{E}\left[|X|^t\right]\right)^{1/\alpha t}$, we conclude that $\mathsf{E}\left[|X|^{1/\alpha}\right] < \infty$ and hence $n\mathsf{P}(|X| \ge \varepsilon n^\alpha) \to 0$ as $n \to \infty$ for all $\varepsilon > 0$. Therefore, condition (13.24) with an arbitrary $N \ge 1$ follows from the convergence of series (13.14) for all $\varepsilon > 0$.

Now we choose $N = 2^j$ such that $N(\alpha t - 1) > \theta + 2$. Since $\int_{|x|<n^\alpha} |x|^t \, dF(x) \le \mathsf{E}\left[|X|^t\right]$, condition (13.25) holds for $r = t$. Hence Theorem 13.6 follows from Lemma 13.4 in the case of a symmetric distribution.

Now let the distribution function of random variables $\{X_n\}$ be general (not necessarily F is symmetric). Consider the sequence of symmetrizations $\{X_n^{(s)}\}$ and note that all the assumptions of Theorem 13.6 also hold for the sequence $\{X_n^{(s)}\}$. The part of Theorem 13.6 proved above implies that

$$\sum_{n=1}^{\infty} w_n \mathsf{P}(|S_n^{(s)}| \ge \varepsilon n^\alpha) < \infty \qquad \text{for all} \quad \varepsilon > 0.$$

The symmetrization inequalities (see Sect. A.18) yield

$$\sum_{n=1}^{\infty} w_n \mathsf{P}(|S_n - \operatorname{med}(S_n)| \ge \varepsilon n^\alpha) < \infty \qquad \text{for all} \quad \varepsilon > 0,$$

where $\operatorname{med}(S_n)$ is a median of the random variable S_n. Since $\operatorname{med}(S_n)/n^\alpha \to 0$, the convergence of series (13.5) follows for all $\varepsilon > 0$. □

Proof of Corollary 13.3 We check the assumptions of Theorem 13.6. Condition (13.20) clearly holds for $\theta > \max\{0, \rho\}$. Let $0 < \gamma < 1+\rho$. Since $n^{1+\rho-\gamma} L(n) \to \infty$ as $n \to \infty$, relation (13.21) implies that $\sum n^\gamma \mathsf{P}(|X| \ge n^\alpha) < \infty$. This means that $\mathsf{E}\left[|X|^t\right] < \infty$ for $t = (1+\gamma)/\alpha$, that is, condition (13.19) holds.

Finally, the convergence of series (13.14) for all $\varepsilon > 0$ follows from (13.21), since L is a regularly varying function. Indeed, fix $\varepsilon > 0$ and, for a given $n \in \mathbf{N}$, choose $m \in \mathbf{N}$ such that $m^\alpha \le \varepsilon n^\alpha < (m+1)^\alpha$. Thus

$$\begin{aligned}\sum_{n=1}^{\infty} n w_n \mathsf{P}(|X| \ge \varepsilon n^\alpha) &\le \sum_{m=1}^{\infty} \mathsf{P}(|X| \ge m^\alpha) \sum_{n: m^\alpha \le \varepsilon n^\alpha < (m+1)^\alpha} n w_n \\ &\le \sum_{m=1}^{\infty} O(1) m w_m \mathsf{P}(|X| \ge m^\alpha),\end{aligned}$$

since L is a regularly varying function. This proves the convergence of series (13.5) for all $\varepsilon > 0$ and therefore Corollary 13.3 follows from Theorem 13.6. □

Proof of Theorem 13.7 We check the assumptions of Lemmas 13.4 and 13.5. Put $t = 2$ and $\varphi(n) = n^\alpha$ and choose $N = 2^j$, $j \ge 0$, such that $N(2\alpha - 1) > \theta$. Then (13.22) implies that

$$\sum_{n=m}^{\infty} \frac{w_n}{n^{N(2\alpha-1)}} = O\left(\frac{W_m}{m^{N(2\alpha-1)+1}}\right).$$

We derive from (ii) that $n\mathsf{P}(|X| \ge \varepsilon n^\alpha) \to 0$ as $n \to \infty$ for all $\varepsilon > 0$. Thus condition (13.24) follows from assumption (i).

Next we prove that (13.25) also holds. Put

$$\sigma^2(n) = \int_{|x|<\varphi(n)} x^2\, dF(x), \qquad \Delta\left[\sigma^{2N}(n)\right] = \sigma^{2N}(n) - \sigma^{2N}(n-1).$$

It is clear that

$$\begin{aligned}\sum_{n=1}^{\infty} w_n \left[n^{1-2\alpha}\sigma^2(n)\right]^N &= \sum_{n=1}^{\infty} \frac{w_n}{n^{N(2\alpha-1)}} \sum_{m\le n} \Delta\left[\sigma^{2N}(m)\right] \\ &= \sum_{m\ge 1} \Delta\left[\sigma^{2N}(m)\right] \sum_{n\ge m} \frac{w_n}{n^{N(2\alpha-1)}} \\ &= O(1) \sum_{m\ge 1} \Delta\left[\sigma^{2N}(m)\right] \frac{W_m}{m^{N(2\alpha-1)+1}}.\end{aligned}$$

Since $\Delta\left[\sigma^{2N}(m)\right] \le N\left(\sigma^2(m) - \sigma^2(m-1)\right)\sigma^{2(N-1)}(m)$, we conclude that

$$\sum_{n=1}^{\infty} w_n \left[n^{1-2\alpha}\sigma^2(n)\right]^N \le O(1) \sum_{m\ge 1} \int_{\varphi(m-1)\le|x|<\varphi(m)} x^2\, dF \frac{W_m \sigma^{2(N-1)}(m)}{m^{N(2\alpha-1)+1}}.$$

Condition (ii) yields $\sup_m \sigma^2(m)/m^{2\alpha-1} < \infty$. Thus

$$\begin{aligned}\sum_{n=1}^{\infty} w_n \left[n^{1-2\alpha}\sigma^2(n)\right]^N &\le O(1) \sum_{m\ge 1} \frac{W_m}{m^{2\alpha}} \int_{\varphi(m-1)\le|x|<\varphi(m)} x^2\, dF \\ &\le O(1) \sum_{m\ge 1} W_m \mathsf{P}(m - 1 \le |X|^{1/\alpha} < m),\end{aligned}$$

and we establish (13.25) from (13.23). Now Theorem 13.7 follows from Lemmas 13.4 and 13.5. □

Proof of Corollary 13.4 It is clear that condition (13.20) holds for all $\theta > 0$. We are going to check condition (13.22) for $\theta > 1$. Indeed,

$$\sum_{n=m}^{\infty} \frac{w_n}{n^\theta} \le w_m \sum_{n=m}^{\infty} \frac{1}{n^\theta} \le \text{const} \frac{w_m}{m^{\theta-1}} \le \text{const} \frac{W_m}{m^{\theta+1}}.$$

The property $W_m = O(W_{m-1})$ is also clear. Now Corollary 13.4 follows from Theorem 13.7. □

13.3 Sufficient Conditions for Complete Convergence of Sums with Random Indices

The sufficient conditions introduced in this section allow us to prove relation (13.6) for sums with random indices. Namely, we study the sufficient conditions for

$$\sum_{n=1}^{\infty} n^{\alpha r-2} \mathsf{P}(|S_{\nu_n}| \ge \varepsilon \nu_n^\alpha) < \infty, \tag{13.28}$$

where $\{\nu_n, n \ge 1\}$ are random variables assuming only positive integer values. The weight coefficients are $w_n = n^{\alpha r-2}$.

If the indices $\{\nu_n\}$ are non-random, then relation (13.28) holds only for $\alpha > \frac{1}{2}$ (the proof is similar to that of assumption (i) in Theorem 13.3). In contrast, the restriction $\alpha > 1/2$ is not generally speaking necessary for (13.28) if the indices are random. Below is the corresponding example.

Example 13.1 Assume that $\alpha > 0$. Let $\{X_n, n \ge 1\}$ be a sequence of independent Bernoulli random variables, that is

$$\mathsf{P}(X_n = \pm 1) = \frac{1}{2}.$$

It is well known that

$$\mathsf{P}(\, S_n = 0 \text{ i. o.}) = 1.$$

We define the sequence $\{\nu_n, n \ge 1\}$ as follows: $\nu_1 = 1$ and

$$\nu_n = \min\{k\colon k > \nu_{n-1} \text{ and } S_k = 0\}, \qquad n \ge 2.$$

It is clear that the sequence $\{\nu_n, n \ge 1\}$ is well defined for almost all elementary random events. For other elementary random events we set $\nu_n = \nu_{n-1} + 1$. For all $\varepsilon > 0$, we have

$$\mathsf{P}(\, |S_{\nu n}| \ge \varepsilon \nu_n^\alpha \,) = 0.$$

Thus (13.28) holds for all $\varepsilon > 0$. We highlight again that the number α in this example is arbitrary (for example, it can be less than $\frac{1}{2}$).

Moreover, any sequence of positive numbers (in place of ν_n^α) fits the same construction.

13.3.1 Gut's Theorem

The proof of relation (13.28) relies on the assumption that the law of large numbers in the sense of complete convergence holds for random variables $\{\nu_n\}$.

Theorem 13.8 (A. Gut) *Let $\{\nu_n, n \geq 1\}$ be integer-valued random variables that assume only positive values. Fix some numbers $\varepsilon > 0$ and $A > \varepsilon$ and a non-negative random variable ν such that $\mathsf{P}(\nu \geq A) = 1$.*

(a) Let $\alpha r > 1,\ \alpha > 1/2$. Assume that condition (13.7) *holds. If*

$$\sum_{n=1}^{\infty} n^{\alpha r-2}\mathsf{P}\left(\left|\frac{\nu_n}{n} - \nu\right| \geq \varepsilon\right) < \infty, \tag{13.29}$$

then condition (13.28) *also holds.*

(b) Let $\alpha r > 1,\ \alpha > 1/2$. Assume that

$$\mathsf{E}\left[|X|^{1/\alpha}\log^+|X_1|\right] < \infty \text{ and additionally } \mathsf{E}[X_1] = 0 \text{ in the case of } \alpha \leq 1. \tag{13.30}$$

If condition (13.29) *holds for some random variable ν with the same properties as in case* (a), *then condition* (13.28) *also holds.*

Remark 13.5 The restrictions imposed on the numbers α and r are quite natural. If $\alpha r < 1$, then series (13.28) converges for all $\varepsilon > 0$ for arbitrary sequences of random variables $\{X_n, n \geq 1\}$ and $\{\nu_n, n \geq 1\}$. Note also that the proof of the cases $\alpha r = 1$ and $\alpha r > 1$ are quite different.

Remark 13.6 A disadvantage of Theorem 13.8 is that the moment condition (13.7) is the same for all sequences $\{\nu_n\}$, while optimal conditions in the case of non-random indices essentially depends on how the sequence behaves at infinity. Let, for example, $\nu_n = \left[n^\beta\right]$, $\beta \geq 1$. Condition (13.7) is optimal in this case only for $\beta = 1$. If $\beta > 1$, then this condition is quite far from being optimal. Moreover, condition (13.29) does not hold at all in this case for every random variable ν, that is, Theorem 13.8 cannot be used at all if $\nu_n = \left[n^\beta\right]$, $\beta > 1$. Indeed, for every random variable ν for which Gut's condition holds we have

$$\mathsf{P}\left(\left|\frac{\nu_n}{n} - \nu\right| \geq \varepsilon\right) \geq \mathsf{P}\left(\left[n^\beta\right] - n\nu \geq n\varepsilon\right) \geq \mathsf{P}\left(\left[n^\beta\right] \geq n(A+\varepsilon)\right).$$

The latter "probability" equals 1 for sufficiently large n. Since $\alpha r > 1$, condition (13.29) does not hold for any random variable ν.

The following is the optimal result for $\nu_n = [n^\beta]$, $\beta \geq 1$.

Proposition 13.1 *Let $\alpha r > 1$, $\alpha > 1/2$. Assume that $\nu_n = [n^\beta]$ almost surely for some $\beta > 0$. Then condition* (13.28) *is equivalent to the following set of conditions*:

$$\mathsf{E}\left[|X|^\theta\right] < \infty \quad \text{and} \quad \mathsf{E}[X] = 0 \text{ if } \alpha \leq 1, \tag{13.31}$$

where $\theta = (\alpha r - 1 + \beta)/\alpha\beta$.

According to condition (13.7), the existence of the moment of order r is assumed in Theorem 13.8. It is clear that $\theta < r$ for $\beta > 1$, which supports what we have said above that condition (13.7) is not optimal in this case.

Remark 13.7 One should distinguish between the cases $\beta > 1, \beta = 1$, and $\beta < 1$ for the sequence $\nu_n = \left[n^\beta\right]$. In the first case, $\{\nu_n, n \geq 1\}$ is a subset of positive integers, while it coincides with $\mathbf{N}$ in the second case. In the case of $\beta < 1$, the sequence $\{\nu_n, n \geq 1\}$ contains all positive integers with some multiplicity and consists of the "blocks" $B_m = \{n : \left[n^\beta\right] = m\}$, $m \geq 1$. Clearly $\operatorname{card}(B_m) \asymp m^{1/\beta}$. In the latter case, Proposition 13.1 follows from Theorem 13.2.

13.3.2 The Case $\alpha r > 1$

The following result is an extension of Theorem 13.8. We assume in Theorem 13.9 that the sequence $\left\{\nu_n/n^\beta\right\}$ is bounded (in the sense of complete convergence) for some $\beta > 0$.

Theorem 13.9 *Let $\alpha r > 1$, $\alpha > 1/2$. Assume that condition* (13.19) *holds for some $t > 1/\alpha$ and, additionally, that* (13.18) *holds if $\alpha \leq 1$. Let*

$$\sum_{n=1}^{\infty} n^{\alpha r - 2} \mathsf{P}(\nu_n < cn^\beta) < \infty. \tag{13.32}$$

If, for some $\varepsilon_0 > 0$,

$$\sum_{n=1}^{\infty} n^{\alpha r - 2} \mathsf{P}\left(\max_{k \leq \nu_n} |X_k| \geq \varepsilon_0 \nu_n^\alpha\right) < \infty \tag{13.33}$$

with some constants $\beta > 0$ and $c > 0$, then there exists a number $\varepsilon_1 = O(\varepsilon_0) > 0$ such that

$$\sum_{n=1}^{\infty} n^{\alpha r - 2} \mathsf{P}(|S_{\nu_n}| \geq \varepsilon \nu_n^\alpha) < \infty \tag{13.34}$$

for all $\varepsilon \geq \varepsilon_1$.

The convergence of series (13.34) for *all* $\varepsilon > 0$ is obtained in the following result.

Theorem 13.10 *Let* $\alpha r > 1,\ \alpha > 1/2$. *Let* $\beta > 0, c > 0$, *and* $\theta = (\alpha r - 1 + \beta)/\alpha\beta$. *Assume that conditions* (13.31) *and* (13.32) *hold. Then the series on the right-hand side of* (13.34) *converges for all* $\varepsilon > 0$.

The simplest case of Theorem 13.10 corresponds to $\beta = 1$.

Corollary 13.5 *Let* $\alpha r > 1,\ \alpha > 1/2$. *Let condition* (13.7) *hold and*

$$\sum_{n=1}^{\infty} n^{\alpha r-2} \mathsf{P}(\nu_n < cn) < \infty \quad \textit{for some} \quad c > 0. \tag{13.35}$$

Then the series on the left-hand side of (13.34) *converges for all* $\varepsilon > 0$.

Remark 13.8 The results obtained in Theorem 13.8 and Corollary 13.5 are quite different. While Corollary 13.5 asserts that series (13.34) converges for *all* $\varepsilon > 0$, Theorem 13.8 states the same only for *some* $\varepsilon > 0$.

We show that case (a) in Theorem 13.8 follows from Corollary 13.5. More precisely, we show that condition (13.35) follows from (13.29). Indeed, let condition (13.29) hold for some $\varepsilon_0 > 0$ and let a number $A > \varepsilon_0$ and random variable ν be chosen as in Theorem 13.8. Then

$$\mathsf{P}\left(\left|\frac{\nu_n}{n} - \nu\right| \ge \varepsilon_0\right) \ge \mathsf{P}(\nu_n \le n\nu - n\varepsilon_0) \ge \mathsf{P}(\nu_n < cn)$$

for all $0 < c < A - \varepsilon_0$, that is, condition (13.35) follows from (13.29).

Note further that Corollary 13.5 does not coincide with case (a) in Theorem 13.8. An appropriate example is constructed as follows. Let $\nu_n = [n^\beta]$, $\beta > 1$. In this case,

$$\mathsf{P}(\nu_n < cn) = 0 \quad \text{for all} \quad c > 0$$

if n is sufficiently large. Thus condition (13.35) holds.

Further, if ν is an arbitrary random variable, then

$$\mathsf{P}\left(\left|\frac{\nu_n}{n} - \nu\right| \ge \varepsilon\right) \ge \mathsf{P}(\nu \le n^{\beta-1} - 1 - \varepsilon) \ge \frac{1}{2}$$

for all $\varepsilon > 0$ if n is sufficiently large. This clearly indicates that condition (13.29) is not satisfied for any random variable ν and that Corollary 13.5 is a more general statement than Theorem 13.8. This especially concerns Theorems 13.9 and 13.10.

13.3.3 The Baum–Katz Theorem

Theorem 13.11 (L. Baum and M. Katz) *Let $\{X_n\}$ be a sequence of independent identically distributed random variables.* (a)*Let $\alpha r > 1$, $\alpha > 1/2$. If condition* (13.7) *holds, then*

$$\sum_{n=1}^{\infty} n^{\alpha r-2} \mathsf{P}\left(\sup_{j\geq n} \frac{|S_j|}{j^{\alpha}} \geq \varepsilon\right) < \infty \tag{13.36}$$

for all $\varepsilon > 0$. Moreover, the converse statement is also true.

(b) *If $\alpha > 1/2$ and condition* (13.30) *holds, then*

$$\sum_{n=1}^{\infty} \frac{1}{n} \mathsf{P}\left(\sup_{j\geq n} \frac{|S_j|}{j^{\alpha}} \geq \varepsilon\right) < \infty \tag{13.37}$$

for all $\varepsilon > 0$. Moreover, the converse statement is also true.

Theorem 13.11 can easily be derived from Corollary 13.5. Put

$$\nu_n = \min\left\{k \geq n : \frac{|S_k|}{k^{\alpha}} = \sup_{j\geq n} \frac{|S_j|}{j^{\alpha}}\right\}. \tag{13.38}$$

Since $\theta\alpha > 1$ in condition (13.31), the Marcinkiewicz–Zygmund strong law of large numbers implies that $S_j/j^{\alpha} \to 0$ almost surely and thus ν_n is well defined almost surely for all $n \geq 1$. For those elementary random events ω where $\nu_n(\omega)$ is not defined by equality (13.38) we put $\nu_n = n$.

If $0 < c < 1$, then $\mathsf{P}(\nu_n < cn) = 0$, that is, condition (13.35) holds and relation (13.36) follows for all $\varepsilon > 0$ from Corollary 13.5.

Remark 13.9 Yet another application of Corollary 13.5 is to the following sequence $\{\nu_n, n \geq 1\}$: let $0 < c < 1$ and

$$\nu_n = \max\left\{k : cn < k \leq n \text{ and } |S_k| = \max_{cn<j\leq n} |S_j|\right\}.$$

Since $\mathsf{P}(\nu_n < cn) = 0$, Corollary 13.5 implies that

$$\sum_{n=1}^{\infty} n^{\alpha r-2} \mathsf{P}\left(\max_{cn<k\leq n} |S_k| \geq \varepsilon n^{\alpha}\right) < \infty \quad \text{for all } \varepsilon > 0. \tag{13.39}$$

In fact, one can also choose $c = 0$. Moreover, condition (13.39) is equivalent to condition (13.7) (see, for example, [57]).

13.3.4 The Case $\alpha r = 1$

Theorem 13.12 *Assume that* $\{X_n\}$ *is a sequence of independent identically distributed random variables. Let* $\alpha > 1/2$ *and* $\alpha r = 1$. *If conditions* (13.30) *and* (13.32) *hold for some* $\beta > 0$ *and* $c > 0$, *then*

$$\sum_{n=1}^{\infty} \frac{1}{n} \mathsf{P}(|S_{\nu_n}| \geq \varepsilon \nu_n^{\alpha}) < \infty \tag{13.40}$$

for all $\varepsilon > 0$.

Remark 13.10 Part (b) of Theorem 13.8 follows from Theorem 13.12, since condition (13.32) is less restrictive than condition (13.29) (see Remark 13.8). We should highlight that relation (13.40) holds for *all* $\varepsilon > 0$ under the assumptions of Theorem 13.12.

Remark 13.11 The moment condition (13.30) in Theorem 13.12 cannot be improved in general. For the sake of simplicity, assume that X_n are symmetric random variables. Define the sequence $\{\nu_n, n \geq 1\}$ by equality (13.38). It is easy to see that $\nu_n \geq n$ and thus condition (13.32) holds for $c = 1$ and $\beta = 1$. On the other hand,

$$\sum_{n=1}^{\infty} \frac{1}{n} \mathsf{P}(|S_{\nu_n}| \geq \varepsilon \nu_n^{\alpha}) = \sum_{n=1}^{\infty} \frac{1}{n} \mathsf{P}\left(\sup_{j \geq n} \left| \frac{S_j}{j^{\alpha}} \right| \geq \varepsilon \right).$$

According to Theorem 13.11, the convergence of the series on the right-hand side is equivalent to condition (13.30).

Sufficient conditions for the complete convergence of the sequence $\{S_{\nu_n}/\nu_n^{\alpha}\}$ are given in the following assertion.

Corollary 13.6 *Let* $\{X_n\}$ *be a sequence of independent identically distributed random variables. Let* $\alpha > 1/2$. *Assume that*

$$\mathsf{E}\left[|X|^{2/\alpha}\right] < \infty \quad \textit{and} \quad \mathsf{E}[X] = 0 \quad \textit{in the case of} \quad \alpha \leq 1. \tag{13.41}$$

(a) *If*

$$\sum_{n=1}^{\infty} \mathsf{P}(\nu_n < cn^{\beta}) < \infty \tag{13.42}$$

for some $c > 0$ *and* $\beta \geq 1$, *then the sequence* $\{S_{\nu_n}/\nu_n^{\alpha}, n \geq 1\}$ *converges completely to* 0, *that is*

$$\sum_{n=1}^{\infty} \mathsf{P}(|S_{\nu_n}| \geq \varepsilon \nu_n^{\alpha}) < \infty \tag{13.43}$$

for all $\varepsilon > 0$.

(*b*) *If condition* (13.42) *holds for some* $c > 0$ *and* $0 < \beta < 1$, *then the sequence* $\{S_{\nu_n}/\nu_n^\alpha, n \ge 1\}$ *is bounded in the sense of complete convergence, that is, there exists a number* $\varepsilon_1 \ge 0$ *such that condition* (13.43) *holds for* $\varepsilon > \varepsilon_1$ (*we do not claim here that* $\varepsilon_1 = 0$ *as in part* (a) *above*).

Proof of Theorem 13.9 In view of condition (13.32), we need to show that there exists a number $\varepsilon_1 > 0$ such that

$$\sum_{n=1}^{\infty} n^{\alpha r-2}\mathsf{P}(|S_{\nu n}| \ge \varepsilon\nu_n^\alpha, \nu_n \ge cn^\beta) < \infty \tag{13.44}$$

for all $\varepsilon \ge \varepsilon_1$. Let $0 < \gamma < 1$ and let q be a positive integer such that

$$\gamma\alpha t > 1, \qquad \alpha r - 2 + \beta(q - q\gamma\alpha t + 1) < -1.$$

We introduce the random events

$$\begin{aligned} A_n^1 &= \left\{\omega\colon \max_{k\le\nu_n} |X_k| \ge \varepsilon\nu_n^\alpha/q\right\}, \\ A_n^2 &= \left\{\omega : \text{there are at least } q \text{ indices } k_i \le \nu_n \text{ such that } |X_{k_i}| \ge \varepsilon\nu_n^{\gamma\alpha}\right\}, \\ A_n^3 &= \{\omega\colon |\zeta_{\nu_n}| \ge \varepsilon\nu_n^\alpha/q\}, \end{aligned}$$

where $\zeta_m = \sum_{k=1}^m X_{mk}$ and $X_{mk} = X_k\mathbb{I}(|X_k| < \varepsilon m^{\gamma\alpha})$. To prove (13.44), it is necessary to show that

$$\sum_{n=1}^{\infty} n^{\alpha r-2}\mathsf{P}(A_n^i,\ \nu_n \ge cn^\beta) < \infty \tag{13.45}$$

for $i = 1, 2, 3$ and $\varepsilon \ge \varepsilon_1$, since $\{\omega\colon |S_{\nu_n}| \ge \varepsilon\nu_n^\alpha\} \subseteq A_n^1 \cup A_n^2 \cup A_n^3$. The proof of this inclusion follows the lines of that in Lemma 13.4 and thus is omitted here. Condition (13.45) for $i = 1$ and $\varepsilon \ge \varepsilon_0 q$ follows directly from condition (13.33). Put $\varepsilon_1 = \varepsilon_0 q$. Now we show that condition (13.45) holds for $i = 2$ and $i = 3$ for *all* $\varepsilon > 0$ if q is a sufficiently large number. To prove (13.45) for $i = 2$ we introduce the following random events

$$B(k_1, \dots, k_q, m, n) = \{\omega : \nu_n = m \quad \text{and} \quad |X_{k_1}| \ge \varepsilon m^{\gamma\alpha}, \dots, |X_{k_q}| \ge \varepsilon m^{\gamma\alpha}\}$$

and the "cube without diagonals" in the space $\mathbf{N}^d$:

$$[1, m]_q = \{k_1, \dots, k_q : k_1 \le m, \dots, k_q \le m \text{ and } k_i \ne k_j,\ \text{if } i \ne j\}.$$

It is clear that

$$\mathsf{P}(A_n^2,\ \nu_n = m) \le \sum_{[1,m]_q} \int\limits_{B(k_1,\dots,k_q,m,n)} d\mathsf{P}$$
$$\le \frac{1}{\varepsilon^{qt}} \sum_{[1,m]_q} \frac{1}{m^{q\gamma\alpha t}} \int\limits_{B(k_1,\dots,k_q,m,n)} |X_{k_1}|^t \dots |X_{k_q}|^t \, d\mathsf{P},$$

where the summation in $\sum_{[1,m]_q}$ involves all multi-indices $(k_1, \dots, k_q) \in [1, m]_q$. Since the total number of elements of the set $[1, m]_q$ is less than m^q,

$$\mathsf{P}(A_n^2, \nu_n \ge cn^\beta) \le \frac{(\mathsf{E}\left[|X|^t\right])^q}{\varepsilon^{qt}} \sum_{m=[cn^\beta]}^{\infty} m^{q-q\gamma\alpha t}.$$

Now we deduce that

$$\sum_{n=1}^{\infty} n^{\alpha r-2} \mathsf{P}(A_n^2, \nu_n \ge cn^\beta) \le O(1) \sum_{n=1}^{\infty} n^{\alpha r-2} \sum_{m=[cn^\beta]}^{\infty} m^{q-q\gamma\alpha t} \le O(1) \sum_{n=1}^{\infty} n^b < \infty$$

according to the defining properties of the number q, where

$$b = \alpha r - 2 + \beta(q - q\gamma\alpha t + 1).$$

Thus condition (13.45) holds for $i = 2$ and all $\varepsilon > 0$.

Finally, we consider the case where $i = 3$. Fix $\varepsilon > 0$. We show first that

$$\lim_{k\to\infty} \frac{\mathsf{E}\left[\zeta_k\right]}{k^\alpha} = 0. \tag{13.46}$$

Without loss of generality, one can assume that $t < 1$ for $\alpha > 1$. If $t < 1$, then

$$|\mathsf{E}\left[\zeta_k\right]| \le k \int\limits_{|x|<\varepsilon k^{\gamma\alpha}} |x|\, dF(x) \le \varepsilon^{1-t} \mathsf{E}\left[|X|^t\right] \cdot k^{1+(1-t)\gamma\alpha}.$$

A similar bound holds for $\alpha \le 1$, too, since $t > 1$ and $\mathsf{E}\left[X\right] = 0$ in this case, whence

$$|\mathsf{E}\left[\zeta_k\right]| = k \left| \int\limits_{|x|\ge\varepsilon k^{\gamma\alpha}} x\, dF(x) \right| \le \varepsilon^{1-t} \mathsf{E}\left[|X|^t\right] \cdot k^{1+(1-t)\gamma\alpha}.$$

Since $\gamma < 1$ and $1 < \gamma\alpha t$, we conclude that $1 + (1-t)\gamma\alpha < \alpha$. This proves (13.46). Thus for all $\delta < \varepsilon/q$ and sufficiently large k,

$$\mathsf{P}(A_n^3,\ \nu_n = k) \le \mathsf{P}(|\zeta_k - \mathsf{E}[\zeta_k]| \ge \delta k^\alpha, \quad \nu_n = k).$$

Now we choose $v \ge \max\{2, t\}$. The precise value of v will be specified later. The latter bound implies that

$$\mathsf{P}(A_n^3,\ \nu_n = k) \le \left(\frac{1}{\delta k^\alpha}\right)^v \int\limits_{\{\nu_n = k\}} |\zeta_k - \mathsf{E}[\zeta_k]|^v\, d\mathsf{P} \le O(1)\frac{\mathsf{E}\left[|\zeta_k - \mathsf{E}[\zeta_k]|^v\right]}{k^{\alpha v}}.$$

Thus

$$\begin{aligned}\sum_{n=1}^\infty n^{\alpha r-2}\mathsf{P}(A_n^3, \nu_n \ge cn^\beta) &\le O(1)\sum_{n=1}^\infty n^{\alpha r-2}\sum_{k=[cn^\beta]}^\infty \frac{\mathsf{E}\left[|\zeta_k - \mathsf{E}[\zeta_k|^v]\right]}{k^{\alpha v}}\\ &\le O(1)\sum_{k=1}^\infty \frac{\mathsf{E}\left[|\zeta_k - \mathsf{E}[\zeta_k]|^v\right]}{k^b},\end{aligned} \tag{13.47}$$

where $b = \alpha v - (\alpha r - 1)/\beta$. We use Rosenthal's inequality (2.11) to prove a bound for $\mathsf{E}\left[|\zeta_k - \mathsf{E}[\zeta_k]|^v\right]$:

$$\begin{aligned}\mathsf{E}\left[|\zeta_k - \mathsf{E}[\zeta_k]|^v\right] &\le O(1)\sum_{i=1}^k \mathsf{E}\left[|X_{ki} - \mathsf{E}[X_{ki}]|^v\right] + O(1)\Bigg(\sum_{i=1}^k \mathsf{var}[X_{ki}]\Bigg)^{v/2}\\ &\le O(1)\left(k\mathsf{E}\left[|X_{k1}|^v\right] + k|\mathsf{E}[X_{k1}]|^v + \left(k\mathsf{E}\left[X_{k1}^2\right]\right)^{v/2}\right).\end{aligned} \tag{13.48}$$

Each term on the right-hand side of (13.48) is estimated separately:

$$\begin{aligned}&\mathsf{E}\left[|X_{k1}|^v\right] \le O(1)k^{(v-t)\gamma\alpha}; \qquad |\mathsf{E}[X_{k1}]|^v \le \begin{cases} O(1), & t \ge 1,\\ O(1)k^{(1-t)\gamma\alpha v}, & t < 1;\end{cases}\\ &\qquad\qquad \left(\mathsf{E}\left[X_{k1}^2\right]\right)^{v/2} \le \begin{cases} O(1), & t \ge 2,\\ O(1)k^{(2-t)\gamma\alpha v/2}, & t < 2.\end{cases}\end{aligned} \tag{13.49}$$

The variables $O(1)$ in these bounds do not depend on k but they do depend on ε. Choose v so large that $b > 1$ (recall that $b = \alpha v - (\alpha r - 1)/\beta$) and

$$\begin{aligned} 1 + (v - t)\gamma\alpha &< b - 1,\\ 1 + \max\{0, 1 - t\}\gamma\alpha v &< b - 1,\\ \frac{v}{2} + \frac{\max\{0, 2 - t\}}{2}\gamma\alpha v &< b - 1.\end{aligned} \tag{13.50}$$

The first of these restrictions can be achieved by the choice of the number v in view of $\gamma < 1$, the second restriction is satisfied, since $\gamma \max\{0, 1-t\} < 1$, and the third one holds by $\gamma \max\{0, 2-t\} < 2$. For such a number v we estimate each term of the series on the right-hand side of (13.47) with the help of (13.48) and then use inequality (13.49). Considering restrictions (13.50), the series on the right-hand side of (13.47) converges and condition (13.45) holds for $i = 3$.

The proof of Theorem 13.9 is complete. □

Proof of Theorem 13.10 First we note that $\theta > \frac{1}{\alpha}$ and thus both conditions (13.19) and (13.18) hold for $t = \theta$. It remains to show that condition (13.33) holds for all $\varepsilon > 0$ in order to apply Theorem 13.9. Let $\varepsilon > 0$. Then

$$\sum_{n=1}^{\infty} n^{\alpha r-2}\mathsf{P}\left(\max_{k\le\nu_n}|X_k| \ge \varepsilon\nu_n^{\alpha}\right)$$
$$\le \sum_{n=1}^{\infty} n^{\alpha r-2}\mathsf{P}\left(\nu_n < cn^{\beta}\right) + \sum_{n=1}^{\infty} n^{\alpha r-2}\mathsf{P}\left(\max_{k\le\nu_n}|X_k| \ge \varepsilon\nu_n^{\alpha},\, \nu_n \ge cn^{\beta}\right).$$

The first series on the right-hand side converges by condition (13.32). To prove the convergence of the second series we put

$$p(k,n,j) = \mathsf{P}(|X_k| \ge \varepsilon\nu_n^{\alpha},\, \nu_n = j).$$

Then we have

$$\sum_{n=1}^{\infty} n^{\alpha r-2}\mathsf{P}\left(\max_{k\le\nu_n}|X_k| \ge \varepsilon\nu_n^{\alpha},\, \nu_n \ge cn^{\beta}\right)$$
$$= \sum_{n=1}^{\infty} n^{\alpha r-2} \sum_{j\ge cn^{\beta}} \mathsf{P}\left(\max_{k\le\nu_n}|X_k| \ge \varepsilon\nu_n^{\alpha},\, \nu_n = j\right)$$
$$\le \sum_{n=1}^{\infty} n^{\alpha r-2} \sum_{j\ge cn^{\beta}} \left[\sum_{k<cn^{\beta}} p(k,n,j) + \sum_{cn^{\beta}\le k\le j} p(k,n,j)\right].$$

Further,

$$\sum_{n=1}^{\infty} n^{\alpha r-2} \sum_{j\ge cn^{\beta}} \sum_{k<cn^{\beta}} p(k,n,j) \le \sum_{n=1}^{\infty} n^{\alpha r-2} \sum_{k<cn^{\beta}} \mathsf{P}(|X_k| \ge \varepsilon\nu_n^{\alpha},\, \nu_n \ge cn^{\beta})$$
$$\le \sum_{n=1}^{\infty} n^{\alpha r-2} \sum_{k<cn^{\beta}} \mathsf{P}(|X_k| \ge \varepsilon c^{\alpha} n^{\alpha\beta})$$
$$\le c\sum_{n=1}^{\infty} n^{\alpha r+\beta-2}\mathsf{P}(|X_1| \ge \varepsilon c^{\alpha} n^{\alpha\beta}).$$

According to Proposition A.23, condition (13.31) implies the convergence of the latter series for all $\varepsilon > 0$.

Moreover,

$$\begin{aligned}\sum_{n=1}^{\infty} n^{\alpha r-2} \sum_{j \ge cn^{\beta}} \sum_{cn^{\beta} \le k \le j} p(k,n,j) &\le \sum_{n=1}^{\infty} n^{\alpha r-2} \sum_{k \ge cn^{\beta}} \mathsf{P}(|X_k| \ge \varepsilon \nu_n^{\alpha}, \nu_n \ge k) \\ &\le \sum_{n=1}^{\infty} n^{\alpha r-2} \sum_{k \ge cn^{\beta}} \mathsf{P}(|X_k| \ge \varepsilon k^{\alpha}) \\ &\le 1 + \sum_{k=1}^{\infty} \mathsf{P}(|X_1| \ge \varepsilon k^{\alpha}) \sum_{n: n \le (k/c)^{1/\beta}} n^{\alpha r-2}.\end{aligned}$$

The asymptotic behavior of the inner sum is defined by formula (A.69), whence

$$\sum_{n=1}^{\infty} n^{\alpha r-2} \sum_{j \ge cn^{\beta}} \sum_{cn^{\beta} \le k \le j} p(k,n,j) \le O(1) \sum_{k=1}^{\infty} k^{(\alpha r-1)/\beta} \mathsf{P}(|X_1| \ge \varepsilon k^{\alpha}).$$

The convergence on the right-hand side follows from condition (13.31) by Proposition A.23. Therefore, condition (13.33) holds and thus Theorem 13.10 follows from Theorem 13.9. □

Proof of Corollary 13.5 It is clear that condition (13.35) is a particular case of condition (13.32) corresponding to $\beta = 1$. In this case, conditions (13.7) and (13.31) coincide. Hence Corollary 13.5 follows directly from Theorem 13.10. □

Proof of Theorem 13.12 Since

$$\mathsf{P}(|S_{\nu_n}| \ge \nu_n^{\alpha}\varepsilon) \le \mathsf{P}(\nu_n < cn^{\beta}) + \mathsf{P}(|S_{\nu_n}| \ge \nu_n^{\alpha}\varepsilon, \nu_n \ge cn^{\beta})$$

and

$$\mathsf{P}(|S_{\nu_n}| \ge \nu_n^{\alpha}\varepsilon, \nu_n \ge cn^{\beta}) \le \mathsf{P}\left(\sup_{k \ge cn^{\beta}} \left| \frac{S_k}{k^{\alpha}} \right| \ge \varepsilon \right),$$

we need to show that

$$\sum_{n=1}^{\infty} \frac{1}{n} \mathsf{P}\left(\sup_{k \ge cn^{\beta}} \left| \frac{S_k}{k^{\alpha}} \right| \ge \varepsilon \right) < \infty \text{ for all } \varepsilon > 0 \tag{13.51}$$

in view of (13.32). Now we estimate the left-hand side of (13.51) separately for different values of β.

The case $0 < \beta < 1$. After simple algebra we obtain

$$\sum_{n=1}^{\infty}\frac{1}{n}\mathsf{P}\Bigl(\sup_{k\geq cn^{\beta}}\Bigl|\frac{S_k}{k^{\alpha}}\Bigr|\geq\varepsilon\Bigr)=\sum_{j=0}^{\infty}\sum_{n:j\leq cn^{\beta}<j+1}\frac{1}{n}\mathsf{P}\Bigl(\sup_{k\geq cn^{\beta}}\Bigl|\frac{S_k}{k^{\alpha}}\Bigr|\geq\varepsilon\Bigr)$$

$$\leq O(1)\left[1+\sum_{j=1}^{\infty}\frac{1}{j^{1/\beta}}\Bigl((j+1)^{1/\beta}-j^{1/\beta}\Bigr)\mathsf{P}\Bigl(\sup_{k\geq j}\Bigl|\frac{S_k}{k^{\alpha}}\Bigr|\geq\varepsilon\Bigr)\right] \tag{13.52}$$

$$\leq O(1)\left[1+\sum_{j=1}^{\infty}\frac{1}{j}\mathsf{P}\Bigl(\sup_{k\geq j}\Bigl|\frac{S_k}{k^{\alpha}}\Bigr|\geq\varepsilon\Bigr)\right].$$

The cases $\beta > 1$ and $\beta = 1, c \geq 1$. Bound (13.52) is obvious.

The case $\beta = 1, 0 < c < 1$. For $\delta = \varepsilon c^{\alpha}/2$, we have

$$\mathsf{P}\Bigl(\sup_{k\geq cn}\Bigl|\frac{S_k}{k^{\alpha}}\Bigr|\geq\varepsilon\Bigr)\leq\mathsf{P}\Bigl(\max_{cn\leq k<n}|S_k|\geq\delta n^{\alpha}\Bigr)+\mathsf{P}\Bigl(\sup_{k\geq n}\Bigl|\frac{S_k}{k^{\alpha}}\Bigr|\geq\varepsilon/2\Bigr).$$

Relation (13.30) and the Marcinkiewicz–Zygmund strong law of large numbers (see Corollary 9.2 for $d = 1$) imply that $\operatorname{med}(S_n)/n^{\alpha} \to 0$ (recall that $\operatorname{med}(\xi)$ is a median of a random variable ξ). Thus

$$\frac{1}{n^{\alpha}}\max_{1\leq k\leq n}|\operatorname{med}(S_k)|\to 0,\quad\text{whence}\quad\frac{1}{n^{\alpha}}\max_{cn\leq k<n}|\operatorname{med}(S_k-S_n)|\to 0.$$

Therefore $\max_{cn\leq k<n}|\operatorname{med}(S_k - S_n)| < n^{\alpha}\delta/2$ for sufficiently large n. Now we apply the Lévy inequality for non-symmetric random variables (Corollary 2.3) and obtain that

$$\mathsf{P}\Bigl(\max_{cn\leq k<n}|S_k|\geq\delta n^{\alpha}\Bigr)\leq\mathsf{P}\Bigl(\max_{cn\leq k<n}|S_k-\operatorname{med}(S_k-S_n)|\geq\delta n^{\alpha}/2\Bigr)$$

$$\leq 2\mathsf{P}(|S_n|\geq\delta n^{\alpha}/2)\leq 2\mathsf{P}\Bigl(\sup_{k\geq n}\Bigl|\frac{S_k}{k^{\alpha}}\Bigr|\geq\delta/2\Bigr)$$

for sufficiently large n. Therefore,

$$\sum_{n=1}^{\infty}\frac{1}{n}\mathsf{P}\Bigl(\sup_{k\geq[cn^{\beta}]}\Bigl|\frac{S_k}{k^{\alpha}}\Bigr|\geq\varepsilon\Bigr)\leq O(1)\left[1+\sum_{n=1}^{\infty}\frac{1}{n}\mathsf{P}\Bigl(\sup_{k\geq n}\Bigl|\frac{S_k}{k^{\alpha}}\Bigr|\geq\varepsilon c_1\Bigr)\right]$$

irrespective of the values of β and c, where $c_1 = \min\{1/2, c^{\alpha}/4\}$.

According to Theorem 13.11, the convergence of the latter series follows from condition (13.30). Thus condition (13.51) and, as a result, Theorem 13.12 are proved. □

Proof of Corollary 13.6 We check the assumptions of Theorem 13.10 with $r = \frac{2}{\alpha}$ for case (a). Since $(\alpha r - 1 + \beta)/\alpha\beta \leq 2/\alpha$, condition (13.31) follows from (13.41) and (13.32) coincides with (13.42). Thus case (a) of Corollary 13.6 follows from Theorem 13.10.

For case (b), we check the assumptions of Theorem 13.9 with $r = \frac{2}{\alpha}$. Indeed, condition (13.19) with $t = \frac{2}{\alpha}$ and (13.31) follow from (13.41), while (13.32) coincides with (13.42). We have shown in the proof of Theorem 13.9 that condition (13.33) follows from (13.31) and (13.32). Thus case (b) of Corollary 13.6 follows from Theorem 13.9. □

13.4 Complete Convergence of Sums with Non-random Indices

Assume that ν_n are non-random indices. The main restriction of Sect. 13.3 imposed on the sequence $\{\nu_n, n \geq 1\}$ is the existence of a number $c > 0$ for which condition (13.32) holds. This condition for non-random $\{\nu_n\}$ is equivalent to

$$\text{there exists a } c > 0 \text{ such that} \quad \sum_{n \in A} n^{\alpha r - 2} < \infty, \quad \text{where } A = \{n : \nu_n < cn^{\beta}\}.$$

The latter condition follows from

$$\liminf_{n \to \infty} \frac{\nu_n}{n^{\beta}} > 0. \tag{13.53}$$

In turn, condition (13.33) for non-random indices $\{\nu_n\}$ follows from

$$\sum_{n=1}^{\infty} n^{\alpha r - 2} \nu_n \mathsf{P}(|X| \geq \varepsilon_0 \nu_n^{\alpha}) < \infty \quad \text{for some } \varepsilon_0 > 0, \tag{13.54}$$

since

$$\mathsf{P}\left(\max_{k \leq m} |X_k| \geq z \right) = 1 - (1 - \mathsf{P}(|X| \geq z))^m \leq m\mathsf{P}(|X| \geq z)$$

for all $m \geq 1$ and $z \geq 0$.

The above consideration shows that Theorem 13.9 implies the following result.

Corollary 13.7 *Let $\{X_n\}$ be a sequence of independent identically distributed random variables. Let $\alpha > 1/2$, $\alpha r > 1$, and let $\{\nu_n\}$ be a sequence of non-random numbers. Assume that conditions* (13.53) *and* (13.54) *hold for some constants $\beta > 0$ and $\varepsilon_0 > 0$. Moreover, let condition* (13.19) *be satisfied for some $t > 1/\alpha$ and additionally let condition* (13.18) *hold if $\alpha \leq 1$. Then there exists a number $\varepsilon_1 > 0$ such that relation* (13.34) *holds for all $\varepsilon > \varepsilon_1$.*

For the sake of simplicity, we assume that the sequence $\{\nu_n\}$ is non-decreasing. Let τ_k be the number of solutions of the equation $\nu_n = k$. We further assume that

$$\tau_k < \infty \quad \text{for all} \quad k \ge 1.$$

In most cases, it is convenient to change condition (13.53) to its analog written in terms of the sequence $\{\tau_k\}$. Such a condition is involved in the following result.

Lemma 13.6 *Condition* (13.53) *is equivalent to*

$$\limsup_{m\to\infty} \frac{\tau_1 + \cdots + \tau_m}{m^{1/\beta}} < \infty. \tag{13.55}$$

Proof of Lemma 13.6 First we note that $\tau_1 + \cdots + \tau_m$ is the number of members of the sequence $\{\nu_n\}$ that do not exceed m. Thus

$$\nu_n > m \iff \tau_1 + \cdots + \tau_m \le n - 1. \tag{13.56}$$

If condition (13.53) holds, then there exists a constant $c > 0$ such that $\nu_n \ge cn^\beta$ for all $n \ge 1$. Thus $\nu_n > m$, where $m \stackrel{\text{def}}{=} \left[cn^\beta\right] - 1$ and (13.56) implies $\tau_1 + \cdots + \tau_m \le n - 1$. Since

$$\frac{n-1}{m^{1/\beta}} \to \frac{1}{c},$$

relation (13.55) holds.

Conversely, if condition (13.55) holds, then there exists a constant $c < \infty$ such that $\tau_1 + \cdots + \tau_m \le cm^{1/\beta}$, $m \ge 1$. Thus $\tau_1 + \cdots + \tau_m \le n - 1$, where $n \stackrel{\text{def}}{=} \left[cm^{1/\beta}\right] + 2$ and we conclude from (13.56) that $\nu_n > m$. Since

$$\frac{m}{n^\beta} \to \frac{1}{c},$$

relation (13.53) is proved. □

For simplicity, we assume in what follows that $\alpha r = 2$. The general case of $\alpha r \ge 1$ can be considered similarly.

Lemma 13.7 *Let* $X, \{X_n\}$ *be independent identically distributed random variables. If* $\alpha r = 2$, *then condition* (13.54) *for non-random indices* $\{\nu_n\}$ *is equivalent to*

$$\sum_{n=1}^{\infty} n\tau_n \mathsf{P}(|X| \ge n^\alpha \varepsilon_0) < \infty \textit{ for some } \varepsilon_0 > 0. \tag{13.57}$$

13.4.1 Series with Repeating Terms

Now we are going to provide some sufficient conditions for the convergence of series (13.5) whose weight coefficients are positive integers. Passing to the integer parts, analogous results can be derived for the general case of weight coefficients. Another problem for the case of a subsequence of positive integers is considered in Sect. 13.4.2.

Corollary 13.8 *Let $\{X_n\}$ be a sequence of independent identically distributed random variables. Let $\alpha > 1/2$ and let $\{\tau_k\}$ be a sequence of positive integers. Assume that condition* (13.55) *holds for some $\beta > 0$.*

(a) If condition (13.57) *holds and additionally $\mathsf{E}[X] = 0$ in the case of $\alpha \le 1$, then there exists a number ε_1 such that the series*

$$\sum_{n=1}^{\infty} \tau_n \mathsf{P}(|S_n| \ge n^{\alpha}\varepsilon) \tag{13.58}$$

converges for all $\varepsilon > \varepsilon_1$.

(b) If series (13.58) *converges for all $\varepsilon > 0$, then*

$$\sum_{n=1}^{\infty} n\tau_n \mathsf{P}(|X| \ge n^{\alpha}\varepsilon) < \infty \quad \textit{for all } \varepsilon > 0 \tag{13.59}$$

and $\mathsf{E}[X] = 0$ if $\alpha \le 1$.

Proof of Corollary 13.7 (a) Since $\tau_n \ge 1$ for all $n \ge 1$, relation (13.57) implies that $\mathsf{E}\left[|X|^{2/\alpha}\right] < \infty$ (see Proposition A.25 with $\nu_1 = 1$, $\nu_2 = \alpha$). Hence conditions (13.18) and (13.19) hold for $t = \frac{2}{\alpha}$. We construct a sequence of positive integers $\{\nu_n\}$ by using the following rule:

$$\{\nu_1, \nu_2, \nu_3, \dots\} = \left\{ \underbrace{1, \dots, 1}_{\tau_1}, \underbrace{2, \dots, 2}_{\tau_2}, \dots, \underbrace{m, \dots, m}_{\tau_m}, \dots \right\}.$$

By Lemmas 13.6 and 13.7, conditions (13.53) and (13.54) hold for the sequence $\{\nu_n, n \ge 1\}$ introduced above. Thus Corollary 13.7 with $\alpha r = 2$ implies that there exists a number ε_1 such that

$$\sum_{n=1}^{\infty} \mathsf{P}(|S_{\nu_n}| \ge \varepsilon \nu_n^{\alpha}) < \infty \quad \text{for all} \quad \varepsilon \ge \varepsilon_1.$$

Therefore Corollary 13.8 is proved for case (a).

(b) If series (13.58) converges for all $\varepsilon > 0$, then the following law of large numbers

$$\frac{S_n}{n^{\alpha}} \xrightarrow{\mathsf{P}} 0, \qquad n \to \infty,$$

holds, since $\tau_n \ge 1$. This yields $n\mathsf{P}(|X| \ge \varepsilon n^\alpha) \to 0$ as $n \to \infty$ for an arbitrary $\varepsilon > 0$. Now condition (13.59) implies Theorem 13.5. Finally, condition $\mathsf{E}[X] = 0$ follows from Theorem 13.2, since $\tau_n \ge 1$. □

The simplest case of Corollary 13.8 corresponds to $\tau_k = 1$, $k \ge 1$, that is, to $\nu_n = n$, $n \ge 1$. The result below describes the complete convergence to 0 of the sequence S_n/n^α.

Corollary 13.9 *Let $\{X_n\}$ be a sequence of independent identically distributed random variables. Let $\alpha r = 2$. Condition* (13.16) *is equivalent to the complete convergence to 0 of the sequence S_n/n^α, that is, condition* (13.16) *is equivalent to the convergence of the series $\sum \mathsf{P}(|S_n| \ge \varepsilon n^\alpha)$ for all $\varepsilon > 0$.*

Proof of Corollary 13.9 Proposition A.25 with $\nu_1 = 1$ and $\nu_2 = \alpha$ means that condition (13.57) is equivalent to the convergence of the series $\sum n\mathsf{P}(|X| \ge \varepsilon n^\alpha)$ for all $\varepsilon > 0$. Thus Corollary 13.9 follows from Corollaries 13.2 and 13.8. □

13.4.2 The Case of a Subsequence

Let $\{\nu_n\}$ be a subsequence of positive integers, that is, $\nu_n < \nu_{n+1}$ for all $n \ge 1$. We study the convergence of the series

$$\sum_{k=1}^{\infty} \nu_k^{\alpha r-2} \mathsf{P}(|S_{\nu_k}| \ge \varepsilon \nu_k^\alpha). \tag{13.60}$$

Corollary 13.7 can be used to investigate the convergence of series (13.60) for all $\varepsilon > 0$. In this case, condition (13.53) holds for $\beta = 1$, since $\nu_n \ge n$. The only condition of Corollary 13.7 to be verified is the "moment" condition (13.54).

Theorem 13.13 *Let $\{X_n\}$ be independent identically distributed random variables and let $\{\nu_n\}$ be a subsequence of positive integers. Assume that $\alpha > 1/2$, $\alpha r \ge 1$, condition* (13.19) *holds, and additionally $\mathsf{E}[X] = 0$ in the case of $\alpha \le 1$. The convergence of series* (13.60) *for all $\varepsilon > 0$ is equivalent to the following condition*

$$\sum_{k=1}^{\infty} \nu_k^{\alpha r-1} \mathsf{P}(|X| \ge \varepsilon \nu_k^\alpha) \quad \textit{for all } \varepsilon > 0.$$

The most studied case is the simplest one, corresponding to $\alpha = 1$, $\alpha r = 2$.

Corollary 13.10 *Let $\{X_n, n \ge 1\}$ be independent identically distributed random variables and let $\{\nu_n, n \ge 1\}$ be a subsequence of positive integers. Assume that*

$$\mathsf{E}\left[|X|^t\right] < \infty \quad \textit{for some } t > 1. \tag{13.61}$$

Then the following two conditions are equivalent:

$$\sum_{k=1}^{\infty} \mathsf{P}(|S_{\nu_k}| \geq \varepsilon \nu_k) < \infty \tag{13.62}$$

for all $\varepsilon > 0$; *and*

$$\sum_{k=1}^{\infty} \nu_k \mathsf{P}(|X| \geq \varepsilon \nu_k) < \infty \tag{13.63}$$

for all $\varepsilon > 0$.

13.4.3 The Asmussen–Kurtz Conjecture

Corollary 13.10 contains the criterion for the convergence for subsequences under the additional condition that a moment of order higher than 1 exists (see (13.61)). If $\nu_k = k$, then this additional condition is negligible in view of (13.63), which is equivalent to the existence of the second moment. Nevertheless, condition (13.61) is too restrictive if a sequence $\{\nu_n, n \geq 1\}$ grows fast enough. For example, if $\nu_k = 2^k$, then condition (13.63) is equivalent to the existence of the first moment and (13.61) is too restrictive in this case.

It seems reasonable to conjecture that if ν_k increases faster, then (13.63) transforms in an even weaker condition than the existence of the first moment. Surprisingly, condition (13.63) remains equivalent to the existence of the first moment even for very quickly growing sequences $\{\nu_n\}$. This is true, for example, for $\nu_n = n!$. Asmussen and Kurtz [12] conjecture that the condition $\mathsf{E}[|X|] < \infty$ is necessary for the law of large numbers in the sense of complete convergence for subsequences.

Remark 13.12 Gut [135] discards the Asmussen–Kurtz conjecture by constructing the following example. The law of large numbers for symmetric random variables $\{X_n\}$ follows from the condition $x\mathsf{P}(X \geq x) \to 0$ as $x \to \infty$ where X is a copy of X_n (see Theorem 4.4). We assume that this condition holds but the first moment $\mathsf{E}[|X|]$ does not exist. If a subsequence $\{\nu_k\}$ is chosen such that

$$\mathsf{P}\left(|S_{\nu_k}| \geq \nu_k \varepsilon^k\right) \leq \frac{1}{2^k}, \qquad \varepsilon_k \stackrel{\text{def}}{=} \frac{1}{2^k},$$

then (13.62) holds for all $\varepsilon > 0$.

We will extend this result and show that, if one does not restrict to a given class of subsequences, then any a priori moment condition is not necessary for (13.62). The following example of Deli shows that, for every positive unbounded function ψ such that

$$\mathsf{E}[\psi(|X|)] = \infty, \tag{13.64}$$

there exists a distribution function F and a subsequence $\{\nu_n, n \geq 1\}$ for which condition (13.62) is satisfied for all $\varepsilon > 0$.

Choosing $\psi(x) = x$ in Deli's example below we get another counterexample to the Asmussen–Kurtz conjecture.

13.4.4 Deli's Example

Let ψ be a non-decreasing unbounded function defined on $[0, \infty)$. Then we construct recursively two sequences of positive integers $\{n_k\}$ and $\{m_k\}$. The first members of these subsequences are chosen such that

$$n_1 = 1, \qquad \psi(m_1) > 1.$$

If, for some $k \geq 1$, the numbers $n_1, \dots, n_k$ and $m_1, \dots, m_k$ are defined, then we choose n_{k+1} to satisfy the following three properties

(a) $\quad n_{k+1} > (k+1) \sum_{i=1}^{k} \frac{m_i^2}{\psi(m_i)}$,

(b) $\quad n_{k+1} > (k+1)^2$,

(c) $\quad n_{k+1} > m_k$.

Then we choose m_{k+1} such that

(d) $\quad m_{k+1} > n_k$,

(e) $\quad \psi(m_{k+1}) > n_{k+1}^2$,

(f) $\quad \psi(m_{k+1}) > 2\psi(m_k)$.

The property (f) implies that the series

$$\sum_{k=1}^{\infty} \frac{1}{\psi(m_k)}$$

converges; denote its sum by C. Next we introduce a random variable X with the distribution

$$\mathsf{P}(X = \pm m_k) = \frac{C}{2\psi(m_k)}, \qquad k \geq 1.$$

It is clear that the distribution function F of the random variable X is symmetric and that

$$\mathsf{E}\left[\psi(|X|)\right] = \infty.$$

Moreover,

$$
\begin{aligned}
n_k \mathsf{P}(|X| \geq n_k) &= n_k \sum_{i: m_i \geq n_k} \mathsf{P}(|X| = m_i) \\
&\stackrel{(c)}{\leq} n_k \sum_{i=k}^{\infty} \mathsf{P}(|X| = m_i) = C n_k \sum_{i=k}^{\infty} \frac{1}{\psi(m_i)} \\
&\stackrel{(e)}{\leq} C n_k \sum_{i=k}^{\infty} \frac{1}{n_i^2} \leq C \sum_{i=k}^{\infty} \frac{1}{n_i} \\
&\stackrel{(b)}{\leq} C \sum_{i=k}^{\infty} \frac{1}{i^2} \to 0
\end{aligned}
$$

and

$$
\frac{1}{n_k} \int_{|x|<n_k} x^2 \, dF = \frac{1}{n_k} \sum_{i: m_i < n_k} m_i^2 \mathsf{P}(|X| = m_i) \stackrel{(f)}{=} \frac{C}{n_k} \sum_{i=1}^{k-1} \frac{m_i^2}{\psi(m_i)} \to 0
$$

as $k \to \infty$. Let $\{X_n\}$ be a sequence of independent random variables with the distribution function F. Since

$$
\begin{aligned}
\mathsf{P}(|S_{n_k}| \geq n_k \varepsilon) &\leq n_k \mathsf{P}(|X| \geq n_k) + \mathsf{P}\left(\left| \sum_{i \leq n_k} X_i \, \mathbb{I}_{\{|X_i| < n_k\}} \right| \geq n_k \varepsilon \right) \\
&\leq n_k \mathsf{P}(|X| \geq n_k) + \frac{1}{n_k \varepsilon^2} \int_{|x|<n_k} x^2 \, dF,
\end{aligned}
$$

property (a) implies that

$$
\frac{S_{n_k}}{n_k} \xrightarrow{\mathsf{P}} 0.
$$

Completing the construction, we choose an arbitrary sequence of real numbers $\varepsilon_i \downarrow 0$. Given $i \geq 1$, define an index k_i such that

$$
\mathsf{P}(|S_{\nu_i}| \geq \varepsilon_i \nu_i) \leq \frac{1}{2^i}
$$

for $\nu_i \stackrel{\text{def}}{=} n_{k_i}$. This relation proves (13.62) for all $\varepsilon > 0$.

13.5 Complete Convergence of Multi-Indexed Sums

Let $\{\tau_k\}$ be the Dirichlet sequence (see Sect. A.10). Recall that $\tau_k = \tau_k(d)$ is the number of solutions of the equation $n_1 \cdots n_d = k$. The rough asymptotics in the Dirichlet divisors problem (relation (A.41)) implies that condition (13.55) holds for all $\beta < 1$. By Remark A.5, condition (13.57) is equivalent to the following one

$$\mathsf{E}\left[|X|^{2/\alpha}\left(\log^+|X|\right)^{d-1}\right] < \infty \tag{13.65}$$

whatever $\varepsilon_0 > 0$.

The following result is a straightforward consequence of Corollary 13.8.

Corollary 13.11 *Let X, $\{X(\mathbf{n}), \mathbf{n} \in \mathbf{N}^d\}$ be independent identically distributed random variables and let $\alpha > 1/2$ and $d \geq 1$. If condition* (13.65) *holds and additionally $\mathsf{E}[X] = 0$ for the case of $\alpha \leq 1$, then*

$$\sum_{n_1=1}^{\infty} \cdots \sum_{n_d=1}^{\infty} \mathsf{P}(|S(n_1, \dots, n_d)| \geq (n_1 \cdots n_d)^{\alpha}\varepsilon) < \infty \quad \textit{for all } \varepsilon > 0, \tag{13.66}$$

that is, the field $S(\mathbf{n})/|\mathbf{n}|^{\alpha}$ converges to 0 in the sense of complete convergence.

The assumptions of Corollary 13.11 are, in fact, necessary and sufficient. Below we consider multi-indexed series with more general weights.

Let $w(x)$ be a regularly function of order $\rho \geq -1$ varying at infinity, that is, $w(x) = x^{\rho}L(x)$, where $L(x)$ is a slowly varying function. For $\alpha > \frac{1}{2}$, put

$$G(x) = \sum_{n \leq x} n^{\rho+1}L(n)\left(\log^+ n\right)^{d-1}. \tag{13.67}$$

Theorem 13.14 *Let X, $\{X(\mathbf{n}), \mathbf{n} \in \mathbf{N}^d\}$ be independent identically distributed random variables. Let $w(x) = x^{\rho}L(x)$, where $\rho \geq -1$ and $L(\cdot)$ is a measurable slowly varying function. Assume that there exists a number $x_0 > 0$ such that*

$$\inf_{x > x_0} L(x)\left(\log^+ x\right)^{d-1} > 0. \tag{13.68}$$

Let X, $\{X(\mathbf{n}), \mathbf{n} \in \mathbf{N}^d\}$ be independent identically distributed random variables and let $\alpha > 1/2$. The series

$$\sum_{\mathbf{n} \in \mathbf{N}^d} w(|\mathbf{n}|)\mathsf{P}(|S(\mathbf{n})| \geq \varepsilon|\mathbf{n}|^{\alpha}) \tag{13.69}$$

converges for all $\varepsilon > 0$ if and only if

$$\mathsf{E}\left[G\left(|X|^{1/\alpha}\right)\right] < \infty, \tag{13.70}$$

$$\mathsf{E}[X] = 0, \qquad \text{if } \alpha \le 1. \tag{13.71}$$

Remark 13.13 The asymptotic behavior of the function G with $\rho > -1$ can be written as follows

$$G(x) \asymp x^{\rho+2} L(x) \left(\log^+ x\right)^{d-1}, \qquad x \to \infty.$$

Thus condition (13.70) is equivalent in this case to the existence of the moment

$$\mathsf{E}\left[|X|^{(\rho+2)/\alpha} L\left(|X|^{1/\alpha}\right)\left(\log^+ |X|\right)^{d-1}\right].$$

Note that (13.68) is a superfluous assumption in this case. If $\rho = -1$, then the function G is slowly varying (see [334]). Theorem 13.14 in this case becomes false if one drops condition (13.68).

Corollary 13.12 *Let X, $\{X(\mathbf{n}), \mathbf{n} \in \mathbf{N}^d\}$ be independent identically distributed random variables. Let $\alpha > \frac{1}{2}$, $\alpha r \ge 1$. The series*

$$\sum_{\mathbf{n} \in \mathbf{N}^d} |\mathbf{n}|^{\alpha r - 2} \mathsf{P}(|S(\mathbf{n})| \ge \varepsilon |\mathbf{n}|^{\alpha}) \tag{13.72}$$

converges for all $\varepsilon > 0$ if and only if

$$\mathsf{E}\left[|X|^r \left(\log^+ |X|\right)^{d-1}\right] < \infty \text{ and } \mathsf{E}[X] = 0 \text{ in the case of } \alpha \le 1.$$

Proof of Theorem 13.14 Put $w_n = \tau_n n^{\rho} L(n)$, where $\{\tau_n\}$ is the Dirichlet sequence. Let $\{S_n, n \ge 1\}$ be a sequence associated with the field $\{S(\mathbf{n}), \mathbf{n} \in \mathbf{N}^d\}$, that is, $S_n = \sum_{k=1}^{n} X_k$, where $\{X_n, n \ge 1\}$ is a sequence of independent copies of the random variables $X(\mathbf{n})$. Further, let $\{X_n^{(s)}\}$ be symmetrizations for the sequence $\{X_n, n \ge 1\}$, while $S_n^{(s)}$ are cumulative sums of the random variables $X_k^{(s)}$.

First we prove that conditions (13.70)–(13.71) imply the convergence of series (13.69) for all $\varepsilon > 0$. Using the rough asymptotics in the Dirichlet divisors problem (see bound (A.41)) we derive condition (13.22) for $\theta > \rho + 2$. The property $W_n = O(W_{n-1})$ is also clear. It is easy to check that (13.70) and (13.68) imply assumptions (i)–(ii) of Theorem 13.7 and that (13.71) implies assumption (iii). Thus Theorem 13.7 proves that series (13.5) converges for all $\varepsilon > 0$ which is equivalent to the convergence of series (13.69).

Now we prove the converse, namely we prove that if series (13.69) converges for all $\varepsilon > 0$, then conditions (13.70)–(13.71) are satisfied. In doing so we use Theorem 13.5. All the assumptions of this theorem are clearly satisfied except condition (13.15). In particular, series $\sum w_n$ diverges in view of (13.68).

Thus we concentrate on the proof of condition (13.15). By convergence (13.69), we conclude that

$$\sum_{n=1}^{\infty} w_n \mathsf{P}(|S_n| \ge \varepsilon n^{\alpha}) < \infty$$

for all $\varepsilon > 0$. Since $\mathsf{P}(|S_n^{(s)}| \ge x) \le 2\mathsf{P}(|S_n| \ge x/2)$ for all $x > 0$, we establish

$$\sum_{n=1}^{\infty} w_n \mathsf{P}(|S_n^{(s)}| \ge \varepsilon n^{\alpha}) < \infty \quad \text{for all } \varepsilon > 0.$$

By the Lévy inequality for symmetric random variables (Corollary 2.4),

$$\begin{aligned}\sum_{n=1}^{\infty} w_n \mathsf{P}(|S_n^{(s)}| \ge \varepsilon n^{\alpha}) &= w_1 \mathsf{P}(|X^{(s)}| \ge \varepsilon) + \sum_{n=1}^{\infty} \sum_{2^{n-1}<m\le 2^n} w_m \mathsf{P}(|S_m^{(s)}| \ge \varepsilon m^{\alpha}) \\ &\ge \frac{1}{2}\sum_{n=1}^{\infty} \mathsf{P}(|S_{2^{n-1}}^{(s)}| \ge \varepsilon 2^{n\alpha}) \sum_{2^{n-1}<m\le 2^n} w_m. \end{aligned} \tag{13.73}$$

Since L is a slowly varying function, we have $\min_{2^{n-1}<m\le 2^n} L(m) \ge \frac{1}{2} L\left(2^{n-1}\right)$ for sufficiently large n (we may assume without loss of generality that this property holds for all n). Thus

$$\sum_{2^{n-1}<m\le 2^n} w_m \ge \frac{1}{2} L\left(2^{n-1}\right) 2^{(n-1)\rho}\left(T_{2^n} - T_{2^{n-1}}\right),$$

where T is the Dirichlet function (see Sect. A.10). The function T is regularly varying of order 1 (see (A.32)) and hence there exists a constant $c > 0$ such that $T_{2^n} - T_{2^{n-1}} \ge cT_{2^{n-1}} \ge c2^n n^{d-1}$, $n \ge 1$. Therefore

$$\begin{aligned}\sum_{n=1}^{\infty} w_n \mathsf{P}(|S_n^{(s)}| \ge \varepsilon n^{\alpha}) &\ge \text{const} \sum_{n=1}^{\infty} 2^{n\rho} L(2^n)(T_{2^n} - T_{2^{n-1}}) \mathsf{P}(|S_{2^n}^{(s)}| \ge \varepsilon_1 2^{n\alpha}) \\ &\ge \text{const} \sum_{n=1}^{\infty} 2^{n(\rho+1)} L(2^n) n^{d-1} \mathsf{P}(|S_{2^n}^{(s)}| \ge \varepsilon_1 2^{n\alpha}).\end{aligned}$$

Condition (13.68) implies that $\sum_{n=1}^{\infty} \mathsf{P}(|S_{2^n}^{(s)}| \ge \varepsilon 2^{n\alpha}) < \infty$ for all $\varepsilon > 0$, since $\rho \ge -1$. An argument similar to that used to prove (13.73) establishes

$$\sum_{n=1}^{\infty} \frac{1}{n} \mathsf{P}(|S_n^{(s)}| \ge \varepsilon n^{\alpha}) < \infty.$$

Now Theorem 13.2 with $r = \frac{1}{\alpha}$ implies the existence of the moment $\mathsf{E}\left[|X^{(s)}|^{1/\alpha}\right] < \infty$. Then using Remark 13.2, we derive condition (13.15) for $X^{(s)}$ and then for X, too. Note that $\mathsf{E}[|X|] < \infty$ if $\alpha \le 1$. Then Theorem 13.5 proves

$$\sum_{n=1}^{\infty} \tau_n n^{\rho+1} L(n) \mathsf{P}(|X^{(s)}| \ge n^\alpha) < \infty.$$

By Proposition A.26, condition (13.70) follows.

As mentioned above, the expectation exists if $\alpha \le 1$; let $\mu = \mathsf{E}[X_1]$. The sufficient part of the theorem just proved implies that

$$\sum_{n=1}^{\infty} w_n \mathsf{P}(|S_n - n\mu| \ge \varepsilon n^\alpha) < \infty.$$

The convergence of series (13.69) yields $\sum w_n \mathsf{P}(|S_n| \ge \varepsilon n^\alpha) < \infty$, whence we conclude that $\sum w_n \mathsf{P}(n|\mu| \ge \varepsilon n^\alpha) < \infty$ for all $\varepsilon > 0$ which is the case only if $\mu = 0$, since $\sum w_n = \infty$.

The proof of Theorem 13.14 is complete. □

13.5.1 More General Multi-Indexed Series

Now we study more general series as compared to (13.69), namely we no longer assume that the values of the weight function $w(\mathbf{n})$ are the same for multi-indices $\mathbf{n}$ with identical $|\mathbf{n}|$. Thus we consider the series

$$\sum_{\mathbf{n} \in \mathbf{N}^d} w(\mathbf{n}) \mathsf{P}(|S(\mathbf{n})| \ge \varepsilon |\mathbf{n}|^\alpha). \tag{13.74}$$

Putting

$$w_k = \sum_{\mathbf{n}: |\mathbf{n}| = k} w(\mathbf{n}),$$

series (13.74) reduces to the form of (13.5). If the function $w(\mathbf{n})$ possesses a certain property of regularity, then checking the assumptions of Theorem 13.7 is quite easy. For example, let $w(\mathbf{n}) = n^{\alpha r_1 - 2} \dots n^{\alpha r_d - 2}$. The assumptions of Theorem 13.7 imposed on W_n can be checked by using representation (A.42). Using Proposition A.20, it is not difficult to reduce condition (13.14) to a moment condition. Indeed, the following result is proved in this way.

Corollary 13.13 *Let X, $\{X(\mathbf{n}), \mathbf{n} \in \mathbf{N}^d\}$ be independent identically distributed random variables. Let $\alpha > 1/2$, $r = \max\{r_1, \dots, r_d\} > 0$, and $c = \operatorname{card}\{i : r_i = r\}$. Also let $w(\mathbf{n}) = n^{\alpha r_1 - 2} \dots n^{\alpha r_d - 2}$. The set of conditions*

$$\mathsf{E}\left[|X|^r \left(\log^+ |X|\right)^{c-1}\right] < \infty,$$
$$\mathsf{E}[X] = 0 \quad \textit{in the case of} \quad \alpha \le 1,$$

is necessary and sufficient for the convergence of series (13.74) *for all* $\varepsilon > 0$.

The convergence of the series, an analog of the Baum–Katz series,

$$\sum_{\mathbf{n}\in\mathbf{N}^d} n_1^{\alpha_1 r_1 - 2} \dots n_d^{\alpha_d r_d - 2} \mathsf{P}(|S(\mathbf{n})| \ge \varepsilon n_1^{\alpha_1} \dots n_d^{\alpha_d}) \tag{13.75}$$

is also easy to study.

Corollary 13.14 *Let* X, $\{X(\mathbf{n}), \mathbf{n} \in \mathbf{N}^d\}$ *be independent identically distributed random variables. Let the numbers* $\alpha_1, \dots, \alpha_d$ *and* $r_1, \dots, r_d$ *be such that:*

(i) $\min\{\alpha_1, \dots, \alpha_d\} > \frac{1}{2}$;
(ii) $\min\{\alpha_1 r_1, \dots, \alpha_d r_d\} \ge 1$.

Put

$$r = \max\{r_1, \dots, r_d\}, \qquad q = \operatorname{card}\{k : r_k = r\}, \qquad \alpha = \min\{\alpha_1, \dots, \alpha_d\}.$$

Series (13.75) *converges for all* $\varepsilon > 0$ *if and only if*

$$\mathsf{E}\left[|X|^r \left(\log^+ |X|\right)^{q-1}\right] < \infty$$

in the case of $\alpha < 1$; *or if and only if*

$$EX = 0, \qquad \mathsf{E}\left[|X|^r \left(\log^+ |X|\right)^{q-1}\right] < \infty$$

in the case of $\alpha \ge 1$.

Proof of Corollary 13.14 Put

$$f(\mathbf{n}) = n_1^{\alpha_1 r_1 - 2} \dots n_d^{\alpha_d r_d - 2}, \qquad g(\mathbf{n}) = n_1^{\alpha_1} \dots n_d^{\alpha_d}.$$

The convergence of series (13.75) for all $\varepsilon > 0$ is equivalent to the convergence for all $\varepsilon > 0$ of the series

$$\sum_{k=1}^{\infty} w_k \mathsf{P}(|S_k| \ge \varepsilon k),$$

where

$$w_k = \sum_{\mathbf{n}:[g(\mathbf{n})]=k} f(\mathbf{n}).$$

It is easy to check that $w_k \asymp k^r \left(\log^+ k\right)^{q-1}$ as $k \to \infty$. □

13.6 The Asymptotic Behavior of Series of Large Deviation Probabilities with Respect to a Small Parameter

According to the commonly accepted terminology, series of type (13.4) are constituted by *probabilities of large deviations* if $b_n = n^{\alpha}$, $\alpha > 1/2$. In the literature one can find many investigations of series of probabilities of *moderate* and *small* deviations corresponding to the sequences $b_n = \sqrt{n}\,(\log n)^{\alpha}$ and $b_n = \sqrt{n}\,(\log\log n)^{\alpha}$, $\alpha > 0$, respectively.

The study of the convergence of series of large, moderate, and small deviations is traditional for the topic of limit theorems in probability theory. A new direction was introduced by C. C. Heyde who considered series (13.4) for $w_n = 1, b_n = n$ and found its asymptotic behavior as $\varepsilon \downarrow 0$.

Theorem 13.15 (C. C. Heyde) *Let $\{X_n\}$ be independent identically distributed random variables. Assume that*

$$\mathsf{E}[X_1] = 0, \qquad \mathsf{E}\left[X_1^2\right] = \sigma^2 < \infty. \tag{13.76}$$

Then

$$\lim_{\varepsilon\downarrow 0} \varepsilon^2 \sum_{n=1}^{\infty} \mathsf{P}(|S_n| \geq \varepsilon n) = \sigma^2. \tag{13.77}$$

Recall (see Theorem 13.2) that condition (13.76) is necessary and sufficient for the convergence of the series on the left-hand side of (13.77) for all $\varepsilon > 0$.

13.6.1 Asymptotic Behavior of Multi-Indexed Series

The law of large numbers in the sense of complete convergence for independent identically distributed variables $\{X(\mathbf{n}), \mathbf{n} \in \mathbf{N}^d\}$ is, by definition, equivalent to the following condition

$$Q(\varepsilon) \stackrel{\text{def}}{=} \sum_{\mathbf{n}\in\mathbf{N}^d} \mathsf{P}\left(|S(\mathbf{n})| \geq \varepsilon|\mathbf{n}|\right) < \infty \quad \text{for all} \quad \varepsilon > 0. \tag{13.78}$$

By Corollary 13.12, the following condition

$$\mathsf{E}[X] = 0, \qquad \mathsf{E}\left[X^2 \left(\log^+ |X|\right)^{d-1}\right] < \infty \tag{13.79}$$

is necessary and sufficient for (13.78).

We study the asymptotic behavior of the function $Q(\varepsilon)$ over the parameter $\varepsilon \downarrow 0$.

Theorem 13.16 *Assume that condition* (13.79) *holds and* $\mathsf{E}\left[X^2\right] = \sigma^2, 0 < \sigma^2 < \infty$. *Then*

$$\lim_{\varepsilon\downarrow 0} \frac{\varepsilon^2 Q(\varepsilon)}{\left(\log^+ \varepsilon^{-2}\right)^{d-1}} = \frac{\sigma^2}{(d-1)!}.$$

Proof of Theorem 13.16 We consider the case of $d > 1$, since Theorem 13.16 coincides with Theorem 13.15 for $d = 1$. We will use several functions $Q(\varepsilon)$ in the proof that are constructed according to the rule (13.78) but for different distribution functions. To indicate the distribution function used to construct the function $Q(\varepsilon)$ we sometimes write it as a subscript. Hence if each of the random variables $\{X(\mathbf{n}), \mathbf{n} \in \mathbf{N}^d\}$ have the distribution function F, then we write $Q_F(\varepsilon)$. If the random variables are standard Gaussian, then we write $Q_\Phi(\varepsilon)$. It is sometimes convenient to write $Q_X(\varepsilon)$ instead of $Q(\varepsilon)$ to indicate that the random variables $\{X(\mathbf{n}), \mathbf{n} \in \mathbf{N}^d\}$ are used to construct the function Q.

We restrict ourselves to the case $\sigma^2 = 1$ in the proof, since the general case follows from this particular one. Indeed, define random variables $Y(\mathbf{n}) = X(\mathbf{n})/\sigma$. Since $\mathsf{E}[Y] = 0$ and $\mathsf{var}[Y] = 1$, we establish from Theorem 13.16 with $\sigma^2 = 1$ that

$$\frac{1}{(d-1)!} = \lim_{\varepsilon\downarrow 0} \frac{\varepsilon^2 Q_Y(\varepsilon)}{\left(\log^+ \varepsilon^{-2}\right)^{d-1}} = \lim_{\varepsilon\downarrow 0} \frac{\varepsilon^2 Q_X(\sigma\varepsilon)}{\left(\log^+ \varepsilon^{-2}\right)^{d-1}},$$

that is

$$\lim_{\varepsilon\downarrow 0} \frac{\varepsilon^2 Q_X(\varepsilon)}{\left(\log^+ \varepsilon^{-2}\right)^{d-1}} = \frac{\sigma^2}{(d-1)!}.$$

So, we concentrate on the case $\sigma^2 = 1$. It is clear that Theorem 13.16 follows from the following two relations:

$$\lim_{\varepsilon\downarrow 0} \frac{\varepsilon^2 Q_\Phi(\varepsilon)}{\left(\log^+ \varepsilon^{-2}\right)^{d-1}} = \frac{1}{(d-1)!}, \tag{13.80}$$

$$\lim_{\varepsilon\downarrow 0} \frac{\varepsilon^2 |Q_F(\varepsilon) - Q_\Phi(\varepsilon)|}{\left(\log^+ \varepsilon^{-2}\right)^{d-1}} = 0. \tag{13.81}$$

To prove (13.80) we rewrite the series by using the Dirichlet sequence:

$$Q_\Phi(\varepsilon) = \sum_{k=1}^{\infty} \tau_k \mathsf{P}\left(|Z| \geq \varepsilon\sqrt{k}\right),$$

where Z is a standard Gaussian random variable. Changing the order of summation we get

$$Q_\Phi(\varepsilon) = \sum_{n=1}^{\infty} T_n \mathsf{P}\left(\sqrt{n} \leq |Z| < \varepsilon\sqrt{n+1}\right),$$

where $T_n = \tau_1 + \cdots + \tau_n$. Then we use representation (A.32), for the sequence $\{T_n\}$: there exists a λ with $0 < \lambda < 1$ such that

$$Q_\Phi(\varepsilon) = \sum_{n=1}^{\infty} \left[n\mathscr{P}(\log^+ n) + o(n^\lambda) \right] \mathsf{P}\left(\varepsilon\sqrt{n} \le |Z| < \varepsilon\sqrt{n+1} \right), \tag{13.82}$$

where $\mathscr{P}(\cdot)$ is a polynomial of degree $d-1$ whose leading coefficient equals $\frac{1}{(d-1)!}$ (see (A.32)). Since

$$\varepsilon^2 \sum_{n=1}^{\infty} n^\lambda \mathsf{P}\left(\varepsilon\sqrt{n} \le |Z| < \varepsilon\sqrt{n+1} \right)$$
$$= \varepsilon^{2-2\lambda} \sum_{n=1}^{\infty} (\varepsilon\sqrt{n})^{2\lambda} \int\limits_{\varepsilon\sqrt{n} \le |x| < \varepsilon\sqrt{n+1}} d\Phi(x) \le 2\varepsilon^{2-2\lambda} \int_0^\infty x^{2\lambda}\, d\Phi(x)$$

and all moments of Gaussian random variables are finite,

$$\lim_{\varepsilon \downarrow 0} \frac{\varepsilon^2}{\left(\log^+ \varepsilon^{-2}\right)^{d-1}} \sum_{n=1}^{\infty} o(n^\lambda) \mathsf{P}\left(\varepsilon\sqrt{n} \le |Z| < \varepsilon\sqrt{n+1} \right) = 0.$$

Let $0 \le k \le d-1$. Then, $(\log n)^k = \sum_{i=0}^k C_k^i \left(\log \varepsilon^{-2}\right)^i \left(\log \varepsilon^2 n\right)^{k-i}$ for all $n \ge 1$ and $\varepsilon > 0$. Thus

$$\varepsilon^2 \sum_{n=1}^{\infty} n\,(\log n)^k\, \mathsf{P}\left(\varepsilon\sqrt{n} \le |Z| < \varepsilon\sqrt{n+1} \right)$$
$$= \sum_{i=0}^{k} C_k^i \left(\log \varepsilon^{-2}\right)^i \left[\sum_{n=1}^{\infty} \left(\varepsilon^2 n\right) \left(\log \varepsilon^2 n\right)^{k-i} \mathsf{P}\left(\varepsilon\sqrt{n} \le |Z| < \varepsilon\sqrt{n+1} \right) \right] \tag{13.83}$$
$$\le \sum_{i=0}^{k} C_k^i \left(\log \varepsilon^{-2}\right)^i \int_0^\infty x^2 \left(\log^+ x^2\right)^{k-i} d\Phi(x)$$

for $0 < \varepsilon < 1$. This bound implies that

$$\lim_{\varepsilon \downarrow 0} \frac{\varepsilon^2}{\left(\log^+ \varepsilon^{-2}\right)^{d-1}} \sum_{n=1}^{\infty} n\,(\log n)^k\, \mathsf{P}\left(\varepsilon\sqrt{n} \le |Z| < \varepsilon\sqrt{n+1} \right) = 0$$

for $0 \le k < d-1$. The latter limit for $k = d-1$ can be written as a sum of d limits corresponding to the terms with $0 \le i \le d-1$ (see equality (13.83)). Moreover, only one of these limits (namely, the one corresponding to $i = d-1$) is non-zero:

$$\lim_{\varepsilon\downarrow 0} \frac{\varepsilon^2}{\left(\log \varepsilon^{-2}\right)^{d-1}} \sum_{n=1}^{\infty} \left(\log \varepsilon^{-2}\right)^{d-1} n\,(\log n)^{d-1}\,\mathsf{P}\left(\varepsilon\sqrt{n} \le |Z| < \varepsilon\sqrt{n+1}\right)$$
$$= \lim_{\varepsilon\downarrow 0} \varepsilon^2 \sum_{n=1}^{\infty} n\mathsf{P}\left(\varepsilon\sqrt{n} \le |Z| < \varepsilon\sqrt{n+1}\right) = 1$$

by Theorem 13.15. Now representation (13.82) implies equality (13.80) in view of equality (A.40).

Next we prove (13.81). We start by proving that

$$\lim_{\varepsilon\downarrow 0} \frac{\varepsilon^2}{\left(\log^+ \varepsilon^{-2}\right)^{d-1}} \sum_{n\le[\varepsilon^{-2}m]} \tau_n |\mathsf{P}(|S_n| \ge \varepsilon n) - 2\Phi(-\varepsilon\sqrt{n})| = 0 \tag{13.84}$$

for all $m \ge 1$. Put

$$\Delta_n = \sup_{x\in\mathbf{R}} \left|\mathsf{P}(|S_n| \ge x\sqrt{n}) - 2\Phi(-x)\right|.$$

To estimate Δ_n we apply the Petrov bound for the rate of convergence in the central limit theorem.

Theorem 13.17 (V.V. Petrov) *Let $X_1, \dots, X_n$ be independent random variables such that $\mathsf{E}\left[X_j\right] = 0$ and $\mathsf{E}\left[X_j^2 g(X_j)\right] < \infty$ for $j = 1, \dots, n$ and some function g satisfying the following conditions*:

(a) g is non-negative even and non-decreasing in the domain $x > 0$;
(b) $\frac{x}{g(x)}$ is a non-decreasing function in the domain $x > 0$.

Put $S_n = \sum_{j=1}^n X_j$ and

$$\sigma_j^2 = \mathsf{E}\left[X_j^2\right], \qquad B_n = \sum_{j=1}^n \sigma_j^2, \qquad F_n(x) = \mathsf{P}\left(S_n < xB_n^{1/2}\right).$$

Then

$$\sup_{x\in\mathbf{R}} |F_n(x) - \Phi(x)| \le \frac{A}{B_n g\left(B_n^{1/2}\right)} \sum_{j=1}^n \mathsf{E}\left[X_j^2 g(X_j)\right]. \tag{13.85}$$

Theorem 13.17 with $g(x) = (\log |x|)^{d-1}$ implies that

$$\Delta_n \le \frac{\text{const}}{(\log n)^{d-1}}.$$

The latter inequality together with Remark A.5 proves equality (13.84), since $d > 1$ and

$$\sum_{n=1}^{N} \frac{\tau_n}{\left(\log^+ n\right)^{d-1}} \asymp N.$$

Now we show that

$$\lim_{m\to\infty} \limsup_{\varepsilon\downarrow 0} \frac{\varepsilon^2}{\left(\log^+ \varepsilon^{-2}\right)^{d-1}} \sum_{n>[\varepsilon^{-2}m]} \tau_n |\mathsf{P}(|S_n| \ge \varepsilon n) - 2\Phi(-\varepsilon\sqrt{n})| = 0. \tag{13.86}$$

We derive relation (13.86) from the following two equalities

$$\lim_{m\to\infty} \limsup_{\varepsilon\downarrow 0} \frac{\varepsilon^2}{\left(\log^+ \varepsilon^{-2}\right)^{d-1}} \sum_{n>[\varepsilon^{-2}m]} \tau_n \mathsf{P}(|S_n| \ge \varepsilon n) = 0, \tag{13.87}$$

$$\lim_{m\to\infty} \limsup_{\varepsilon\downarrow 0} \frac{\varepsilon^2}{\left(\log^+ \varepsilon^{-2}\right)^{d-1}} \sum_{n>[\varepsilon^{-2}m]} \tau_n \Phi(-\varepsilon\sqrt{n}) = 0. \tag{13.88}$$

Equality (13.88) is in fact a particular case of (13.87), since

$$2\Phi(-\varepsilon\sqrt{n}) = \mathsf{P}(|Z_1 + \cdots + Z_n| \ge \varepsilon n),$$

where Z_j are independent standard Gaussian random variables for which condition (13.79) holds. It remains to prove (13.87).

We apply the Buldygin–Hoffman-Jørgensen inequality (13.26) with $j = 1$ and for random variables with finite second moments:

$$\begin{aligned} \mathsf{P}(|S_n| \ge \varepsilon n) &= O(1)\left[n\mathsf{P}(|X_1| \ge \varepsilon n/3) + (\mathsf{P}(|S_n| \ge \varepsilon n/3))^2\right] \\ &= O(1)\left[n\mathsf{P}(|X_1| \ge \varepsilon n/3) + \frac{81\sigma^4}{n^2\varepsilon^4}\right]. \end{aligned}$$

Thus (13.87) follows from

$$\lim_{m\to\infty} \limsup_{\varepsilon\downarrow 0} \frac{\varepsilon^{-2}}{\left(\log^+ \varepsilon^{-2}\right)^{d-1}} \sum_{n>[\varepsilon^{-2}m]} \frac{\tau_n}{n^2} = 0, \tag{13.89}$$

$$\lim_{m\to\infty} \limsup_{\varepsilon\downarrow 0} \frac{\varepsilon^2}{\left(\log^+ \varepsilon^{-2}\right)^{d-1}} \sum_{n>[\varepsilon^{-2}m]} n\tau_n \mathsf{P}(|X| \ge \varepsilon n) = 0. \tag{13.90}$$

Equality (13.89) follows from Remark A.4. Equality (13.90) is a consequence of Remark A.5 and the following estimates

$$\begin{aligned}
&\sum_{n>[\varepsilon^{-2}m]} n\tau_n \mathsf{P}(|X| \ge \varepsilon n) \\
&\qquad \le \sum_{n>[\varepsilon^{-2}m]} \mathsf{P}(\varepsilon n \le |X| < \varepsilon(n+1)) \sum_{k=1}^{n} k\tau_k \\
&\qquad \le \text{const} \sum_{n>[\varepsilon^{-2}m]} n^2 (\log n)^{d-1} \mathsf{P}(\varepsilon n \le |X| < \varepsilon(n+1)),
\end{aligned}$$

whence we conclude that

$$\begin{aligned}
&\limsup_{\varepsilon\downarrow 0} \frac{\varepsilon^2}{\left(\log^+ \varepsilon^{-2}\right)^{d-1}} \sum_{n>[\varepsilon^{-2}m]} n\tau_n \mathsf{P}(|X| \ge \varepsilon n) \\
&\quad \le O(1) \limsup_{\varepsilon\downarrow 0} \frac{1}{\left(\log^+ \varepsilon^{-2}\right)^{d-1}} \int_{|x|\ge\varepsilon[\varepsilon^{-2}m]} x^2 \left(\log^+ \frac{|x|}{\varepsilon}\right)^{d-1} dF(x) = 0.
\end{aligned}$$

To prove that the latter limit equals zero we use the binomial formula for

$$\left(\log^+ |x| + \log^+ \frac{1}{\varepsilon}\right)^{d-1}$$

and take into account that $\varepsilon\left[\varepsilon^{-2}m\right] \to \infty$ as $\varepsilon \downarrow 0$ for all $m \ge 1$. The proof of Theorem 13.16 is complete. □

The method used to prove Theorem 13.16 applies to a more precise result, namely to get a better asymptotics of the function $Q_F(\varepsilon)$. The asymptotic expansion of this function involve the coefficients of the Dirichlet polynomial that have a closed form only in the case of $d = 2$. This explains why we provide the corresponding result only for $d = 2$.

Theorem 13.18 *Let X, $\{X(m,n)\}$ be a field of independent identically distributed random variables such that*

$$\mathsf{E}[X] = 0, \qquad \mathsf{E}\left[X^2\right] = 1, \qquad \mathsf{E}\left[X^2 \log^+ |X|\right] < \infty.$$

Then

$$\lim_{\varepsilon\downarrow 0} \left(\varepsilon^2 Q(\varepsilon) - \log^+ \varepsilon^{-2}\right) = \gamma - 1 + \log 2,$$

where the function $Q(\varepsilon)$ is defined by equality (13.78) *with $\alpha = 2$ and where γ is the Euler–Mascheroni constant,*

$$\gamma = \lim_{m\to\infty} \left(\sum_{k=1}^{m} \frac{1}{k} - \log m\right).$$

A more careful analysis allows us to obtain the asymptotic behavior of the series related to the Marcinkiewicz–Zygmund law of large numbers.

Theorem 13.19 *Let* X, $\{X(m,n)\}$ *be a field of independent identically distributed random variables such that*

$$\mathsf{E}[X]=0, \qquad \mathsf{E}\left[X^2\right]=1.$$

Let $1 \le u, v < 2$. *Put*

$$D_v=\frac{2^{\frac{v}{2-v}}}{\sqrt{\pi}}\Gamma\left(\frac{2+v}{2(2-v)}\right), \quad D_v'=\frac{2^{\frac{v}{2-v}}}{\sqrt{\pi}}\Gamma'\left(\frac{2+v}{2(2-v)}\right),$$

where $\Gamma(\cdot)$ *is the* Γ*-function. Let*

$$Q(\varepsilon)=\sum_{m=1}^{\infty}\sum_{n=1}^{\infty}\mathsf{P}\left(|S(m,n)|\ge \varepsilon m^{1/u}n^{1/v}\right).$$

For the sake of definiteness assume that $u \le v$. *If*

$$\mathsf{E}\left[|X|^{2v}\right]<\infty \quad \text{for} \quad u<v$$

or

$$\mathsf{E}\left[|X|^{2v}\log^+|X|\right]<\infty \quad \text{for} \quad u=v,$$

then

$$\lim_{\varepsilon\downarrow 0}\varepsilon^{\frac{2v}{2-v}}Q(\varepsilon)=D_v\zeta\left(\frac{v(2-u)}{u(2-v)}\right)$$

in the case of $u<v$; *or*

$$\lim_{\varepsilon\downarrow 0}\left[\varepsilon^{\frac{2v}{2-v}}Q(\varepsilon)-D_v\log\varepsilon^{-\frac{2v}{2-v}}\right]$$
$$=D_v\left(2\gamma-1+\log 2^{\frac{2v}{2-v}}\right)+D_v'\frac{2v}{2-v}$$

in the case of $u=v$, *where* $\zeta(\cdot)$ *is the* ζ*-function,* $\zeta(z)=\sum_{n\ge 1}n^{-z}$.

The proof of Theorems 13.18 and 13.19 is similar to that of Theorem 13.16.

Note that the asymptotic behavior of the function $Q(\varepsilon)$ is different for the cases of $u=v$ and $u\ne v$: not only are the constants different on the right-hand sides, the rate of convergence as $\varepsilon\downarrow 0$ is also different.

13.6.2 Rate of Convergence in Heyde's Theorem

We are going to study the rate of convergence in Theorem 13.15, that is, the rate of convergence of

$$\varepsilon^2 \sum_{n=1}^{\infty} \mathsf{P}(|S_n| \geq \varepsilon n) \to \sigma^2, \qquad \varepsilon \downarrow 0.$$

First we solve the problem for Gaussian random variables. Put

$$\lambda(\varepsilon) = \sum_{n=1}^{\infty} \mathsf{P}(|S_n| \geq \varepsilon n).$$

Lemma 13.8 *Assume that* $\{X_n, n \geq 1\}$ *are independent Gaussian random variables for which condition* (13.76) *holds with* $\sigma^2 \neq 0$. *Then*

$$\lim_{\varepsilon \downarrow 0} \left(\frac{\sigma^2}{\varepsilon^2} - \lambda(\varepsilon) \right) = \frac{1}{2}.$$

We see that Theorem 13.15 can be improved for Gaussian random variables. If one additionally assumes that some higher moments exist, then it is possible to improve the result. Nevertheless, the same rate of convergence as in the Gaussian case is not available in general.

Theorem 13.20 *Assume that*

$$\mathsf{E}[X_1] = 0, \qquad \mathsf{E}\left[X_1^2\right] = \sigma^2 \neq 0, \qquad \mathsf{E}\left[|X_1|^3\right] < \infty. \tag{13.91}$$

Then

$$\lim_{\varepsilon \downarrow 0} \varepsilon^{3/2} \left(\lambda(\varepsilon) - \frac{\sigma^2}{\varepsilon^2} \right) = 0. \tag{13.92}$$

In contrast to the Gaussian case, an exact asymptotic behavior is not known in general. Nevertheless, equality (13.92) better describes the rate of convergence in Theorem 13.15.

The proof of Theorem 13.20 is based on Lemma 13.8 and on the following nonuniform bound for the rate of convergence in the central limit theorem.

Theorem 13.21 (S.V. Nagaev) *Let* $X_1, X_2, \ldots, X_n$ *be independent identically distributed random variables such that*

$$\mathsf{E}[X_1] = 0, \qquad \mathsf{E}\left[X_1^2\right] = \sigma^2 > 0, \qquad \mathsf{E}\left[|X_1|^3\right] < \infty.$$

Then, for all $x \in \mathbf{R}$,

$$F_n(x) - \Phi(x)| \leq \frac{A\rho}{\sqrt{n}(1 + |x|)^3}, \tag{13.93}$$

where $F_n(x) = \mathsf{P}(S_n < x\sigma\sqrt{n})$, $S_n = X_1 + \cdots + X_n$, *and where* A *is a universal constant,* $\rho = \mathsf{E}\left[|X_1|^3\right]/\sigma^3$. *One can put, for example,* $A = 0.4784$ *in inequality* (13.93).

If all assumptions of Theorem 13.20 hold, then by Nagaev's inequality

$$\left|\mathsf{P}(|S_n| \geq \varepsilon n) - 2\Phi\left(-\varepsilon\sqrt{n}\right)\right| \leq A\frac{\mathsf{E}\left[|X_1|^3\right]}{\sigma^3}\frac{1}{\sqrt{n}\left(1+\varepsilon\sqrt{n}\right)^3} \tag{13.94}$$

for all $\varepsilon > 0$ and $n \geq 1$.

Proof of Lemma 13.8 First we show that

$$\lim_{\delta\downarrow 0}\sqrt{\delta}\sum_{n=1}^{\infty}\frac{e^{-\delta n}}{\sqrt{n}} = \sqrt{\pi}. \tag{13.95}$$

Indeed, for all $n \geq 1$,

$$\sqrt{\frac{\delta}{n}}e^{-\delta n} = \frac{1}{\sqrt{\delta n}}e^{-\delta n}\int_{\delta(n-1)}^{\delta n} dx \leq \int_{\delta(n-1)}^{\delta n}\frac{e^{-x}}{\sqrt{x}}\,dx,$$

whence

$$\limsup_{\delta\downarrow 0}\sqrt{\delta}\sum_{n=1}^{\infty}\frac{e^{-\delta n}}{\sqrt{n}} \leq \int_0^{\infty}\frac{e^{-x}}{\sqrt{x}}\,dx.$$

Analogously, for all $n \geq 1$,

$$\sqrt{\frac{\delta}{n}}e^{-\delta n} \geq \int_{\delta n}^{\delta(n+1)}\frac{e^{-x}}{\sqrt{x}}\,dx$$

and thus

$$\liminf_{\delta\downarrow 0}\sqrt{\delta}\sum_{n=1}^{\infty}\frac{e^{-\delta n}}{\sqrt{n}} \geq \int_0^{\infty}\frac{e^{-x}}{\sqrt{x}}\,dx.$$

Equality (13.95) follows from the known identity

$$\int_0^{\infty}\frac{e^{-x}}{\sqrt{x}}\,dx = \Gamma\left(\frac{1}{2}\right) = \sqrt{\pi}$$

(see Corollary 1, Sect. 12.14, in [416]).

Now we turn to the proof of Lemma 13.8. Clearly, we can restrict attention to the case $\sigma^2 = 1$. In this case,

$$\lambda(\varepsilon) = \sqrt{\frac{2}{\pi}} \sum_{n=1}^{\infty} \int_{\varepsilon\sqrt{n}}^{\infty} e^{-t^2/2}\, dt = \sqrt{\frac{2}{\pi}} \sum_{n=1}^{\infty} n \int_{\varepsilon\sqrt{n}}^{\varepsilon\sqrt{n+1}} e^{-t^2/2}\, dt.$$

Since

$$1 = \sqrt{\frac{2}{\pi}} \sum_{n=0}^{\infty} \int_{\varepsilon\sqrt{n}}^{\varepsilon\sqrt{n+1}} t^2 e^{-t^2/2}\, dt,$$

we conclude that

$$1-\varepsilon^2\lambda(\varepsilon) = \sqrt{\frac{2}{\pi}} \sum_{n=1}^{\infty} \int_{\varepsilon\sqrt{n}}^{\varepsilon\sqrt{n+1}} \left(t^2 - \varepsilon^2 n\right) e^{-t^2/2}\, dt + \sqrt{\frac{2}{\pi}} \int_0^{\varepsilon} t^2 e^{-t^2/2}\, dt. \quad (13.96)$$

Put

$$I_n = \sqrt{\frac{2}{\pi}} \int_{\varepsilon\sqrt{n}}^{\varepsilon\sqrt{n+1}} \left(t^2 - \varepsilon^2 n\right) dt.$$

Evaluating the integral I_n, we get

$$I_n = \sqrt{\frac{2}{\pi}} \cdot \frac{\varepsilon^3}{3} \cdot \left(\sqrt{n+1} - 2n\left(\sqrt{n+1} - \sqrt{n}\right)\right).$$

Moreover,

$$e^{-\varepsilon^2(n+1)/2} I_n \le \int_{\varepsilon\sqrt{n}}^{\varepsilon\sqrt{n+1}} \left(t^2 - \varepsilon^2 n\right) e^{-t^2/2}\, dt \le e^{-\varepsilon^2 n/2} I_n.$$

The latter result together with (13.96) implies that

$$\lim_{\varepsilon\downarrow 0} \left(\frac{1}{\varepsilon^2} - \lambda(\varepsilon)\right) = \lim_{\varepsilon\downarrow 0} \frac{1}{\varepsilon^2} \sum_{n=1}^{\infty} e^{-\varepsilon^2 n/2} I_n.$$

Now we evaluate the limit on the right-hand side.

The Taylor formula with Peano remainder applied to the function $\sqrt{x+1}$ in a neighborhood of the point $x = 0$ implies that

$$\sqrt{x+1} = 1 + \frac{x}{2} - \frac{x^2}{8} + o\left(x^3\right), \qquad 0 \le x \le 1.$$

Thus

$$\sqrt{n+1}-\sqrt{n}=\sqrt{n}\left(\sqrt{1+(1/n)}-1\right)=\frac{1}{2n^{1/2}}-\frac{1}{8n^{3/2}}+o\left(\frac{1}{n^{5/2}}\right)$$

and

$$\sqrt{n+1}-2n\left(\sqrt{n+1}-\sqrt{n}\right)=\frac{3}{4\sqrt{n}}+O\left(\frac{1}{n^{3/2}}\right).$$

Therefore

$$I_n=\sqrt{\frac{2}{\pi}}\varepsilon^3\left[\frac{1}{4\sqrt{n}}+O\left(\frac{1}{n^{3/2}}\right)\right],$$

whence

$$\frac{1}{\varepsilon^2}\sum_{n=1}^{\infty}e^{-\varepsilon^2 n/2}I_n=\frac{\varepsilon}{4}\sqrt{\frac{2}{\pi}}\sum_{n=1}^{\infty}\frac{e^{-\varepsilon^2 n/2}}{\sqrt{n}}+O\left(\varepsilon\sum_{n=1}^{\infty}\frac{e^{-\varepsilon^2 n/2}}{n^{3/2}}\right). \tag{13.97}$$

The second term on the right-hand side of (13.97) is uniformly bounded with respect to ε, since

$$\sup_{\varepsilon\ge 0}\sum_{n=1}^{\infty}\frac{e^{-\varepsilon^2 n/2}}{n^{3/2}}\le\sum_{n=1}^{\infty}\frac{1}{n^{3/2}}<\infty.$$

Putting $\delta=\varepsilon^2/2$ in the first term on the right-hand side of (13.97) we get

$$\lim_{\varepsilon\downarrow 0}\frac{1}{\varepsilon^2}\sum_{n=1}^{\infty}e^{-\varepsilon^2 n/2}I_n=\frac{1}{2\sqrt{\pi}}\lim_{\delta\downarrow 0}\sqrt{\delta}\sum_{n=1}^{\infty}\frac{e^{-\delta n}}{\sqrt{n}}.$$

Now Lemma 13.8 follows from equality (13.95). □

Proof of Theorem 13.20 Recalling Lemma 13.8 we need to prove that

$$\lim_{\varepsilon\downarrow 0}\varepsilon^{3/2}\sum_{n=1}^{\infty}\left|\mathsf{P}(|S_n|\ge\varepsilon n)-2\Phi\left(-\varepsilon\sqrt{n}\right)\right|=0. \tag{13.98}$$

Let $0<\delta_1<\delta_2$ be fixed. Put $m_1=\left[\delta_1\varepsilon^{-3/2}\right]$ and $m_2=\left[\delta_2\varepsilon^{-3/2}\right]$. Then

$$\sum_{n=1}^{m_1}\left|\mathsf{P}(|S_n|\ge\varepsilon n)-2\Phi\left(-\varepsilon\sqrt{n}\right)\right|\le m_1. \tag{13.99}$$

Now inequality (13.94) implies that

$$\sum_{n=m_1+1}^{m_2} \left|\mathsf{P}(|S_n| \geq \varepsilon n) - 2\Phi\left(-\varepsilon\sqrt{n}\right)\right|$$
$$\leq 2A\frac{\mathsf{E}\left[|X_1|^3\right]}{\sigma^3} \sum_{n=m_1+1}^{m_2} \frac{1}{\sqrt{n}} \leq 2A\frac{\mathsf{E}\left[|X_1|^3\right]}{\sigma^3} \frac{m_2 - m_1}{\sqrt{m_1}}. \tag{13.100}$$

Finally, we apply inequality (13.94) once more:

$$\sum_{n=m_2+1}^{\infty} \left|\mathsf{P}(|S_n| \geq \varepsilon n) - 2\Phi\left(-\varepsilon\sqrt{n}\right)\right|$$
$$\leq \frac{2A\mathsf{E}\left[|X_1|^3\right]}{\varepsilon^3\sigma^3} \sum_{n=m_2+1}^{\infty} \frac{1}{n^2} \leq 2A\frac{\mathsf{E}\left[|X_1|^3\right]}{\sigma^3} \frac{1}{\varepsilon^3 m_2}. \tag{13.101}$$

Combining inequalities (13.99)–(13.101) we obtain

$$\limsup_{\varepsilon\downarrow 0} \varepsilon^{3/2} \sum_{n=1}^{\infty} \left|\mathsf{P}(|S_n| \geq \varepsilon n) - 2\Phi\left(-\varepsilon\sqrt{n}\right)\right| \leq \delta_1 + 2A\frac{\mathsf{E}\left[|X_1|^3\right]}{\sigma^3} \cdot \frac{1}{\delta_2}.$$

Passing to the limit in the latter relation as $\delta_1 \to 0$ and $\delta_2 \to \infty$ we prove equality (13.98). □

13.7 Comments

Complete convergence was introduced by Hsu and Robbins [174]. General results concerning complete convergence can be found in the book by Dugue [78], where the author calls it *almost sure convergence in a restricted sense.*

Hsu and Robbins [174] proved the sufficiency of condition (13.2) in Theorem 13.1. The necessity of this condition for (13.3) is proved by Erdös [84,85]. The case $\alpha = r = 1$ in Theorem 13.2 is considered by Spitzer [377]. Other values of α and r were considered by Katz [192] and later by Baum and Katz [22].

Section 13.1 An inequality similar to (13.10) is mentioned in the book [57], Sect. 10.2, Exercise 12. Theorem 13.3 is proved in [228]. Theorem 13.5 was found by Pruss [344] (the same result is obtained in [228] under a more restrictive assumption than (13.15)). Lemma 13.2 is proved by Ledoux and Talagrand [278]. Lemma 13.3 is, in fact, from [344].

Section 13.2 Corollaries 13.1 and 13.3 are proved by Heyde and Rohatgi [162] for the case of increasing continuous functions L. Other results of Sect. 13.2 are obtained in [228].

Section 13.3 Complete convergence for sums with random indices is studied by Szynal [391], Csörgő and Révész [65]. Theorem 13.8 is proved by Gut [134]. The other results of Sect. 13.3 are proved in [224,233]. Complete convergence for random fields with random multi-indices is considered by Lagodowski and Rychlik [275,270].

Section 13.4 The convergence of series (13.60) for subsequences is investigated by Asmussen and Kurtz [12]. Less restrictive conditions for the convergence of series (13.60) are discussed by Gut [135] and Kuczmaszewska and Szynal [268]. Theorem 13.13 is established in [233].

Section 13.5 Corollary 13.11 for $\alpha = 1$ is proved by Smythe [374]. Theorem 13.14 is given in [228]. Corollary 13.12 is obtained by Gut [129]. Corollaries 13.13 and 13.14 can be found in [221].

Section 13.6 Some analogs of relation (13.77) have been obtained for many other series of type (13.4). A rather complete bibliography and a survey of results concerning (13.77) is given by Sirazhdinov and Gafurov [370]. The proof of Theorem 13.17 can also be found in the monograph [336] (Chap. 5, Theorem 5).

Theorem 13.19 is proved in [223]. The asymptotic behavior of series (13.72) is investigated by Gut and Spataru [140]. Dil'man [74] considers the case $d > 1$.

Theorem 13.20 is proved in [229]. Some related results have recently been obtained by Gut and Steinebach [145, 146, 147].

S.V. Nagaev's Theorem 13.21 is proved in [325]. This result corresponds to a similar uniform bound due to Berry [23] and Esseen [86]. The upper bound of the constant $A \leq 0.4784$ mentioned in Theorem 13.21 is due to Korolev and Shevtsova [259]. The lower bound $A \geq 0.4097$ was found by Esseen [87].

Complete convergence for dependent random fields is studied by Cai [42].

Appendix A
Auxiliary Definitions and Results

This appendix contains the main results and definitions concerning sequences that depend on several indices used throughout the book. Such a sequence is called a *field*.

A.1 The Space of Multiple Indices

Let $\mathbf{N}$ be the set of positive integers, $\mathbf{N} = \{1, 2, \dots\}$. If $d \geq 2$, then $\mathbf{N}^d$ denotes the space of d-dimensional vectors $\mathbf{n} = (n_1, \dots, n_d)$, each of whose coordinates $n_1, \dots, n_d$ belongs to $\mathbf{N}$. Therefore

$$\mathbf{N}^d = \underbrace{\mathbf{N}\times \cdots \times\mathbf{N}}_{d \text{ times}}.$$

For the sake of consistency, we sometimes also write $\mathbf{N}^1$ instead of $\mathbf{N}$.

Adding to the space $\mathbf{N}^d$ the set of vectors $\mathbf{n} = (n_1, \dots, n_d)$ with non-negative coordinates such that $n_1 \dots n_d = 0$, we obtain the spaces

$$\mathbf{N}_0^d = \underbrace{\mathbf{N}_0\times \cdots \times\mathbf{N}_0}_{d \text{ times}},$$

where $\mathbf{N}_0 = \{0\} \cup \mathbf{N}$.

For every $\mathbf{n} \in \mathbf{N}^d$, we set $|\mathbf{n}| = n_1 \cdots n_d$. We also use the notation $\mathbf{0} = (0, \dots, 0)$, $\mathbf{1} = (1, \dots, 1)$, $\mathbf{2} = (2, \dots, 2)$, etc.

A.1.1 Coordinatewise Ordering

In the space $\mathbf{N}^d$, we introduce the coordinatewise ordering $\leq$, namely if $\mathbf{m}, \mathbf{n} \in \mathbf{N}^d$ and $\mathbf{m} = (m_1, \dots, m_d)$, $\mathbf{n} = (n_1, \dots, n_d)$, then we write $\mathbf{m} \leq \mathbf{n}$ if and only if

O. Klesov, *Limit Theorems for Multi-Indexed Sums of Random Variables*,
Probability Theory and Stochastic Modelling 71, DOI 10.1007/978-3-662-44388-0

$m_i \le n_i$ for all $1 \le i \le d$, that is

$$\mathbf{m} \le \mathbf{n} \Longleftrightarrow m_i \le n_i \quad \text{for all } 1 \le i \le d. \tag{A.1}$$

We write $\mathbf{m} \ne \mathbf{n}$ if there exists at least one index i, $1 \le i \le d$, such that $m_i \ne n_i$, that is

$$\mathbf{m} \ne \mathbf{n} \Longleftrightarrow m_i \ne n_i \quad \text{for some } 1 \le i \le d.$$

Further, we write $\mathbf{m} < \mathbf{n}$ if $\mathbf{m} \le \mathbf{n}$ and $\mathbf{m} \ne \mathbf{n}$. Similarly, $\mathbf{m} \not\le \mathbf{n}$, if there exists an i, $1 \le i \le d$, such that $m_i > n_i$, that is

$$\mathbf{m} \not\le \mathbf{n} \Longleftrightarrow m_i > n_i \quad \text{for some } 1 \le i \le d.$$

Now it is clear how to define the notation where the sign "$>$" is used in place of "$<$", namely

$$\begin{aligned} \mathbf{m} \ge \mathbf{n} &\Longleftrightarrow \mathbf{n} \le \mathbf{m}, \\ \mathbf{m} > \mathbf{n} &\Longleftrightarrow \mathbf{n} < \mathbf{m}, \\ \mathbf{m} \not\ge \mathbf{n} &\Longleftrightarrow \mathbf{n} \not\le \mathbf{m}. \end{aligned}$$

Note that the relation $\mathbf{m} \not\le \mathbf{n}$ does not necessarily imply that $\mathbf{m} > \mathbf{n}$ if $d > 1$, that is, the space $\mathbf{N}^d$, $d > 1$, is not completely ordered with respect to the ordering $\le$. This is the reason why there are so many essential changes in the theory of summation of random variables with multiple indices as compared to the classical case of $d = 1$.

A.1.2 Linear Ordering

Along with the coordinatewise ordering in the space $\mathbf{N}^d$, one can introduce a complete ordering $\preccurlyeq$ of its elements. This can be done in many ways. One such ordering is called the lexicographical order:

$$\begin{aligned} \mathbf{m} \preccurlyeq \mathbf{n} \Longleftrightarrow\ & |\mathbf{m}| < |\mathbf{n}|, \\ & \text{or } |\mathbf{m}| = |\mathbf{n}|, \ \text{but } m_1 < n_1, \\ & \text{or } |\mathbf{m}| = |\mathbf{n}|, \ m_1 = n_1, \ \text{but } m_2 < n_2, \\ & \dots\dots\dots\dots\dots\dots\dots\dots\dots\dots \\ & \text{or } |\mathbf{m}| = |\mathbf{n}|, \ m_1 = n_1, \dots, m_{d-2} = n_{d-2}, \ \text{but } m_{d-1} < n_{d-1}, \\ & \text{or } \mathbf{m} = \mathbf{n}. \end{aligned}$$

For example, the ordering $\preccurlyeq$ for $d = 2$ orders the elements $\mathbf{N}^2$ as follows

$$(1,1) \preccurlyeq (1,2) \preccurlyeq (2,1) \preccurlyeq (1,3) \preccurlyeq (3,1) \preccurlyeq (1,4) \preccurlyeq (2,2) \preccurlyeq (4,1) \preccurlyeq \dots.$$

Throughout the book, the phrase "*consider the complete ordering in the space* $\mathbf{N}^d$", means that we are going to consider the linear ordering $\preccurlyeq$ defined above.

A.2 Fields. Convergence of Fields

For every element $\mathbf{n} \in \mathbf{N}^d$, let a real number $x(\mathbf{n})$ be defined. Each set of real numbers $\{x(\mathbf{n}), \mathbf{n} \in \mathbf{N}^d\}$ is called a *field*. It is convenient to extend the domain of a field from $\mathbf{N}^d$ to $\mathbf{N}_0^d$ by letting $x(\mathbf{n}) = 0$ if at least one of the coordinates of the vector $\mathbf{n}$ equals zero, that is, $x(\mathbf{n}) = 0$ for all elements $\mathbf{n} \in \mathbf{N}_0^d \setminus \mathbf{N}^d$.

The concept of a limit is the main concept of classical calculus. Limits in the case of fields can be defined in various ways. We usually use the following two definitions.

A.2.1 Min-convergence

Definition A.1 We say that a field $\{x(\mathbf{n}), \mathbf{n} \in \mathbf{N}^d\}$ lim(min)-converges to a number x_0 if, for all positive ε, there exists a positive integer n_0 such that $|x(\mathbf{n}) - x_0| < \varepsilon$ for all elements $\mathbf{n} = (n_1, \dots, n_d) \in \mathbf{N}^d$ whose coordinates are such that

$$n_1 \geq n_0, \dots, n_d \geq n_0,$$

or, in other words, if

$$\min\{n_1, \dots, n_d\} \geq n_0.$$

Similarly, we say that a field $\{x(\mathbf{n}), \mathbf{n} \in \mathbf{N}^d\}$ lim(min)-converges to $+\infty$ (or, to $-\infty$) if, given an arbitrary $E > 0$, there exists a positive integer n_0 such that

$$x(\mathbf{n}) > E \quad \big(\text{or, } x(\mathbf{n}) < -E\big)$$
$$\text{for all } \mathbf{n} = (n_1, \dots, n_d) \in \mathbf{N}^d \text{ with } n_1 \geq n_0, \dots, n_d \geq n_0.$$

We say that the limit $\lim(\min) x(\mathbf{n})$ does not exists if the property indicated in Definition A.1 does not hold for any number x_0.

If a field $\{x(\mathbf{n}), \mathbf{n} \in \mathbf{N}^d\}$ converges to x_0 in the sense of Definition A.1, then we write $\lim(\min) x(\mathbf{n}) = x_0$. There is another way to indicate lim(min)-convergence which is even more widely used, namely

$$\lim_{\substack{n_1 \to \infty \\ \dots \\ n_d \to \infty}} x(\mathbf{n}) = x_0.$$

On the other hand, the notation $\lim(\min) x(\mathbf{n}) = x_0$ is shorter. Moreover, this notation is compatible with the one we use to indicate another type of convergence which we now introduce.

A.2.2 Max-convergence

Definition A.2 We say that a field $\{x(\mathbf{n}), \mathbf{n} \in \mathbf{N}^d\}$ lim(max)-converges to a number x_0 if, given an arbitrary positive ε, there exists a positive integer n_0 such that $|x(\mathbf{n}) - x_0| < \varepsilon$ for all elements $\mathbf{n} = (n_1, \dots, n_d) \in \mathbf{N}^d$ whose coordinates satisfy the following condition

$$\text{there exists } i,\ 1 \le i \le d, \text{ such that } n_i \ge n_0,$$

that is

$$\max\{n_1, \dots, n_d\} \ge n_0.$$

Similarly, we say that a numerical field $\{x(\mathbf{n}), \mathbf{n} \in \mathbf{N}^d\}$ lim(max)-converges to $+\infty\big(\text{or, to } -\infty\big)$ if, given an arbitrary $E > 0$, there exists a positive integer n_0 such that

$$x(\mathbf{n}) > E \qquad \big(\text{or, } x(\mathbf{n}) < -E\big)$$

$$\text{for all } \mathbf{n} = (n_1, \dots, n_d) \in \mathbf{N}^d \text{ such that } \max\{n_1, \dots, n_d\} \ge n_0.$$

We say that the limit lim(max)$x(\mathbf{n})$ does not exists if the property indicated in Definition A.2 does not hold for any x_0.

If a field $\{x(\mathbf{n}), \mathbf{n} \in \mathbf{N}^d\}$ converges to x_0 in the sense of Definition A.2, then we write lim(max)$x(\mathbf{n}) = x_0$ and say that the field $\{x(\mathbf{n}), \mathbf{n} \in \mathbf{N}^d\}$ is *totally* convergent to x_0. The notation lim(max)$x(\mathbf{n}) = x_0$ is equivalent to

$$\lim_{\max\{n_1,\dots,n_d\}\to\infty} x(\mathbf{n}) = x_0.$$

Both Definitions A.1 and A.2 rely on the same inequality $|x(\mathbf{n}) - x_0| < \varepsilon$; but the sets of elements $\mathbf{n}$ for which it holds are different for these two modes of convergence. The difference between Definitions A.1 and A.2 is exhibited in Fig. A.1 for the case of $\mathbf{N}^2$. Figure A.1 exhibits the exceptional sets for both modes of convergence where the defining inequality $|x(\mathbf{n}) - x_0| < \varepsilon$ does not hold.

Denote the sets of elements of $\mathbf{N}^d$ where the defining inequality holds by $\mathfrak{N}_{\min}$ and $\mathfrak{N}_{\max}$ for convergences lim(min) and lim(max), respectively. It is clear that

$$\mathfrak{N}_{\min} = \{\mathbf{n}\colon\ \min(n_1, \dots, n_d) \ge n_0\},$$
$$\mathfrak{N}_{\max} = \{\mathbf{n}\colon\ \max(n_1, \dots, n_d) \ge n_0\}.$$

This notation makes clear the first part of the following assertion.

Proposition A.1 *If* lim(max)$x(\mathbf{n}) = x_0$, *then* lim(min)$x(\mathbf{n}) = x_0$. *The converse statement is not true.*

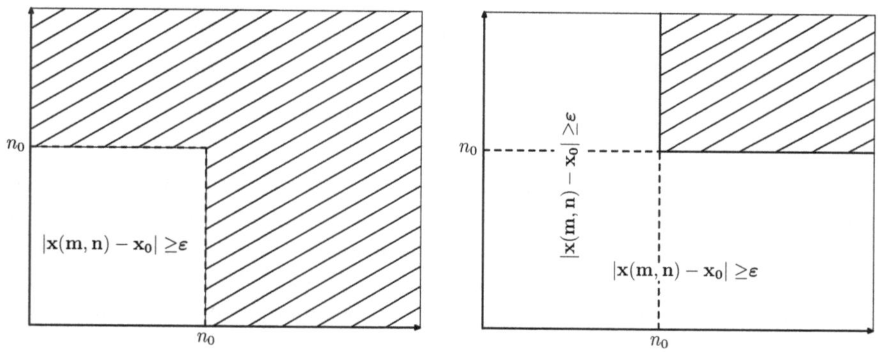

Fig. A.1 Exceptional sets for lim(max)-convergence (*left*) and lim(min)-convergence (*right*)

Proof of Proposition A.1. As mentioned above, the first part of Proposition A.1 is obvious, since $\mathfrak{N}_{\min} \subseteq \mathfrak{N}_{\max}$.

The second part of Proposition A.1 is proved by the following example of a field

$$x(\mathbf{n}) = \frac{1}{n_1} + \cdots + \frac{1}{n_d},$$

for which $\lim(\min) x(\mathbf{n}) = 0$, but $\lim(\max) x(\mathbf{n})$does not exist. □

The following result allows us to prove the equivalence of lim(min)- and lim(max)-convergences for some cases. Let $1 \leq c < d$ and let $\mathfrak{S} = \{i_1, \ldots, i_c\}$ be a subset of the first d positive integers $\{1, \ldots, d\}$. Also let $\mathfrak{N} = \{n_i^*, i \notin \mathfrak{S}\}$ be a family of fixed positive integers. We write

$$(\mathfrak{S}, \mathfrak{N})\text{-}\lim(\min) x(\mathbf{n}) = 0, \tag{A.2}$$

if, given an arbitrary $\varepsilon > 0$, there exists an $n_0 \geq 1$ such that $|x(\mathbf{n})| < \varepsilon$ for all multi-indices $\mathbf{n} \in \mathbf{N}^d$ with $\mathbf{n} = (n_1, \ldots, n_d)$ and

$$\begin{aligned} n_i &= n_i^*, \quad i \notin \mathfrak{S}, \\ n_i &\geq n_0, \quad i \in \mathfrak{S}. \end{aligned}$$

Relation (A.2) means that the field $\{x(\mathbf{n})\}$ lim(min)-converges in the space $\mathbf{N}^{\operatorname{card}(\mathfrak{S})}$ if the coordinates n_i, $i \notin \mathfrak{S}$, of multi-indices $\mathbf{n}$ are fixed.

Proposition A.2 *A field* $\{x(\mathbf{n})\}$ *is such that* $\lim(\max) x(\mathbf{n}) = 0$ *if and only if* $\lim(\min) x(\mathbf{n}) = 0$ *and equality* (A.2) *holds for an arbitrary subset* $\mathfrak{S} \subset \{1, \ldots, d\}$, $1 \leq \operatorname{card}(\mathfrak{S}) < d$, *and for an arbitrary set* $\mathfrak{N}$ *of positive integers* n_i^*, $i \notin \mathfrak{S}$.

For example, Proposition A.2 in the case $d = 2$ means that

$$\lim(\max)x(m,n) = x_0 \iff \begin{cases} \lim(\min)x(m,n) = x_0, \\ \lim\limits_{m\to\infty} x(m,n) = x_0 \ \forall n \ge 1, \\ \lim\limits_{n\to\infty} x(m,n) = x_0 \ \forall m \ge 1. \end{cases}$$

Proof of Proposition A.2. Let $\lim(\min)x(\mathbf{n}) = 0$ and condition (A.2) hold for an arbitrary subset $\mathfrak{S} \subset \{1,\ldots,d\}$, $\mathfrak{S} \ne \varnothing$. If the lim(max)-limit does not exist, then there are an infinite set of increasing element $\mathbf{n}_k$, $k \ge 1$, and a positive number δ for which $|x(\mathbf{n}_k)| \ge \delta$, $k \ge 1$. Let $\mathbf{n}_k = (n_{1k},\ldots,n_{dk})$. Then there exists an i, $1 \le i \le d$, for which the sequence $\{n_{ik}, k \ge 1\}$ is bounded (otherwise the lim(min)-convergence does not hold).

Now denote by $\mathfrak{S}$ the subset of $\{1,\ldots,d\}$ such that $n_{ik} \not\to \infty$, $i \in \mathfrak{S}$. As mentioned above, the set $\mathfrak{S}$ is non-empty. Moreover, starting from some $k \ge 1$, the numbers n_{ik}, $i \in \mathfrak{S}$, do not change. Note that $\mathfrak{S} \ne \{1,\ldots,d\}$ and $\mathfrak{S} \ne \varnothing$. Considering (A.2) one can assert that the inequality $|x(\mathbf{n}_k)| \ge \delta$ does not hold for an infinite number of elements $\mathbf{n}_k$. This completes the proof of Proposition A.2. □

Remark A.1 The lim(max)-convergence is not popular in the theory of functions. This can partially be explained by the fact that, in many situations, the convergence holds only for trivial cases. Nevertheless, it is the lim(max)-convergence that is a natural type of convergence for limit theorems of multi-indexed sums of random variables.

In order to understand this phenomenon in full detail, consider some random events $\mathfrak{E}_\alpha$ that depend on a parameter α of a general nature. As is well known, various questions of probability theory lead to consideration of upper limits of a family of random events that are defined as follows

$$\limsup\,[\mathfrak{E}_\alpha] = \{\omega \in \Omega\colon \omega \text{ belongs to an infinite number of events } \mathfrak{E}_\alpha\}.$$

The probability $\mathsf{P}(\limsup\,[\mathfrak{E}_\alpha])$ plays an important role in various problems of probability theory. Recall that it can only equal either 1 or 0 if the random events E_α are independent.

Let, for example, indices α be elements of the space $\mathbf{N}^d$ and $\mathfrak{E}_\alpha = \{\omega\colon |X_\alpha| \ge \varepsilon\}$, where X_α are random variables and where ε is a positive number. Assume that $\mathsf{P}(\limsup\,[\mathfrak{E}_\alpha]) = 0$ for all $\varepsilon > 0$, then $X_\alpha \to 0$ almost surely in the sense of lim(max)-convergence.

It is worth mentioning that the Borel–Cantelli lemma also leads to lim(max)-convergence. Indeed, the Borel–Cantelli lemma involves the series $\sum \mathsf{P}(\mathfrak{E}_\alpha)$. The convergence of this series indicates that $\mathsf{P}(\limsup\,[\mathfrak{E}_\alpha]) = 0$ which also is related to lim(max)-convergence.

A.2.3 *f*-Convergence

Both modes of convergence discussed above are particular cases of the following general concept.

Definition A.3 Let a function f of d arguments be defined for all $\mathbf{n} \in \mathbf{N}^d$. We say that a field $\{x(\mathbf{n}), \mathbf{n} \in \mathbf{N}^d\}$ is f-convergent to x_0 if, given an arbitrary positive number ε, there exists a positive integer n_0 such that

$$|x(\mathbf{n}) - x_0| < \varepsilon \text{ for all } \mathbf{n} = (n_1, \dots, n_d) \in \mathbf{N}^d \text{ for which } f(\mathbf{n}) \geq n_0.$$

The definition of infinite limits is similar.

If a field $\{x(\mathbf{n}), \mathbf{n} \in \mathbf{N}^d\}$ converges to x_0 in the sense of Definition A.3, then we write $\lim(f)\ x(\mathbf{n}) = x_0$. This notation clarifies our preceding notation $\lim(\max)$ and $\lim(\min)$.

We do not discuss the question of identifying the class of functions f for which Definition A.3 is correct. We only note that if f is non-decreasing with respect to each of its arguments, then the above definition is meaningful.

Note also that, in the class of monotone functions, the $\lim(\min)$-convergence is the weakest, while the $\lim(\max)$-convergence is the strongest.

Remark A.2 Not only the functions max or min are of interest for f-convergence. For example, when studying the functions of several arguments f-convergence is often considered for $f(x_1, \dots, x_d) = \sqrt{x_1^2 + \dots + x_d^2}$. For this particular function, we use a special notation for the f-convergence, namely $\|\cdot\|$-convergence. In what follows, we also deal with vol-convergence, which corresponds to the function $f(x_1, \dots, x_d) = x_1 \dots x_d$. Note that the three types of convergence, namely max-, $\|\cdot\|$-, and vol-convergence are equivalent in the space $\mathbf{N}^d$, since for $\mathbf{n} \in \mathbf{N}^d$

$$\max\{n_1, \dots, n_d\} \leq \sqrt{n_1^2 + \dots + n_d^2} \leq \sqrt{d} \max\{n_1, \dots, n_d\},$$
$$\max\{n_1, \dots, n_d\} \leq n_1 \dots n_d \leq (\max\{n_1, \dots, n_d\})^d.$$

Another approach is used when studying the convergence of subsequences in $\mathbf{N}^d$.

Definition A.4 Let A be a certain subset of $\mathbf{N}^d$, $A \subseteq \mathbf{N}^d$. Let a function f depend on d arguments and be defined for all $\mathbf{n} \in \mathbf{N}^d$. We say that a field $\{x(\mathbf{n}), \mathbf{n} \in \mathbf{N}^d\}$ f-converges to x_0 in the set A if, given an arbitrary positive ε, there exists a positive integer n_0 such that

$$|x(\mathbf{n}) - x_0| < \varepsilon \text{ for all } \mathbf{n} = (n_1, \dots, n_d) \in A \text{ with } f(\mathbf{n}) \geq n_0.$$

The definition of infinite limits is similar.

If a field $\{x(\mathbf{n}), \mathbf{n} \in \mathbf{N}^d\}$ converges to x_0 in the sense of Definition A.4, then we write $\lim(f, A)\ x(\mathbf{n}) = x_0$. Using this notation, it is clear that $\lim(f) = \lim(f, \mathbf{N}^d)$.

The superior limit and inferior limit for f-convergence is defined in the same way as in the case $d = 1$. We say that a number x_0 is a limit point of a field $\{x(\mathbf{n})\}$ for the f-convergence if there exists an infinite set A such that $\lim(f, A) = x_0$.

Let $L_f(\{x(\mathbf{n})\})$ denote the set of limit points of a field $\{x(\mathbf{n}), \mathbf{n} \in \mathbf{N}^d\}$ for the f-convergence.

Definition A.5 If a function f that depends on d arguments is defined for every $\mathbf{n} \in \mathbf{N}^d$ and if $\{x(\mathbf{n}), \mathbf{n} \in \mathbf{N}^d\}$ is a field, then the number $\sup L_f(\{x(\mathbf{n})\})$ is called the superior limit of the field $\{x(\mathbf{n}), \mathbf{n} \in \mathbf{N}^d\}$ for the f-convergence. The superior limit is denoted by $\limsup(f)\ x(\mathbf{n})$.

Definition A.6 If a function f that depends on d arguments is defined for every $\mathbf{n} \in \mathbf{N}^d$ and if $\{x(\mathbf{n}), \mathbf{n} \in \mathbf{N}^d\}$ is a field, then the number $\inf L_f(\{x(\mathbf{n})\})$ is called the inferior limit for the field $\{x(\mathbf{n}), \mathbf{n} \in \mathbf{N}^d\}$ for the f-convergence. The inferior limit is denoted by $\liminf(f)\ x(\mathbf{n})$.

Turning to the field $\{x(\mathbf{n}), \mathbf{n} \in \mathbf{N}^d\}$ we see that $L_{\min}(\{x(\mathbf{n})\}) = \{0\}$ for the field defined in the proof of Proposition A.1, that is

$$\limsup(\min)x(\mathbf{n}) = \liminf(\min)x(\mathbf{n}) = 0,$$

which means the existence of the $\lim(\min)$-limit. On the other hand, this field is such that $\liminf(\max)x(\mathbf{n}) = 0$ and $\limsup(\max)x(\mathbf{n}) = d - 1$, which means that the $\lim(\max)$-limit exists for this field only in the case of $d = 1$.

A.3 Cauchy Convergence Criteria

The following assertion is useful when studying the convergence of fields (it is called the *Cauchy criterion* if $d = 1$). The Cauchy convergence criterion allows us to reduce the consideration of the convergence of a field to the problem of the convergence of its subfields (subsequences if $d = 1$) for $d > 1$.

Proposition A.3 *Let a function f that depends on d arguments be defined for all $\mathbf{n} \in \mathbf{N}^d$. The following two conditions are equivalent for all fields* $\{x(\mathbf{n}), \mathbf{n} \in \mathbf{N}^d\}$:

(a) $\lim(f)\ x(\mathbf{n}) = x_0$;
(b) $\lim_{k\to\infty} x(\mathbf{n}_k) = x_0$ *for every sequence* $\{\mathbf{n}_k, k \geq 1\}$ *such that* $\mathbf{n}_k \leq \mathbf{n}_{k+1}$, $k \geq 1$, *and*

$$\lim_{k\to\infty} f(\mathbf{n}_k) = \infty.$$

Proof of Proposition A.3. Let condition (a) hold, that is, $\lim(f)\ x(\mathbf{n}) = x_0$. Consider an arbitrary monotone and unbounded sequence $\{\mathbf{n}_k, k \geq 1\}$, that is, $\mathbf{n}_k \leq \mathbf{n}_{k+1}$, $k \geq 1$, and $\lim_{k\to\infty} f(\mathbf{n}_k) = +\infty$. Then condition (a) implies that, for every $\varepsilon > 0$, there exists a positive integer n_0 such that $|x(\mathbf{n}) - x_0| < \varepsilon$ for all elements $\mathbf{n} \in \mathbf{N}^d$ such that $f(\mathbf{n}) \geq n_0$. It is clear that there exists a k_0 for which $f(\mathbf{n}_k) \geq n_0$ if $k \geq k_0$, whence $|x(\mathbf{n}_k) - x_0| < \varepsilon$ for such numbers k and this is equivalent to condition (b).

Now let condition (b) hold but assume that condition (a) does not hold. Then there is a number $\varepsilon > 0$ and some infinite set of elements $\{\mathbf{m}_k, k \geq 1\}$ with $f(\mathbf{m}_k) \to \infty$

such that $|x(\mathbf{m}_k) - x_0| \geq \varepsilon$. In the set $\{\mathbf{m}_k\}$, we choose a monotone sequence $\{\mathbf{m}'_k, k \geq 1\}$. The existence of such a sequence for $d > 1$ is not as obvious as in the case $d = 1$. To explain why such a sequence exists for $d > 1$ we first choose a subsequence $\{\mathbf{m}_k^{(1)}\}$, increasing with respect to the first coordinate. Then we choose a further subsequence $\{\mathbf{m}_k^{(2)}\}$ of $\{\mathbf{m}_k^{(1)}\}$, increasing with respect to the second coordinate. Repeating this procedure d times we finally obtain a subsequence $\{\mathbf{m}_k^{(d)}\}$ which is increasing with respect to the last coordinate. Now it is obvious that this sequence is monotone with respect to each coordinate. For short, we denote the latter sequence by $\{\mathbf{m}'_k, k \geq 1\}$.

It is clear that $\lim_{k\to\infty} f(\mathbf{m}'_k) = +\infty$, $\mathbf{m}'_k \leq \mathbf{m}'_{k+1}$, and that $|x(\mathbf{m}'_k) - x_0| \geq \varepsilon$. This means that condition (b) does not hold. This contradiction proves that condition (a) is valid and that Proposition A.3 is proved. □

As explained above, the functions $f = \max$ and $f = \min$ are the most useful in the study of convergence of fields.

Corollary A.1 *The following two conditions are equivalent for every field* $\{x(\mathbf{n}), \mathbf{n} \in \mathbf{N}^d\}$ *of real numbers*:

(a) $\lim(\max)x(\mathbf{n}) = x_0$;
(b) $\lim_{k\to\infty} x(\mathbf{n}_k) = x_0$ *for all sequences* $\{\mathbf{n}_k, k \geq 1\}$ *such that* $\mathbf{n}_k \leq \mathbf{n}_{k+1}$, $k \geq 1$, *and*

$$\lim_{k\to\infty} \max(n_{1k}, \dots, n_{dk}) = +\infty.$$

Corollary A.2 *The following two conditions are equivalent for every field* $\{x(\mathbf{n}), \mathbf{n} \in \mathbf{N}^d\}$ *of real numbers*:

(a) $\lim(\min)x(\mathbf{n}) = x_0$;
(b) $\lim_{k\to\infty} x(\mathbf{n}_k) = x_0$ *for all sequences* $\{\mathbf{n}_k, k \geq 1\}$ *such that* $\mathbf{n}_k \leq \mathbf{n}_{k+1}$, $k \geq 1$, *and*

$$\lim_{k\to\infty} \min(n_{1k}, \dots, n_{dk}) = +\infty.$$

A.4 Multi-indexed Sums and Series

Let $\{x(\mathbf{n}), \mathbf{n} \in \mathbf{N}^d\}$ be a certain field. The numbers

$$s(\mathbf{n}) = \sum_{\mathbf{k}\leq\mathbf{n}} x(\mathbf{k}), \quad \mathbf{n} \in \mathbf{N}^d, \tag{A.3}$$

are called *multi-indexed sums* or *rectangular sums* constructed from the field $\{x(\mathbf{n}), \mathbf{n} \in \mathbf{N}^d\}$. The members of the field $\{s(\mathbf{n}), \mathbf{n} \in \mathbf{N}^d\}$ are also called *partial sums of the field* $\{x(\mathbf{n}), \mathbf{n} \in \mathbf{N}^d\}$.

Definition A.7 The limit of the field $\{s(\mathbf{n}), \mathbf{n} \in \mathbf{N}^d\}$ in the sense of lim(min)-convergence (if it exists) is called the *sum of the multi-indexed series* constructed from the field $\{x(\mathbf{n}), \mathbf{n} \in \mathbf{N}^d\}$. The sum of a multi-indexed series is denoted by

$$\sum_{\mathbf{n}\in\mathbf{N}^d} x(\mathbf{n}) \stackrel{\text{def}}{=} \lim(\min) s(\mathbf{n}), \tag{A.4}$$

where $s(\mathbf{n})$ is defined by equality (A.3). If the limit in (A.4) exists, then we say that the multi-indexed series

$$\sum_{\mathbf{n}\in\mathbf{N}^d} x(\mathbf{n}) = \sum x(\mathbf{n})$$

converges. If the lim(min)-limit for $\{s(\mathbf{n}), \mathbf{n} \in \mathbf{N}^d\}$ does not exist, then we say that the multi-indexed series diverges.

A special name is used for $S(\mathbf{n})$ in the case when $d = 2$: we refer to "*double*" (sums and series) in this case instead of "multiple" or "multi-indexed" (sums and series).

Many properties inherent in the case of converging series for $d = 1$ disappear for $d > 1$. For example, the well-known necessary condition of a convergent series no longer holds if $d > 1$. Even a weaker form of the necessary condition does not hold, namely the terms $x(\mathbf{k})$ of a converging series are not necessarily bounded. Moreover, the partial sums in (A.3) of a convergent series are also not necessarily bounded. We exhibit these two "anomalies" in the following example for $d = 2$. The construction we show in this example is often used (with corresponding changes) in the other chapters of this book.

Example A.1 Let

$$x(k,l) = \begin{cases} (-1)^{l-1}, & k \ge 1,\ l \le 2, \\ 0, & k \ge 1,\ l > 2; \end{cases} \quad \text{that is } s(m,n) = \begin{cases} m, & n = 1, \\ 0, & n > 1. \end{cases}$$

Thus the double series $\sum_{k,l} x(k,l)$ converges but

$$\sup_{m,n\ge 1} |s(m,n)| = \infty.$$

A minor change of terms yields unbounded terms:

$$x(k,l) = \begin{cases} (-1)^{l-1}k, & k \ge 1,\ l \le 2, \\ 0, & k \ge 1,\ l > 2; \end{cases} \quad \text{that is } s(m,n) = \begin{cases} \frac{m(m+1)}{2}, & n = 1, \\ 0, & n > 1. \end{cases}$$

Again, the double series $\sum_{k,l} x(k,l)$ converges but

$$\sup_{m,n\ge 1} |x(m,n)| = \infty.$$

Despite the "anomalies" mentioned above, some properties of convergent series are valid for $d > 1$, too. We show this in the following assertion.

Proposition A.4 *If multi-indexed series* (A.4) *converges, then*

$$\lim(\min)x(\mathbf{n}) = 0.$$

We prove Proposition A.4 in Sect. A.5 below after we introduce the definition of increments of a multi-indexed series.

Remark A.3 It seems reasonable to consider the convergence of multi-indexed series not only for $f = \min$ (as in Definition A.3) but also for other functions f.

It turns out, surprisingly, that other f-convergences, if they are not equivalent to min-convergence, may essentially restrict the area of converging series. For example, choose the function $f = \max$ and define the sum of a multi-indexed series as the limit $\lim(\max)s(\mathbf{n})$. Then if a series converges and the terms $x(\mathbf{n})$ are non-negative, then $x(\mathbf{n}) = 0$ for all $\mathbf{n} \neq \mathbf{1} = (1, \dots, 1)$. This explains why we use lim(min)-convergence when studying the convergence of multi-indexed series.

A.5 Increments of a Field

Definition A.8 The collection $\{\Delta[x(\mathbf{n})], \mathbf{n} \in \mathbf{N}^d\}$ of real numbers is called the field of *increments* constructed from a field $\{x(\mathbf{n}), \mathbf{n} \in \mathbf{N}^d\}$ if

$$x(\mathbf{n}) = \sum_{\mathbf{k} \le \mathbf{n}} \Delta[x(\mathbf{k})] \quad \text{for all } \mathbf{n} \in \mathbf{N}^d. \tag{A.5}$$

We also say that $\Delta[x(\mathbf{n})]$ is the increment of the field $\{x(\mathbf{n}), \mathbf{n} \in \mathbf{N}^d\}$ at element $\mathbf{n} \in \mathbf{N}^d$ (for short, $\Delta[x(\mathbf{n})]$ is an increment of $\{x(\mathbf{n})\}$).

The field of increments exists and is unique for every field $\{x(\mathbf{n}), \mathbf{n} \in \mathbf{N}^d\}$. Recall that we agreed above that $x(\mathbf{n}) = 0$ for any field $\{x(\mathbf{n}), \mathbf{n} \in \mathbf{N}^d\}$ if at least one of the coordinates of $\mathbf{n}$ equals 0.

Proposition A.5 *The field of increments* $\{\Delta[x(\mathbf{n})], \mathbf{n} \in \mathbf{N}^d\}$ *for an arbitrary field* $\{x(\mathbf{n}), \mathbf{n} \in \mathbf{N}^d\}$ *is uniquely defined as follows*

$$\Delta[x(\mathbf{n})] = \sum_{(\delta_1, \dots, \delta_d) \in \mathcal{E}_d} (-1)^{\delta_1 + \dots + \delta_d} x(n_1 - \delta_1, \dots, n_d - \delta_d), \tag{A.6}$$

where $\mathcal{E}_d$ *denotes the set of all vectors of dimension d, each of whose coordinates equals either 0 or 1. The sum on the right-hand side of* (A.6) *is considered with respect to all* 2^d *possible combinations of numbers* $\delta_1, \dots, \delta_d \in \{0, 1\}$.

One can view Δ as an operator (which is also called the *full increment operator* or *backwards increment operator*) acting at each field of real numbers $\{x(\mathbf{n}), \mathbf{n} \in \mathbf{N}^d\}$.

The following bound is a trivial consequence of the latter representation

$$\max_{\mathbf{k}\le\mathbf{n}} |\Delta [S(\mathbf{k})]| \le 2^d \max_{\mathbf{k}\le\mathbf{n}} |S(\mathbf{k})|. \tag{A.7}$$

Note that this bound holds for an arbitrary field $\{S(\mathbf{n})\}$.

Proof of Proposition A.5. Proposition A.5 can easily be proved by induction. Equality (A.6) for $d = 1$ means that $\Delta [x_n] = x_n - x_{n-1}$, $n > 1$, and $\Delta [x_1] = x_1$ (since $x_0 = 0$ by assumption). Thus property (A.5) obviously holds for $d = 1$.

Assume that representation (A.6) holds for spaces $\mathbf{N}^k$ with $k \le d - 1$. Our goal is to prove it for the space $\mathbf{N}^d$.

Representation (A.6) can be rewritten in the following equivalent form

$$\Delta [x(\mathbf{n})] = \sum_{\delta_1=0}^{1} \cdots \sum_{\delta_d=0}^{1} (-1)^{\delta_1+\cdots+\delta_d} x(n_1 - \delta_1, \ldots, n_d - \delta_d).$$

Now we define a collection of operators $\Delta_{\{i_1,\ldots,i_r\}}$ as follows:

$$\Delta_{\{i_1,\ldots,i_r\}} [x(\mathbf{n})] = \sum_{\delta_{i_1}=0}^{1} \cdots \sum_{\delta_{i_r}=0}^{1} (-1)^{\delta_{i_1}+\cdots+\delta_{i_r}} x(\mathbf{n} - \boldsymbol{\delta}_{\{i_1,\ldots,i_r\}}),$$

where the element $\boldsymbol{\delta}_{\{i_1,\ldots,i_r\}} = (\delta_1, \ldots, \delta_d)$ has zero coordinates δ_i for indices $i \notin \{i_1, \ldots, i_r\}$. Note that the right-hand side of (A.6) equals $\Delta_{\{1,\ldots,d\}} [x(\mathbf{n})]$. Moreover, $\Delta_{\{1,\ldots,d\}} = \Delta_{\{1\}}\Delta_{\{2,\ldots,d\}}$. Using the linearity of the operator $\Delta_{\{1\}}$ and the inductive assumption

$$\sum_{k_2=1}^{n_2} \cdots \sum_{k_d=1}^{n_d} \Delta_{\{2,\ldots,d\}} [x(k_1, k_2, \ldots, k_d)] = x(k_1, n_2, \ldots, n_d)$$

we get

$$\begin{aligned}
\sum_{\mathbf{k}\le\mathbf{n}} \Delta_{\{1,\ldots,d\}} [x(\mathbf{k})] &= \sum_{k_1=1}^{n_1} \sum_{k_2=1}^{n_2} \cdots \sum_{k_d=1}^{n_d} \Delta_{\{1\}} \left[\Delta_{\{2,\ldots,d\}} [x(k_1, k_2, \ldots, k_d)]\right] \\
&= \sum_{k_1=1}^{n_1} \Delta_{\{1\}} \left[\sum_{k_2=1}^{n_2} \cdots \sum_{k_d=1}^{n_d} \Delta_{\{2,\ldots,d\}} [x(k_1, k_2, \ldots, k_d)] \right] \\
&= \sum_{k_1=1}^{n_1} \Delta_{\{1\}} [x(k_1, n_2, \ldots, n_d)] = x(\mathbf{n}).
\end{aligned}$$

This completes the proof. □

Proof of Proposition A.4. Proposition A.5 implies that every term $x(\mathbf{n})$ is a linear combination of partial sums $s(\mathbf{k})$. Note also that every coordinate of the element $\mathbf{k}$ in this combination either equals the corresponding coordinate of the element $\mathbf{n}$ or differs from it by 1 or by -1. Moreover, exactly 2^{d-1} sums in this combination have the coefficient -1, while the other 2^{d-1} sums have the coefficient $+1$. This proves Proposition A.4. □

Definition A.9 Starting with a field of real numbers $\{x(\mathbf{n}), \mathbf{n} \in \mathbf{N}^d\}$ we define the field $\{\nabla[x(\mathbf{n})], \mathbf{n} \in \mathbf{N}^d\}$ as follows:

$$\nabla[x(\mathbf{n})] = \sum_{\delta_1,\dots,\delta_d=0}^{1} (-1)^{\delta_1+\dots+\delta_d} x(n_1+\delta_1,\dots,n_d+\delta_d),$$

where the summation on the right-hand side is considered with respect to all 2^d possible combinations of the numbers $\delta_1,\dots,\delta_d \in \{0,1\}$. We also view ∇ as an operator (called the *forward increment operator*) acting on each field $\{x(\mathbf{n})\}$.

The properties of the operator ∇ are similar to the properties of the operator Δ. We also mention a relationship between these two operators

$$\nabla[x(\mathbf{n})] = (-1)^d \Delta[x(\mathbf{n}+\mathbf{1})]. \tag{A.8}$$

The operators ∇ and Δ come together in the so-called *partial summation formula for multi-indexed sums.*

Proposition A.6 *Let elements $\mathbf{m} \le \mathbf{n} \in \mathbf{N}^d$ be fixed. Assume that the following property holds for two fields $\{a(\mathbf{k}), \mathbf{k} \in \mathbf{N}^d\}$ and $\{b(\mathbf{k}), \mathbf{k} \in \mathbf{N}^d\}$: given $\boldsymbol{\varepsilon} \in \mathcal{E}_d$,*

$$a(\mathbf{k})b(\mathbf{k}+\boldsymbol{\varepsilon}) = 0 \tag{A.9}$$

for all $\mathbf{k} \in \Pi(\mathbf{m},\mathbf{n}) \setminus \Pi(\mathbf{m}-\boldsymbol{\varepsilon},\mathbf{n}-\boldsymbol{\varepsilon})$ and $\mathbf{k} \in \Pi(\mathbf{m}-\boldsymbol{\varepsilon},\mathbf{n}-\boldsymbol{\varepsilon}) \setminus \Pi(\mathbf{m},\mathbf{n})$, where

$$\Pi(\mathbf{m},\mathbf{n}) = \{\mathbf{k}\colon \mathbf{m} \le \mathbf{k} \le \mathbf{n}\} \quad \textit{and} \quad \Pi(\mathbf{m}-\boldsymbol{\varepsilon},\mathbf{n}-\boldsymbol{\varepsilon}) = \{\mathbf{k}\colon \mathbf{m}-\boldsymbol{\varepsilon} \le \mathbf{k} \le \mathbf{n}-\boldsymbol{\varepsilon}\}.$$

Then

$$\sum_{\mathbf{m}\le\mathbf{k}\le\mathbf{n}} \Delta[a(\mathbf{k})]\, b(\mathbf{k}) = \sum_{\mathbf{m}\le\mathbf{k}\le\mathbf{n}} a(\mathbf{k})\nabla[b(\mathbf{k})]. \tag{A.10}$$

Condition (A.9) for $d=1$ means that

$$a_{m-1}b_m = 0, \quad a_{n-1}b_n = 0.$$

If property (A.9) does not hold for $d=1$, then equality (A.10) becomes

$$\sum_{k=m}^{n} \Delta[a_k]b_k = \sum_{k=m}^{n} a_k\nabla[b_k] + a_{m-1}b_m - a_{n-1}b_n.$$

A similar correction of equality (A.10) is also needed in the general case of $d > 1$, if assumption (A.9) is dropped.

Proof of Proposition A.6. First we use representation (A.6):

$$\sum_{\mathbf{m}\le\mathbf{k}\le\mathbf{n}} \Delta\left[a(\mathbf{k})\right] b(\mathbf{k}) = \sum_{\delta_1,\dots,\delta_d=0}^{1} (-1)^{\delta_1+\dots+\delta_d} \sum_{\mathbf{m}\le\mathbf{k}\le\mathbf{n}} a(\mathbf{k}-\boldsymbol{\delta}) b(\mathbf{k}).$$

Then (A.9) implies that

$$\sum_{\mathbf{m}\le\mathbf{k}\le\mathbf{n}} a(\mathbf{k}-\boldsymbol{\delta}) b(\mathbf{k}) = \sum_{\mathbf{m}-\boldsymbol{\delta}\le\mathbf{k}\le\mathbf{n}-\boldsymbol{\delta}} a(\mathbf{k}) b(\mathbf{k}+\boldsymbol{\delta}) = \sum_{\mathbf{m}\le\mathbf{k}\le\mathbf{n}} a(\mathbf{k}) b(\mathbf{k}+\boldsymbol{\delta})$$

for an arbitrary vector $\boldsymbol{\delta} = (\delta_1, \dots, \delta_d) \in \mathscr{E}_d$. Thus

$$\sum_{\mathbf{m}\le\mathbf{k}\le\mathbf{n}} \Delta\left[a(\mathbf{k})\right] b(\mathbf{k}) = \sum_{\mathbf{m}\le\mathbf{k}\le\mathbf{n}} a(\mathbf{k}) \sum_{\delta_1,\dots,\delta_d=0}^{1} (-1)^{\delta_1+\dots+\delta_d} b(\mathbf{k}+\boldsymbol{\delta})$$

and this proves Proposition A.6. □

A.6 Monotonicity in the Space of Multiple Indices

There are several ways to define the monotonicity for $d > 1$. Each of the definitions inherits some aspect of monotonicity for $d = 1$ and each of them is a "natural" generalization of this property. Below we discuss "increasing" fields; the notion of "decreasing" fields can obviously be deduced from here.

Definition A.10 A field $\{x(\mathbf{n}), \mathbf{n} \in \mathbf{N}^d\}$ is called *increasing* if

$$x(\mathbf{m}) \le x(\mathbf{n}) \quad \text{for all } \mathbf{m} \le \mathbf{n}. \tag{A.11}$$

Definition A.11 A field $\{x(\mathbf{n}), \mathbf{n} \in \mathbf{N}^d\}$ is called a *field with non-negative increments* if

$$\Delta\left[x(\mathbf{n})\right] \ge 0 \quad \text{for all } \mathbf{n} \in \mathbf{N}^d.$$

Each of these definitions for $d = 1$ coincides with the usual definition of monotonicity. Definition A.10 for $d = 1$ is reduced to the condition that $x_n \le x_{n+1}$, while Definition A.11 can be written as $x_n - x_{n-1} \ge 0$. Nevertheless, these definitions are different if $d > 1$.

Proposition A.7 *Every field with non-negative increments is increasing. The converse statement is false for $d > 1$.*

Proof of Proposition A.7. The definition of increment (see representation (A.5)) implies that $x(\mathbf{m}) \le x(\mathbf{n})$ for all $\mathbf{m} \le \mathbf{n}$ if $\Delta[x(\mathbf{n})] \ge 0$ for all $\mathbf{n} \in \mathbf{N}^d$. This proves the first part of Proposition A.7.

We prove the second part of Proposition A.7 by example. Without loss of generality, we restrict attention to the case $d = 2$. Put

$$x(1,1) = 0, \quad x(1,2) = x(2,1) = 1, \quad x(2,2) = \sqrt{2}$$

and $x(m,n) = m + n$ for other m and n. It is clear that the field $\{x(m,n); m,n \ge 1\}$ is increasing but it is not a field with non-negative increments, since $\Delta[x(2,2)] = x(2,2) - x(1,2) - x(2,1) + x(1,1) = 2 - 2\sqrt{2} < 0$, while $\Delta[x(1,2)] > 0$, $\Delta[x(2,1)] > 0$ and $\Delta[x(1,1)] \ge 0$. □

The example constructed in the proof of Proposition A.7 shows that both inequalities $\Delta[x(\mathbf{n})] < 0$ and $\Delta[x(\mathbf{n})] > 0$ may occur in the general case of an increasing field. In a similar manner, one can construct an increasing field for which each of these inequalities may happen infinitely often.

The partial summation formula (A.10) for the space $\mathbf{N}^d$ is often useful when estimating multi-indexed sums. One of the approaches to proving the so-called *Hájek–Rényi type* inequalities in probability theory relies on estimates of this type.

Proposition A.8 *Let $\{X(\mathbf{n}), \mathbf{n} \in \mathbf{N}^d\}$ be a field of real numbers and let $\{S(\mathbf{n}), \mathbf{n} \in \mathbf{N}^d\}$ be their partial sums. Assume that $\{b(\mathbf{n}), \mathbf{n} \in \mathbf{N}^d\}$ is a field with non-negative increments, that is $\Delta[b(\mathbf{n})] \ge 0$ for all $\mathbf{n} \in \mathbf{N}^d$. Put*

$$T(\mathbf{n}) = \sum_{\mathbf{k} \le \mathbf{n}} \frac{X(\mathbf{k})}{b(\mathbf{k})}.$$

Then

$$\max_{\mathbf{n} \le \mathbf{m}} \left| \frac{S(\mathbf{n})}{b(\mathbf{n})} \right| \le 2^d \max_{\mathbf{n} \le \mathbf{m}} |T(\mathbf{n})| \tag{A.12}$$

for all $\mathbf{m} \in \mathbf{N}^d$.

Proof Recall that we agreed to extend the domain of each field to $\mathbf{N}^d \cup \mathbf{N}_0^d$. For $\{T(\mathbf{n})\}$, this means that $T(\mathbf{n}) = 0$ if at least one of the coordinates of an element $\mathbf{n}$ equals 0. Proposition A.6 implies that

$$\left| \frac{S(\mathbf{n})}{b(\mathbf{n})} \right| = \frac{1}{b(\mathbf{n})} \left| \sum_{\mathbf{k} \le \mathbf{n}} \Delta[T(\mathbf{k})] \cdot b(\mathbf{k}) \right| = \frac{1}{b(\mathbf{n})} \left| \sum_{\mathbf{k} \le \mathbf{n}} T(\mathbf{k}) \nabla \left[b'(\mathbf{k}) \right] \right|$$

for all $\mathbf{n} \in \mathbf{N}^d$, where

$$b'(\mathbf{k}) = \begin{cases} b(\mathbf{k}), & \mathbf{k} \le \mathbf{n}, \\ 0, & \mathbf{k} \not\le \mathbf{n}. \end{cases}$$

Thus

$$\left|\frac{S(\mathbf{n})}{b(\mathbf{n})}\right| \le \max_{\mathbf{k}\le\mathbf{n}} |T(\mathbf{k})| \cdot \frac{1}{b(\mathbf{n})} \sum_{\mathbf{k}\le\mathbf{n}} \left|\nabla\left[b'(\mathbf{k})\right]\right|,$$

whence

$$\max_{\mathbf{n}\le\mathbf{m}} \left|\frac{S(\mathbf{n})}{b(\mathbf{n})}\right| \le \max_{\mathbf{k}\le\mathbf{m}} |T(\mathbf{k})| \cdot \max_{\mathbf{n}\le\mathbf{m}} \frac{1}{b(\mathbf{n})} \sum_{\mathbf{k}\le\mathbf{n}} \left|\nabla\left[b'(\mathbf{k})\right]\right|. \tag{A.13}$$

Since not all increments of the field $\{b'(\mathbf{n})\}$ are non-negative, one cannot immediately use equality (A.8) and derive bound (A.12) from it.

To complete the proof of bound (A.12), we introduce the following notation: if $\ell = 1$, then $I(\varepsilon, \ell) = \{1\}$ for $\varepsilon \in \{0, 1\}$; if $\ell > 1$, then

$$I(\varepsilon, \ell) = \begin{cases} \{1, \dots, \ell - 1\}, & \text{for } \varepsilon = 0, \\ \{\ell\}, & \text{for } \varepsilon = 1. \end{cases}$$

Let $\mathbf{n} = (n_1, \dots, n_d)$ and $\boldsymbol{\varepsilon} = (\varepsilon_1, \dots, \varepsilon_d) \in \mathscr{E}_d$. Put

$$\zeta(\boldsymbol{\varepsilon}) = \sum_{k_1 \in I(\varepsilon_1, n_1)} \cdots \sum_{k_d \in I(\varepsilon_d, n_d)} \left|\nabla\left[b'(\mathbf{k})\right]\right|$$

and note that $\zeta(\mathbf{1}) = \nabla\left[b'(\mathbf{n})\right]$. Moreover, in the general case

$$\sum_{\mathbf{k}\le\mathbf{n}} \left|\nabla\left[b'(\mathbf{k})\right]\right| = \sum_{\boldsymbol{\varepsilon}\in\mathscr{E}_d} \zeta(\boldsymbol{\varepsilon}). \tag{A.14}$$

Now we estimate each term $\zeta(\boldsymbol{\varepsilon})$, $\boldsymbol{\varepsilon} \in \mathscr{E}_d$, separately.

The case $\varepsilon_1 = \dots = \varepsilon_d = 1$. In this case, $\zeta(\boldsymbol{\varepsilon}) = \left|\Delta\left[b'(\mathbf{n}+\mathbf{1})\right]\right| = b(\mathbf{n})$ according to (A.8) and to the definition of the field $\{b'(\mathbf{n})\}$.

The case $\varepsilon_1 = \dots = \varepsilon_d = 0$. According to the definition of the field $\{b'(\mathbf{n}), \mathbf{n} \in \mathbf{N}^d\}$, we have

$$\zeta(\boldsymbol{\varepsilon}) = \sum_{\mathbf{k}\le\mathbf{n}-\mathbf{1}} \left|\nabla\left[b'(\mathbf{k})\right]\right| = \sum_{\mathbf{k}\le\mathbf{n}-\mathbf{1}} \Delta\left[b(\mathbf{k}+\mathbf{1})\right]$$

by (A.8) and in view of the non-negativity of increments of the field $\{b(\mathbf{n}), \mathbf{n} \in \mathbf{N}^d\}$. Further,

$$\zeta(\boldsymbol{\varepsilon}) = \sum_{\mathbf{2}\le\mathbf{k}\le\mathbf{n}} \Delta\left[b(\mathbf{k})\right] \le \sum_{\mathbf{k}\le\mathbf{n}} \Delta\left[b(\mathbf{k})\right] = b(\mathbf{n})$$

again by the non-negativity of the increments $\{\Delta\left[b(\mathbf{n})\right]\}$.

The case $\varepsilon_1 = \cdots = \varepsilon_c = 0$, $\varepsilon_{c+1} = \cdots = \varepsilon_d = 1$. We assume that $1 \le c < d$. Let ∇_c be the operator similar to ∇ but with the difference that ∇_c acts on the space $\mathbf{N}^c$ with respect to the first c coordinates. Then

$$\zeta(\boldsymbol{\varepsilon}) = \sum_{k_1 \le n_1 - 1} \cdots \sum_{k_c \le n_c - 1} \left| \nabla_c \left[b(k_1, \ldots, k_c, n_{c+1}, \ldots, n_d) \right] \right|.$$

As in the preceding case, we prove that $\zeta(\boldsymbol{\varepsilon}) \le b(\mathbf{n})$.

The general term $\zeta(\boldsymbol{\varepsilon})$ of the sum in (A.14) is considered similarly by symmetry. Hence

$$\zeta(\boldsymbol{\varepsilon}) \le b(\mathbf{n}) \quad \text{for every } \boldsymbol{\varepsilon} \in \mathscr{E}_d.$$

Since $\operatorname{card}(\mathscr{E}_d) = 2^d$, equality (A.14) implies that

$$\sum_{\mathbf{k} \le \mathbf{n}} \left| \nabla \left[b'(\mathbf{k}) \right] \right| \le 2^d b(\mathbf{n}) \quad \text{for all } \mathbf{n} \in \mathbf{N}^d.$$

Therefore Proposition A.8 follows from inequality (A.14). □

A.7 Kronecker's Lemma

One of the most often used results of the classical calculus is the so-called *Kronecker's lemma*. Unfortunately, this result for $d > 1$ does not hold in general. On the other hand, it can be proved for non-negative terms even in the case of $d > 1$.

Proposition A.9 *Let* $x(\mathbf{n}) \ge 0$ *for all* $\mathbf{n} \in \mathbf{N}^d$. *Assume that a field* $\{b(\mathbf{n}), \mathbf{n} \in \mathbf{N}^d\}$ *is increasing and is unbounded, that is,* $b(\mathbf{m}) \le b(\mathbf{n})$ *for all* $\mathbf{m} \le \mathbf{n}$ *and* $\lim(\max) b(\mathbf{n}) = \infty$. *If the multi-indexed series*

$$\sum_{\mathbf{n} \in \mathbf{N}^d} \frac{x(\mathbf{n})}{b(\mathbf{n})}$$

converges, then

$$\lim(\max) \frac{1}{b(\mathbf{n})} \sum_{\mathbf{k} \le \mathbf{n}} x(\mathbf{k}) = 0.$$

Proof of Proposition A.9. The proof of Proposition A.9 for $d > 1$ is the same as in the case $d = 1$. Indeed, we have for any fixed n_0,

$$\frac{1}{b(\mathbf{n})} \sum_{\mathbf{k} \le \mathbf{n}} x(\mathbf{k}) = \frac{1}{b(\mathbf{n})} \sum_{|\mathbf{k}| < n_0} x(\mathbf{k}) + \frac{1}{b(\mathbf{n})} \sum_{\mathbf{k} \le \mathbf{n}, |\mathbf{k}| \ge n_0} x(\mathbf{k}).$$

The lim(max)-limit of the first term is equal to 0, since n_0 is fixed and the field $b(\mathbf{n})$ is unbounded. The second term is estimated by the tail of a convergent series of non-negative terms

$$\frac{1}{b(\mathbf{n})} \sum_{\mathbf{k}\le\mathbf{n}, |\mathbf{k}|\ge n_0} x(\mathbf{k}) \le \sum_{|\mathbf{k}|\ge n_0} \frac{x(\mathbf{k})}{b(\mathbf{k})}.$$

Since n_0 is arbitrary, Proposition A.9 is proved. □

Example A.2 We are going to show that Proposition A.9 fails if one drops the assumption that the $x(\mathbf{n})$ are non-negative. We restrict our attention to the case $d = 2$ and $b(m, n) = mn$. Note, however, that a similar example can be constructed in the same way for all $d > 1$ and almost all fields $b(m, n)$. Note also that the field $b(m, n) = mn$ is increasing and, moreover, its increments are non-negative. Set

$$x(k, l) = \begin{cases} -l, & k = 1, \\ 2l, & k = 2, \\ 0, & k > 2. \end{cases}$$

For all $m \ge 2$ and $n \ge 1$, we have

$$\sum_{k=1}^{m}\sum_{l=1}^{n} \frac{x(k, l)}{b(k, l)} = 0, \qquad \frac{1}{b(m, n)} \sum_{k=1}^{m}\sum_{l=1}^{n} x(k, l) = \frac{n + 1}{2m},$$

whence we conclude that the double series $\sum_{k,l=1}^{\infty} x(k, l)/b(k, l)$ converges to 0. On the other hand, neither the lim(max)- nor the lim(min)-limits exist for the field $(1/b(m, n)) \sum_{k=1}^{m} \sum_{l=1}^{n} x(k, l)$.

A.8 Martikainen's Characteristic of a Field

Let $\{b(\mathbf{n}), \mathbf{n} \in \mathbf{N}^d\}$ be a field of non-negative numbers for which the series $\sum_{\mathbf{n}\in\mathbf{N}^d} b(\mathbf{n})$ diverges. With the field $\{b(\mathbf{n})\}$, we associate a function $\mathscr{B} = \mathscr{B}_b$ of subsets of $\mathbf{N}^d$:

$$\mathscr{B}(D) = \mathscr{B}_b(D) = \sum_{\mathbf{n}\in D} b(\mathbf{n}), \quad D \subseteq \mathbf{N}^d.$$

For $\Pi(\mathbf{n}) = \{\mathbf{k} \in \mathbf{N}^d : \mathbf{k} \le \mathbf{n}\}$, we also write $\mathscr{B}(\mathbf{n}) = \mathscr{B}(\Pi(\mathbf{n}))$.

Now let $x > 0$. Introduce the sets

$$D(x) = \{\mathbf{n} \in \mathbf{N}^d : \mathscr{B}(\mathbf{n}) \le x\}.$$

The function $\chi(x)$ defined by

$$\chi(x) = \chi_b(x) = \mathscr{B}(D(x)) = \sum_{\mathbf{n}:\, \mathscr{B}(\mathbf{n}) \le x} b(\mathbf{n})$$

is called *Martikainen's characteristic of the field* $\{b(\mathbf{n}), \mathbf{n} \in \mathbf{N}^d\}$.

The asymptotic behavior of Martikainen's characteristic for $d = 1$ (almost) does not depend on the sequence $\{b_n, n \ge 1\}$ from which it is constructed. More precisely,

$$\limsup_{x\to\infty} \frac{\chi(x)}{x} = 1,$$

since $\chi(x) \le x$ and $\chi(B_n) = B_n$, where $B_n = \sum_{k \le n} b_k$. If the sequence $\{b_n, n \ge 1\}$ is bounded, then lim substitutes lim sup in the latter relation. This property disappears for $d > 1$.

Example A.3 Let a field $\{b(\mathbf{n}), \mathbf{n} \in \mathbf{N}^d\}$ be "concentrated" at the axes, that is $b(\mathbf{1}) = 1$; $b(\mathbf{n}) = 1$ if exactly one coordinate of an element $\mathbf{n} \ne \mathbf{1}$ equals 1, and $b(\mathbf{n}) = 0$ otherwise. Then $\mathscr{B}(\mathbf{n}) = n_1 + \cdots + n_d - d + 1$ and $\chi(x) = d[x] - d + 1$, where the symbol $[\cdot]$ denotes the integer part of a real number. Thus the asymptotic behavior of Martikainen's characteristic for this field is the same as in the case $d = 1$ (up to a multiplicative constant), namely

$$\lim_{x\to\infty} \frac{\chi(x)}{x} = d.$$

On the other hand, let $b(\mathbf{n}) = 1$ for all $\mathbf{n} \in \mathbf{N}^d$. Then $\mathscr{B}(\mathbf{n}) = |\mathbf{n}|$ and $\chi(x)$ has the same asymptotic behavior as the Dirichlet function (see A.32), that is

$$\lim_{x\to\infty} \frac{\chi(x)}{x(\log x)^{d-1}} = \frac{1}{(d-1)!}.$$

The above examples show that the asymptotic behavior of Martikainen's characteristic in the case of $d > 1$ varies even in the class of bounded fields $\{b(\mathbf{n}), \mathbf{n} \in \mathbf{N}^d\}$. One may nevertheless expect there to be a universal "dominating" function $\psi_d = \psi$ in the case of $d > 1$ such that

$$\lim_{x\to\infty} \frac{\chi(x)}{\psi(x)} = 0$$

for all fields $\{b(\mathbf{n}), \mathbf{n} \in \mathbf{N}^d\}$. Surprisingly, even this property does not hold for $d > 1$. In the following example, given a function $\psi(x)$, we construct a field $\{b(\mathbf{n}), \mathbf{n} \in \mathbf{N}^d\}$ such that

$$\limsup_{x\to\infty} \frac{\chi(x)}{\psi(x)} = \infty. \tag{A.15}$$

Example A.4 It is clear that one need only consider the case $d = 2$. Without loss of generality we may assume that ψ is a non-negative non-decreasing unbounded function.

Let $\varphi(x) = x(\psi(x) + a)$, where a number a is chosen such that $\psi(1) + a \geq 3$. We introduce the sequence of positive integers $\{n_k, k \geq 1\}$ as follows:

$$\begin{aligned} &n_1 = 2, \quad n_2 = [\varphi(2)], \\ &n_k = [\varphi(n_1 + \cdots + n_{k-1} - k + 2)], \quad k > 2. \end{aligned}$$

Since $\varphi(x) \geq 3x$ for all $x \geq 1$, one can prove by induction that $n_1 + \cdots + n_{k-1} \geq k(k-1)$, $k \geq 1$, whence $n_k \to \infty$ as $k \to \infty$. Now we put $G_k = \{(i, j) \in N^2 : i + j = n_k\}$, $k \geq 1$. Note that $\operatorname{card}(G_k) = n_k - 1$. Let the field $\{b(i, j)\}$ be defined by

$$b(i, j) = \begin{cases} 1, & \text{if } (i, j) \in \bigcup_{k\geq 1} G_k, \\ 0, & \text{otherwise.} \end{cases}$$

If $(m, n) \in G_k$ and $k \geq 2$, then

$$\begin{aligned} B(m, n) &= \sum_{i=1}^{m} \sum_{j=1}^{n} b(i, j) \leq \operatorname{card}(G_1) + \cdots + \operatorname{card}(G_{k-1}) + 1 \\ &= n_1 + \cdots + n_{k-1} - k + 2. \end{aligned}$$

Further, if $x = n_1 + \cdots + n_{k-1} - k + 2$, then $D(x) \supseteq \{(i, j) : i + j \leq n_k\} \stackrel{\text{def}}{=} M_k$, whence $\mathscr{B}(D(x)) \geq \mathscr{B}(M_k) = n_1 + \cdots + n_{k-1} + n_k - k = x + [\varphi(x)] - 2$. This relation implies (A.15).

The above phenomenon yields several consequences in the theory of limit theorems for random fields. Namely, a universal function exists for $d = 1$ that describes the asymptotic behavior of cumulative sums of independent random variables. In contrast, such a function does not exist for any $d > 1$.

Example A.5 (*continuation of Example* A.4) Put $G = \bigcup_{k\geq 1} G_k$. Without loss of generality we assume that $\psi(1) > 1$. Under this assumption, one may choose $a = 2$ in the definition of the function φ. Let a field $\{X(m, n)\}$ of independent random variables be such that $X(m, n) = 0$ for $(m, n) \notin G$. Given an element $(m, n) \in G$, let $X(m, n)$ be a random variable that does not depend on any other random variable and has the distribution

$$\mathsf{P}(X(m,n) = \psi(B(m,n))) = \mathsf{P}(X(m,n) = -\psi(B(m,n))) = \frac{1}{2\psi(B(m,n))},$$

$$\mathsf{P}(X(m,n) = 0) = 1 - \frac{1}{\psi(B(m,n))}.$$

It is clear that $\mathsf{E}\,[X(i,j)] = 0$ and $\mathsf{E}\left[X^2(i,j)\right] = b(i,j)$ for all $i,j \geq 1$. Is it possible that the random variables just defined satisfy the following relation

$$\lim(\max) \frac{S(m,n)}{\psi(B(m,n))} = 0 \quad \text{a. s.?} \tag{A.16}$$

If relation (A.16) holds, then

$$\lim(\max) \frac{X(m,n)}{\psi(B(m,n))} = 0 \quad \text{a. s.,}$$

since $X(m,n) = \Delta\,[S(m,n)]$ and ψ is an increasing function. This implies that

$$\mathsf{P}(|X(m,n)| \geq \psi(B(m,n)) \text{ i. o.}) = 0.$$

The random events $\{\omega\colon |X(m,n)| \geq \psi(B(m,n))\}$ are independent and thus the Borel–Cantelli lemma for independent events implies

$$\sum_{m=1}^{\infty}\sum_{n=1}^{\infty} \mathsf{P}(|X(m,n)| \geq \psi(B(m,n))) < \infty. \tag{A.17}$$

On the other hand,

$$\sum_{m=1}^{\infty}\sum_{n=1}^{\infty} \mathsf{P}(|X(m,n)| \geq \psi(B(m,n))) = \sum_{k=1}^{\infty} \sum_{(m,n)\in G_k} \frac{1}{\psi(B(m,n))}.$$

Since

$$\begin{aligned}\sum_{(m,n)\in G_k} \frac{1}{\psi(B(m,n))} &\geq \sum_{(m,n)\in G_k} \frac{1}{\psi(n_1 + \cdots + n_{k-1} - k + 2)} \\ &= \frac{n_k - 1}{\psi(n_1 + \cdots + n_{k-1} - k + 2)} \\ &= \frac{[\varphi(n_1 + \cdots + n_{k-1} - k + 2)] - 1}{\psi(n_1 + \cdots + n_{k-1} - k + 2)} \geq 1\end{aligned}$$

for sufficiently large k, condition (A.17) cannot be true.

Now it is not hard to define a field of independent random variables $\{X(m,n)\}$ such that

$$\limsup_{m,n} \max \frac{|S(m,n)|}{\psi(B(m,n))} = \infty \quad \text{a. s.} \tag{A.18}$$

To define such a field, one needs to choose an arbitrary increasing function ψ_1 such that $\psi_1(x)/\psi(x) \to \infty$, $x \to \infty$, and then construct the field $\{X(m,n)\}$ by using the above procedure for ψ_1 rather than for ψ. Then

$$\limsup_{\max} \left| \frac{S(m,n)}{\psi_1(m,n)} \right| > 0,$$

since (A.16) does not hold, and thus relation (A.18) is indeed valid.

Definition A.12 Denote by Ψ_c the set of positive increasing functions $\psi(x)$ such that

$$\sum_{n=n_0}^{\infty} \frac{1}{n\psi(n)} < \infty, \tag{A.19}$$

where n_0 is the first positive integer for which $\psi(n_0) > 0$.

Further, denote by Ψ_d the set of positive increasing functions $\psi(x)$ for which the series on the right-hand side of (A.19) diverges.

Next, denote by Ψ^c the set of positive increasing functions $\psi(x)$ for which

$$x\psi\left(\frac{1}{x}\right) \downarrow 0, \quad x \downarrow 0, \tag{A.20}$$

and the series on the right-hand side of (A.19) converges.

Finally, denote by Ψ^d the set of positive increasing functions $\psi(x)$ for which condition (A.20) holds and the series on the right-hand side of (A.19) diverges.

A.9 Regularly Varying Functions

A function f defined for $x \ge 0$ is called *regularly varying* if $f(x) > 0$ for all $x \ge 0$ and the limit

$$\lim_{x\to\infty} \frac{f(cx)}{f(x)}$$

exists for all $c > 0$. If f is a measurable and regularly varying function, then

$$\lim_{x\to\infty} \frac{f(cx)}{f(x)} = c^{\rho}$$

for some number $\rho \in \mathbf{R}$ (this number is called the *index* of the function f). Throughout this book we consider only measurable regularly varying functions f. The set of regularly varying functions with index ρ is denoted by $\mathcal{RV}_\rho$.

A.9.1 Monotone Versions

The theory of regularly varying functions significantly simplifies if one restricts to the case of monotone functions. It is also true that every regularly varying function with index $\rho \neq 0$ is asymptotically equivalent to some monotone function which is also regularly varying and possesses the same index as the function f.

Proposition A.10 *Let f be a regularly varying function with index $\rho \neq 0$.*

(i) *If $\rho > 0$, then there exists an increasing function $f^{\uparrow}$ for which*

$$f(t) \sim f^{\uparrow}(t), \quad t \to \infty.$$

This function $f^{\uparrow}$ is regularly varying with the same index ρ.

(ii) *If $\rho < 0$, then there exists a decreasing function $f^{\downarrow}$ for which*

$$f(t) \sim f^{\downarrow}(t), \quad t \to \infty.$$

This function $f^{\downarrow}$ is regularly varying with the same index ρ.

If $\rho = 0$, then f is called a *slowly varying* function. It is not true in general that, given a slowly varying function, there is a monotone version of it. What can be derived from Proposition A.10 is that, for an arbitrary slowly varying function f and for all $\rho > 0$, there are monotone regularly varying functions $f^{\uparrow} \in \mathcal{RV}_\rho$ and $f^{\downarrow} \in \mathcal{RV}_{-\rho}$ such that

$$x^{\rho} f(x) \sim f^{\uparrow}(x), \quad x \to \infty,$$
$$x^{-\rho} f(x) \sim f^{\downarrow}(x), \quad x \to \infty.$$

From here, one can easily deduce that if f is slowly varying and $\varepsilon > 0$, then

$$\lim_{x \to \infty} x^{\varepsilon} f(x) = \infty,$$
$$\lim_{x \to \infty} x^{-\varepsilon} f(x) = 0.$$

All these results follow from the following assertion describing the asymptotic behavior of a slowly varying function.

Proposition A.11 *Let L be a slowly varying function and* $\gamma > 0$. *Then*

$$\sup_{t \geq x} t^{-\gamma} L(t) \sim x^{-\gamma} L(x), \tag{A.21}$$

$$\inf_{t \geq x} t^{-\gamma} L(t) \sim x^{-\gamma} L(x). \tag{A.22}$$

A.9.2 Karamata's Theorems

One of the main achievements of the theory of regularly varying functions is establishing the asymptotic properties of integrals.

Proposition A.12 (Karamata's theorem: direct half) *Let* f *be a locally integrable regularly varying function with index* ρ.

(i) *If* $\sigma + \rho + 1 > 0$, *then*

$$\frac{1}{x^{\sigma+1} f(x)} \int_0^x t^\sigma f(t)\, dt \to \frac{1}{\sigma + \rho + 1}, \quad x \to \infty. \tag{A.23}$$

(ii) *If* $\sigma + \rho + 1 < 0$, *then*

$$\frac{1}{x^{\sigma+1} f(x)} \int_x^\infty t^\sigma f(t)\, dt \to -\frac{1}{\sigma + \rho + 1}, \quad x \to \infty. \tag{A.24}$$

Proposition A.13 (Karamata's theorem: inverse half) *Let* f *be a positive locally integrable function. If relation* (A.23) *or* (A.24) *holds for some* ρ *and* $\sigma + \rho + 1 \neq 0$, *then* f *is regularly varying with index* ρ; *moreover* $\sigma + \rho + 1 > 0$ *in the case of* (A.23) *and* $\sigma + \rho + 1 < 0$ *in the case of* (A.24).

Proposition A.14 *Let* f *be a regularly varying function of index* ρ, *that is* $f \in \mathcal{RV}_\rho$. *Put* $f_k = f(k)$.

(i) *If* $\sigma > \rho + 1$, *then*

$$\sum_{k=n}^{\infty} \frac{f_k}{k^\sigma} \sim \frac{f_n}{n^{\sigma-1}} (\sigma - \rho), \quad n \to \infty. \tag{A.25}$$

(ii) *If* $\sigma < \rho$, *then*

$$\sum_{k=1}^{n} \frac{f_k}{k^\sigma} \sim \frac{f_n}{n^{\sigma-1}} (\rho - \sigma), \quad n \to \infty. \tag{A.26}$$

A.9.3 Inverses of Regularly Varying Functions

It is often important to determine the asymptotic behavior of inverse functions. For regularly varying functions this can easily be done, since their inverses are also regularly varying.

Proposition A.15 *Let f be a continuous monotone regularly varying function with index $\rho \neq 0$. Then its inverse function also is a regularly varying function with index $1/\rho$, that is*

$$f^{-1}(x) = x^{1/\rho} L(x),$$

where L is a slowly varying function.

Proposition A.15 is not valid for slowly varying functions. A counterexample is presented by the function $f(x) = \log x$.

A.9.4 Uniform Convergence

One of the two fundamental facts about slowly varying functions is the so-called *uniform convergence theorem*.

Proposition A.16 (uniform convergence theorem) *Let f be a slowly varying function. Then, for all $0 < a < b < \infty$,*

$$\lim_{x\to\infty} \sup_{a\le c\le b} \left| \frac{f(cx)}{f(x)} - 1 \right| = 0.$$

Proposition A.16 extends in an obvious way to the case of regularly varying functions with non-zero index.

We show how Proposition A.16 is used in the theory of regularly varying functions by proving Proposition A.14.

Proof of Proposition A.14. We restrict ourselves to the case of part (i) (part (ii) is proved similarly).

Indeed, we have $f_k = k^\rho L(k)$, where $L(\cdot)$ is a slowly varying function. Let $\sigma > \rho + 1$. Considering $0 < \gamma < \sigma - \rho - 1$ we deduce from Proposition A.11 that

$$\sum_{k=n}^{\infty} \frac{f_k}{k^\sigma} \le \left(\sup_{x\ge n} \frac{L(x)}{x^\gamma} \right) \sum_{k=n}^{\infty} \frac{1}{k^{\sigma-\rho-\gamma}} \sim \frac{L(n)}{n^\gamma} \cdot \frac{\sigma-\rho-\gamma}{n^{\sigma-\rho-1-\gamma}} = \frac{f_n}{n^{\sigma-1}}(\sigma-\rho-\gamma),$$

whence

$$\limsup_{n\to\infty} \frac{n^{\sigma-1}}{f_n} \sum_{k=n}^{\infty} \frac{f_k}{k^\sigma} \le \sigma - \rho - \gamma.$$

Similarly we prove

$$\liminf_{n\to\infty} \frac{n^{\sigma-1}}{H_n} \sum_{k=n}^{\infty} \frac{H_k}{k^\sigma} \geq \sigma - \rho - \gamma$$

by using (A.22) instead of (A.21). □

Now we are going to provide an estimate for the so-called *Dirichlet convolution* that will be used when estimating the Dirichlet polynomial below.

Proposition A.17 *Assume that $L(x)$ is a slowly varying function such that $L(x) \geq c_0 > 0$ for all $x \geq 1$. Then*

$$F(x) \overset{\text{def}}{=} \sum_{i=1}^{x} \frac{1}{i} L\left(\frac{x}{i}\right)$$

is a slowly varying function such that $F(x) \geq c_0$ for all $x \geq 1$ and moreover $F(x) \to \infty$ as $x \to \infty$.

Proof of Proposition A.17. Considering the term corresponding to the index $i = 1$ we prove that $F(x) \geq L(x) \geq c_0$ for all $x \geq 1$. Moreover,

$$F(x) \geq c_0 \sum_{i \leq x} \frac{1}{i} \sim \log x, \quad x \to \infty,$$

in view of the asymptotic behavior of the harmonic series (A.46).

Now let $\lambda > 1$. Then

$$F(\lambda x) = \sum_{i \leq x} \frac{1}{i} L\left(\frac{\lambda x}{i}\right) + \sum_{x < i \leq \lambda x} \frac{1}{i} L\left(\frac{\lambda x}{i}\right) \overset{\text{def}}{=} F_1(x) + F_2(x).$$

For the second term, we get by (A.46)

$$F_2(x) \leq \sup_{\frac{1}{\lambda} \leq t < 1} L(t) \sum_{x < i \leq \lambda x} \frac{1}{i} \asymp \log \lambda + O\left(\frac{1}{x}\right).$$

Thus $F_2(x) = O(1)$, $x \to \infty$.

Further, fix $\varepsilon > 0$ and choose $x_0 > 1$ such that

$$\left| \frac{L(\lambda x)}{L(x)} - 1 \right| < \varepsilon, \quad x \geq x_0.$$

Such a number x_0 exists since $L(\cdot)$ is a slowly varying function. Then, for $x \geq x_0$, we obtain

$$F_1(x) = \sum_{i \le x, \frac{x}{i} \le x_0} \frac{1}{i} L\left(\frac{\lambda x}{i}\right) + \sum_{i \le x, \frac{x}{i} > x_0} \frac{1}{i} L\left(\frac{\lambda x}{i}\right) \stackrel{\text{def}}{=} F_{11}(x) + F_{12}(x).$$

As above, relation (A.46) implies that

$$F_{11}(x) \le \sup_{t \le \lambda x_0} L(t) \sum_{\frac{x}{x_0} \le i \le x} \frac{1}{i} \asymp \log x_0 + O\left(\frac{1}{x}\right).$$

Hence $F_{11}(x) = O(1)$, $x \to \infty$.

Further,

$$\begin{aligned} F_{12}(x) &= \sum_{i \le x, \frac{x}{i} > x_0} \frac{1}{i} \left(\frac{L(\lambda x/i)}{L(x/i)} - 1\right) L\left(\frac{x}{i}\right) + \sum_{i \le x, \frac{x}{i} > x_0} \frac{1}{i} L\left(\frac{x}{i}\right) \\ &\stackrel{\text{def}}{=} F_{121}(x) + F_{122}(x). \end{aligned}$$

We estimate the terms separately:

$$|F_{121}(x)| \le \sum_{i \le x, \frac{x}{i} > x_0} \frac{1}{i} \left|\frac{L(\lambda x/i)}{L(x/i)} - 1\right| L\left(\frac{x}{i}\right) \le \varepsilon \sum_{i \le x} \frac{1}{i} L\left(\frac{x}{i}\right) = \varepsilon F(x),$$

$$F_{122}(x) = F(x) - \sum_{i \le x, \frac{x}{i} \le x_0} \frac{1}{i} L\left(\frac{x}{i}\right).$$

The last term, as shown when estimating $F_{11}(\cdot)$, is bounded. Therefore $F(\lambda x) = F(x) + F_{121}(x) + O(1)$, whence

$$\left|\frac{F(\lambda x)}{F(x)} - 1\right| \le \varepsilon + O\left(\frac{1}{F(x)}\right).$$

Since $F(x) \to \infty$, $x \to \infty$, and ε is an arbitrary number,

$$\lim_{x \to \infty} \frac{F(\lambda x)}{F(x)} = 1.$$

□

A.10 Dirichlet's Divisor Problem

Let k be a fixed number. A set of positive integers $\{n_1, \ldots, n_d\}$ is called a solution of the Dirichlet equation if

$$n_1 \cdots n_d = k. \tag{A.27}$$

Now let τ_k, $k \geq 1$, be the number of positive integer solutions of Eq. (A.27). We call $\{\tau_k\}$ the *Dirichlet sequence* constructed for the space $\mathbf{N}^d$. Further, let $T_k = \tau_1 + \cdots + \tau_k$, $k \geq 2$. Clearly, T_k is the number of positive integer solutions of the inequality

$$n_1 \cdots n_d \leq k.$$

A.10.1 The Case d = 2

One of the classical problems in number theory is to study the asymptotic behavior of τ_k and T_n. The sequence $\{\tau_k, k \geq 1\}$ varies irregularly but T_n admits some nice estimates. In 1849 P. Dirichlet studied the case $d = 2$ and proved

$$T_n = n \log n + (2\gamma - 1)n + o\left(\sqrt{n}\right), \tag{A.28}$$

where γ is the Euler–Mascheroni constant:

$$\gamma = \lim_{n\to\infty} \left[\sum_{k=1}^{n} \frac{1}{k} - \log n\right], \quad \gamma \approx 0.577. \tag{A.29}$$

One of the best earlier results for $d = 2$ is due to G. Voronoi: let $\lambda_0 = \frac{1}{3}$, then

$$T_n = n \log n + (2\gamma - 1)n + o\left(n^{\lambda}\right) \tag{A.30}$$

for all $\lambda > \lambda_0$. The problem of finding the minimal λ_0 possessing this property is called the *Dirichlet divisor problem* in number theory. The precise value of this number λ_0 is not known yet. What is presently known is that

$$\lambda_0 \leq \frac{131}{416}. \tag{A.31}$$

A.10.2 The Case d > 2

One can prove for the general case of $d > 1$ that there exists a certain polynomial $\mathscr{P}(z) = a_0 + a_1 z + \cdots + a_{d-1} z^{d-1}$ of degree $d - 1$ such that

$$T_n = n\mathscr{P}(\log n) + o\left(n^{\lambda}\right) \tag{A.32}$$

for some $\lambda < 1$. The problem of finding the minimal λ_0 such that representation (A.32) holds for all $\lambda > \lambda_0$ is called the *generalized Dirichlet divisor problem*. The optimal λ_0 in the generalized Dirichlet divisor problem for dimension d is denoted by λ_d. It is known that

$$\lambda_3 \leq \frac{43}{96}, \quad \lambda_4 \leq \frac{1}{2}, \quad \lambda_5 \leq \frac{11}{20}. \tag{A.33}$$

The classical Landau bound holds for an arbitrary dimension:

$$\lambda_d \leq \frac{d-1}{d+1}, \quad d \geq 2. \tag{A.34}$$

Moreover, a lower bound is also known for the Dirichlet divisor problem:

$$\lambda_d \geq \frac{d-1}{2d}, \quad d \geq 4. \tag{A.35}$$

The Hardy–Titchmarsh conjecture asserts that

$$\lambda_d = \frac{d-1}{2d}, \quad d \geq 2. \tag{A.36}$$

The best of the presently known results in the generalized Dirichlet divisor problem (with respect to the order of growth with d) is

$$\lambda_d \leq 1 - \frac{c}{d^{2/3}} \tag{A.37}$$

for some constant c.

Equality (A.36) is equivalent to the well-known *Lindelöf conjecture* on the ζ function:

$$\zeta\left(\frac{1}{2} + it\right) = O(t^{\varepsilon}), \quad t \to \infty, \tag{A.38}$$

for an arbitrary $\varepsilon > 0$. Unfortunately it is not known whether relation (A.38) is true or not. However, it is known that (A.38) holds for

$$\varepsilon > \frac{32}{205}. \tag{A.39}$$

The Lindelöf conjecture is equivalent to the following result on zeros of the ζ function, which is also called Lindelöf's conjecture.

Lindelöf's conjecture *Let $\sigma \in (1/2, 1)$ be fixed. The number of solutions of the equation $\zeta(s) = 0$ (that is, the number of zeros of the ζ function) that belong to the set*

$$\mathfrak{Re}\,(s) > \sigma, \quad T < \mathfrak{Im}\,(s) < T + 1,$$

is $o(\log T)$.

Lindelöf's conjecture (A.38) is weaker than the Riemann hypothesis, in particular relation (A.36) follows from the Riemann hypothesis.

The Riemann hypothesis *The real part of any non-trivial zero of the ζ function is equal to $\frac{1}{2}$.*

The trivial zeros of the ζ function are $s = -2, -4, -6, \ldots$.

A.10.3 The Coefficients of Dirichlet's Polynomial

The coefficients of the polynomial $\mathscr{P}$ can explicitly be written in terms of some integral transforms of the Γ function. The evaluation is messy and the expressions are quite complicated. Only the leading coefficient a_{d-1} can be evaluated comparatively easily:

$$a_{d-1} = \frac{1}{(d-1)!}. \tag{A.40}$$

Some "rough" bounds for the generalized Dirichlet divisor problem can be obtained by elementary methods: namely, for $n \geq 1$,

$$C_1 n(\log n)^{d-1} \leq T_n \leq C_2 n(\log n)^{d-1} \tag{A.41}$$

for some constants $0 < C_1 < C_2$ that do not depend on n but do depend on d.

Several problems rely on the fact that T_n is a regularly varying function of index one. The corresponding representation for T_n involves a slowly varying function which is asymptotically equivalent to $(\log^+ x)^{d-1}$ as $x \to \infty$. This result easily follows from (A.32) and the fact itself seems to be trivial. However, this property can be proved independently of (A.32) and is much simpler. In order to prove it we say that a point of the space $\mathbf{R}^d$ is an *integer point* if all its coordinates are positive integers.

Now we introduce the *Dirichlet function* $D(x)$, $x \geq 1$, as the number of integer points in the domain $\{t_1 \geq 1, \ldots, t_d \geq 1 \colon t_1 \ldots t_d \leq x\}$. It is obvious that $D(n) = T_n$ for positive integer n.

Proposition A.18 *The function $D(x)$ is regularly varying with index one, that is*

$$D(x) = xL(x), \tag{A.42}$$

where $L(x)$ is a certain slowly varying function.

Below we provide an elementary proof of Proposition A.18 that is independent of representation (A.32). Moreover, we obtain (A.40) in the course of the proof.

Prior to proving (A.40)–(A.42) we obtain some elementary bounds for the harmonic series needed in the sequel. Let

$$H_n = \sum_{k=1}^{n} \frac{1}{k}.$$

The "rough" bounds for the harmonic series are straightforward:

$$\log(n+1) \le H_n \le 1 + \log n, \quad n \ge 2. \tag{A.43}$$

To check (A.43) one only needs to sum the inequalities

$$\frac{1}{n-1} \le \int_{n-1}^{n} \frac{1}{x}\, dx \le \frac{1}{n}, \quad n \ge 2.$$

More precise bounds for H_n can be obtained with the help of the Euler–Maclaurin formula.

Proposition A.19 (Euler–Maclaurin formula) *If a function $f(x)$ is continuous and has a continuous derivative $f'(x)$ in the interval $[\ell_1, \ell_2]$, where $\ell_1 < \ell_2$ are integers, then*

$$\sum_{\ell_1 \le i \le \ell_2} f(i) = \int_{\ell_1}^{\ell_2} f(x)\, dx + f(\ell_1) + \int_{\ell_1}^{\ell_2} \rho(x) f'(x)\, dx, \tag{A.44}$$

where $\rho(x) = [x] - x$.

The precise asymptotic behavior of the harmonic series, known already to L. Euler, follows from the next result with $\alpha = 0$.

Lemma A.1 *The limit*

$$\lim_{x\to\infty} \left(\sum_{i=1}^{[x]} \frac{1}{i} \log^{\alpha} i - \frac{1}{\alpha+1} (\log x)^{\alpha+1} \right)$$

exists for all $\alpha \ge 0$.

Proof of Lemma A.1 We apply the Euler–Maclaurin formula with $\ell_1 = 1$, $\ell_2 = [x]$ and $f(x) = (1/x)(\log x)^{\alpha}$. Then $f'(x) = x^{-2}\left(\alpha(\log x)^{\alpha-1} - (\log x)^{\alpha}\right)$ and thus

$$\sum_{i=1}^{[x]} \frac{(\log i)^{\alpha}}{i} = \int_{1}^{[x]} \frac{(\log t)^{\alpha}}{t}\, dt + \psi(1) - \int_{1}^{[x]} \rho(t) \frac{\alpha(\log t)^{\alpha-1} - (\log t)^{\alpha}}{t^2}\, dt, \tag{A.45}$$

where

$$\psi(x) = \begin{cases} 0, & \alpha > 0, \\ \frac{1}{x}, & \alpha = 0. \end{cases}$$

Therefore

$$\sum_{i=1}^{[x]} \frac{(\log i)^\alpha}{i} = \int_0^{\log[x]} u^\alpha \, du + \psi(1) + \int_1^\infty \rho(t) \frac{\alpha(\log t)^{\alpha-1} - (\log t)^\alpha}{t^2} \, dt + o(1),$$

since $|\rho(t) \leq \frac{1}{2}$ and the integral $\int_1^\infty \frac{(\log t)^\alpha - \alpha(\log t)^{\alpha-1}}{t^2} \, dt$ converges absolutely. The latter representation proves Lemma A.1. □

The reasoning given above can be extended to give the precise value of the limit and to estimate the rate of convergence. We restrict ourselves to the case of $\alpha = 0$. Then

$$\sum_{i=1}^{[x]} \frac{1}{i} = \log[x] + 1 - \int_1^\infty \frac{t - [t]}{t^2} \, dt + \int_{[x]}^\infty \frac{t - [t]}{t^2} \, dt.$$

We also use the following "rough" asymptotic behavior of the harmonic series:

$$\sum_{i=1}^{m} \frac{1}{i} = \log m + \gamma + O\left(\frac{1}{m}\right), \tag{A.46}$$

where the Euler–Mascheroni constant γ is defined by

$$\gamma = 1 - \int_1^\infty \frac{t - [t]}{t^2} \, dt.$$

Proof of formula (A.41) The function T_k can be estimated from the above by the number of integer points in the set

$$\left\{ \mathbf{n} \in \mathbf{N}^d : n_1 \leq k, n_2 \leq \left[\frac{k}{n_1}\right], \ldots, \right.$$

$$\left. n_{d-1} \leq \left[\frac{k}{n_1 \ldots n_{d-2}}\right], n_d \leq \left[\frac{k}{n_1 \ldots n_{d-1}}\right] \right\},$$

where the symbol $[x]$ stands for the integer part of a real number x. Thus

$$T_k \leq \sum_{n_1=1}^{k} \sum_{n_2=1}^{\left[\frac{k}{n_1}\right]} \cdots \sum_{n_{d-1}=1}^{\left[\frac{k}{n_1 \dots n_{d-2}}\right]} \sum_{n_d=1}^{\left[\frac{k}{n_1 \dots n_{d-1}}\right]} \mathbb{I}(\mathbf{n})$$

$$\leq \sum_{n_1=1}^{k} \sum_{n_2=1}^{\left[\frac{k}{n_1}\right]} \cdots \sum_{n_{d-1}=1}^{\left[\frac{k}{n_1 \dots n_{d-2}}\right]} \left[\frac{k}{n_1 n_2 \dots n_{d-1}}\right],$$

where $\mathbb{I}(\mathbf{n}) = 1$ for all $\mathbf{n} \in \mathbf{N}^d$. Hence

$$T_k \leq \sum_{n_1=1}^{k} \sum_{n_2=1}^{k} \cdots \sum_{n_{d-1}=1}^{k} \frac{k}{n_2 \dots n_d} \leq k \left(\sum_{i=1}^{k} \frac{1}{i} \right)^{d-1} \leq C_2 k (\log k)^{d-1}$$

in view of the upper bound (A.43).

To prove the lower bound in (A.41) we use induction. For $d = 2$ and $k > 1$, we have

$$T_k = \sum_{n_1=1}^{k} \sum_{n_2=1}^{\left[\frac{k}{n_1}\right]} \mathbb{I}(n_1, n_2) = \sum_{n_1=1}^{k} \left[\frac{k}{n_1}\right] \geq \sum_{n_1=1}^{k} \left(\frac{k}{n_1} - 1\right)$$
$$\geq \text{const}\, k \log k - k = O(k \log k)$$

in view of the first inequality in (A.43). Therefore the lower bound in (A.41) is proved for $d = 2$. Assume that the lower bound in (A.41) is true for dimensions less than or equal to $d - 1 \geq 1$; we will prove it for the dimension d.

Let $T_k^{(d-1)}$ be the number of integer solutions $\{n_1, \dots, n_{d-1}\}$ of the inequality $n_1 \dots n_{d-1} \leq k$ for a given $k \geq 1$. Then

$$T_k = T_k^{(d)} = \operatorname{card}\left\{ \mathbf{n}\colon n_1 \leq k, n_2 \dots n_d \leq \left[\frac{k}{n_1}\right] \right\},$$

$$T_k^{(d)} = \sum_{n_1=1}^{k} T^{(d-1)}_{\left[\frac{k}{n_1}\right]}.$$

By the inductive assumption,

$$T_k \geq O(1) \sum_{n_1=1}^{k} \left[\frac{k}{n_1}\right] \left(\log \left[\frac{k}{n_1}\right]\right)^{d-2} \geq O(1) \sum_{m=1}^{\left[\frac{k}{2}\right]} \frac{k}{2m} \left(\log \frac{k}{2m}\right)^{d-2}$$
$$= O(k) \sum_{m=1}^{\left[\frac{k}{2}\right]} \frac{1}{m} (\log k - \log 2m)^{d-2}.$$

The sum on the right-hand side is $O\left(k\,(\log k)^{d-1}\right)$ in view of Lemma A.1 with $\alpha = 0, 1, \ldots, d-2$. This proves the lower bound in (A.41) for dimension d. □

Proof of formula (A.40) We need the following auxiliary result.

Lemma A.2 *Let* $C_m^n = \frac{m!}{n!(m-n)!}$. *Then*

$$\sum_{l=0}^{d-2} C_{d-2}^l (-1)^{d-2-l} \frac{1}{d-1-l} = \frac{1}{d-1}.$$

Proof of Lemma A.2 It is clear that

$$\begin{aligned}
\sum_{l=0}^{d-2} C_{d-2}^l (-1)^{d-2-l} \frac{1}{d-1-l} &= \sum_{k=1}^{d-1} C_{d-2}^{d-1-k} (-1)^{k-1} \frac{1}{k} \\
&= \frac{1}{d-1} \sum_{k=1}^{d-1} C_{d-1}^k (-1)^{k-1} \\
&= \frac{1}{d-1} \left(1 - \sum_{k=0}^{d-1} C_{d-1}^k (-1)^k\right) \\
&= \frac{1}{d-1} \left(1 - (1-1)^{d-1}\right) = \frac{1}{d-1}.
\end{aligned}$$

Lemma A.2 is proved. □

We return to the proof of (A.40). Let $d > 1$. Put $T^{(d)}(x) = \operatorname{card}\{\mathbf{n}\colon |\mathbf{n}| \le x\}$ for $x > 0$. If $|\mathbf{n}|_{2,\ldots,d} = n_2 \ldots n_d$, then

$$T^{(d)}(x) = \operatorname{card}\left\{\mathbf{n} \in \mathbf{N}^d\colon n_1 \le x,\, |\mathbf{n}|_{2,\ldots,d} \le \frac{x}{n_1}\right\}.$$

The function $T^{(1)}(x)$ looks simpler if $d = 1$, namely $T^{(1)}(x) = [x]$. Hence

$$T^{(d)}(x) = \sum_{n_1=1}^{[x]} T^{(d-1)}(x/n_1). \tag{A.47}$$

The latter result is used to prove the representation

$$T^{(d)}(x) = \frac{1}{(d-1)!} x (\log x)^{d-1} + O\left(x (\log x)^{d-2}\right), \quad x \to \infty. \tag{A.48}$$

Here and in what follows we write $\log^k x$ instead of $(\log x)^k$.

To prove representation (A.48) we use induction. First, if $d = 1$, then

$$T^{(1)}(x) = \sum_{k_1=1}^{[x]} \mathbb{I}_{k_1} = [x] = x + O(1) = x + O\left(\frac{x}{\log x}\right), \quad x \to \infty.$$

Next, Lemma A.1 with $\alpha = 0$ implies for $d = 2$ that

$$T^{(2)}(x) = \sum_{n_1=1}^{[x]} T^{(1)}\left(\frac{x}{n_1}\right) = \sum_{n_1=1}^{[x]} \left(\frac{x}{n_1} + O(1)\right) = x \log x + O(x), \quad x \to \infty.$$

Assume that representation (A.48) is true for dimensions less than d; our aim is to prove (A.48) for dimension d. By (A.47) and by the inductive assumption,

$$T^{(d)}(x) = \frac{1}{(d-2)!} \sum_{i=1}^{[x]} \frac{x}{i} \left(\log \frac{x}{i}\right)^{d-2} + \sum_{i=1}^{[x]} O\left(\frac{x}{i}\left(\log \frac{x}{i}\right)^{d-3}\right), \quad x \to \infty.$$

The second sum is estimated with the help of Lemma A.1 with $\alpha = 0$:

$$\sum_{i=1}^{[x]} O\left(\frac{x}{i}\left(\log \frac{x}{i}\right)^{d-3}\right) = O\left(x(\log x)^{d-3}\right) \sum_{i=1}^{[x]} \frac{1}{i} = O\left(x(\log x)^{d-2}\right), \quad x \to \infty.$$

To complete the proof, one needs to show that

$$\sum_{i=1}^{[x]} \frac{1}{i} \left(\log \frac{x}{i}\right)^{d-2} = \frac{1}{d-1} (\log x)^{d-1} + O\left((\log x)^{d-2}\right), \quad x \to \infty. \tag{A.49}$$

Indeed, by the binomial theorem

$$\begin{aligned} \sum_{i=1}^{[x]} \frac{1}{i} \left(\log \frac{x}{i}\right)^{d-2} &= \sum_{i=1}^{[x]} \frac{1}{i} (\log x - \log i)^{d-2} \\ &= \sum_{l=0}^{d-2} C_{d-2}^{l} (-1)^{d-2-l} (\log x)^{l} \sum_{i=1}^{[x]} \frac{1}{i} (\log)^{d-2-l}. \end{aligned}$$

For the sum with respect to i we use Lemma A.1 for $\alpha = d - 2 - l$:

$$\sum_{i=1}^{[x]} \frac{1}{i} (\log i)^{d-2-l} = \frac{1}{d-1-l} (\log x)^{d-1-l} + O(1), \quad x \to \infty.$$

Thus, as $x \to \infty$,

$$\sum_{i=1}^{[x]} \frac{1}{i}\left(\log \frac{x}{i}\right)^{d-2} = (\log x)^{d-1} \sum_{l=0}^{d-2} C_{d-2}^{l} \frac{(-1)^{d-2-l}}{d-1-l} + O\left((\log x)^{d-2}\right).$$

Lemma A.2 implies equalities (A.49) and (A.48). Now relation (A.40) follows from (A.48). □

Proof of formula (A.42) Note that $L(x) \geq \frac{1}{2}$ for all $x \geq 1$. Indeed, if $1 \leq x < 2$, then $D(x) = 1 \geq \frac{1}{2}x$. Otherwise, if $x \geq 2$, then $D(x) = [x] \geq \frac{1}{2}x$.

First we treat the case of $d = 2$. In this case,

$$D(x) = \sum_{i=1}^{[x]} \left[\frac{x}{i}\right] = \sum_{i \leq \frac{x}{2}} \left[\frac{x}{i}\right] + \sum_{\frac{x}{2} < i \leq x} \left[\frac{x}{i}\right] \stackrel{\text{def}}{=} D_1(x) + D_2(x). \tag{A.50}$$

For the first term in (A.50), we have

$$D_1(x) = \sum_{i \leq \frac{x}{2}} \frac{x}{i} \eta_i(x), \quad \eta_i(x) = \left[\frac{x}{i}\right] \frac{i}{x}.$$

It is clear that $\frac{1}{2} \leq \eta_i(x) \leq 1$. Further,

$$D_1(x) = x \sum_{i \leq \frac{x}{2}} \frac{1}{i} \eta_i(x) = x \log x \cdot G(x),$$

$$G(x) = \frac{1}{\log x} \sum_{i \leq \frac{x}{2}} \frac{1}{i} \eta_i(x).$$

Now we show that $G(x)$ approaches 1 as $x \to \infty$. It is clear that

$$G(x) = \frac{1}{x \log x} \sum_{i \leq \frac{x}{2}} \left[\frac{x}{i}\right].$$

Using the asymptotic behavior of the harmonic series (A.46) we conclude that

$$G(x) \leq \frac{1}{x \log x} \sum_{i \leq \frac{x}{2}} \frac{x}{i} = \frac{1}{\log x}\left(\log \frac{x}{2} + \gamma + O\left(\frac{1}{x}\right)\right), \tag{A.51}$$

$$G(x) \geq \frac{1}{x \log x} \sum_{i \leq \frac{x}{2}} \left(\frac{x}{i} - 1\right) = \frac{1}{\log x}\left(\log \frac{x}{2} + \gamma + O\left(\frac{1}{x}\right)\right) - \frac{1}{2 \log x}.$$

The latter two inequalities prove that $\lim_{x\to\infty} G(x) = 1$. Therefore

$$D_1(x) = G(x)x \log x$$

and, moreover, the limit of $G(x)$ exists as $x \to \infty$. This means that $D_1(x)$ is a regularly varying function of index one.

For the term $D_2(x)$, we have

$$D_2(x) = \sum_{\frac{x}{2}<i\le x} \left[\frac{x}{i}\right] = [x] - \left[\frac{x}{2}\right] = x \cdot \frac{[x]-[x/2]}{x},$$

that is, $D_2(x)$ also is a regularly varying function of index one. Finally, $D(x) = D_1(x) + D_2(x)$ is a regularly varying function of index one, since the sum of two slowly varying functions is also a slowly varying function.

Our next aim is to prove (A.42) in the general case of $d > 2$. For this, we again use induction. To avoid a misleading notation, we introduce the dimension of the space as a superscript for the Dirichlet function (A.42). More precisely, we write $D(x) = D^{(d)}(x)$. Similarly, we write $L^{(d)}(x)$ instead of $L(x)$. Then

$$D^{(d)}(x) = \sum_{i=1}^{x} D^{(d-1)}\left(\frac{x}{i}\right).$$

Assume that (A.42) holds for dimension $d-1$ and $L^{(d-1)}$ is such that $L^{(d-1)}(x) \ge \frac{1}{2}$ for all $x \ge 1$. We are going to prove (A.42) for dimension d. Indeed,

$$D^{(d)}(x) = x \sum_{i=1}^{x} \frac{1}{i} L^{(d-1)}\left(\frac{x}{i}\right).$$

Proposition A.17 implies that the function

$$L^{(d)}(x) \stackrel{\text{def}}{=} \sum_{i=1}^{x} \frac{1}{i} L^{(d-1)}\left(\frac{x}{i}\right)$$

is slowly varying and such that $L^{(d)}(x) \ge \frac{1}{2}$ for all $x \ge 1$. □

A.11 Some Applications of Dirichlet's Divisor Function

The asymptotic behavior of the Dirichlet function is used when evaluating multi-indexed series whose terms depend on $|\mathbf{n}|$. For example, we consider the truncated versions $Y(\mathbf{n}) = X(\mathbf{n}) \mathbb{I}(|X(\mathbf{n})| \le |\mathbf{n}|)$ of random variables $X(\mathbf{n})$ in the proof of the strong law of large numbers for identically distributed random variables in order to deal with random variables possessing second moments. This allows us to prove first the strong law of large numbers for truncated versions $Y(\mathbf{n})$. Then we use the Borel–Cantelli lemma to prove that $\mathsf{P}(X(\mathbf{n}) \neq Y(\mathbf{n}) \text{ i. o.}) = 0$ and finally conclude that the strong law of large numbers also holds for random variables $X(\mathbf{n})$ (see Chap. 9). The second step (application of the Borel–Cantelli lemma) reduces to the proof of the convergence of the series

$$\sum_{\mathbf{n}\in\mathbf{N}^d} \mathsf{P}(|X(\mathbf{n})| \ge |\mathbf{n}|) \tag{A.52}$$

by using the upper bound in the Dirichlet divisor problem (A.41). Namely, denoting by X a copy of the random variables $X(\mathbf{n})$ we get

$$\begin{aligned}\sum_{\mathbf{n}\in\mathbf{N}^d} \mathsf{P}(|X(\mathbf{n})| \ge |\mathbf{n}|) &= \sum_{k=1}^{\infty} \tau_k \mathsf{P}(|X| \ge k)\\ &= \sum_{k=1}^{\infty} T_k \mathsf{P}(k \le |X| < k+1)\\ &\le \text{const} \sum_{k=1}^{\infty} k \left(\log^+ k\right)^{d-1} \mathsf{P}(k \le |X| < k+1)\\ &\le \text{const}\, \mathsf{E}\left[|X| \left(\log^+ |X|\right)^{d-1}\right].\end{aligned}$$

Using the lower bound instead of the upper bound in (A.41), we similarly prove the converse inequalities. This reasoning yields the following result.

Lemma A.3 *Let* $\{X(\mathbf{n}), \mathbf{n} \in \mathbf{N}^d\}$ *be independent identically distributed random variables. The convergence of series* (A.52) *is equivalent to the existence of the following moment*

$$\mathsf{E}\left[|X| \left(\log^+ |X|\right)^{d-1}\right].$$

A.11.1 Feller's Condition

The Dirichlet function T_n admits the following bound

$$\sum_{k=n}^{\infty} \frac{T_k}{k^3} = O\left(\frac{T_n}{n^2}\right) \tag{A.53}$$

which is often used in the proof of limit theorems for random fields. The proof of relation (A.53) is based on representation (A.42) and Proposition A.14. We call (A.53) the *Feller type condition* and discuss it in more detail in Chap. 9.

Remark A.4 Condition (A.53) can conveniently be rewritten in the following equivalent form

$$\sum_{k=n}^{\infty} \frac{\tau_k}{k^2} = O(1) \frac{\left(\log^+ n\right)^{d-1}}{n}. \tag{A.54}$$

Indeed,

$$\sum_{k=n}^{\infty} \frac{\tau_k}{k^2} = -\frac{T_{n-1}}{n^2} + \sum_{k=n}^{\infty} T_k \left(\frac{1}{k}^2 - \frac{1}{(k+1)^2}\right).$$

This implies that (A.53) and (A.54) are equivalent in view of the rough bounds (A.41) for the Dirichlet function.

A.12 A Bound for the Number of Integer Points in Certain Domains

We write $a(\mathbf{n}) \asymp b(\mathbf{n})$ if there are two constants $0 < c_1 \le c_2 < \infty$ such that

$$c_1 b(\mathbf{n}) \le a(\mathbf{n}) \le c_2 b(\mathbf{n}).$$

The symbol $\{\mathbb{I}(\mathbf{n})\}$ denotes the field such that $\mathbb{I}(n) = 1$ for all $\mathbf{n} \in \mathbf{N}^d$.

Proposition A.20 *Let* $t_1, \ldots, t_d > 0$;

$$t = \min(t_1, \ldots, t_d); \quad c = \operatorname{card}(i : t_i = t).$$

Put $A_k = \operatorname{card}(\mathbf{n} \colon n_1^{t_1} \cdots n_d^{t_d} \le k)$. *Then*

$$A_k \asymp k^{1/t} (\log k)^{c-1}. \tag{A.55}$$

Proof of Proposition A.20. If $c = d$, then $t_1 = \cdots = t_d$ and $A_k = T_{[k^{1/t}]}$. Using the lower and upper bounds (A.41) in the Dirichlet divisor problems we prove (A.55).

Now let $c < d$ (this automatically implies that $d \ge 2$). Without loss of generality one can assume that $t = t_1 = \cdots = t_c$. First we prove an upper bound for (A.55). Put

$$\begin{aligned} s &= \min\{t_{c+1}, \ldots, t_d\}, \\ A'_x &= \operatorname{card}(\mathbf{n} \colon (n_1 \cdots n_c)^t (n_{c+1} \cdots n_d)^s \le x). \end{aligned} \tag{A.56}$$

Also let $|\mathbf{n}|_{1,\ldots,c} = n_1 \ldots n_c$, $|\mathbf{n}|_{c+1,\ldots,d} = n_{c+1} \ldots n_d$. It is clear that $A_k \le A'_k$ and

$$A'_k = \sum{}' \sum{}'' \mathbb{I}(\mathbf{n}), \tag{A.57}$$

where $\sum'$ denotes the sum with respect to indices $n_{c+1}, \ldots, n_d$ such that $|\mathbf{n}|^s_{1,\ldots,c} \le k$ and the symbol $\sum''$ denotes the sum with respect to the indices $n_1, \ldots, n_c$ that satisfy the condition $|\mathbf{n}|^t_{1,\ldots,c} \le k/|\mathbf{n}|^s_{c+1,\ldots,d}$. Using the upper bound in (A.41) for dimension c we have

$$\begin{aligned} A'_k &\le \text{const} \sum{}' \left(\frac{k}{|\mathbf{n}|^s_{c+1,\ldots,d}}\right)^{1/t} \left(\log \frac{k}{|\mathbf{n}|^s_{c+1,\ldots,d}}\right)^{c-1} \\ &\le \text{const}\, k^{1/t} (\log k)^{c-1} \sum_{n_{c+1}\ge 1} \cdots \sum_{n_d \ge 1} \frac{1}{(n_{c+1} \ldots n_d)^{s/t}}. \end{aligned}$$

This result proves the inequality needed, since the latter multi-indexed sums converges for $s > t$.

To prove the lower bound we follow the same approach for $s = \max\{t_{c+1}, \ldots, t_d\}$. Let A'_k be defined according to (A.56) and note that $A'_k \le A_k$. Applying representation (A.57) and the lower bound in (A.41) for dimension c we obtain

$$A'_k \ge \text{const} \sum{}' \left(\frac{k}{\mathbf{n}|^s_{c+1,\ldots,d}}\right)^{1/t} \left(\log \frac{k}{\mathbf{n}|^s_{c+1,\ldots,d}}\right)^{c-1} \ge \text{const}\, k^{1/t} (\log k)^{c-1}.$$

Here we took into account only one term with $n_{c+1} = \cdots = n_d = 1$ in the sum $\sum'$. The proof of Proposition A.20 is complete. □

Proposition A.21 *Let* $\theta \in \mathbf{R}$, $t_1, \ldots, t_d > 0$;

$$t = \min(t_1, \ldots, t_d); \quad c = \operatorname{card}(i\colon t_i = t).$$

Put $A_k = \operatorname{card}(\mathbf{n}\colon n_1^{t_1} \cdots n_d^{t_d} (\log |\mathbf{n}|)^{t\theta} \le k)$. *Then*

$$A_k \asymp k^{1/t} (\log k)^{c-1-\theta}. \tag{A.58}$$

Proof of Proposition A.21. Let $R(x) = x(\log x)^{t\theta}$ and denote by $R^{-1}(\cdot)$ its inverse function. Both functions R and R^{-1} are increasing in the interval $(e^{t|\theta|}, \infty)$. Put

$$A'(x) = \left\{\mathbf{n} \in \mathbf{N}^d\colon R\left(n_1^{t_1} \ldots n_d^{t_d}\right) \le x\right\}.$$

If $s = \max\{t_1, \ldots, t_d\}$, then

$$A'\left(\frac{k}{s^{t\theta}}\right) \le A_k \le A'\left(\frac{k}{t^{t\theta}}\right).$$

This means that we only need to show that $A'(x) \asymp x^{1/t} (\log x)^{c-1-\theta}$. Note further that $A'(x) \asymp A''(x)$ for all $a \ge 1$, where $A''(x) = \{\mathbf{n} \in \mathbf{N}^d\colon n_1^{t_1} \ldots n_d^{t_d} \le x,$

$n_d^{t_d} \geq a\}$. The functions R and R^{-1} are monotone in the interval $(e^{t|\theta|}, \infty)$, whence we conclude that, for $x \geq e^{t|\theta|}$,

$$\begin{aligned} A''(x) &= \left\{\mathbf{n} \in \mathbf{N}^d : n_1^{t_1} \ldots n_d^{t_d} \leq R^{-1}(x), n_d^{t_d} \geq e^{t|\theta|}\right\} \\ &\asymp \left\{\mathbf{n} \in \mathbf{N}^d : n_1^{t_1} \ldots n_d^{t_d} \leq R^{-1}(x)\right\}. \end{aligned}$$

Proposition A.20 implies

$$A'(x) \asymp \left(R^{-1}(x)\right)^{1/t} \left(\log R^{-1}(x)\right)^{c-1}.$$

Moreover,

$$R^{-1}(x) \asymp \frac{x}{(\log x)^{t\theta}}.$$

Combining the latter two results we complete the proof of Proposition A.21. □

Proposition A.22 *Let* $t_1, \ldots, t_d > 0$. *Put*

$$A_k = \operatorname{card}(\mathbf{n} : n_1^{t_1} + \cdots + n_d^{t_d} \leq k).$$

Then

$$A_k \asymp k^{\frac{1}{t_1} + \cdots + \frac{1}{t_d}}. \tag{A.59}$$

Proof of Proposition A.22. Relation (A.59) is obvious for $d = 1$. Assume that relation (A.59) holds for all dimensions less than d; now our goal is to prove relation (A.59) for dimension d. It is clear that

$$A_k = \sum_{n_d \leq k^{1/t_d}} \sum_{\substack{n_1, \ldots, n_{d-1}: \\ n_1^{t_1} + \cdots + n_{d-1}^{t_{d-1}} \leq k - n_d^{t_d}}} \mathbb{I}(n_1, \ldots, n_{d-1}, n_d). \tag{A.60}$$

By the inductive assumption,

$$A_k \asymp \sum_{i : i \leq k^{1/t_d}} (k - i^{t_d})^{\frac{1}{t_1} + \cdots + \frac{1}{t_{d-1}}}.$$

To prove the lower bound, we note that

$$A_k \geq \text{const} \sum_{i \leq (k/2)^{1/t_d}} (k - i^{t_d})^{\frac{1}{t_1} + \cdots + \frac{1}{t_{d-1}}}$$

$$\geq \text{const} \left(\frac{k}{2}\right)^{\frac{1}{t_1}+\cdots+\frac{1}{t_{d-1}}} \sum_{i \leq (k/2)^{1/t_d}} \mathbb{I}_i \geq \text{const}\, k^{\frac{1}{t_1}+\cdots+\frac{1}{t_d}}.$$

For the upper bound, we put

$$a = t_d, \qquad b = \frac{1}{t_1} + \cdots + \frac{1}{t_{d-1}},$$
$$\psi_j = \{i : j-1 < i^a \leq j\}, \qquad j \geq 1.$$

It is clear that $\psi_1 + \cdots + \psi_k \asymp k^{1/a}$. Hence (A.60) and the inductive assumption imply that

$$A_k \leq \text{const} \sum_{i \leq k^{1/a}} \left(k - i^a\right)^b \leq \text{const} \sum_{1 \leq j \leq k} \psi_j (k-j+1)^b \leq \text{const}\, k^b \sum_{1 \leq j \leq k} \psi_j$$
$$\leq \text{const}\, k^{b+\frac{1}{a}}.$$

This completes the proof. □

A.13 A criterion for the Existence of a Moment of a Random Variable

Let X be a random variable. The series

$$\sum_{n=1}^{\infty} w_n \mathsf{P}(|X| \geq \varphi_n)$$

appears quite often in the proofs of limit theorems of probability theory for various sequences $\{w_n\}$ and $\{\varphi_n\}$. The simplest case is represented by $w_n = n^\nu$ and $\varphi_n = n$:

$$\sum_{n=1}^{\infty} n^\nu \mathsf{P}(|X| \geq \varepsilon n). \tag{A.61}$$

One of the applications of series (A.61) is that its convergence indicates the existence of a certain moment of the random variable X.

Proposition A.23 *Let* $\nu > -1$.

(i) *If series* (A.61) *converges for some* $\varepsilon > 0$, *then*

$$\mathsf{E}\left[|X|^{\nu+1}\right] < \infty. \tag{A.62}$$

(ii) *Conversely, if condition* (A.62) *holds, then series* (A.61) *converges for all* $\varepsilon > 0$.

Below are some particular cases of Proposition A.23:

$$\mathsf{E}\left[|X|\right] < \infty \iff \sum_{n=1}^{\infty} \mathsf{P}(|X| \geq n) < \infty, \tag{A.63}$$

$$\mathsf{E}\left[X^2\right] < \infty \iff \sum_{n=1}^{\infty} n\mathsf{P}(|X| \geq n) < \infty. \tag{A.64}$$

The case of $\nu = -1$ differs from all others.

Proposition A.24 (i) *If the series*

$$\sum_{n=1}^{\infty} \frac{1}{n} \mathsf{P}(|X| \geq \varepsilon n) \tag{A.65}$$

converges for some $\varepsilon > 0$, *then*

$$\mathsf{E}\left[\log^{+} |X|\right] < \infty. \tag{A.66}$$

(ii) *Conversely, if condition* (A.66) *holds, then series* (A.65) *converges for all* $\varepsilon > 0$.

Proposition A.23 is easy to extend to the case of other sequences $\{\varphi_n\}$.

Proposition A.25 *Let* $\nu_1 > -1$, $\nu_2 > 0$.

(i) *If the series*

$$\sum_{n=1}^{\infty} n^{\nu_1} \mathsf{P}(|X| \geq \varepsilon n^{\nu_2}) \tag{A.67}$$

converges for some $\varepsilon > 0$, *then*

$$\mathsf{E}\left[|X|^{(\nu_1+1)/\nu_2}\right] < \infty. \tag{A.68}$$

(ii) *Conversely, if condition* (A.68) *holds, then series* (A.67) *converges for all* $\varepsilon > 0$.

Proof of Proposition A.23. Put $Y = X/\varepsilon$ and rewrite series (A.61) as follows

$$\sum_{k=1}^{\infty} \mathsf{P}(k \leq |Y| < k+1) \sum_{n=1}^{k} n^{\nu}.$$

Using the asymptotics

$$\sum_{n=1}^{k} n^{\nu} \sim \frac{k^{\nu+1}}{\nu+1}, \quad k \to \infty, \tag{A.69}$$

(see Proposition A.14) we prove that the convergence of the latter series is equivalent to the convergence of the series

$$\sum_{k=1}^{\infty} k^{\nu+1} \mathsf{P}(k \le |Y| < k+1).$$

In turn, the latter series converges if and only if condition (A.62) holds. Note that the asymptotics (A.69) can also be obtained from the Lagrange mean value theorem

$$(n+1)^{\nu+1} - n^{\nu+1} = \frac{\xi_n^{\nu}}{\nu+1}, \quad n \le \xi_n \le n+1. \qquad \square$$

Consider a more general result than Proposition A.23.

Proposition A.26 *Let g be a regularly varying function of index ρ. Let $\{\tau_k\}$ be the Dirichlet divisor sequence for the space $\mathbf{N}^d$ (see* Sect. A.10*). The series*

$$\sum_{k=1}^{\infty} \tau_k g(k) \mathsf{P}(|X| \ge k) \tag{A.70}$$

converges if and only if

$$\mathsf{E}\,[G(|X|)] < \infty, \tag{A.71}$$

where

$$G(x) = \sum_{n \le x} g(n) \left(\log^+ n\right)^{d-1}. \tag{A.72}$$

Proof It is clear that series (A.70) converges and condition (A.71) holds if $\rho < -1$. In what follows we consider the case of $\rho \ge -1$. Note also that G is a regularly varying function.

Given a positive integer $N \ge 1$ choose n such that $2^{n-1} < N \le 2^n$. Without loss of generality we assume that $g(1) = 0$. Since g is a regularly varying function,

$$\sum_{k=1}^{N} \tau_k g(k) \le \sum_{k=1}^{2^n} \tau_k g(k) = \sum_{i=1}^{n} \sum_{k=2^{i-1}+1}^{2^i} \tau_k g(k) \le \text{const} \sum_{i=1}^{n} g(2^i) T_{2^i}.$$

Applying the rough asymptotics (A.41) in the Dirichlet problem, we obtain

$$\sum_{k=1}^{N} \tau_k g(k) \le \text{const} \sum_{i=1}^{n} g(2^i) 2^i \left(\log^+ 2^i\right)^{d-1} \le \text{const} \sum_{i=1}^{n} g(2^i) \sum_{k=2^{i-1}+1}^{2^i} \left(\log^+ k\right)^{d-1}.$$

Since g is a regularly varying function,

$$\sum_{k=1}^{N} \tau_k g(k) \le \text{const} \sum_{i=1}^{n} g(2^i)2^i \left(\log^+ 2^i\right)^{d-1} \le \text{const} \sum_{i=1}^{2^n} g(k) \left(\log^+ k\right)^{d-1}$$
$$= \text{const}\, G(2N) \le \text{const}\, G(N).$$

The above argument can be applied in reverse order (for other constants, however). Therefore

$$\sum_{k=1}^{N} \tau_k g(k) \asymp G(N).$$

The proof of Proposition A.26 is completed after the following transformation of series (A.70):

$$\sum_{k=1}^{\infty} \tau_k g(k) \mathsf{P}(|X| \ge k) = \sum_{n=1}^{\infty} \mathsf{P}(n \le |X| < n+1) \sum_{k=1}^{n} \tau_k g(k)$$

by recalling again that G is regularly varying. □

Remark A.5 It is shown in the proof of Proposition A.26 that

$$\sum_{k=1}^{n} \tau_k g(k) \asymp \sum_{k=1}^{n} g(k) \left(\log^+ k\right)^{d-1}$$

if a function g is regularly varying. The asymptotic behavior of the function G can be described in other terms if $\rho > -1$:

$$G(x) \asymp x g(x) \left(\log^+ x\right)^{d-1} \asymp g(x) T(x),$$

where T is the Dirichlet function for the space $\mathbf{N}^d$.

A.14 An Auxiliary Proposition for Chung's Theorem

Proposition A.27 *Let X be a random variable with the distribution function $F(x)$, b a non-random non-negative constant, $g(x)$ a non-decreasing positive even function. Assume that $\mathsf{E}\,[g(X)] < \infty$ and that at least one of the following conditions holds:*

(i) *$x/g(x)$ is non-decreasing for $x > 0$;*
(ii) *$x/g(x)$ and $g(x)/x^2$ are non-increasing for $x > 0$ and $\mathsf{E}\,[X] = 0$;*
(iii) *$g(x)/x^2$ is non-increasing for $x > 0$ and X is a symmetric random variable.*

Then

$$\mathsf{P}(|X| \geq b) \leq \frac{\mathsf{E}\,[g(X)]}{g(b)}, \tag{A.73}$$

$$\left| \int\limits_{|x|<b} x\,dF(x) \right| \leq \frac{b}{g(b)} \mathsf{E}\,[g(X)], \tag{A.74}$$

$$\int\limits_{|x|<b} x^2\,dF(x) \leq \frac{b^2}{g(b)} \mathsf{E}\,[g(X)]. \tag{A.75}$$

Proof of Proposition A.27. First of all, note that condition (A.73) is a direct consequence of the Chebyshev–Markov inequality for the non-decreasing function $g(x)$.

If the function $g(x)$ satisfies condition (i), then

$$\frac{x^2}{b^2} \leq \frac{g^2(x)}{g^2(b)} \leq \frac{g(x)}{g(b)}$$

in the domain $|x| < b$, whence

$$\frac{x^2}{g(x)} \leq \frac{b^2}{g(b)}.$$

The latter inequality is obvious in the same domain if $g(x)$ satisfies either condition (ii) or (iii). Thus

$$\int\limits_{|x|<b} x^2\,dF(x) \leq \frac{b^2}{g(b)} \int\limits_{|x|<b} g(x)\,dF(x) \leq \frac{b^2}{g(b)} \mathsf{E}\,[g(X)]$$

and this proves (A.75).

Further, if condition (i) holds, then

$$\left| \int\limits_{|x|<b} x\,dF(x) \right| \leq \frac{b}{g(b)} \mathsf{E}\,[g(X)].$$

In the case of condition (ii), we have

$$\begin{aligned} \left| \int\limits_{|x|<b} x\,dF(x) \right| &= \left| \int\limits_{|x|\geq b} x\,dF(x) \right| \\ &\leq \frac{b}{g(b)} \int_{|x|\geq b} g(x)\,dF(x) \leq \frac{b}{g(b)} \mathsf{E}\,[g(X)]. \end{aligned}$$

Finally, $\int_{|x|<b} x\,dF(x) = 0$ in the case (iii), since X is a symmetric random variable and thus inequality (A.74) is obvious.

The proof of Proposition A.27 is complete. □

The simplest case of Proposition A.27 corresponds to the function $g(x) = |x|^r$.

Corollary A.3 *Let X be a random variable with the distribution function $F(x)$, $0 < r \le 2$. Assume that $\mathsf{E}\left[|X|^r\right] < \infty$ and $\mathsf{E}[X] = 0$ if $1 \le r \le 2$. Then, for an arbitrary non-random non-negative constant b,*

$$\mathsf{P}(|X| \ge b) \le \frac{\mathsf{E}\left[|X|^r\right]}{b^r},$$

$$\left|\int_{|x|<b} x\,dF(x)\right| \le \frac{\mathsf{E}\left[|X|^r\right]}{b^{r-1}},$$

$$\int_{|x|<b} x^2\,dF(x) \le \frac{\mathsf{E}\left[|X|^r\right]}{b^{r-2}}.$$

A.15 The Strong Law of Large Numbers for a Subsequence

Proposition A.28 *Let $\{\xi_i, i \ge 1\}$ be independent identically distributed random variables, $\zeta_j = \xi_1 + \cdots + \xi_j$, $j \ge 2$; $\zeta_1 = \xi_1$. If*

$$\lim_{k\to\infty} \frac{\zeta_{k^2}}{k^2} = 0 \quad \text{a. s.},$$

then the expectation $\mathsf{E}[\xi_1]$ exists and moreover $\mathsf{E}[\xi_1] = 0$.

Note that Kolmogorov's strong law of large numbers for independent identically distributed random variables (see Theorem 9.2) asserts that if $\zeta_k/k \to 0$ almost surely, then the expectation $\mathsf{E}[\xi_1]$ exists and moreover $\mathsf{E}[\xi_1] = 0$. Proposition A.28 allows one to make the same conclusion under a weaker assumption.

Proof The assumption of the theorem implies that

$$\frac{\zeta_{(k+1)^2} - \zeta_{k^2}}{k^2} = \frac{\zeta_{(k+1)^2}}{(k+1)^2} \cdot \frac{(k+1)^2}{k^2} - \frac{\zeta_{k^2}}{k^2} \to 0 \quad \text{a. s.}$$

Since random variables $\zeta_{(k+1)^2} - \zeta_{k^2}$ are independent, the Borel–Cantelli lemma implies that

$$\sum_{k=1}^{\infty} \mathsf{P}(|\zeta_{(k+1)^2} - \zeta_{k^2}| \ge \varepsilon k^2) < \infty$$

for all $\varepsilon > 0$. Further, the random variables $\zeta_{(k+1)^2} - \zeta_{k^2}$ have the same distributions as ζ_{2k+1}. Thus, for all $\varepsilon > 0$,

$$\sum_{k=1}^{\infty} \mathsf{P}(|\zeta_{2k+1}| \geq \varepsilon k^2) < \infty, \tag{A.76}$$

whence we get $\zeta_{2k+1}/k^2 \to 0$ almost surely and thus $(\xi_{2k} + \xi_{2k+1})/k^2 \to 0$ almost surely. By the Borel–Cantelli lemma, for all $\varepsilon > 0$,

$$\sum_{k=1}^{\infty} \mathsf{P}(|\xi_{2k} + \xi_{2k+1}| \geq \varepsilon k^2) < \infty.$$

Since $\{\omega : |\xi_{2k}(\omega) + \xi_{2k+1}(\omega)| \geq \varepsilon k^2\} \supseteq \{\omega : |\xi_{2k+1}(\omega)| \geq 2\varepsilon k^2, |\xi_{2k}(\omega)| \leq \varepsilon k^2\}$, we derive from the independence and identical distribution that, for sufficiently large (thus, for all) $\delta > 0$,

$$\sum_{k=1}^{\infty} \mathsf{P}(|\xi_{2k+1}| \geq \delta k^2) < \infty.$$

This relation together with (A.76) proves that, for all $\varepsilon > 0$,

$$\sum_{k=1}^{\infty} \mathsf{P}(|\zeta_{2k}| \geq \varepsilon k^2) < \infty.$$

Combining this result with (A.76) we get, for all $\varepsilon > 0$,

$$\sum_{k=1}^{\infty} \mathsf{P}(|\zeta_{k}| \geq \varepsilon k^2) < \infty.$$

To complete the proof we use Proposition A.29 with $\alpha = 2$ and $t = 1$. □

Proposition A.29 (L. Baum and M. Katz) *Let $\{\xi_k\}$ be independent identically distributed random variables and let $\{\zeta_k\}$ be their partial sums. Let $\alpha t > 1$ and $\alpha > \frac{1}{2}$. The convergence for all $\varepsilon > 0$ of the series*

$$\sum_{k=1}^{\infty} k^{\alpha t - 2} \mathsf{P}(|\zeta_k| \geq \varepsilon k^{\alpha})$$

is equivalent to $\mathsf{E}\left[|\xi_1|^t\right] < \infty$ and $\mathsf{E}[\xi_1] = 0$ if $t \geq 1$.

A.16 Some Classical Results of Measure Theory

Consider a field of random events $\{E(\mathbf{n}), \mathbf{n} \in \mathbf{N}^d\}$. We use the notation $\{E(\mathbf{n}) \text{ i. o.}\}$ for the random event constituted from elementary random events $\omega \in \Omega$ for which the inclusion $\omega \in E(\mathbf{n})$ holds infinitely often, that is $\omega \in E(\mathbf{n})$ for infinitely many multi-indices $\mathbf{n} \in \mathbf{N}^d$. This random event is often called the *upper limit* for the field of random events $\{E(\mathbf{n}), \mathbf{n} \in \mathbf{N}^d\}$ and is denoted by $\limsup[E(\mathbf{n})]$. The probability of the event $\limsup[E(\mathbf{n})]$ can sometimes be evaluated by using the following generalization of the Borel–Cantelli lemma for the space $\mathbf{N}^d$.

Proposition A.30 (Borel–Cantelli lemma) *If*

$$\sum_{\mathbf{n}\in\mathbf{N}^d} \mathsf{P}(E(\mathbf{n})) < \infty,$$

then $\mathsf{P}(E(\mathbf{n}) \text{ i. o.}) = 0$. *On the other hand, if random events* $E(\mathbf{n})$, $\mathbf{n} \in \mathbf{N}^d$, *are jointly independent and*

$$\sum_{\mathbf{n}\in\mathbf{N}^d} \mathsf{P}(E(\mathbf{n})) = \infty,$$

then $\mathsf{P}(E(\mathbf{n}) \text{ i. o.}) = 1$.

Proof of Proposition A.30. We use the complete ordering $\preccurlyeq$ to place random events $E(\mathbf{n})$, $\mathbf{n} \in \mathbf{N}^d$, one by one. Now we apply the usual Borel–Cantelli lemma for the sequence of random events. □

Remark A.6 Proposition A.30 is often used to prove that a certain field converges almost surely. Let $\{X(\mathbf{n}), \mathbf{n} \in \mathbf{N}^d\}$ be a field of random variables and let $E(\mathbf{n}) = \{\omega\colon |X(\mathbf{n})| \ge \varepsilon\}$ for $\varepsilon > 0$. If $\sum \mathsf{P}(E(\mathbf{n})) < \infty$ for all $\varepsilon > 0$, then the probability that an infinite number of events $E(\mathbf{n})$ occur is equal to 0, whence $\lim(\max) X(\mathbf{n}) = 0$ almost surely. Note that the Borel–Cantelli lemma implies lim(max)-convergence from which lim(min)-convergence follows. This explain why, along with the "natural and classical" lim(min)-convergence, we also study lim(max)-convergence.

Remark A.7 Proposition A.30 can be used to evaluate the superior and inferior limits of a field of random variables. For example, if random events $E(\mathbf{n})$ are defined as in the preceding remark via random variables $X(\mathbf{n})$ and the series $\sum \mathsf{P}(E(\mathbf{n}))$ converges for $\varepsilon > \varepsilon_0$, then $\limsup_{\max} X(\mathbf{n}) \le \varepsilon_0$ almost surely.

Moreover, if random variables $X(\mathbf{n})$ are jointly independent and the multi-indexed series $\sum \mathsf{P}(E(\mathbf{n}))$ diverges for $\varepsilon < \varepsilon_0$, then $\liminf_{\max} X(\mathbf{n}) \ge \varepsilon_0$ almost surely.

The main results on the interchange of limits and integrals for fields follow easily from the corresponding results for sequences. The Beppo Levi lemma for $\mathbf{N}^d$ is often useful in proofs concerning the convergence of multi-indexed series.

Proposition A.31 (Beppo Levi lemma) *Let a field of random variables* $\{S(\mathbf{n}), \mathbf{n} \in \mathbf{N}^d\}$ *be monotone, that is*

$$S(\mathbf{m}) \le S(\mathbf{n}) \quad \textit{for all } \mathbf{m} \le \mathbf{n}.$$

Assume that the expectations $\mathsf{E}[S(\mathbf{n})]$ *exist and are uniformly bounded, that is, for some constant* $C > 0$,

$$\mathsf{E}[S(\mathbf{n})] \le C \quad \textit{for all } \mathbf{n} \in \mathbf{N}^d.$$

Then $S = \lim(\min) S(\mathbf{n})$ *exists and moreover*

$$\mathsf{E}[S] \le \lim(\min) \mathsf{E}[S(\mathbf{n})].$$

Proposition A.32 (another Beppo Levi lemma) *Let* $\{X(\mathbf{n}), \mathbf{n} \in \mathbf{N}^d\}$ *be a field of non-negative random variables. Assume that*

$$\sum_{\mathbf{n} \in \mathbf{N}^d} \mathsf{E}[X(\mathbf{n})] < \infty.$$

Then the series $\sum_{\mathbf{n} \in \mathbf{N}^d} X(\mathbf{n})$ *converges almost surely and moreover*

$$\mathsf{E}\left[\sum_{\mathbf{n} \in \mathbf{N}^d} X(\mathbf{n})\right] = \sum_{\mathbf{n} \in \mathbf{N}^d} \mathsf{E}[X(\mathbf{n})].$$

Propositions A.31 and A.32 are not equivalent for $d > 1$ (in contrast to the case of $d = 1$). This can be explained by the difference between the notions of monotone fields and fields with non-negative increments (see Sect. A.5). Indeed, $\{S(\mathbf{n}), \mathbf{n} \in \mathbf{N}^d\}$ is monotone in Proposition A.31, while it has non-negative increments in Proposition A.32, since $\Delta[S(\mathbf{n})] = X(\mathbf{n})$.

We make use of the following generalization of Fatou's lemma for the space $\mathbf{N}^d$ several times.

Proposition A.33 (Fatou lemma) *Let random variables* $\{S(\mathbf{n}), \mathbf{n} \in \mathbf{N}^d\}$ *be non-negative and have finite expectations. Moreover, let* $\mathsf{E}[S(\mathbf{n})] \le C$ *for some constant* C. *If*

$$\lim(\min) S(\mathbf{n}) = S \quad \textit{a. s.}$$

(*or* $\lim(\max) S(\mathbf{n}) = S$ *a. s.*) *almost surely, then the expectation exists for* S, *too, and moreover*

$$\mathsf{E}[S] \le \liminf(\min) \mathsf{E}[S(\mathbf{n})]$$

(*correspondingly,* $\mathsf{E}[S] \le \liminf(\max) \mathsf{E}[S(\mathbf{n})]$).

Another generalization of a classical result useful for the space $\mathbf{N}^d$ is the Lebesgue dominated convergence theorem for random fields.

Proposition A.34 (Lebesgue dominated convergence theorem) *Let $|S(\mathbf{n})| \le Y$ almost surely and let the expectation of the random variable Y exist. If*

$$\lim(\min) S(\mathbf{n}) = S \quad \text{or} \lim(\max) S(\mathbf{n}) = S$$

in probability, then

$$\lim(\min) \mathsf{E}\,[S(\mathbf{n})] = \mathsf{E}\,[S] \quad \text{or} \lim(\max) \mathsf{E}\,[S(\mathbf{n})] = \mathsf{E}\,[S]\,.$$

Since the convergence in probability follows from the almost sure convergence, Proposition A.34 holds for the almost sure convergence, too.

A.17 Identifying a Field with a Subsequence

Some problems for random fields $\{S(\mathbf{n})\}$ deal with a subset of multi-indices rather than with the whole space $\mathbf{N}^d$. The simplest subset of this kind is presented by a subsequence of elements $\mathbf{n}_k = (n_{1k}, \dots, n_{dk})$, $k \ge 1$. If all d sequences $\{n_{1k}, k \ge 1\}, \dots, \{n_{dk}, k \ge 1\}$ are increasing, then it is convenient to view the collection $\{S(\mathbf{n}_k), k \ge 1\}$ as a usual sequence of sums of random variables.

For example, let the random variables $X(\mathbf{n})$ have identical distribution and put $m_k = |\mathbf{n}_k| = n_{1k} \dots n_{dk}$. We construct an auxiliary sequence $\{\zeta_j, j \ge 1\}$ as follows. For every $k \ge 1$, we write the random variables $X(\mathbf{n})$ with indices such that $\mathbf{n} \le \mathbf{n}_k$ and $\mathbf{n} \not\le \mathbf{n}_{k-1}$ in an arbitrary order after the random variables $X(\mathbf{n})$, $\mathbf{n} \le \mathbf{n}_{k-1}$. The random variables are numbered according the order they are written and are denoted by ξ_i, $m_{k-1} < i \le m_k$. Then we put $\zeta_j = \xi_1 + \dots + \xi_{m_j}$. It is clear that $S(\mathbf{n}_k) = \zeta_{m_k}$ for all $k \ge 1$ and moreover the random variables ξ_i and $X(\mathbf{n})$ have the same distribution.

Using such an "embedding" of a field into a sequence, one can study weak convergence. Moreover, this sometimes allows us to get new results for almost sure convergence by using the corresponding results for subsequences. For example, let $\mathsf{E}\,[X(\mathbf{n})] = 0$ and $\mathsf{E}\left[X^2(\mathbf{n})\right] = 1$ for all $\mathbf{n} \in \mathbf{N}^d$. Then

$$\limsup_{k\to\infty} \frac{S(k,\dots,k)}{\sqrt{2k^d \log\log k}} = 1 \quad \text{a. s.},$$
$$\liminf_{k\to\infty} \frac{S(k,\dots,k)}{\sqrt{2k^d \log\log k}} = -1 \quad \text{a. s.}$$

This result is a consequence of the above "embedding" and the law of the iterated logarithm for cumulative sums of independent identically distributed random variables.

A.17.1 A Sequence Associated with a Field

Assume that the random variables $\{X(\mathbf{n}), \mathbf{n} \in \mathbf{N}^d\}$ are independent and have identical distribution. Along with the field $\{X(\mathbf{n})\}$ it is often useful to consider a sequence of independent random variables $\{Y_k, k \geq 1\}$ that have the same distribution as each of the random variables $X(\mathbf{n})$. Such a sequence $\{Y_k, k \geq 1\}$ is said to be *associated with the field* $\{X(\mathbf{n}), \mathbf{n} \in \mathbf{N}^d\}$. If $\zeta_k = Y_1 + \cdots + Y_k$, then

$$\zeta_k \stackrel{d}{=} S(\mathbf{n}) \quad \text{if} \quad k = |\mathbf{n}|.$$

The choice of an associated sequence is not unique at all. One can, for example, choose any of the "coordinate" sequences, say $Y_k \stackrel{\text{def}}{=} X(k, 1, \ldots, 1)$.

A.18 Symmetrization Inequalities

Definition A.13 Let Y be a random variable. The random variable $Y^{(s)} = Y - Y'$ is called the *symmetrization* of Y if Y' is an independent copy of Y.

Definition A.14 Let Y be a random variable. Each number m such that

$$\mathsf{P}(Y \geq m) \geq \frac{1}{2} \quad \text{and } \mathsf{P}(Y \leq m) \geq \frac{1}{2}$$

is called a *median of the random variable* Y and denoted by $\operatorname{med}(Y)$.

Let $\{Y_n, n \geq 1\}$ be an arbitrary sequence of random variables, $\operatorname{med}(Y_n)$ a median of the random variable Y_n, and $\{Y_n^{(s)}, n \geq 1\}$ the sequence of symmetrizations $\{Y_n, n \geq 1\}$. Then, for all $x > 0$,

$$\mathsf{P}\left(\sup_{n\geq 1} |Y_n - \operatorname{med}(Y_n)| \geq x\right) \leq 2\mathsf{P}\left(\sup_{n\geq 1} |Y_n^{(s)}| \geq x\right). \tag{A.77}$$

This result can be rewritten for $d > 1$, too.

Proposition A.35 (symmetrization inequalities) *Let* $\{Y(\mathbf{n}), \mathbf{n} \in \mathbf{N}^d\}$ *be a field of random variables, and* $\operatorname{med}(Y(\mathbf{n}))$ *and* $Y^{(s)}(\mathbf{n})$ *be a median and symmetrization of the random variable* $Y(\mathbf{n})$. *Then, for all* $x > 0$,

$$\begin{aligned} \mathsf{P}\left(\sup\nolimits_{\mathbf{n}\in\mathbf{N}^d} Y(\mathbf{n}) - \operatorname{med}(Y(\mathbf{n})) \geq x\right) &\leq 2\mathsf{P}\left(\sup\nolimits_{\mathbf{n}\in\mathbf{N}^d} Y^{(s)}(\mathbf{n}) \geq x\right), \\ \mathsf{P}\left(\sup\nolimits_{\mathbf{n}\in\mathbf{N}^d} |Y(\mathbf{n}) - \operatorname{med}(Y(\mathbf{n}))| \geq x\right) &\leq 2\mathsf{P}\left(\sup\nolimits_{\mathbf{n}\in\mathbf{N}^d} |Y^{(s)}(\mathbf{n})| \geq x\right). \end{aligned} \tag{A.78}$$

A.19 A Formula for the Expectation

Let ξ be a non-negative random variable. Since

$$\xi = \int_0^\infty \mathbb{I}_{\{\xi \geq u\}}\, du$$

for all $\omega \in \Omega$, Fubini's theorem implies that

$$\mathsf{E}[\xi] = \int_0^\infty \mathsf{E}\left[\mathbb{I}_{\{\xi \geq u\}}\right] du = \int_0^\infty \mathsf{P}(\xi \geq u)\, du \tag{A.79}$$

if one of the integrals exists.

Let $r > 0$. We apply the latter equality for the random variable ξ^r and then make a change of variables:

$$\mathsf{E}[\xi^r] = \int_0^\infty \mathsf{P}(\xi^r \geq u)\, du = \int_0^\infty \mathsf{P}\left(\xi \geq u^{1/r}\right) du = r \int_0^\infty u^{r-1} \mathsf{P}(\xi \geq u)\, du.$$

A.19.1 Mallows's Bound for a Median

Let μ, a, and σ^2 be the median, expectation, and variance of a random variable X, respectively. Then Chebyshev's inequality implies that

$$|\mu - a| \leq \sqrt{2}\sigma.$$

A better bound for the deviation between the median and expectation of a random variable is known, namely

$$|\mu - a| \leq \sigma. \tag{A.80}$$

Indeed,

$$|\mu - a| \leq \mathsf{E}[|X - \mu|] \leq \mathsf{E}[|X - a|] \leq \sqrt{\mathsf{var}[X]}.$$

The second inequality is explained by the following property of a median:

$$\mathsf{E}[|X - \mu|] \leq \mathsf{E}[|X - c|] \qquad \text{for all } c \in \mathbf{R}. \tag{A.81}$$

The inequality, in fact, is strict for all numbers $c \in \mathbf{R}$ that are not medians of X. If $c \in \mathbf{R}$ is an arbitrary median of X, then the above inequality becomes an equality.

To prove property (A.81) note that (A.79) implies

$$\mathsf{E}[|X - c_2|] - \mathsf{E}[|X - c_1|] = \int_{c_1}^{c_2} \big(\mathsf{P}(X \le t) - \mathsf{P}(X \ge t)\big)\, dt$$

for all $c_1, c_2 \in \mathbf{R}$.

By μ_1 we denote the minimal median of the random variable X and by μ_2 we denote its maximal median. Since $\mathsf{P}(X \le t) - \mathsf{P}(X \ge t) = 0$ for all $\mu_1 < t < \mu_2$, we get

$$\int_{\mu_1}^{\mu_2} \big(\mathsf{P}(X \le t) - \mathsf{P}(X \ge t)\big)\, dt = 0$$

and thus $\mathsf{E}[|X - \mu|]$ is the same for all medians μ.

If $c < \mu_1$, then, for all $c \le t < \mu_1$,

$$\mathsf{P}(X \ge t) \ge \mathsf{P}(X \ge \mu_1) \ge \frac{1}{2}$$

and thus $\mathsf{P}(X \le t) < \frac{1}{2}$ (otherwise t is also a median). Therefore, $\mathsf{P}(X \le t) - \mathsf{P}(X \ge t) < 0$, whence

$$\mathsf{E}[|X - c|] - \mathsf{E}[|X - \mu_1|] = \int_{\mu_1}^{c} \big(\mathsf{P}(X \le t) - \mathsf{P}(X \ge t)\big)\, dt > 0.$$

If $c > \mu_2$, then $\mathsf{P}(X \le t) \ge \mathsf{P}(X \le \mu_2) \ge \frac{1}{2}$ for an arbitrary $\mu_2 < t \le c$, whence $\mathsf{P}(X \ge t) < \frac{1}{2}$ (otherwise t is also a median). Therefore, $\mathsf{P}(X \le t) - \mathsf{P}(X \ge t) > 0$ in this case, whence

$$\mathsf{E}[|X - c|] - \mathsf{E}[|X - \mu_2|] = \int_{\mu_2}^{c} \big(\mathsf{P}(X \le t) - \mathsf{P}(X \ge t)\big)\, dt > 0.$$

A.20 Comments

Section A.1 The orders of the type (A.1) are studied in the theory of spaces with incomplete orders. The order $\preccurlyeq$ is also called *lexicographical* (see [354]).

Section A.2 The convergence in the sense of Definition A.1 was used by A.-L. Cauchy at the beginning of the 19th century (see the remark on p. 26 in [416] and Chap. 5 in the book by Bromwich [30]). Pringsheim [342] and Hardy [152] systematically studied this type of convergence at the beginning of the 20th century. The modern state of art can be found in the monographs by Yanushauskas [426, 427].

Section A.3 Cauchy's convergence criterion for the convergence of a double series has been studied by O. Stolz in 1884.

Section A.4 Another definition of the convergence of multi-indexed series is given in the paper by Deo [68], where the limit of partial sums is considered for lim(max)-convergence. This definition of convergence is not acceptable for $d \geq 2$. For example, if a series $\sum x(\mathbf{n})$ absolutely converges in the sense of Deo's definition, then one can ensure that all its terms are equal to 0 (except, probably, $x(\mathbf{1})$). This definition is not acceptable even in the case of non-absolutely convergent multi-indexed series, since it imposes a strong restriction on the terms of a series.

Section A.6 Definition A.10 is used in the theory of partially ordered sets, where the functions satisfying condition (A.11) are called *isotone* or *order preserving functions* (see [354]). Definition A.11 means that the field $\{x(\mathbf{n})\}$ "generates" a measure in the space $\mathbf{N}^d$.

Section A.7 Proposition A.9 is proved in Kronecker's paper: C. R. Acad. Sci. Paris 103 (1886), pp. 980–987. Moricz in [321] proves an analog of Kronecker's lemma for $d > 1$ under the so-called "regular" convergence of a multi-indexed series that is more restrictive as compared to the usual convergence.

Section A.8 The characteristic $\chi(x)$ of a field was introduced by Martikainen in [308].

Example A.5 is interesting in view of a result of Petrov [336] for the case $d = 1$:

$$\lim_{n\to\infty} \frac{S_n}{\psi(B_n)} = 0 \quad \text{a. s.}$$

for every increasing function $\psi(x) \in \Psi_c$. Example A.5 shows that an analog of this property does not hold for any $d > 1$ and for any function $\psi(x)$.

Section A.9 The definition of a regularly varying function and the main results concerning these functions are due to Karamata [189]. Proposition A.10 is proved by Karamata [189] (also see Theorem 1.5.3 in [26]; an analogous result for a more general class of functions is obtained in [37]).

Proposition A.12 is known as *Karamata's theorem, direct half* (Theorem 1.5.11 in [26]); Proposition A.13 is known as *Karamata's theorem, converse half* (Theorem 1.6.11 in [26]). The case of $\sigma + \rho + 1 = 0$ is considered by Parameswaran [334] (in this case, one needs to assume additionally that $\int_0^\infty t^\sigma f(t)\,dt < \infty$ in part (ii)).

Proposition A.15 is a particular case of Lemma 1.8 in [368] (see also Theorem 1.5.12 in [26]). Related results for non-monotone and discontinuous functions are discussed in [37]. One of the candidates for f^{-1} is the so-called generalized inverse function; an even more general notion of *asymptotically quasi-inverse functions* is discussed in [37]. Proposition A.11 is statement 4° in the book [368, p. 19]; Proposition A.16 coincides with Theorem 1.1 in [368].

Section A.10 For reader's convenience, we give below some references concerning the Dirichlet divisor problem. Bound (A.30) is obtained by Voronoi [403], inequality (A.31) is proved by Huxley [176]. Estimates (A.33) for $\lambda = 3, 4, 5$ can be found in Kolesnik [251] and Heath-Brown [158].

Bound (A.34) for general $d \geq 2$ is proved by Landau [277]. The lower bound (A.35) is proved by Hardy [153]. Inequality (A.37) is obtained by Karatsuba [190]. Lindelöf conjecture appears in [296]. The proof that the Lindelöf conjecture is equivalent to relation (A.36) can be found in [399]. The lower bound (A.39) for ε in the Lindelöf conjecture is obtained in [177].

The Euler–Maclaurin formula in the form of (A.44) is stated in [11].

Section A.11 Rough estimates for the Dirichlet function have been applied many times for the proof of the strong law of large numbers for random variables with multi-indices. Smythe [373] obtained similar bounds by estimating the r-tuple integral

$$\int \cdots \int_{\substack{x_1 \geq 1, \ldots, x_r \geq 1 \\ x_1 \ldots x_r \leq N}} dx_1 \ldots dx_r.$$

Relation (A.53) is considered in [221].

Section A.12 Evaluation of the number of integer points in subsets of $\mathbf{N}^d$ and determining the asymptotic behavior of the number of points in growing sets is a traditional topic in number theory (it suffices to mention the Gauss circle problem) and combinatorics. For almost all our purposes, "rough" bounds are sufficient (see the monographs by Gelfond and Linnik [122] and Krätzel [260] for finer results).

Section A.13 The assertion that the function G defined by (A.72) is regularly varying coincides with Karamata's theorem for $\rho > -1$ (see Proposition A.14) or with Parameswaran's theorem [334] for $\rho = -1$.

Section A.14 Proposition A.27 is obtained in the paper by Chung [63] for a partial case. A generalization of Chung's result is given in the book by Petrov [338, Theorem 9, §2, Chap. VI].

Section A.15 Proposition A.29 is proved by Baum and Katz [22].

Section A.16 These results can be found, for example, in the book by Riesz and Szökefalvi-Nagy [353] for $d = 1$.

Section A.18 The proof of inequality A.77 is exhibited, for example, in Petrov's monograph [336, Lemma 12, Chap. IX].

Section A.19 The method of the proof of inequality (A.80) given above is hinted in the textbook by Strook [386, Problem 1.4.23, p. 43 therein].

References

1. A. Abay, Renewal theorems for random walks in multidimensional time. Mathematica Slovaca **49**(3), 371–380 (1999)
2. A. de Acosta, Inequalities for B-valued random vectors with applications to the strong law of large numbers. Ann. Probab. **9**(1), 157–161 (1981)
3. A. Adler, One-sided limit theorems for sums of multidimensional indexed random variables. J. Theory Probab. **5**(4), 335–341 (1992)
4. A. Adler, The strong law of large numbers for multidimensionally indexed random variables. Bull. Inst. Math. Acad. Sinica **23**(4), 335–341 (1995)
5. A. Adler, A weak law for randomly stopped sums of multidimensionally indexed random variables. Stoch. Anal. Appl. **15**(4), 463–472 (1997)
6. A. Adler, Exact strong laws for multidimensionally indexed random variables. J. Multivar. Anal. **77**(1), 73–83 (2001)
7. A. Adler, A. Rosalsky, R.L. Taylor, Strong laws of large numbers for weighted sums of random elements in normed linear spaces. Int. J. Math. Math. Sci. **12**(3), 507–530 (1989)
8. R.J. Adler, J. Taylor, *Random Fields and Geometry*, Springer Monographs in Mathematics (Springer, New York, 2007)
9. C. Aisleitner, On the law of the iterated logarithm for the discrepancy of sequences $\{n_{\mathbf{k}}x\}$ with multidimensional indices. Unif. Distrib. Theory **2**(2), 89–104 (2007)
10. K.S. Alexander, R.A. Pyke, A uniform central limit theorem for set-indexed partial-sum processes with finite variance. Ann. Probab. **14**(2), 582–597 (1986)
11. T.M. Apostol, An elementary view of Euler's summation formula. Amer. Math. Mon. **106**(5), 409–418 (1999)
12. S. Asmussen, T.G. Kurtz, Necessary and sufficient conditions for complete convergence in the law of large numbers. Ann. Probab. **8**(1), 176–172 (1980)
13. J.-M. Azaïs, M. Wshebor, *Level Sets and Extrema of Random Processes and Fields* (Wiley, Hoboken, 2009)
14. H.A. Azarnoosh, Marcinkiewicz-type strong law of large numbers for double arrays of negatively dependent random variables. J. Sci. Islam. Repub. Iran **13**(3), 255–261, 290 (2002)
15. T.A. Azlarov, N.A. Volodin, Laws of large numbers for identically distributed Banach space valued random variables. Teor. Veroyatnost. i Primenen. **26**(3), 584–590 (1981) (English transl. in Theory Probab. Appl. **26**(3), 573–580 (1982))
16. B. von Bahr, C.G. Esseen, Inequalities for r-th absolute moment of a sum of random variables, $1 < r < 2$. Ann. Math. Stat. **36**(1), 299–300 (1965)
17. Y. Bakhtin, A.V. Bulinski, Moment inequalities for sums of dependent multiindexed random variables. Fundam. Prikl. Mat. **3**(4), 1101–1108 (1997) (Russian)

O. Klesov, *Limit Theorems for Multi-Indexed Sums of Random Variables*,
Probability Theory and Stochastic Modelling 71, DOI 10.1007/978-3-662-44388-0

18. A.D. Barbour, Tail sums of convergent series of independent random variables. Proc. Camb. Philos. Soc. **75**, 361–364 (1974)
19. J. Barone, A. Novikoff, A history of the axiomatic formulation of probability from Borel to Kolmogorov I. Arch. Hist. Exact Sci. **18**(2), 123–190 (1978)
20. R. Bass, R. Pyke, Functional law of the iterated logarithm and uniform central limit theorem for partial-sum processes indexed by sets. Ann. Probab. **12**(1), 13–34 (1984)
21. R. Bass, R. Pyke, A strong law of large numbers for partial sum processes indexed by sets. Ann. Probab. **12**(1), 268–271 (1984)
22. L.E. Baum, M. Katz, Convergence rates in the law of large numbers. Trans. Amer. Math. Soc. **120**(1), 108–123 (1965)
23. A.C. Berry, The accuracy of the Gaussian approximation to the sum of independent variates. Trans. Amer. Math. Soc. **49**(1), 122–136 (1941)
24. A. Bikyalis, Estimates of the remainder term in the central limit theorem. Liet. Mat. Rinkinys **6**, 323–346 (1966) (Russian)
25. N.H. Bingham, Variants on the law of the iterated logarithm. Bull. Lond. Math. Soc. **18**(5), 433–467 (1986)
26. N.H. Bingham, C.M. Goldie, J.L. Teugels, *Regular Variation*, Encyclopedia of Mathematics and Its Applications, vol. 27 (Cambridge University Press, Cambridge, 1987)
27. S. Bochner, Partial ordering in the theory of martingales. Ann. Math. **62**(1), 162–169 (1955)
28. É. Borel, Les probabilités dénombrables et leurs applications arithmétiques. Rendiconti del Circolo Matematico di Palermo **27**, 247–271 (1909)
29. V.M. Borodikhin, On a generalization of Nagaev-Fuk inequalities to a class of random fields. Sibirsk. Mat. Zh. **36**(6), 1275–1282 (1995) (English transl. in Siberian Math. J. **36**(6), 1101–1107 (1995))
30. T.J. I'a. Bromwich, *An Introduction to the Theory of Infinite Series*, 2nd edn., revised with the assistance of T. M. Macrobert. (Macmillan and Company, London, 1926)
31. H.D. Brunk, The strong law of large numbers. Duke Math. J. **15**(1), 181–195 (1948)
32. G. Budianu, On the law of the iterated logarithm for tail sums of random variables. Studii si Cercetari Mat. **33**, 149–158 (1981)
33. V.V. Buldygin, On random series in Banach spaces. PhD thesis, Institute of Mathematics, Academy of Science of Ukraine, Kiev, 1973 (Russian)
34. V.V. Buldygin, *The Convergence of Random Elements in Topological Spaces* (Naukova dumka, Kiev, 1980) (Russian)
35. V.V. Buldygin, The strong law of large numbers and the convergence to zero of Gaussian sequences, Teor. Veroyatnost. i Mat. Statist. **19**, 33–41, 156–157 (1978) (English transl. in Theor. Probab. Math. Stat. **19**, 33–41 (1978))
36. V.V. Buldygin, K.-H. Indlekofer, O.I. Klesov, J.G. Steinebach, Asymptotics of renewal processes: some recent developments. Ann. Univ. Sci. Bp. Sect. Comp. **28**, 107–139 (2008)
37. V.V. Buldygin, K.-H. Indlekofer, O.I. Klesov, J.G. Steinebach, *Pseudo Regularly Varying Functions and Generalized Renewal Processes* (TBiMC, Kyiv, 2012) (Ukrainian)
38. V.V. Buldygin, S.A. Solntsev, *Functional Methods in Problems of the Summation of Random Variables* (Naukova dumka, Kiev, 1989) (Russian)
39. V.V. Buldygin, S.A. Solntsev, *Asymptotic Behaviour of Linearly Transformed Sums of Random Variables* (Kluwer, Dordrecht, 1997)
40. M. Bulst, Große und kleine Abweichungen für mehrparametrige Partialsummenprocessen und ein Strassensches gesetz des iterierten Logrithmus. Math. Nachr. **99**, 337–354 (1980)
41. D.L. Burkholder, Successive conditional expectations of an integrable function. Ann. Math. Stat. **33**(3), 887–893 (1962)
42. C. Guang Hui, Complete convergence for ρ-mixing random fields. J. Zhejiang Univ. Sci. Ed. **30**(4), 380–383, 396 (2003) (Chinese, English summary)
43. R. Cairoli, Une inégalité pour martingales 'a indices multiples et ses applications, in *Séminaire des Probabilités* Vol. IV. Lecture Notes in Mathematics (Springer, Berlin, 1970), pp. 1–27
44. R. Cairoli, R.C. Dalang, *Sequential Stochastic Optimization*, Wiley Series in Probability and Statistics: Probability and Statistics (A Wiley-Interscience Publication, Wiley, New York, 1996)

45. R. Cairoli, J.-P. Gabriel, *Arrêt de certaines suites multiples de variables aléatoires indépendantes*. Lecture Notes in Math, Vol. 721 (Springer, Berlin, 1979), pp. 174–198
46. R. Cairoli, J.B. Walsh, Stochastic integrals in the plane. Acta Math. **134**, 111–183 (1975)
47. Ping-yan Chen, The law of the iterated logarithm for independent random variables with multidimensional parameters and its application. J. Math. Res. Expos. **24**(1), 55–63 (2004)
48. P. Chen, S. Gan, On moments of sup $|Sn/n^{1/r}|$ and sup $|Sn/\sqrt{n\overline{\text{loglog}n}}|$. Acta Math. Scientia **23**(A), 573–582 (2003) (Chinese, English summary)
49. P. Chen, Gan S, On moments of the maximum of normed partial sums of ρ-mixing random variables. Stat. Probab. Lett. **78**(10), 1215–1221 (2008)
50. P. Chen, R.G. Antonini, T.-C. Hu, A. Volodin, Limiting behaviour of moving average processes under ρ-mixing assumption. Note di Matematica **30**(1), 17–23 (2010)
51. Ping-yan Chen, Zhi-yong Shan, On Chover's LIL for stable random fields. J. Math. (Wuhan) **20**(2), 227–230 (2000)
52. H. Chernoff, H. Teicher, Limit distributions of the minimax of independent identically distributed random variables. Trans. Amer. Math. Soc. **116**, 474–491 (1965)
53. B.D. Choi, S.H. Sung, On the integrability of sup $|S_n/n^{1/r}|$ for $1 < r < 2$. Stat. Probab. Lett. **5**(4), 267–272 (1987)
54. B.D. Choi, S.H. Sung, On moment conditions for the supremum of normed sums. Stoch. Process. Appl. **26**(1), 99–106 (1987)
55. J. Chover, A law of the iterated logarithm for stable summands. Proc. Amer. Math. Soc. **47**(3), 441–443 (1966)
56. Y.S. Chow, H. Teicher, Iterated logarithm laws for weighted averages. Z. Wahrsch. verw. Gebiete **26**(2), 87–94 (1973)
57. Y.S. Chow, H. Teicher, *Probability Theory: Independence, Interchangeability Martingales*, 3rd edn. Springer Texts in Statistics (Springer, New York, 1997)
58. Y.S. Chow, H. Robbins, D. Siegmund, *Great Expectations: The Theory of Optimal Stopping* (Houghton Mifflin Co., Boston, 1971)
59. T.C. Christofides, Rate of convergence in the strong law of large numbers for U-statistics based on a multidimensionally indexed array of random variables. Stat. Probab. Lett. **9**(3), 267–272 (1990)
60. T.C. Christofides, R.J. Serfling, Maximal inequalities for multidimensionally indexed submartingale arrays. Ann. Probab. **18**(2), 630–641 (1990)
61. T.C. Christofides, R.J. Serfling, Maximal inequalities and convergence results for generalized U-statistics. J. Stat. Plann. Inf. **24**(3), 271–286 (1990)
62. T.C. Christofides, R.J. Serfling, U-statistics on a lattice of i.i.d. random variables. Stat. Probab. Lett. **40**(3), 293–303 (1998)
63. K.L. Chung, Note on some strong laws of large numbers. Amer. J. Math. **69**(1), 189–192 (1947)
64. R.M. Crujeiras, I. Van Keilegom, Least squares estimation of nonlinear spatial trends. Comp. Stat. Data Anal. **54**(2), 452–465 (2010)
65. M. Csörgő, P. Révész, *Strong Approximations in Probability and Statistics*. Probability and Mathematical Statistics (Academic Press Inc, New York, London, 1981)
66. B. Davis, Stopping rules for $\frac{Sn}{n}$ and the class L log L. Z. Wahrsch. verw. Gebiete **17** (2), 147–150 (1971)
67. P. Deheuvels, On the Erdös-Rényi theorem for random fields and sequences and its relationships with the theory of runs and spacings. Z. Wahrsch. verw. Gebiete **70**(1), 91–115 (1985)
68. C.M. Deo, Strong laws of large numbers for weakly stationary random fields. Sankhyā Ser. A **40**(1), 19–27 (1978)
69. D.L. Deng, The law of the iterated logarithm for tail sums of B-valued random variables. Acta Sci. Nat. Univ. Jilin. **3**, 1–10 (1994) (Chinese, English summary)
70. D.L. Deng, Convergence rates for probabilities of moderate deviation for sums of random variables indexed by Z_+^d. Acta Math. Hungar. **67**(1–2), 131–149 (1995)

71. D.L. Deng, Complete convergence and convergence rates in Marcinkiewicz law of large numbers for random variables indexed by Z_+^d. Math. Appl. (Wuhan) **9**(4), 441–448 (1996)
72. D.L. Deng, Wittmann's law of iterated logarithm for tail sums of B-valued random variables. J. Theoret. Probab. **17**(2), 367–385 (2004)
73. S.W. Dharmadhikari, K. Jogdeo, Bounds on moments of certain random variables. Ann. Math. Stat. **40**(4), 1506–1508 (1969)
74. S.V. Dil'man, Asymptotics in the Baum-Katz formula for random fields, Matem. Zametki **79**(5), 674–680 2006 (English transl. in Math. Notes **79**(5–6), 625–631 (2006))
75. W. Doeblin, Sur l'ensemble de puissance d'une loi de probabilité. Studia Math. **9**(1), 71–96 (1940)
76. J.L. Doob, Renewal theory from the point of view of the theory of probability. Trans. Amer. Math. Soc. **63**(3), 422–438 (1948)
77. J.L. Doob, *Stochastic Processes*. Wiley Classics Library. (A Wiley Interscience Publication, Wiley, New York, 1990) (Reprint of the 1953 original)
78. D. Dugue, *Traité de Statistique Théorique et Appliquée* (Masson, Paris, 1958)
79. N. Dunford, An individual ergodic theorem for non-commutative transformations. Acta Sci. **14**(1), 1–4 (1951)
80. L.V. Dung, T. Ngamkham, N.D. Tien, A.I. Volodin, Marcinkiewicz-Zygmund type law of large numbers for double arrays of random elements in Banach spaces. Lobachevskii J. Math. **30**(4), 337–346 (2009)
81. L.V. Dung, N.D. Tien, Mean convergence theorems and weak laws of large numbers for double arrays of random elements in Banach spaces. Bull. Korean Math. Soc. **47**(3), 467–482 (2010)
82. V.A. Egorov, The strong law of large numbers and the law of the iterated logarithm for a sequence of independent random variables, Teor. Veroyatnost. i Primenen. **15**(3), 520–527 (1970) (English transl. in Theory Probab. Appl. **15**(3), 509–514 (1970))
83. V.A. Egorov, On Kolmogorov's theorem on the law of the iterated logarithm, Vestnik Leningrad. Univ. Mat. Meh. Astronom. **13**(3), 140–142, 153 (1972) (Russian)
84. P. Erdös, On a theorem of Hsu and Robbins. Ann. Math. Stat. **20**(2), 286–291 (1949)
85. P. Erdös, Remark on my paper "On a theorem of Hsu and Robbins". Ann. Math. Stat. **21**(1), 138–138 (1950)
86. C.-G. Esseen, On the Liapunoff limit of error in the theory of probability, in *Arkiv för matematik, astronomi och fysik*, Vol. A28 (Almquist & Wiksell, Stockholm, 1942), pp. 1–19
87. C.-G. Esseen, A moment inequality with an application to the central limit theorem. Skand. Aktuarietidskr. **39**, 160–170 (1956)
88. C.-G. Esseen, S. Janson, On moment conditions for normed sum of independent variables and martingale differences. Stoch. Process. Appl. **19**(1), 173–182 (1985)
89. N. Etemadi, On some classical results in probability theory. Sankhyā Ser. A **47**(2), 215–221 (1975)
90. N. Etemadi, An elementary proof of the strong law of large numbers. Z. Wahrsch. verw. Gebiete **55**(1), 119–122 (1981)
91. N. Etemadi, On the maximal inequalities for the average of pairwise i.i.d. random variables. Comm. Stat. A Theory Method **13**(22), 2749–2756 (1984)
92. N. Etemadi, Maximal inequalities for partial sums of independent random vectors with multi-dimensional time parameters. Comm. Stat. A Theory Methods **12**(20), 3909–3923 (1991)
93. I. Fazekas, *Marcinkiewicz Strong Law of Large Numbers for B-Valued Random Variables with Multidimensional Indices*. Proceedings of III Pannonian Symposium Mathematical Statistics. (Viségrad, Hungary, 1982), pp. 53–61
94. I. Fazekas, Convergence rates in the Marcinkiewicz strong law of large numbers for Banach space valued random variables with multidimensional indices. Publ. Math. **32**(3–4), 203–209 (1985)
95. I. Fazekas, Burkholder's inequality for multiindex martingales. Annales Mathematicae et Informaticae **32**, 45–51 (2005)

96. I. Fazekas, A.N. Chuprunov, Almost sure limit theorems for the Pearson statistic. Teor. Veroyatnost. i Primenen. **48**(1), 162–169 (2003) (English transl. in Theory Probab. Appl. **48**(1), 140–147 (2004))
97. I. Fazekas, O. Klesov, Limit laws for sums of independent random variables on sets. Theory Stoch. Process. **2(18)**(1–2), 137–149 (1996)
98. I. Fazekas, O. Klesov, A new approach to the strong laws of large numbers. Teor. Veroyatnost. i Primenen. **42**(2), 401–402 (1997)
99. I. Fazekas, O. Klesov, A general approach to the strong laws of large numbers. Teor. Veroyatnost. i Primenen. **45**(3), 568–583 (2000) (English transl. in Theory Probab. Appl. **45**(3), 436–449 (2002))
100. I. Fazekas, T. Tómács, Strong law of large numbers for pairwise independent random variables with multidimensional indices. Publ. Math. **53**(1–2), 149–161 (1998)
101. W. Feller, Úber das Gesetz der grossen Zahlen. Acta. Scient. Math. Szeged **8**(4), 191–201 (1937)
102. W. Feller, On the integral equation of renewal theory. Ann. Math. Stat. **12**(3), 243–267 (1941)
103. W. Feller, The general form of the so-called law of the iterated logarithm. Trans. Amer. Math. Soc. **54**(3), 373–402 (1943)
104. W. Feller, *An Introduction to Probability Theory and Its Applications*, Vol. II, 2nd edn. (Wiley, New York, London, Sydney, 1971)
105. X. Fernique, Régularité de processus gaussiens. Invent. Math. **12**(4), 304–320 (1971)
106. X. Fernique, *Regularité des trajectoires des fonctions aéatoires gaussiennes*. Lecture Notes in Mathematics, Vol. 480 (Springer, Berlin, 1975), pp. 1–96
107. G.M. Fikhtengol'ts, *A Course on Differential and Integral Calculus*, Vol. II (Nauka, Moscow, 1969) (Russian)
108. B. de Finetti, Sulla funzione a incremento aleatorio. Atti Acad. Naz. Lincei. Rend. Cl. Sci. Fis. Mat. Nat. **6**(10), 163–168, 325–329, 548–553 (1929)
109. D. Freedman, A remark on the law of the iterated logarithm. Ann. Math. Stat. **38**(2), 598–600 (1967)
110. A.N. Frolov, Strong limit theorems for increments of random fields. Theory Stoch. Process. **8(24)**(1), 89–97 (2002)
111. A.N. Frolov, On the asymptotic behavior of increments of random fields. Zap. Nauchn. Sem. S.-Peterburg. Otdel. Mat. Inst. Steklov. (POMI) **298**(6), 191–207, 319 (2003) (English transl. in J. Math. Sci. (N. Y.) **128**(1), 2604–2613 (2005))
112. D.H. Fuk, S.V. Nagaev, Probabilistic inequalities for sums of independent random variables. Teor. Veroyatnost. i Primenen. **16**(4), 660–675 (1971) (English transl. in Theory Probab. Appl. **16**(4), 643–660. Letter to the Editors, **21**(4), 896 (1976))
113. M. Fukamiya, On dominated ergodic theorem in L_p, $p \geq 1$. Tôhoku Math. J. **46**, 150–153 (1939)
114. J.P. Gabriel, Loi des grands nombres, séries et martingales á deux indices. C. R. Acad. Sci. Paris Sér. A **279**, 169–171 (1974)
115. J.P. Gabriel, An inequality for sums of independent random variables indexed by finite dimensional filtering sets and its application to convergence of series. Ann. Probab. **5**(5), 779–786 (1977)
116. J.P. Gabriel, Martingales with a countable filtering index set. Ann. Probab. **5**(6), 888–898 (1977)
117. M.U. Gafurov, M.K. Kholmuradov, *Convergence of Series of Probabilities of Large Deviations for Sums of Random Variables with Multidimensional Indices*. Probability Distributions and Mathematical Statistics (Fan, Tashkent, 1986), pp. 114–131, 490 (Russian)
118. J. Galambos, K.-H. Indlekofer, I. Kátai, A renewal theorem for random walks in multidimensional. Trans. Amer. Math. Soc. **300**(2), 759–769 (1987)
119. J. Galambos, I. Kátai, A note on random walks in multidimensional time. Math. Proc. Camb. Phil. Soc. **99**(1), 163–170 (1986)
120. J. Galambos, I. Kátai, Some remarks on random walks in multidimensional time, in *Proceedings of 5th Pannonian Symposium Mathematical Statistical* (Visegrad, Hungary, 1985; Reidel, Dordrecht, 1986), pp. 65–74

121. S. Gan, P. Chen, Some limit theorems for weighted sums of random variable fields. Wuhan Univ. J. Nat. Sci. **11**(2), 323–327 (2006)
122. A.O. Gel'fond, Y.V. Linnik, *Elementary Methods in the Analytic Theory of Numbers* (Fizmatlit, Moscow, 1962) (English transl. translated by D.E. Brown (Pergamon Press, New York, 1966))
123. N. Van Giang, On moments of the supremum of normed partial sums of random variables indexed by N^k. Acta Math. Hung. **60**(1–2), 73–80 (1992)
124. N. Van Giang, Marcinkiewicz–Zygmund laws for Banach space valued random variables with multidimensional parameters. Teor. Veroyatnost. i Primenen. **40**(1), 213–219 (1995) (English transl. in Theory Probab. Appl. **40**(1), 175–181 (1995))
125. B.V. Gnedenko, Several theorems about powers of distribution functions. Uchenye Zapiski Moskov. Gos. Univ. Matematika **30**, 61–82 (1939) (Russian)
126. B.V. Gnedenko, A.N. Kolmogorov, *Limit Distributions for Sums of Independent Random Variables* (Gostekhizdat, Moscow, 1949); English transl. Translated from the Russian, annotated, and revised by K.L. Chung. With appendices by J.L. Doob, P.L. Hsu. Revised edition, Addison-Wesley Publishing Co., Reading, Mass.-London-Don Mills., Ont., 1968)
127. R.F. Gundy, On the class $L\log L$, martingales, and singular integrals. Studia Math. **33**(2), 109–118 (1969)
128. A. Gut, Convergence of reversed martingales with multidimensional indices. Duke Math. J. **43**(2), 269–275 (1976)
129. A. Gut, Marcinkiewicz laws and convergence rates in the law of large numbers for random variables with multidimensional indices. Ann. Probab. **6**(3), 469–482 (1978)
130. A. Gut, Moments of the maximum of normed partial sums of random variables with multidimensional indices. Z. Wahrsch. verw. Gebiete **46**(2), 205–220 (1979)
131. A. Gut, On the integrability of $\sup|S_n/n|$ for subsequences. Ann. Probab. **7**(6), 1059–1065 (1979)
132. A. Gut, Convergence rates for probabilities of moderate deviations for sums of random variables with multidimansional indices. Ann. Probab. **8**(2), 298–313 (1980)
133. A. Gut, Strong laws for independent identically distributed random variables indexed by a sector. Ann. Probab. **11**(3), 569–577 (1983)
134. A. Gut, Complete convergence and convergence rates for randomly indexed partial sums with an application to the some first passage time. Acta Math. Hung. **42**(3/4), 225–232 (1983)
135. A. Gut, On complete convergence in the law of large numbers for subsequences. Ann. Probab. **13**(4), 1286–1291 (1985)
136. A. Gut, Law of the iterated logarithm for subsequences. Probab. Math. Stat. **7**(1), 27–58 (1986)
137. A. Gut, Convergence rates in the central limit theorem for multidimensionally indexed random variables. Studia Sci. Math. Hung. **37**(3–4), 401–418 (2001)
138. A. Gut, *Stopped Random Walks*, 2nd edn. (Springer, New York, 2009)
139. A. Gut, O. Klesov, J. Steinebach, Equivalences in strong limit theorems for renewal counting processes. Stat. Probab. Lett. **35**(4), 381–394 (1997)
140. A. Gut, A. Spataru, Precise asymptotics in some strong limit theorems for multidimensionally indexed random variables. J. Multivar. Anal. **86**(2), 398–422 (2003)
141. A. Gut, U. Stadtmüller, Laws of the single logarithm for delayed sums of random fields. Bernoulli **14**(1), 249–276 (2008)
142. A. Gut, U. Stadtmüller, Laws of the single logarithm for delayed sums of random fields II. J. Math. Anal. Appl. **346**(2), 403–414 (2008)
143. A. Gut, U. Stadtmüller, An asymmetric Marcinkiewicz–Zygmund LLN for random fields. Stat. Probab. Lett. **79**(8), 1016–1020 (2009)
144. A. Gut, U. Stadtmüller, On the LSL for random fields. J. Theor. Probab. **24**(2), 422–449 (2011)
145. A. Gut, J. Steinebach, Convergence rates in precise asymptotics. J. Math. Anal. Appl. **390**, 1–14 (2012)
146. A. Gut, J. Steinebach, Precise asymptotics—a general approach. Acta Math. Hung. **138**(4), 365–385 (2013)

147. A. Gut, J. Steinebach, Convergence rates in precise asymptotics II. Ann. Univ. Sci. Budapest. Sect. Comp. **39**, 95–110 (2013)
148. L. de Haan, *On Regular Variation and Its Application to the Weak Convergence of Sample Extremes*. Mathematical Centre Tracts, Vol. 32 (Mathematisch Centrum, Amsterdam, 1970)
149. C. Hagwood, A renewal theorem in multidimensional time. Aust. J. Stat. **31**(1), 130–137 (1989)
150. J. Hájek, A. Rényi, Generalization of an inequality of Kolmogorov. Acta Math. Acad. Sci. Hung. **6**(3–4), 281–283 (1955)
151. P. Hall, C.C. Heyde, *Martingale Limit Theory and Its Applications*. Probability and Mathematical Statistics (Academic Press Inc, New York, 1980)
152. G.H. Hardy, On the convergence of certain multiple series. Proc. Lond. Math. Soc. **2**(1), 124–128 (1904)
153. G.H. Hardy, On Dirichlet's divisor problem. Proc. Lond. Math. Soc. **15**(2), 1–25 (1915)
154. G.H. Hardy, J.E. Littlewood, Some problems of Diophantine approximation. Acta Math. **37**(1), 155–190 (1914)
155. G.H. Hardy, J.E. Littlewood, The approximate functional equation in the theory of the zeta-function, with applications to the divisor problems of Dirichlet and Piltz. Proc. Lond. Math. Soc. **21**(2), 39–74 (1922)
156. P. Hartman, A. Wintner, On the law of the iterated logarithm. Amer. J. Math. **64**(2), 273–298 (1941)
157. Shuyuan He, Maximal probability inequality and truncation property for random field and its application. Beijing Daxue Xuebao Ziran Kexue Ban **31**(1), 40–53 (1995)
158. D.R. Heath-Brown, Recent progress in analytic number theory, in *Mean Values of the Zeta-Function and Divisor Problems*, Vol. 1 (Durham, 1979), (Academic Press, London, 1981) pp. 115–119
159. C.C. Heyde, On the converse to the iterated logarithm law. J. Appl. Probab. **5**(1), 210–215 (1968)
160. C.C. Heyde, Some properties of metrics in a study of convergence to normality. Z. Wahrsch. verw. Gebiete **11**(3), 181–192 (1969)
161. C.C. Heyde, A supplement to the strong law of large numbers. J. Appl. Probab. **12**(1), 903–907 (1975)
162. C.C. Heyde, V.K. Rohatgi, A pair of complementary theorems on convergence in the strong law of large numbers. Proc. Camb. Philos. Soc. **63**(1), 73–82 (1967)
163. J. Hoffman-Jørgensen, *Sums of Independent Banach Space Valued Random Variables*. Prepr. Series Vol. 15 (Math. Inst. Århus University, Å rhus, 1972)
164. J. Hoffman-Jørgensen, Sums of independent Banach space valued random variables. Studia Math. **52**(2), 159–186 (1974)
165. D.H. Hong, A.I. Volodin, Marcinkiewicz-type law of large numbers for double arrays. J. Korean Math. Soc. **36**(6), 1133–1143 (1999)
166. D.H. Hong, S.Y. Hwang, Marcinkiewicz-type strong law of large numbers for double arrays of pairwise independent random variables. Int. J. Math. Math. Sci. **22**(1), 171–177 (1999)
167. S. Hu, Some new results for the strong law of large numbers. Acta Math. Sinica (Chin. Ser.) **46**(6), 1123–1134 (2003) (Chinese, English summary)
168. S. Hu, G. Chen, X. Wang, On extending the Brunk–Prokhorov strong law of large numbers for martingale differences. Stat. Probab. Lett. **78**(18), 3187–3194 (2008)
169. Y.J. Hu, Complete convergence theorems for L_p-mixingales. J. Math. Anal. Appl. **290**(1), 271–290 (2004)
170. Sh Hu, X. Li, W. Yang, X. Wang, Maximal inequalities for some dependent sequences and their applications. J. Korean Stat. Soc. **40**(1), 11–19 (2011)
171. N. Van Huan, N. Van Quang, On the strong law of large numbers for twodimensional arrays of blockwise independent and blockwise orthogonal random variables. Probab. Math. Stat. **25**(2), 385–391 (2005)
172. N. Van Huan, N. Van Quang, The Doob inequality and strong law of large numbers for multidimensional arrays in general Banach spaces. Kybernetika **48**(2), 254–267 (2012)

173. N.V. Hung, N.D. Tien, On the almost sure convergence of two parameter martingales and the strong law of large numbers in Banach spaces. Acta Mathematica Vietnamica **17**(1), 127–143 (1992)
174. P.L. Hsu, H. Robbins, Complete convergence and the law of large numbers. Proc. Nat. Acad. Sci. U.S.A. **33**(2), 25–31 (1947)
175. J. Hüsler, Convergence rates in the law of large numbers for random variables on partially ordered sets. Monath. Math. **85**(1), 53–58 (1978)
176. M.N. Huxley, Exponential sums and lattice points III. Proc. Lond. Math. Soc. **87**(3), 591–609 (2003)
177. M.N. Huxley, Exponential sums and the Riemann zeta function V. Proc. Lond. Math. Soc. **90**(1), 1–41 (2005)
178. I.A. Ibragimov, Y.A. Rozanov, *Gaussian Random Processes* (Nauka, Moscow, 1970) (English Transl.: Translated from the Russian by A.B. Aries, *Applications of Mathematics*, Vol. 9 (Springer, Berlin, 1978))
179. K.-H. Indlekofer, I. Kátai, O.I. Klesov, Renewal theorems for some weighted functions. Ann. Univ. Sci. Budapest. Sect. Comp. **34**, 179–194 (2011)
180. K.-H. Indlekofer, O.I. Klesov, Dirichlet's divisors in probability theory. Theory Stoch. Process. **3(19)**(1–2), 208–215 (1997)
181. K.-H. Indlekofer, O.I. Klesov, The asymptotic behavior of the renewal processes constructed from a random walk with a restricted multidimensional time domain. Ann. Univ. Sci. Budapest Sect. Comp. **24**, 209–221 (2004)
182. K.-H. Indlekofer, O.I. Klesov, Strong law of large numbers for multiple sums whose indices belong to a sector with function boundaries. Teor. Veroyatnost. i Primenen. **52**(4), 803–810 (2007) (English transl. in. Theory Probab. Appl. **52**(4), 711–719 (2008))
183. K. Itô, *Stochastic Processes*. Lectures Given at Aarhus University. Reprint of the 1969 original. Edited and with a foreword by O.E. Barndorff-Nielsen, K. Sato, (Springer, Berlin, 2004)
184. B.G. Ivanoff, E. Merzbach, What is a multi-parameter renewal process? Stochastics **78**(6), 411–441 (2006)
185. A.V. Ivanov, N.N. Leonenko, *Statistical Analysis of Random Fields*. With a preface by A.V. Skorokhod (Vyshcha Shkola, Kiev, 1986) (English transl. Mathematics and Its Applications (Soviet Series) (Kluwer Academic Publishers Group, Dordrecht, 1986))
186. N.C. Jain, M.B. Marcus, Integrability of infinite sums of independent vector-valued random variables. Trans. Amer. Math. Soc. **212**(1), 1–36 (1975)
187. Z. Kabluchko, A. Munk, Shao's theorem on the maximum of standartized random walk increments for multidimensional arrays. ESAIM Probab. Stat. **13**, 409–416 (2009)
188. J.-P. Kahane, *Some Random Series of Functions* (D. C. Heath and Co. Raytheon Education Co., Lexington, 1968)
189. J. Karamata, Sur un mode de croissance régulière des fonctions. Mathematica (Cluj) **4**, 38–53 (1930)
190. A.A. Karatsuba, Some problems of contemporary analytic number theorem. Matem. Zametki **17**(2), 341–350 (1974) (English transl. in Math. Notes **17**(2), 195–199 (1974))
191. A.A. Karatsuba, *Basic Analytic Number Theory*, 2nd edn. (Nauka, Moscow, 1983) (English transl. by M.B. Nathanson, Springer, Berlin, 1993)
192. M. Katz, The probability in the tail of a distribution. Ann. Math. Stat. **34**(1), 312–318 (1963)
193. M. Katz, H. Stratton, A limit theorem for sums of random variables indexed by multidimensional indices. Z. Wahrsch. verw. Gebiete **58**(3), 389–396 (1981)
194. R.K. Kendzhaev, *An Estimate of Convergence in the Law of the Iterated Logarithm for Independent Random Variables that are Indexed on a Sector*. Probabilistic Models and Mathematical Statistics (Fan, Tashkent, 1987), pp. 43–52, 174 (Russian)
195. A.Y. Khintchine, Über einen Satz der Wahrscheinlichkeitsrechnung. Fund. Math. **6**, 9–20 (1924)
196. A.Y. Khintchine, Sur la loi des grands nombres. C. R. Acad. Sci. **188**(7), 477–479 (1929)
197. A.Y. Khintchine, Zur Theorie der unbeschränktteilbaren Verteilungsgezetze. Matem. Sb. **2**(1), 79–119 (1937)

198. A.Y. Khintchine, Über Klassenkonvergenz von Verteilungsgezetzen. Izv. Nauchno-Issled. Inst. Mat. Meh. Tomsk. Gos. Univ. **1**, 258–261 (1937)
199. A.Y. Khintchine, Déduction nouvelle d'une formule de M. Paul Lévy. Bull. Univ. d'Etat Moskou Sér. Int. Sect. A. **1**(1), 1–5 (1937)
200. A.Y. Khintchine, *Asymptotische Gesetze der Wahrscheinlichkeitsrechnung* (ONTI, Moskau, 1936) (German transl. Springer, Berlin, 1933)
201. A.Y. Khintchine, *Limit Laws for Sums of Independent Random Variables* (GONTI, Moscow-Leningrad, 1938)
202. A.Y. Khintchine, Two theorems on stochastic processes whose increments are of the same type. Matem. Sb. **3(45)**(3), 577–584 (1938) (Russian)
203. M.K. Kholmuradov, The law of the iterated logarithm for sums of random variables with multidimensional indices. Dokl. Akad. Nauk UzSSR **7**, 3–4 (1985) (Russian)
204. M.K. Kholmuradov, On the existence of generalized moments for the supremum of normalized sums of random variables with multidimensional indices. Uzbek. Mat. Zh. **1**, 84–89 (1995) (Russian)
205. B. Khoshnevisan, *On Sums of iid Random Variables Indexed by N Parameters, Séminaire de Probabilités*. Lecture Notes in Mathematics, Vol. XXXIV. (Springer, Berlin, 2000) pp. 151–156
206. B. Khoshnevisan, *Multiparameter Processes: An Introduction to Random Fields*, Springer Monographs in Mathematics (Springer, New York, 2002)
207. M.J. Klass, On stopping rules and the expected supremum of S_n/a_n and $|S_n|/a_n$. Ann. Prob. **2**(5), 889—905 (1974)
208. O.I. Klesov, On the strong law of large numbers. Dokl. Akad. Nauk Ukrain. SSR Ser. A **12**, 990–992, 1068 (1979) (Russian)
209. O.I. Klesov, A strong law of large numbers for random fields with independent values. Teor. Veroyatnost. i Mat. Statist. **21**, 65–68, 165 (1979) (English transl. in Theory Probab. Math. Stat. **21**, 75–78 (1980))
210. O.I. Klesov, Three series theorem for random fields with independent values. Vestnik Kiev. Univer. Ser. Mat. Mekh. **22**, 35–40 (1980) (Russian)
211. O.I. Klesov, The Hájek–Rényi inequality for random fields and the strong law of large numbers. Teor. Veroyatnost. i Mat. Statist. **22**, 58–66, 163 (1980) (English transl. in Theor. Probability and Math. Statist. **22**, 63–72 (1981))
212. O.I. Klesov, A law of the iterated logarithm for stable summands. Teor. Veroyatnost. i Mat. Statist. **24**, 60–63, 152 (1981) (English transl. in Theor. Probab. Math. Stat. **24**, 67–70 (1982))
213. O.I. Klesov, The strong law of large numbers for homogeneous random fields, Teor. Veroyatnost. i Mat. Statist. **25**, 29–40, 166 (1981) (English transl. in Theor. Probab. Math. Stat. **25**, 31–42 (1982))
214. O.I. Klesov, A remark on the strong law of large numbers. Teor. Veroyatnost. i Mat. Stat. **26**, 69–76, 159 (1982) (English transl. Theor. Probab. Math. Stat. **26**, 79–86 (1982))
215. O.I. Klesov, The law of the iterated logarithm for multiple sums. Teor. Veroyatnost. i Mat. Stat. **27**, 60–67, 158 (1982) (English transl. Theor. Probab. Math. Stat. **27**, 65–72 (1983))
216. O.I. Klesov, The rate of convergence of series of random variables. Ukrain. matem. zh. **35**(3), 309–314 (1983) (English transl. in Ukrain. Math. J. **35**(3), 264–268 (1984))
217. O.I. Klesov, Existence of moments of the supremum of "weighted" multiple sums. Teor. Veroyatnost. i Mat. Stat. **28**, 51–59 (1983) (English transl. Theor. Probab. Math. Stat. **28**, 57–64 (1984))
218. O.I. Klesov, Sets determining the strong law of large numbers. Theory Stoch. Process. **11**, 48–51 (1983) (Russian)
219. O.I. Klesov, Rate of convergence of certain random series. Teor. Veroyatnost. i Mat. Stat. **30**, 81–92, 152 (1984) (English transl. Theor. Probab. Math. Stat **30**, 91–101 (1985))
220. O.I. Klesov, Asymptotic behavior of a series of probabilities. Dokl. Akad. Nauk Ukrain. SSR Ser. A **8**, 15–19 (1984) (Russian)
221. O.I. Klesov, The strong law of large numbers for multiple sums of independent identically distributed random variables. Matem. Zametki **38**(6), 915–930 (1985) (English transl. in Math. Notes **38**(6), 1006–1014 (1986))

222. O.I. Klesov, The law of the iterated logarithm for weighted sums of independent identically distributed random variables. Teor. Veroyatnost. i Primenen. **24**(2), 389–393 (1986) (English transl. in Theory Probab. Appl. **24**(2), 337–342 (1986))
223. O.I. Klesov, *A Limit Theorem for Multiple Sums of Independent Identically Distributed Random Variables*. Probability Distributions and Mathematical Statistics. (Fan, Tashkent, 1986) pp. 265–277) (English transl. in J. Soviet Math. **38**(6), 2321–2326 (1987))
224. O.I. Klesov, Rate of convergence of the sums of a random number of random variables. Teor. Veroyatnost. i Mat. Statist. **39**, 65–71, 127 (1988) (English transl. Theor. Probab. Math. Stat. **39**, 75–81 (1989))
225. O.I. Klesov, A renewal theorem for a random walk with multidimensional time. Ukrain. matem. zh. **43**,(9), 1161–1167 (1991) (English transl. in Ukrain. Math. J. **43**(9), 1089–1094 (1992))
226. O.I. Klesov, The weak law of large numbers for multiple sums of independent identically distributed random variables. in A.N. Shiryaev et al. (ed.) *Proceedings of the Sixth USSR-Japan Symposium, Kiev, USSR*. (World Scientific, Singapore, New Jersey, London, Hong-Kong, 4–10 August 1991) pp. 25–31
227. O.I. Klesov, *On Integrability of* $\sup |S_{n_k}/n_k|$. New Trends in Probability and Statistics, Vol. 1 (Bakuriani, 1990), (VSP, Utrecht, 1991) pp. 38–42
228. O.I. Klesov, Convergence of series of large-deviation probabilities of sums of independent identically distributed random variables. Ukrain. matem. zh. **45**(6), 770–784 (1993) (English transl. in Ukrain. Math. J. **45**(6), 845–862 (1994))
229. O.I. Klesov, On the co nvergence rate in a theorem of Heyde. Teor. Verayanost. i Mat. Stat. **49**, 65–71 (1993) (English transl. Theor. Probab. Math. Stat. **49**, 83–87 (1994))
230. O.I. Klesov, An analogue of Feller's theorem for multiple sums. Matem. Zametki **55**(1), 53–61, 156 (1994) (English transl. in Math. Notes **55**(1–2), 37–42 (1994))
231. O.I. Klesov, The law of large numbers for multiple sums of independent identically distributedrandom variables. Teor. Imovirnost. ta Mat. Statyst. **50**, 76–86 (1994) (English transl. Theor. Probab. Math. Stat. **50**, 77–87 (1995))
232. O.I. Klesov, Almost sure convergence of multiple series of independent random variables. Teor. Veroyatnost. i Primenen. **40**(1), 68–83 (1995) (English transl. in Theory Probab. Appl. **40**(1), 52–65 (1995))
233. O.I. Klesov, Complete convergence for randomly indexed sums of random variables. J. Math. Sci. **76**(2), 2241–2249 (1995)
234. O.I. Klesov, A new method for the strong law of large numbers for random fields. Theory Stoch. Process. **4(20)**(1–2), 122–128 (1998)
235. O.I. Klesov, The law of the iterated logarithm for multiple sums. Teor. Imovirnost. ta Mat. Statyst. **61**, 39–46 (1999) (English transl. Theory Probab. Math. Stat. **61**, 41–48 (2000))
236. O.I. Klesov, The strong law of large numbers for "subsequences" on the plane. Theory Stoch. Process. **6(22)**(1–2), 47–53 (2000)
237. O.I. Klesov, Existence of supremum moments for multiple sums and the strong law of large numbers. Teor. Imovirnost. ta Mat. Statyst. **62**, 27–36 (2000) (English transl. Theory Probab. Math. Stat. **62**, 27–37 (2001))
238. O.I. Klesov, Boundedness of multiple series. Teor. Imovirnost. ta Mat. Statyst. **63**, 90–99 (2000) (English transl. Theor. Probab. Math. Stat. **63**, 99–108 (2001))
239. O.I. Klesov, On the regular convergence of random multiple series. Acta Sci. Math. (Szeged) **75**(3–4), 493–498 (2009)
240. O. Klesov, I. Fazekas, C. Noszály, T. Tómácz, Strong laws of large numbers for sequences and fields. Theory Stoch. Process. **5(21)**(3–4), 91–104 (1999)
241. O. Klesov, A. Rosalsky, A nonclassical law of the iterated logarithm for independent identically distributed square integrable random variables. Stoch. Anal. Appl. **19**(4), 627–642 (2001)
242. O. Klesov, A. Rosalsky, A nonclassical law of the iterated logarithm for independent identically distributed square integrable random variables II. Stoch. Anal. Appl. **20**(4), 839–846 (2002)

243. O. Klesov, Z. Rychlik, The strong law of large numbers on partially ordered sets. Teor. Imovirnost. ta Mat. Statyst. **58**, 31–37 (1998) (English transl. Theor. Probab. Math. Stat. **58**, 35–41 (1999))
244. O. Klesov, Z. Rychlik, J. Steinebach, Strong limit theorems for general renewal processes. Theory Probab. Stat. **21**(2), 329–349 (2001)
245. O. Klesov, J. Steinebach, Asymptotic behavior of renewal processes defined by random walks with multidimensional time. Teor. Imovirnost. ta Mat. Statyst. **56**, 105–111 (1997) (English transl. Theor. Probab. Math. Stat. **56**, 107–113 (1998))
246. O. Klesov, J. Steinebach, The asymptotic behavior of the renewal function constructed from a random walk in multidimensional time with restricted domain. Ann. Univ. Sci. Budapest. Sect. Comp. **22**, 181–192 (2004)
247. K. Knopp, *Theory and Application of Infinite Series*, ed. by R.C. Young. (Blackie, London, 1928) (Translated from the second German edition)
248. Mi-Hwa Ko, The convergence rates in the asymmetric laws of large number for negatively associated random fields. Honam Math. J. **34**(2), 209–217 (2012)
249. Mi-Hwa Ko, On the asymmetric Marcinkiewic z-Zygmund strong law of large numbers for linear random fields. Commun. Stat. Theory Methods **42**(8), 1553–1562 (2013)
250. M.-H. Ko, T.-S. Kim, H.-C. Kim, Strong laws of large numbers for asymptotically quadrant independent random fields. Commun. Korean Math. Soc. **19**(4), 765–773 (2004)
251. G. Kolesnik, On the estimation of multiple exponential sums, *Recent Progress in Analytic-Number Theory*, Vol. 1 (Durham, 1979), (Academic Press, London, New York, 1981), pp. 231–246
252. A.N. Kolmogorov, Une série e Fourier-Lebesgue divergente presque partout. Fund. Math. **4**, 324–328 (1923)
253. A.N. Kolmogoroff, Über das Gesetz des iterierten Logarithmus. Math. Ann. **11**, 126–135 (1929)
254. A.N. Kolmogoroff, Über die Summen durch den Zufall bestimmter unabhängiger Grössen. Math. Ann. **99**, 309–319 (1928); 102, 484–488 (1929)
255. A.N. Kolmogoroff, Sur la loi forte des grandes nombres. C. R. Acad. Sci. Paris. **191**, 910–912 (1930)
256. A.N. Kolmogoroff, Sulla forma generale di un processo stocastico omogeneo. Atti Acad. Naz. Lincei Rend. Cl. Sci. Fis. Mat. Nat. **6**(15), 805–808, 866–869 (1932)
257. A.N. Kolmogoroff, *Foundations of the Theory of Probability*. (GONTI, Moscow, Leningrad, 1936), (Nauka, Moscow, 1974) (English transl, 2nd edn. Chelsea, New York, 1956)
258. A.N. Kolmogoroff, Sur les propriétés des fonctions de concentration de M. P. Lévy. Ann. Inst. Henri Poincaré **16**(1), 27–34 (1958)
259. YuV Korolev, I.G. Shevtsova, On the upper bound for the absolute constant in the Berry–Esseen inequality. Theory Probab. Appl. **54**(4), 638–658 (2010)
260. E. Krätzel, *Lattice Points*. Mathematics and Its Applications, Vol. 33 (East European Series) (Kluwer Academic Publishers Group, Dordrecht, 1988)
261. U. Krengel, *Ergodic theorems. de Gruyter Studies in Mathematics*. With a Supplement by Antoine Brunel, Vol. 6 (Walter de Gruyter & Co., Berlin, 1985)
262. K. Krickeberg, Convergence of martingales with a directed index set. Trans. Amer. Math. Soc. **83**(2), 313–337 (1956)
263. V.M. Kruglov, On the extension of the class of stable distributions. Teor. Veroyatnost. i Primenen. **17**(4), 723–732 (1972) (English transl. in Theory Probab. Appl. **17**(4), 685–694 (1972))
264. N.Yu. Kryzhanovskaya, Moment inequality for sums of multi-indexed dependent random variables. Matem. Zametki **83**(6), 843–856 (2008) (English transl. in Math. Notes **83**(6), 770–782 (2008))
265. L. Kubik, A characteri zation of the class $\mathscr{L}$ of probability distributions. Studia Math. **21**(2), 245–252 (1962)
266. A. Kuczmaszewska, The strong law of large numbers for dependent random variables. Stat. Probab. Lett. **73**(3), 305–314 (2005)

267. A. Kuczmaszewska, On Chung–Teicher type strong law of large numbers for ρ^*-mixing random variables. Discr. Dyn. Nat Soc. Article ID 140548, 1–10 (2008)
268. A. Kuczmaszewska, D. Szynal, On the Hsu–Robbins law of large numbers for subsequences. Bull. Acad. Sci. Polon. Ser. Math. Astronom. Phys. **36**, 69–79 (1988)
269. O. Kurbanmuradov, K. Sabelfeld, Exponential bounds for the probability deviations of sums of random fields. Monte-Carlo Methods Appl. **12**(3–4), 211–229 (2006)
270. Z.A. Lagodowski, Probabilities of moderate deviations for randomly indexed sums of random variables with multidimensional indices. Probab. Math. Stat. **9**(1), 115–123 (1988)
271. Z.A. Lagodowski, Fuk inequalities for random fields of backward martingales. Sibirsk. Mat. Zh. **32**(2), 176–179, 219 (1991) (English transl. in Siberian Math. J. **32**(2), 329–332 (1991))
272. Z.A. Lagodowski, On the Feller strong law of large numbers for fields of B-valued random variables. Math. Pannonica **5**(2), 249–254 (1994)
273. Z.A. Lagodowski, Strong laws of large numbers for $\mathbb{B}$-valued random fields. Discr. Dyn. Nat. Soc. Article ID **485412**, 1–12 (2009)
274. Z.A. Lagodowski, P. Matula, On almost sure limiting behavior of weighted sums of random fields. Acta Math. Hung. **126**(1–2), 16–22 (2010)
275. Z.A. Lagodowski, Z. Rychlik, Complete convergence and convergence rates for randomly indexed sums of random variables with multidimensional indices. Bull. Polish Acad. Sci. Math. **32**(3–4), 219–354 (1985)
276. Z.A. Lagodowski, Z. Rychlik, Convergence rates in the strong law of large numbers for sums of random variables with multidimensional indices. Probab. Math. Stat. **7**(2), 149–158 (1986)
277. E. Landau, Zur Theorie der Riemannschen Zetafunktion. Vierteljahrsschr. Naturf. Ges. Zürich **56**, 125–148 (1911)
278. M. Ledoux, M. Talagrand, *Probability in Banach spaces. Isoperimetry and Processes*. Classics in Mathematics. (Springer, Berlin, 2011) (Reprint of the 1991 edition)
279. N.N. Leonenko, Y.S Mishura, On the invariance principle for multiparameter martingales. Teor. Veroyatnost. i Mat. Stat. **24**, 81–91, 153 (1981) (English transl. Theor. Probab. Math. Stat. **24**, 91–101 (1982))
280. S. Levental, V. Mandrekar, S.A. Chobanyan, Towards Nikishin's theorem on the almost sure convergence of rearrangements of functional series. Funkts. Analiz Prilozhen. **45**(1), 41–55 (2011) (English transl. in Funct. Anal. Appl. **45**(1), 33–45 (2011))
281. P. Lévy, *Calcul des probabilités* (Gauthier-Villars, Paris, 1925)
282. P. Lévy, Sur les intégrales dont les éléments sont des variables aléatoires indépendents. Annali R. Scuola Norm. Sup. Pisa **2**(3), 337–366; (4), 217–218 (1934)
283. P. Lévy, *Théorie de l'addition des variables aléatoires* (Gauthier-Villars, Paris, 1937)
284. R. LePage, Log log law for Gaussian processes. Z. Wahrsch. verw. Gebiete **25**(2), 103–108 (1973)
285. D. Li, Maximum inequalities for B-valued random fields and their application. J. Math. (Wuhan) **10**(4), 371–380 (1990) (Chinese, English summary)
286. D. Li, On moments of the supremum of normed weighted averages. Can. J. Stat. **24**(3), 279–292 (1996)
287. D. Li, M.L. Huang, A note on moments of maximum of Cesáro summation. Stat. Probab. Lett. **38**(1), 73–81 (1998)
288. D. Li, M.B. Rao, X.C. Wang, The law of the iterated logarithm for independent random variables with multidimensional indices. Ann. Probab. **20**(2), 660–674 (1992)
289. D. Li, M.B. Rao, X.C. Wang, On the strong law of large numbers and the law of the logarithm for weighted sums of independent random variables with multidimensional indices. J. Multivar. Anal. **52**(2), 181–198 (1995)
290. D. Li, R.J. Tomkins, Compact laws of the iterated logarithm for B-valued random variables with two-dimensional indices. J. Theory Probab. **11**(2), 443–459 (1998)
291. D. Li, Z.Q. Wu, Law of the iterated logarithm for $\mathbb{B}$-valued random variables with multidimensional indices. Ann. Probab. **17**(2), 760–774 (1989)
292. D. Li, Z.Q. Wu, On the moment of sup $|S_n/|n|^{1/r}|(0 < r < 2)$ and sup $|S_n/\sqrt{|n|\log\log|n|}|$, Acta Math. Sinica **32**, 747–757, (1989) (Chinese, English summary)

293. Z.F. Michael, Li, R.J. Tomkins, Stability of maxima of random variables with multidimensional indices. Extremes **7**(4), 135–147 (2004)
294. Wei-Dong Lin, Some LIL type results on the partial sums and trimmed sums with multidimensional indices. Elect. Comm. Probab. **12**, 221–233 (2007)
295. Z.Y. Lin, Y.K. Choi, K.S. Hwang, Some limit theorems on the increments of a multiparameter fractional Brownian motion. Stoch. Anal. Appl. **19**(4), 499–517 (2001)
296. E. Lindelöf, *Le calcul des résidue et ses applications a la théorie des fonctions* (Gauthier-Villars, Paris, 1905)
297. X. Liu, J. Liu, Moments of the maximum of normed partial sums of ρ^-- mixing random variables. Appl. Math. J. Chin. Univ. **24**(3), 355–360 (2009) (Chinese, English summary)
298. M. Loéve, On almost sure convergence, in *Proceedings of Second Berkeley Symposium on Mathematics Statistics and Probability*. (University of California Press, Berkeley July 31–August 12 1950), pp. 279–303
299. M. Loéve, *Probability Theory*. Graduate Texts in Mathematics, Vol. 46 4th edn. (Springer, New York, 1978)
300. M. Maejima, T. Mori, Some renewal theorems for random walks in multidimensional time. Math. Proc. Camb. Phil. Soc. **95**(1), 149–154 (1984)
301. J. Marcinkiewicz, A. Zygmund, Sur les fonctions indépendantes. Fund. Math. **29**(1), 60–90 (1937)
302. J. Marcinkiewicz, A. Zygmund, Remarque sur la loi du logarithme itéré. Fund. Math. **29**(1), 215–222 (1937)
303. B.J. McCabe, L.A. Shepp, On the supremum of S_n/n. Ann. Math. Stat. **41**(6), 2166–2168 (1970)
304. M.B. Marcus, L.F. Shepp, Continuity of Gaussian processes. Trans. Amer. Math. Soc. **151**, 377–391 (1970)
305. A.I. Martikainen, The law of the iterated logarithm for permuted and weighted sums. Teor. Veroyatnost. i Primenen. **23**(2), 380–383 (1978) (English transl. in Theory Probab. Appl. **23**(2), 362–365 (1978))
306. A.I. Martikainen, A converse to the law of the iterated logarithm for a random walk. Teor. Veroyatnost. i Primenen. **25**(2), 364–366 (1980) (English transl. in Theory Probab. Appl. **25**(2), 361–362 (1980))
307. A.I. Martikainen, The law of the iterated logarithm for a random field. Vestnik Leningrad. Univ. Mat. Mekh. Astronom. **22**, 14–20, 121 (1985) (Russian)
308. A.I. Martikainen, The order of growth of a random field. Matem. Zametki **39**(3), 431–437, 463 (1986) (English transl. in Math. Notes **39**(3–4), 237–240 (1986))
309. A.I. Martikainen, V.V. Petrov, On necessary and sufficient conditions for the law of the iterated logarithm. Teor. Veroyatnost. i Primenen. **22**(1), 16–23 (1977); **22**(2), 442 (1977) (English transl. in Theory Probab. Appl. **22**(1), 18–26 (1977))
310. A.I. Martikainen, V.V. Petrov, On a theorem of Feller. Teor. Veroyatnost. i Primenen. **25**(1), 194–197 (1980) (English transl. in Theory Probab. Appl. **25**(1), 191–193 (1980))
311. I.K. Matsak, On the law of the iterated logarithm in Banach lattices. Teor. Veroyatnost. i Primenen. **44**(4), 865–874 (1999) (English transl. in Theory Probab. Appl. **44**(4), 775–784 (2000))
312. I.K. Matsak, A.M. Plichko, A moment estimate for the supremum of normed sums in the law of the iterated logarithm. Ukrain. matem. zh. **58**(5), 653–665 (2006) (English transl. in Ukrain. Math. J. **58**(5), 737–750 (2006))
313. P. Matula, M. Seweryn, Weighted strong law of large numbers for random variables indexed by a sector. J. Probab. Statist. Article ID 701952, 1–16 (2011)
314. D. Mejzler, On a certain class of infinitely divisible distributions. Israel J. Math. **16**(1), 1–19 (1973)
315. T. Mikosch, On the strong law of large numbers for random fields. Vestnik Leningrad. Univ. Mat. Mekh. Astronom. **19**, 82–85 (1984)
316. T. Mikosch, Almost sure behavior of tail series in functional spaces. Ann. Math. **16**(2), 123–133 (1990)

317. T. Mikosch, R. Norvaisa, Strong law of large numbers for fields of Banach space valued random variables. Probab. Theory Relat. Fields **74**(2), 241–253 (1987)
318. F. Móricz, Moment inequalities for the maximum of partial sums of random fields. Acta Sci. Math. **39**(3–4), 353–366 (1977)
319. F. Móricz, Multiparameter strong laws of large numbers I. Second order moment restrictions. Acta Sci. Math. **40**(1–2), 143–145 (1978)
320. F. Móricz, Strong laws of large numbers for quasistationary random fields. Z. Wahrsch. verw. Gebiete **51**(3), 249–268 (1980)
321. F. Móricz, Generalization of two theorems of Kronecker and of Knopp for multiple numerical series. Analysis **2**(1–4), 49–64 (1982)
322. F. Móricz, U. Stadtmüller, M. Thalmaier, Strong laws for blockwise $\mathscr{M}$-dependent random fields. J. Theory Probab. **21**(3), 660–671 (2008)
323. G.J. Morrow, Approximation of rectangular sums of B-valued random variables. Z. Wahrsch. verw. Gebiete **57**(2), 265–291 (1981)
324. E. Mourier, Eléments aléatoires dans un espace de Banach. Ann. Inst. H. Poincaré **13**(3), 161–244 (1953)
325. S.V. Nagaev, Some limit theorems for large deviations. Teor. Veroyatnost. i Primenen. **10**(2), 231–254 (1965) (English transl. in Theory Probab. Appl. **10**(2), 214–235 (1965))
326. S.V. Nagaev, I.F. Pinelis, Some inequalities for the distributions of sums of independent random variables. Teor. Veroyatnost. i Primenen. **22**(2), 254–263 (1977) (English transl. in Theory Probab. Appl. **10**(2), 248–256 (1965))
327. E. Nam, On the order of growth of convergent series of independent random variables. J. Appl. Math. Stoch. Anal. **2**(2), 159–168 (2004)
328. E. Nam, A. Rosalsky, On the rate of convergence of series of random variables. Teor. Imovirnost. ta Mat. Statyst. **52**, 120–131 (1995) English transl. Theor. Probab. Math. Stat. **52**, 129–140 (1996)
329. C.H. Ndiaye, G.S. Lo, On general strong laws of large numbers for fields of random variables. Ann. Math. Inform. **38**, 3–13 (2011)
330. P. Ney, S. Wainger, The renewal theorem for a random walk in two-dimensional time. Stud. Math. **44**(1), 71–85 (1972)
331. C. Neaderhauser, An almost sure invariance principle for partial sums associated with a random field. Stoch. Process. Appl. **11**(1), 1–10 (1981)
332. C. Noszály, T. Tómács, A general approach to strong laws of large numbers for fields of random variables. Ann. Univ. Sci. Budapest. Sect. Math. **43**, 61–78 (2001)
333. G. Ottaviani, Sulla teoria astratta del calcolo delle probabilitá proposita dal Cantelli. Giorn. Ist. Ital. Attuari **10**, 10–40 (1939)
334. S. Parameswaran, Partition functions whose logarithms are slowly oscillating. Trans. Amer. Math. Soc. **100**(2), 217–240 (1961)
335. S.R. Paranjape, C. Park, Laws of iterated logarirhm for multiparameter Wiener processes. J. Multivar. Anal. **3**(1), 132–136 (1973)
336. V.V. Petrov, *Sums of Independent Random Variables* (Nauka, Moscow, 1972) (English transl. translated from the Russian ed by A.A. Brown, *Ergebnisse der Mathematik und ihrer Grenzgebiete, Band,* Vol. 82 (Springer, New York, 1975))
337. V.V. Petrov, A generalization of a certain inequality of Levy, Teor. Veroyatnost. i Primenen. **20**(1), 140–144 (1975) (English transl. in Theory Probab. Appl. **20**(1), 141–145 (1975))
338. V.V. Petrov, Limit theorems of probability theory, *Sequences of Independent Random Variables* (Nauka, Moscow, 1987) (English transl. in *Oxford Studies in Probability*, Vol. 4 (Oxford Science Publications, The Clarendon Press, Oxford University Press, New York, 1995))
339. W. Pfuhl, J. Steinebach, On precise asymptotics for the Erdös-Rényi increments of random fields. Pub. Inst. Stat. Univ. **2**, 49–66 (1988)
340. G. Pick, *Geometrisches zur Zahlenlehre*. Zeitschrift für Naturwissenschafen, hrsg. vom Naturhistorisch (Vereine "Lot os" in Prague, Prague, 1899), pp. 311–319
341. V.I. Piterbarg, *Asymptotic Metheods in the Theory of Gaussian Processes and Fields*. (MGU Publishing House, Moscow, 1988) (English transl. in *Mathematical Monographs*, Vol. 148. (American Mathematical Society, Providence, RI, 1995))

342. A. Pringsheim, Zur theorie der zweifach unendlichen Zahlenfolgen. Math. Ann. **53**(3), 289–321 (1900)
343. Y.V. Prokhorov, On the strong law of large numbers. Izv. Akad. Nauk SSSR, Ser. Mat. **14**(6), 523–536 (1950) (Russian)
344. A.R. Pruss, Remarks on summability of series formed from deviation probabilities of sums of independent identically distributed random variables. Ukrain. matem. zh. **48**(4), 569–572 (1996) (English transl. in Ukrain. Math. J. **48**(4), 631–635 (1996))
345. R. Pyke, *Partial Sums of Matrix Arrays and Brownian Sheets*. Stochastic Geometry and Stochastic Analysis. (Wiley, New York, 1973) pp. 331–348
346. N.V. Quang, N.V. Huan, On the weak law of large numbers for double arrays of Banach space valued random elements. J. Probab. Stat. Sci. **6**, 125–134 (2008)
347. N.V. Quang, N.V. Huan, On the strong law of large numbers and L_p convergence for double arrays of random elements in p-uniformly smooth Banach spaces. Stat. Probab. Lett. **79**(18), 1891–1899 (2009)
348. N.V. Quang, N.V. Huan, A Hájek-Rényi-type maximal inequality and strong laws of large numbers for multidimensional arrays. J. Inequal. Appl. Article ID 569759, 1–14 (2010)
349. N.V. Quang, N.V. Huan, A characterization of p-uniformly smooth Banach spaces and weak laws of large numbers for d-dimensional adapted arrays. Sankhyā **72-A**(2), 344–358 (2010)
350. N.V. Quang, L.V. Thanh, N.D. Tien, Almost sure convergence for double arrays of $\mathscr{M}$-dependent random elements in Banach spaces. Georgian Math. J. **18**(4), 777–800 (2011)
351. N.V. Quang, N.N. Huy, Weak law of large numbers for adapted double arrays of random variables. J. Korean Math. Soc. **45**, 795–805 (2008)
352. P. Révész, The laws of large numbers, *Probability and Mathematical Statistics*, Vol. 4 (Académiai Kiadó and Academic Press, Budapest and New York, 1967)
353. F. Riesz, B. Szökefalvi-Nagy, *Leçons d'fanalyse fonctionnelle, sixième édition* (Académiai Kiadó, Budapest, 1972)
354. S. Roman, *Lattices and Ordered Sets* (Springer, New York, 2008)
355. A. Rosalsky, On the converse to the iterated logarithm law. Sankhyā **A42**(1–2), 103–108 (1980)
356. A. Rosalsky, Almost certain limiting behavior of the tail series of independent summands. Bull. Inst. Math. Acad. Sinica **11**, 185–208 (1983)
357. A. Rosalsky, L.V. Thanh, Strong and weak laws of large numbers for double sums of independent random elements in Rademacher type p Banach spaces. Stoch. Anal. Appl. **24**(6), 1097–1117 (2006)
358. A. Rosalsky, L.V. Thanh, On almost sure and mean convergence of normed double sums of Banach space valued random elements. Stoch. Anal. Appl. **25**(4), 895–911 (2007)
359. A. Rosalsky, L.V. Thanh, Weak laws of large numbers for double sums of independent random elements in Rademacher type p and stable type p Banach spaces. Nonlinear Anal. **71**(12), e1065–e1071 (2009)
360. H.P. Rosenthal, On the subspaces of $L_p(p > 2)$ spanned by sequences of independent random variables. Israel J. Math. **8**(3), 273–303 (1970)
361. L.V. Rozovskiĭ, A generalization of the Kolmogorov theorem on the law of the iterated logarithm. Teor. Veroyatnost. i Primenen. **42**(1), 134–143 (1997) (English transl. in Theory Probab. Appl. **42**(1), 112–120 (1997))
362. L.V. Rozovskiĭ, A remark on the Kolmogorov law of the iterated logarithm. Teor. Veroyatnost. i Primenen. **47**(1), 166–169 (2002) (English transl. in Theory Probab. Appl. **47**(1), 171–173 (2002))
363. Dae-Hee Ryu, Rényi inequality and strong law of large numbers for AQSI random variables. J. Chungcheong Math. Soc. **23**(4), 835–844 (2010)
364. A.P. Shashkin, Maximal inequality for weakly dependent random fields. Matem. Zametki **75**(5), 773–782 (2004) (English transl. in Math. Notes **75**(5), 717–725 (2004))
365. O.E. Shcherbakova, Asymptotic behavior of increments of random fields. Teor. Imovirnost. ta Mat. Statyst. **68** 158–171 (2003) (English transl. Theor. Probab. Math. Stat. **68**, 173–186 (2004))

366. G. Shorack, R.T. Smythe, Inequalities for the S_k/b_k with $k \in N^r$. Proc. Amer. Math. Soc. **54**(1), 331–336 (1976)
367. Hu Shuhe, Hu Ming, A general approach rate to the strong law of large numbers. Stat. Probab. Lett. **76**(8), 843–851 (2006)
368. E. Seneta, *Regularly Varying Functions*, vol. 508, Lecture Notes in Mathematics (Springer, Berlin, 1976)
369. D. Siegmund, On moments of the maximum of normed partial sums. Ann. Math. Stat. **40**(2), 527–531 (1969)
370. S.K. Sirazhdinov, M.U. Gafurov, *Method of Series in Boundary Problems for Random Walks* (Fan, Tashkent, 1987) (Russian)
371. A.V. Skorohod, *Random Processes With Independent Increments*. (Nauka, Moscow, 1964) (English transl. Mathematics and its Applications (Soviet Series), Vol. 47 (Kluwer Academic Publishers Group, Dordrecht, 1991))
372. W.L. Smith, On the elementary renewal theorem for non-identically distributed variables. Pac. J. Math. **14**(2), 673–699 (1964)
373. R.T. Smythe, Strong law of large numbers for r-dimensional arrays of random variables. Ann. Probab. **1**(1), 164–170 (1973)
374. R.T. Smythe, The sums of independent random variables on the partially ordered sets. Ann. Probab. **2**(5), 906–917 (1974)
375. R.T. Smythe, Ergodic properties of marked point processes in $\mathbf{R}^r$. Ann. Inst. Henri Poincare **11**(2), 109–125 (1975)
376. T.C. Son, D.H. Thang, The Brunk–Prokhorov strong law of large numbers for fields of martingale differences taking values in a Banach space. Stat. Probab. Lett. **83**(8), 1901–1910 (2013)
377. F.L. Spitzer, A combinatorial lemma and its application to probability theory. Trans. Amer. Math. Soc. **82**(2), 323–339 (1956)
378. U. Stadtmüller, Strong limit theorems for increments of random fields, in *Stochastic Geometry, Spatial Statistics and Random Fields*. Lecture Notes in Mathematics Vol. 2068 (Springer, Berlin, 2013), pp. 385–398
379. U. Stadtmüller, L.V. Thanh, On the strong limit theorems for double arrays of $\mathscr{M}$-dependent random variables. Acta Math. Sin. (Engl. Ser.) **27**(10), 1923–1934 (2011)
380. W.L. Steiger, S.K. Zaremba, The converse of the Hartman-Wintner theorem. Z. Wahrsch. verw. Gebiete **22**(3), 193–194 (1972)
381. J. Steinebach, On the increments of partial sum processes with multidimensional indices. Z. Wahrsch. verw. Gebiete **63**(1), 59–70 (1983)
382. G. Stoica, A note on the rate of convergence in the strong law of large numbers for martingales. J. Math. Anal. Appl. **381**(2), 910–913 (2011)
383. W.F. Stout, *Almost Sure Convergence, Probability and Mathematical Statistics*, vol. 24 (Academic Press, New York, London, 1974)
384. V. Strassen, An invariance principle for the law of the iterated logarithm. Z. Wahrsch. verw. Gebiete **3**(3), 211–226 (1964)
385. V. Strassen, A converse to the law of the iterated logarithm. Z. Wahrsch. verw. Gebiete **4**(4), 265–268 (1966)
386. D.W. Strook, *Probability Theory. An Analytic View*, 2nd edn (Cambridge University Press, Cambridge, 2011)
387. K.L. Su, R.L. Taylor, Marcinkiewicz strong laws of large numbers and convergence rates for arrays of independent random elements in Banach spaces. Stoch. Anal. Appl. **10**(2), 223–237 (1992)
388. K.L. Sung, T.C. Hu, A. Volodin, A note on the growth rate in the Fazekas–Klesov general law of large numbers and on the weak law of large numbers for tail series. Publ. Math. Debrecen **73**(1/2), 1–10 (2008)
389. K.L. Sung, A. Volodin, On convergence of series of independent random variables. Bull. Korean Math. Soc. **38**(4), 763–772 (2001)

390. S.H. Sung, Maximal inequalities for dependent random variables and applications. J. Inequal. Appl. Article ID 598319, 1–10 (2008)
391. D. Szynal, On almost complete convergence for the sum of a random number of independent random variables. Bull. Acad. Sci. Polon. Ser. Math. Astronom. Phys. **20**, 571–574 (1972)
392. H. Teicher, A dominated ergodic type theorem. Z. Wahrsch. verw. Gebiete **8**(2), 113–116 (1967)
393. H. Teicher, Some new conditions for the strong law. Proc. Nat. Acad. Sci. U.S.A. **59**(3), 705–707 (1968)
394. H. Teicher, Completion of a dominated ergodic theorem. Ann. Math. Stat. **42**(6), 2156–2158 (1971)
395. L. Van Thanh, Strong law of large numbers and L_p-convergence for double arrays of independent random variables. Acta Math. Vietnamica **30**(3), 225–232 (2005)
396. L. Van Thanh, On the L_p-convergence for multidimensional arrays of random variables. Int. J. Math. Math. Sci. **8**, 1317–1320 (2005)
397. L. Van Thanh, On the strong law of large numbers for D-dimensional arrays of random variables. Elect. Comm. Probab. **12**, 434–441 (2007)
398. N.D. Tien, L. Van Dung, Convergence of double series of random elements in Banach spaces. J. Korean Math. Soc. **42**(5), 1053–1064 (2012)
399. E.C. Titchmarsh, *The Theory of the Riemann Zeta Function*, 2nd edn. (Oxford University Press, Oxford, 1986). (revised)
400. R.J. Tomkins, On the law of the iterated logarithm. Ann. Probab. **6**(1), 162–168 (1978)
401. R.J. Tomkins, Rates of convergence for tail series. J. Stat. Plann. Infer. **43**(1–2), 77–85 (1995)
402. R. Vélez, A new approach to series of independent random variables. Test **10**(2), 405–418 (2001)
403. G.F. Voronoi, Sur un probl'eme du calcul des fonctions asymptotiques. J. Reine Angew. Math. **126**, 241–282 (1903)
404. N.A. Volodin, S.V. Nagaev, A remark on the strong law of large numbers. Teor. Veroyatnost. i Primenen. **22**(4), 829–831 (1977) (English transl. in Theory Probab. Appl. **22**(4), 810–813 (1978))
405. V.G. Vovk, The law of the iterated logarithm for sequences that are random in the sense of Kolmogorov or chaotic. Teor. Veroyatnost. i Primenen. **32**(3), 456–468 (1987) (English transl. in Theory Probab. Appl. **32**(3), 413–425 (1987))
406. X. Wang, S. Hu, Maximal inequalities for demimartingales and their applications. Sci. Chin. Ser. A. Math. **52**(10), 2207–2217 (2011)
407. X. Wang, S. Hu, X. Li, W. Yang, Maximal inequalities and strong law of large numbers for AANA sequences. Commun. Korean Math. Soc. **26**(1), 151–161 (2011)
408. X. Wang, S. Hu, Y. Shen, N. Ling, Strong law of large numbers and growth rate for a class of random variable sequences. Stat. Probab. Lett. **78**(18), 3330–3337 (2011)
409. X. Wang, S. Hu, Y. Shen, W. Yang, Moment inequality for ϕ-mixing sequences and its applications. J. Inequal. Appl. Article ID 379743, 1–12 (2009)
410. X. Wang, S. Hu, Y. Shen, W. Yang, Maximal inequality for ψ-mixing sequences and its applications. Appl. Math. Lett. **23**(10), 1156–1161 (2010)
411. X. Wang, Sh Hu, W. Yang, B.L.S. Prakasa Rao, Maximal inequalities for N-demimartingale and strong law of large numbers. Stat. Probab. Lett. **81**(9), 1348–1353 (2011)
412. X. Wang, S. Hu, W. Yang, Convergence properties for asymptotically almost negatively associated sequence. Discr. Dyn. Nat. Soc. Article ID 218380, 1–15 (2010)
413. X. Wang, Sh Hu, W. Yang, Y. Shen, Hajék-Rényi-type inequalities and laws of large numbers for mixingale array. Math. Inequal. Appl. **14**(3), 621–632 (2011)
414. J.F. Wang, J.S. Jin, Marcinkiewicz laws of large numbers and complete convergence with multidimensional indices. Acta Math. Sci. Ser. A Chin. Ed. **25**(5), 734–743 (2005) Chinese, English summary)
415. T. Watanabe, K. Yamamuro, Limsup behaviors of multi-dimensional selfsimilar processes with independent increments. Alea **7**, 79–116 (2010)

416. E.T. Whittaker, G.N. Watson, An introduction to the general theory of infinite processes and of analytic functions: with an account of the principal transcendental functions, in *A Course of Modern Analysis* 4th edn. (Cambridge University Press, New York, 1963)
417. M. Weiss, On the law of the iterated logarithm. J. Math. Mech. **8**(1), 121–132 (1959)
418. N. Wiener, The ergodic theorem. Duke Math. J. **5**(1), 1–18 (1939)
419. M. Wichura, Inequalities with applications to the weak convergence of random processes with multidimensional time parameter. Ann. Math. Stat. **40**(2), 681–687 (1969)
420. M. Wichura, Some Strassen-type laws of the iterated logarithm for multiparameter stochastic processes with independent increments. Ann. Probab. **1**(2), 272–296 (1973)
421. E. Wong, M. Zakai, Martingales and stochastic integrals for processes with a multidimensional parameter. Z. Wahrsch. verw. Gebiete **29**(2), 109–122 (1974)
422. M.I. Yadrenko, *Spectral Theory of Random Fields*, (Vyshcha Shkola, Kiev, 1980) (English transl. Translation Series in Mathematics and Engineering (Optimization Software Inc, Publications Division, New York, 1983)
423. K. Yamamuro, A law of the iterated logarithm of Chover type for multidimensional Lévy processes. Osaka J. Math. **42**(2), 367–383 (2005)
424. W. Yang, Y. Shen, S. Hu, X. Wang, Hájek-Rényi-type inequality and strong law of large numbers for some dependent sequences. Acta Math. Appl. Sinica Engl. Ser. **28**(3), 495–504 (2012)
425. S. Yang, C. Su, Y. Keming, A general method to the strong law of large numbers and its applications. Stat. Probab. Lett. **78**(6), 794–803 (2008)
426. A.I. Yanushauskas, *Double Series* ("Nauka" Sibirsk. Otdel., Novosibirsk, 1980) (Russian)
427. A.I. Yanushauskas, *Multiple Trigonometric Series* ("Nauka" Sibirsk. Otdel., Novosibirsk, 1986) (Russian)
428. P.A. Yas'kov, A generalization of the Menshov-Rademacher theorem. Matem. Zametki **86**(6), 925–937 (2009) (English transl. in Math. Notes **86**(6), 861–872 (2009))
429. P.A. Yas'kov, Strong convergence for multiple sums of nonorthogonal random variables. Teor. Veroyatnost. i Primenen. **55**(2), 382–386 (2010) (English transl. in Theory Probab. Appl. **55**(2), 351–355 (2011))
430. M. Zakai, Some classes of two-parameter martingales. Ann. Probab. **9**(2), 255–265 (1981)
431. N. Van Zang, N.Z.Tien, Strong law of large numbers for multiparametric independent random variables. Lithuanian Math. J. **31**(1), 103–114 (1991)
432. L.X. Zhang, J.W. Wen, A strong law of large numbers for B-valued random fields. Chin. Ann. Math. Ser. A 22 (2001), no. 2, 205–216. (Chinese, English summary)
433. G.J. Zimmerman, Some sample function properties of the two-parameter Gaussian processes. Ann. Math. Stat **43**(4), 1235–1246 (1972)
434. N.M. Zinchenko, Approximation of sums of random variables with multiindices. Ukrain. matem. zh. **36**(2), 149–154 (1984) (English transl. in Ukrain. Math. J. **36**(2), 133–138 (1984))
435. N.M. Zinchenko, A strong invariance principle for multiple sums of random variables from the domain of attraction of a stable law. Teor. Veroyatnost. i Mat. Stat. **36** 62–75, 140 (1998) (English transl. Theor. Probab. Math. Stat. **36**, 69–81 (1988))
436. N.M. Zinchenko, Asymptotics of stable random fields. Teor. Veroyatnost. i Mat. Stat. **38**, 51–59, 134 (1998) (English transl. Theor. Probab. Math. Stat. **38**, 53–61 (1988))
437. N.M. Zinchenko, On the asymptotic behavior of increments of certain classes of random fields. Teor. Veroyatnost. i Mat. Stat. **48**, 11–18 (1993) (English transl. Theor. Probab. Math. Stat. **48**, 7–11 (1994))
438. N.M. Zinchenko, A modified law of iterated logarithm for stable random variables. Teor. Veroyatnost. i Mat. Stat. **49**, 99–109 (1993) (English transl. Theor. Probab. Math. Stat. **49**, 69–76 (1994))
439. V.M. Zolotarev, *Modern Theory of Summation of Independent Random Variables* (Nauka, Moscow, 1986) (English transl. *Modern Probability and Statistics* (VSP, Utrecht, 1997))
440. A. Zygmund, An individual ergodic theorem for non-commutative transformations. Acta Sci. **14**(2), 103–110 (1951)

Index

O. Klesov, *Limit Theorems for Multi-Indexed Sums of Random Variables*,
Probability Theory and Stochastic Modelling 71, DOI 10.1007/978-3-662-44388-0

H

I

K

L

M

MIX
Papier aus verantwortungsvollen Quellen
Paper from responsible sources
FSC® C105338

If you have any concerns about our products,
you can contact us on
ProductSafety@springernature.com

In case Publisher is established outside the EU,
the EU authorized representative is:
Springer Nature Customer Service Center GmbH
Europaplatz 3, 69115 Heidelberg, Germany

Printed by Libri Plureos GmbH
in Hamburg, Germany